Riti Thapar Kapoor, Maulin P. Shah (Eds.)
BioChar

Also of Interest

Green Pulp and Paper Industry.
Biotechnology for Ecofriendly Processing
Edited by Amit Kumar, Puneet Pathak, Dharm Dutt, 2021
ISBN 978-3-11-059184-2, e-ISBN 978-3-11-059241-2

Industrial Waste.
Characterization, Modification and Applications of Residues
Edited by Herbert Pöllmann, 2021
ISBN 978-3-11-067486-6, e-ISBN 978-3-11-067494-1

Bioelectrochemistry.
Design and Applications of Biomaterials
Edited by Serge Cosnier, 2019
ISBN 978-3-11-056898-1, e-ISBN 978-3-11-057052-6

Photosynthesis.
Biotechnological Applications with Micro-Algae
Edited by Matthias Rögner, 2021
ISBN 978-3-11-071691-7, e-ISBN 978-3-11-071697-9

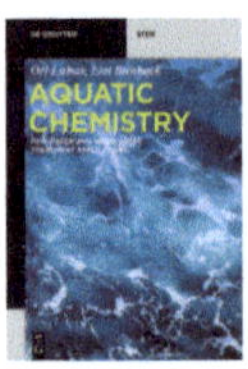

Aquatic Chemistry.
For Water and Wastewater Treatment Applications
Ori Lahav, Liat Birnhack, 2019
ISBN 978-3-11-060392-7, e-ISBN 978-3-11-060395-8

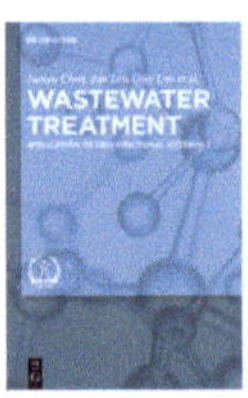

Wastewater Treatment.
Application of New Functional Materials
Jianyu Chen, Jun Luo, Qijin Luo und Zhihua Pang, 2018
ISBN 978-3-11-054278-3, e-ISBN 978-3-11-054438-1

BioChar

Applications for Bioremediation of Contaminated Systems

Edited by
Riti Thapar Kapoor, Maulin P. Shah

DE GRUYTER

Editors

Dr. Riti Thapar Kapoor
Amity Institute of Biotechnology
Amity University
Sector 125
Noida 201313
Uttar Pradesh
India
rititkapoor@gmail.com

Dr. Maulin P. Shah
Environmental Microbiology Lab
Opp. Champapuri Jain Mandir
A/103 Satsang Park
Ankleshwar 393002
Gujarat
India
maulinshah1979@gmail.com

ISBN 978-3-11-073858-2
e-ISBN (PDF) 978-3-11-073400-3
e-ISBN (EPUB) 978-3-11-073406-5

Library of Congress Control Number: 2021944023

Bibliographic information published by the Deutsche Nationalbibliothek
The Deutsche Nationalbibliothek lists this publication in the Deutsche Nationalbibliografie; detailed bibliographic data are available on the Internet at http://dnb.dnb.de.

Cover image: Noppharat05081977 / iStock / Getty Images Plus
Typesetting: Integra Software Services Pvt. Ltd.
Printing and binding: CPI books GmbH, Leck

www.degruyter.com

Contents

List of Authors

Ashwin Raj Suresh
Dept. of Biotechnology
Bannari Amman Institute of Technology
Sathyamangalam
Erode
Tamil Nadu
India
sashwinraj@gmail.com

Moni Philip Jacob Kizhakedathil
Dept. of Biotechnology
Bannari Amman Institute of Technology
Sathyamangalam
Erode
Tamil Nadu
India

Chithra Ashok
Dept. of Biotechnology
Bannari Amman Institute of Technology
Sathyamangalam
Erode
Tamil Nadu
India

Dr. Ashwini A. Waoo
AKS University
Satna
Madhya Pradesh
India
ashwiniwaoo@gmail.com

Charu Vyas
AKS University
Satna
Madhya Pradesh
India

Shikha Kumari
Department of Environmental Science
Maharshi Dayanand University
Rohtak 124001
Haryana
India

Poonam kataria
Department of Environmental Science
Maharshi Dayanand University
Rohtak 124001
Haryana
India

Geeta Dhania
Department of Environmental Science
Maharshi Dayanand University
Rohtak 124001
Haryana
India
geetadhania@rediffmail.com

Harinisri K
Sri Ramakrishna College of Arts and Science
for Women
Coimbatore
Tamil Nadu
India
microdrkh@gmail.com

Dr Thamaraiselvi. B
Sri Ramakrishna College of Arts and Science
for Women
Coimbatore
Tamil Nadu
India
thamaraimicro@srcw.ac.in

Reshmi Gopalakrishnan
Sri Ramakrishna College of Arts and Science
for Women
Coimbatore
Tamil Nadu
India

Manovina M
Sri Ramakrishna College of Arts and Science
for Women
Coimbatore
Tamil Nadu
India

https://doi.org/10.1515/9783110734003-203

Helen La
Department of Civil Engineering
CEERE
University of Calgary
2500 University Drive
NW, Calgary, Alberta
Canada T2N 1N4
hla@ucalgary.ca

J. Patrick A. Hettiaratchi
Department of Civil Engineering
CEERE
University of Calgary
2500 University Drive
NW, Calgary, Alberta
Canada T2N 1N4
jhettiar@ucalgary.ca

Palas Samanta
Department of Environmental Science
Sukanta Mahavidyalaya
University of North Bengal
West Bengal
India
samanta.palas2010@gmail.com

Sukhendu Dey
Department of Environmental Science
The University of Burdwan
West Bengal
India

Jinho Jungc
Division of Environmental Science and
Ecological Engineering
Korea University
Seoul
Republic of Korea

Apurba Ratan Ghosh
Department of Environmental Science
The University of Burdwan
West Bengal
India

Piyush Gupta
Department of Chemistry
SRM Institute of Science and Technology
Delhi-NCR Campus
Modinagar
Ghaziabad 201204
Uttar Pradesh
India
piyushg@srmist.edu.in
piyushg.1977@gmail.com

Namrata Gupta
Department of Chemistry
RBS Engineering Technical Campus
Bichpuri
Agra 283105
Uttar Pradesh
India

Subhakanta Dash
Department of Chemistry
Synergy Institute of Engineering and
Technology
Dhenkanal 759001
Odisha
India

Poulomi Ghosh
Department of Biotechnology
Brainware University
Kolkata 700125
West Bengal
India

Saprativ P. Das
Department of Chemical Engineering
Indian Institute of Technology Bombay
Mumbai 400076
Maharashtra
India
saprativ@iitb.ac.in
saprativdas@gmail.com

Srinithya Ravinuthala
Department of Studies in Biotechnology
University of Mysore
Manasagangothri
Mysore 570006
Karnataka
India

Nithya R.
PG and Research Department of Zoology
Justice Basheer Ahmed Sayeed College for Women
Teynampet
Chennai 600018
Tamil Nadu
India

Saprativ P. Das
Department of Chemical Engineering
Indian Institute of Technology Bombay
Mumbai 400076
Maharashtra
India
saprativ@iitb.ac.in
saprativdas@gmail.com

Mohit Aggarwal
Indian Institute of Technology Bhubaneswar
Bhubaneswar 752050
Odisha
India

Neelancherry Remya
Indian Institute of Technology Bhubaneswar
Bhubaneswar 752050
Odisha
India
remya@iitbbs.ac.in

Sweety Kaur
Archeron Group
Bengaluru 560024
Karnataka
India

Ajay Kumar
School of Bioengineering and Biosciences
Lovely Professional University
Phagwara
Punjab
India

Gurmeen Rakhra
School of Bioengineering and Biosciences
Lovely Professional University
Phagwara
Punjab
India

Richa Arora
Department of Microbiology
Punjab Agricultural University
Ludhiana 141004
Punjab
India
richaarora@pau.edu

Tejveer Singh
Department of Life Sciences
School of Basic Sciences and Research
Sharda University
Greater Noida 201306
Uttar Pradesh
India

Marzuqa Quraishi
Amity Institute of Biotechnology
Amity University
Mumbai 410206
Maharashtra
India

Aindree
Department of Life Sciences
School of Basic Sciences and Research
Sharda University
Greater Noida 201306
Uttar Pradesh
India

Mayur Thakre
Amity Institute of Biotechnology
Amity University
Mumbai 410206
Maharashtra
India

Piyush K Gupta
Department of Life Sciences
School of Basic Sciences and Research
Sharda University
Greater Noida 201306
Uttar Pradesh
India

Aberam Natarajan
Department of Life Sciences
School of Basic Sciences and Research
Sharda University
Greater Noida 201306
Uttar Pradesh
India

Mohit Sahni
Department of Physics
School of Basic Sciences and Research
Sharda University
Greater Noida 201306
Uttar Pradesh
India

Kanupriya
Department of Life Sciences
School of Basic Sciences and Research
Sharda University
Greater Noida 201306
Uttar Pradesh
India

Meenal Gupta
Department of Physics
School of Basic Sciences and Research
Sharda University
Greater Noida 201306
Uttar Pradesh
India

Ankit Kumar
Department of Life Sciences
School of Basic Sciences and Research
Sharda University
Greater Noida 201306
Uttar Pradesh
India

Soumya Pandit
Department of Life Sciences
School of Basic Sciences and Research
Sharda University
Greater Noida 201306
Uttar Pradesh
India

Dr. G. Subbulakshmi
Department of Chemistry
Jain University
Bangalore
Karnataka
India
g.subbulakshmi@jainuniversity.ac.in

Dr. R. Thiruneelakandan
Department of Chemistry, University College
of Engineering, BIT Campus
Anna University,
Tiruchirappalli
Tamil Nadu
India

G. Padmapriya
Department of Chemistry
Jain University
Bangalore
Karnataka
India

Dipita Ghosh
Ecological Restoration Laboratory
Department of Environmental Science and
Engineering
Indian Institute of Technology (ISM)
Dhanbad 826 004
Jharkhand
India

Subodh Kumar Maiti
Ecological Restoration Laboratory
Department of Environmental Science and Engineering
Indian Institute of Technology (ISM)
Dhanbad 826 004
Jharkhand
India
subodh@iitism.ac.in
skmism1960@gmail.com

Dr. Swapnila Roy
Associate Professor (Chemistry)
Natural & Applied Sciences
School of Science & Technology
Glocal University
UP-247121
India
swapnilaroy@gmail.com

Dharm Pal
Department of Chemical Engineering
National Institute of Technology Raipur
Raipur 492010
Chhattisgarh
India

Vandana Gupta
Department of Chemical Engineering
National Institute of Technology Raipur
Raipur 492010
Chhattisgarh
India
vgupta.che@nitrr.ac.in

Vijaya Geetha B.
Department of Biotechnology
Rajalakshmi Engineering College (Autonomous)
Affiliated to Anna University
Thandalam
Chennai 602105
Tamil Nadu
India
vijayageetha.b@rajalakshmi.edu.in

Shreenidhi K.S.
Department of Biotechnology
Rajalakshmi Engineering College (Autonomous)
Affiliated to Anna University
Thandalam
Chennai 602105
Tamil Nadu
India

Vihaa Sriee M.G.
Department of Biotechnology
Rajalakshmi Engineering College (Autonomous)
Affiliated to Anna University
Thandalam
Chennai 602105
Tamil Nadu
India

Rashminiza A.
Department of Biotechnology
Rajalakshmi Engineering College (Autonomous)
Affiliated to Anna University
Thandalam
Chennai 602105
Tamil Nadu
India

Reena Kumari
Department of Mathematics and Statistics
Swami Vivekanand Subharti University
Meerut 250005
Uttar Pradesh
India

Poonam Dhankhar
Department of Biotechnology
IIT Roorkee
Roorkee 247667
Uttarakhand
India
dhankharp16@gmail.com

Vikram Dalal
Department of Anesthesiology
Washington University in St. Louis
St. Louis
MO 63110
USA
vikram.dalal.37@gmail.com

Wan Ting Tee
Department of Chemical and Environmental Engineering
Faculty of Science and Engineering
University of Nottingham Malaysia
43500 Semenyih
Selangor Darul Ehsan
Malaysia

Billie Yan Zhang Hiew
Nanotechnology and Catalysis Research Centre (NANOCAT)
Institute for Advanced Studies
University of Malaya
50603 Kuala Lumpur
Malaysia

Suchithra Thangalazhy-Gopakumar
Department of Chemical and Environmental Engineering
Faculty of Science and Engineering
University of Nottingham Malaysia
43500 Semenyih
Selangor Darul Ehsan
Malaysia
Suchithra.Thangalazhy@nottingham.edu.my

Suyin Gan
Department of Chemical and Environmental Engineering
Faculty of Science and Engineering
University of Nottingham Malaysia
43500 Semenyih
Selangor Darul Ehsan
Malaysia

Lai Yee Lee
Department of Chemical and Environmental Engineering
Faculty of Science and Engineering
University of Nottingham Malaysia
43500 Semenyih
Selangor Darul Ehsan
Malaysia
lai-yee.lee@nottingham.edu.my

A. Saravanan
Department of Biotechnology
Rajalakshmi Engineering College
Chennai
Tamil Nadu
India

S. Jeevanantham
Department of Biotechnology
Rajalakshmi Engineering College
Chennai
Tamil Nadu
India
jeevananthamsbt31@gmail.com

S. Karishma
Department of Biotechnology
Rajalakshmi Engineering College
Chennai
Tamil Nadu
India

Vijyendra Kumar
Department of Chemical Engineering
NIT, Raipur 492010
Chhattisgarh
India
vkumar.che@nitrr.ac.in

Muskaan Kedia
Department of Chemical Engineering
NIT, Raipur 492010
Chhattisgarh
India

A B Soni
Department of Chemical Engineering
NIT, Raipur 492010
Chhattisgarh
India

Bidyut Mazumdar
Department of Chemical Engineering
NIT, Raipur 492010
Chhattisgarh
India

Dharm Pal
Department of Chemical Engineering
NIT, Raipur 492010
Chhattisgarh
India

Gamini Sahu
Faculty of Science
Shri Rawatpura Sarkar University
Raipur 492001
Chhattisgarh
India

Ashwin Raj Suresh, Moni Philip Jacob Kizhakedathil, Chithra Ashok

Removal of the antibiotics from wastewater by biochar

Abstract: Antibiotics are drugs that are used to treat infections caused by microorganisms. Overusage and overexploitation of antibiotics by industries have led to its accumulation in the environment.

This has severely impacted the environment and has been the major reason for the evolution of superbugs. Despite employing several strategies, removal of antibiotics from these contaminated sites has become a Gordian knot. Eco-friendly approaches need to be applied to mitigate this issue. Biochar, produced from the pyrolysis of biomass, may hold the key to curb this issue.

Literature has indicated the use of biochar as a suitable method to tackle the problem due to its high selective adsorption capacity and large surface-to-volume ratio. Use of biochars offer several advantages in environmental and economic aspects. The biochars used for such treatments are often produced from waste agricultural materials and therefore is quite an affordable approach. Also, biochars are degradable and environment friendly. This chapter collates the neoteric data on the consumption of antibiotics, the environmental contamination, and the removal of antibiotics using biochar (BC)-based materials and on the new emerging approaches of BC modification and BC composites in relation to the antibiotic removal from water.

Keywords: Biochar, biomass, antibiotics, adsorption, surface modification, eco-friendly

Introduction

There is no doubt that the discovery of penicillin in 1928 by Alexander Fleming is not only the starting of a new generation of medicine but also the cornerstone of human history. In 1945, research done by Howard Flory and Ernest Chain promotes the commodification of penicillin. This promoted the thorough evolution of antibiotic work during 1940–1970, including commercial use; this period of antibiotic discovery was called the "golden age." Using these antibiotics many lives can be saved. However, the wide usage of antibiotic causes problems because of overuse. Due to the emergence of bacterial strains resistant to penicillin, methicillin is used as the replacement for penicillin in 1959 and ampicillin in 1961. Antibiotics forced the development of more and more new drugs; besides, it has also promoted the

https://doi.org/10.1515/9783110734003-001

steady growth of the cost and consumption of antibiotics. AMR is a serious worldwide health problem. Many organizations, including the WHO, have established this as the main goal to fight against this phenomenon. Antibiotics usage cause hazars for the environment, which is considered as the other major problem [1].

Population growth increases food demand which triggered the changes in the development of antibiotics consumption. In the past few decades (since the 1950s), these drugs have been widely used in livestock, poultry farming, and aquaculture. These are not only used as drugs but also mainly used as growth stimulants and acts as an additive for disease prevention. Many countries such as the United States, Canada, Mexico, and Israel also use antibiotics to grow plants. It is also used as plant protectant to increase yield. Currently, more antibiotics are used in the production of animal-derived foods; approximately 70% of global antibiotics consumption which is 30% more than human consumption of antibiotic drugs. In 2010, the global market of antibiotic production was a total of 100,000 tons from that 63,200 tons were used as medicines for farm animals. In the future, by 2030, the use of veterinary drugs is expected to increase by about 105,600 tons per year [2]. The increase in antibiotic use is reflected in the increase in pollution. This is because 30–90% of the drugs entering the human body are excreted without metabolism [3]. Some antibiotics are now divided as emerging pollutants (EP) or pollutants of concern. EP is a naturally or artificially introduced compound, The presence of EP in the environment is a threat to fauna and flora including humans. Around the world, different types of antibiotics residues have different characteristics, biodegradability, and they also cause toxicity in wastewater, mud, groundwater, soil, surface water, and plants. This leads to drinking toxic water and foods. Many of these compounds are very accumulate and persistent in solid and other soils. The exposure of organisms, as well as humans to antibiotics will affect the microbiome, change the entire ecosystem of microbial balance, reduce drug resistance, and induce antibiotic-resistant bacteria [4]. Therefore, it is important not only to control and suppress antibiotics usage but also is used to remove antibiotics effectively from various environmental substrates mainly in soil and water.

Various methods have been used to remove organic pollutants from environmental matrices usually referred to as a destructive process like biological methods through microbial degradation, and chemical methods through oxidation and precipitation process, and chlorination, and destructive physical process including filtration, sedimentation, and membrane process [5]. Among the above methods, adsorption is particularly common, especially the adsorption related to the removal of antibiotics from the environment which has been confirmed in a large number of scientific publications on the subject of "antibiotic + adsorption" in the SCOPUS database 20 years. The adsorption process takes place at the interface between the liquid phase and the solid-phase adsorbent. This stage of execration is usually the main source of environmental pollution caused by antibiotics [6]. Mainly antibiotics are transferred to solid substrate through some processes, such as soil, sediment, and plants. The adsorption process is relatively cheap, effective, and simple. Therefore, it can be widely used in sewage treatment plants.

There are currently many types of adsorbents available, which can be divided into organic, inorganic, and synthetic adsorbent materials. A suitable desiccant must be selected to remove the impurities of antibiotics from the water which is considered an important process. Activated carbon (AC) has been proven to effectively remove organic and inorganic pollutants. In addition to a wide range of modification options, biochar (BC)-based adsorbents have also become more and more important in recent years. Several scientific articles suggested that BC is a preferable adsorbent that consists of a carbon-rich product obtained by heating the biomass with little or no air at a moderately low temperature (<700 °C) [7]. BC can be made from different types of raw materials which is organic, including waste, such as garbage residual biomass, household waste, agricultural, food waste, and animal waste. Besides, BCs are very cheap when compared to ACs which increases the attractiveness of BCs.

Due to the corresponding physical and chemical properties and their adsorption characteristics which can be further improved by various chemical and physical modifications, BC is used to adsorb many pollutants. There are constant reports of the use of BC from various sources, including industrial BC to adsorb antibiotics. This shows the relevance of production and the use of new cheap adsorbents. However, the environmental pollution caused by such pharmaceutical products, these adsorbents are "green." This article provides a rigorous review of the latest literature data on antibiotic use, related pollution, and examines the use of various BC-based materials to adsorb and remove these types of pollutants. In particular, the mechanism that causes adsorption should be emphasized.

1 General information, definitions, and classifications

The Greek word antibiotics mean "against life" which was first used by Selman Waksman and this substance was produced by microorganisms. These antibiotic substances have the ability to inhibit or kill microorganism growth. Using both semisynthetic and synthetic methods, the antibiotics were produced and they may also have effects like bactericidal and bacteriostatic. Aside from having antibacterial nature, it is also used to treat cancer as a cytostatic and fungal infection. There are around 250 antibiotics grouped in accordance with criteria in animal and human medicine. Antibiotics classification is not an easy thing because of its different chemical structure from single ring compounds to complicated compounds; another way of dividing the antibiotic is the action of mechanism against bacteria. According to this mode of action, five major types were distinguished in the antibiotics. They are cell membrane breakdown, DNA replication and repair, cell wall synthesis, blocking key metabolism pathways, and protein synthesis inhibition [8].

Due to the antimicrobial resistance issue, the popular medicine which is used for the treatment of disease became less effective. That is why antibiotics classification is highly needful due to their safety and effectiveness. According to research, the antibiotics available in the market are active against most of the disease-causing bacteria. The WHO published that quinolones, macrolides, glycopeptides, and cephalosporins are antimicrobials that are widely used for veterinary and human medicines. Ergo, in the future, antibiotics usage will increase [9].

1.1 Antibiotics consumption

It is important to study and monitor the consumption of antibiotics and its environmental contamination because of its usage in the future; with that, pathways of distribution and antibiotic contamination source are also predicted (Figure 1).

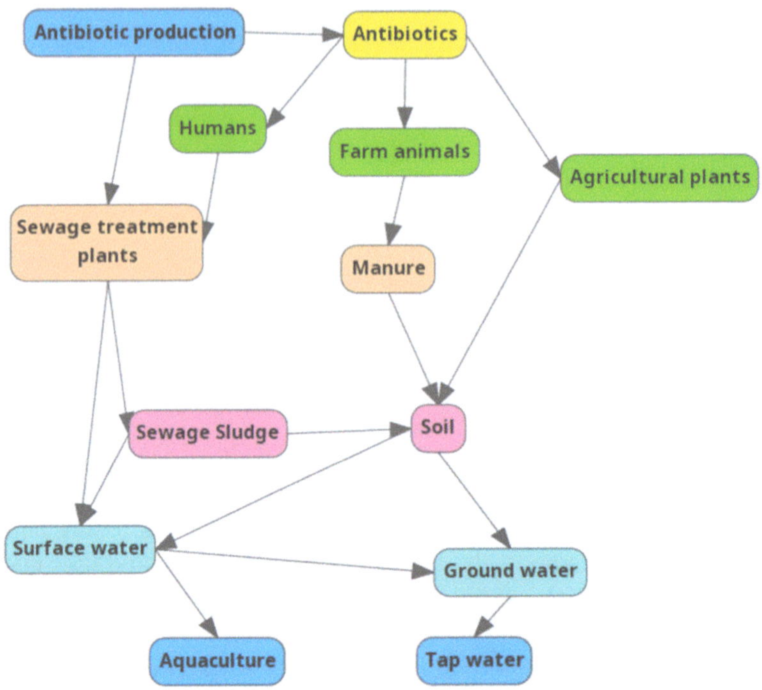

Figure 1: Modes of antibiotic contamination to various water sources.

1.2 Human consumption

Antibiotics used for human consumption are less than one-third of the global market. Consumption of antibiotics increases in the way provided by human medicine guidelines that are throughout daily doses. High-income countries are the major consumers of antibiotics, the consumption rate was recorded as 6% in the defined daily doses. The problem in antimicrobial resistance creates awareness among the people about antibiotics resultant in increasing the rate of consumption by about 114% which also helps to increase the economic development of a country [10]. Recent study by Patrycja Krasucka predicts that the antibiotic consumption rate among the global market will become 200% by 2030.

1.3 Animal consumption

Two-thirds of the global antibiotics will be used for veterinary medicine and they are significant customers of antibiotics. For veterinary medicines, antibiotics play a dominant role in promoting the growth of animals, production of livestock to prevent the disease, and also used for aquaculture. The value of antibiotics in veterinary drugs was estimated, which shows the rise of antibiotic usage of about 42.9 billion. Aureomycin is the first antibiotic used in the 1940s as feed for pigs, calves, and chickens for growth and weight gain. The U.S. Food and Administration states that antibiotics can be used as animal feed additives without the prescription and that European countries used large-scale antibiotics in animal husbandry [11]. For this reason, the antibiotics market share rises higher than 40%. The food production increases lead to the higher use of antibiotics for that growth rate of livestock becoming increase. In developing countries, these trends were particularly found. The estimation of antibiotics consumption in the field of the animal sector will be increased up to 67%. Antibiotic consumption in the production of livestock will account for an increase in the global rate of about 30% in China alone.

According to the data gave by the World Organization for Animal Health, the antimicrobial agents used for livestock production were reduced. As a result, the usage of antibiotics increases. Not only livestock, but aquaculture also plays a major role in the consumption of antibiotics which produce 8% of animal protein. The usage of these substances is restricted in fish farming and it is under the control of regulatory legislation. The use of some antibiotics are prohibited in aquaculture by the authorities due to their specific hazards, for example, an accumulation of these antibiotics in fish will be dangerous for the health of the human food chain [12]. Furthermore, the whole ecosystem was affected by an applied drug, about 70–80% of which will be excreted in the water and cause contamination.

1.4 Antibiotic use in plant agriculture

For the protection of plant products, only a few chemical antibiotics are used on small scale. Among the global production of antibiotics, only 0.4% is used for plant protection. These antibiotics are used for plants only when the disease affects the development of the global economic rate. The bacterial diseases like bacterial spot and fire blight are controlled by using oxytetracycline and streptomycin in the peaches, apple, and others. To overcome a disease called citrus greening in citrus groves, these two antibiotics were sprayed. Apart from fruit tree protection, the gentamycin with that oxytetracycline were both used for vegetable crops and rice crops. The antibiotics used for veterinary and human medicine were sprayed on the plants that cause the microorganism present in the soil to develop resistance gene; this resistance gene will be used to act against the human pathogens [13]. Antifungal chemicals like triazoles and imidazole belong to the azoles group which is also used for the protection of plants, but it is unlike human medicine, although it has a similar chemical structure. The risk of cross-resistance was found in human pathogens because of its chemical similarity to medical azoles.

2 Source and distribution of antibiotics in the environment

A wide range of diversity in human activities results in the entry of antibiotics into the environment. It is important to identify the pathways and risks of contamination when antibiotics and humans are relatively present in the environment, and setting up mutual correlations between them is also equally important [14].

Humans and animals are the main customers of antibiotics and also, they are the important primary source contaminating the environment. After the consumption of antibiotics by human beings, it is metabolized and the rest of the unwanted materials are excreted through urine and feces. These excreted materials further pollute the sewage and wastewater treatment plants by accumulating in the sludge that may also use as a soil improver. Due to the failure of pipelines, these pollutants directly enter the surface of the soil and mix with groundwater. The main source of antibiotics is industrial manufacturing of pharmaceuticals which finally contaminate the environment. The antibiotics are mostly present in natural elements like lakes, wells, soil, and so on, but the concentration of antibiotics is higher than the therapeutic doses recommended. The main pathway to distribute the residues of veterinary antibiotics is to use as manure for a plant source. Tetracycline (TC) and sulfonamides are the major antibiotics found in manures. The antibiotics concentration in manure is higher than the dose prescribed; however, there are some appropriate methods that can lower the concentration in manures [15]. The effective degradation of antibiotics depends on strict conditions such as humidity, oxygen amount, and temperature.

The animal source also contaminates aquaculture. Excreted compounds from fish farming are mixed in water and consumed by plants and aquatic organisms. Subsequently, some amount of pollutant can be transferred to large water bodies such as seas and oceans. Frequently identified sulfonamides in the water bodies are toxic to algae by affecting the photosynthesis pathway. This may affect the aquatic ecosystem balance [16].

Another important pathway to contaminate the environment is plant protection activity. Chemical compounds can enter into soil and water by spraying method cause contamination. These contaminated water and soil are again absorbed and circulated through animals and plants, which links to the food chain of humans. By assessing the food products such as meat, fish, vegetables, and seafood, antibiotic residues are found even in the drinking water. Humans become the secondary consumer of antibiotics by means of devouring the contaminated food varieties [17].

Antibiotics present in the environment are characterized by varying physicochemical properties. As a result of characterization, soils have a higher concentration of antibiotics than water bodies. The half-life of antibiotics will be extended by the temperature reduction and the light absence. Seasonal changes also play a role in the antibiotic drug's half-life. For example, in summer, because of the higher temperature, drug can be degraded easily, also continuous rainfall is responsible for rapid dilution. These drugs are not degraded for several months and years, this results in bioaccumulation [18]. The pollution caused by antibiotics not only affects the insensate elements but also affects living organisms. Whole organisms present in the ecosystem have an impact on antibiotics. Such pollution is the major reason for arising danger like antimicrobial resistance and the spreading of the natural microbiome [19]. Due to these reasons, the removal of antibiotic residues from the environment is most important.

3 Biochar-based materials for the removal of antibiotics from water

Interest in BCs is growing has been increasing in the past 20 years and multidisciplinary fields are also researching the BCs [20]. Some research works are going in this field for the production of BC materials and their practical application. Recent researches are focusing on the risks rises when BC is introduced into the environment. BCs main advantage is to have the possibility to merge in the science field and across life easily. It has applications in the fields of energy production, soil properties improvement, waste management, and mitigation in climate change. BCs continuous reports show that it is sustainable and low cost which is also used for replacing the aromatic carbons. It is also known as "green" and widely used for the removal of contaminants in the environment, adsorption, and immobilization techniques [21]. Due to the functional groups and porous structure presence, it has the ability to remove organic, inorganic, volatile substances, pesticides, and various classes of drugs. Increasing antibiotics consumption will increase the contamination in the environment [22]. This increases the need to produce adsorbents which is effective toward removing the pollutants from the matrices of the environment. Research studies show that the best removers of antibiotics are carbon adsorbents and aromatic carbons. But the ACs are not cost-effective, for that reason the BC is used which is also having the same physiochemical characters as ACs and cost-effective. For this reason, some researchers turn their focus on developing antibiotic contaminants remover using BC.

3.1 Adsorption on biochars

The BC adsorption process can be divided into four stages such as (I) adsorbate molecules transportation from solution to BC; (II) adsorbate diffusion in the liquid film surrounding the BC; (III) diffusion of liquid in the pores; and (IV) adsorbate interaction with BC [23, 24].

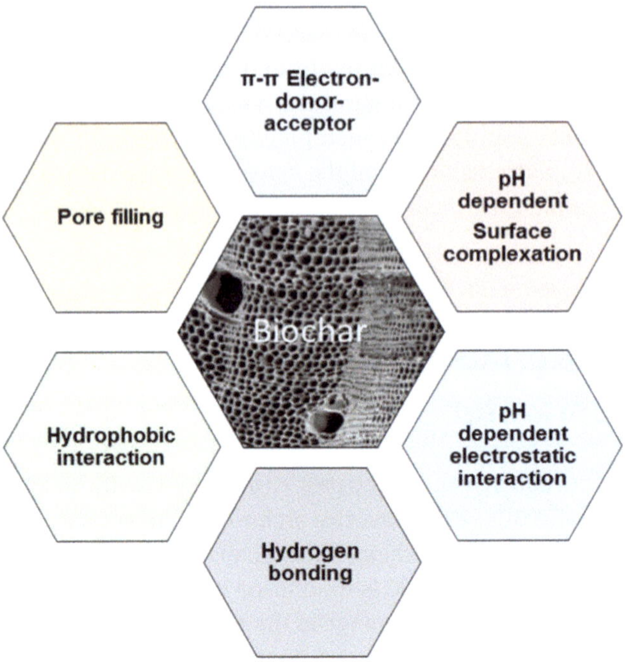

Figure 2: Type of interactions that antibiotics exhibit with biochars.

The mechanism of adsorption depends on the interaction between adsorbent and adsorbate not only this, but various interactions are also involved in the adsorption of antibiotics on BC. The various interactions are electrostatic interaction, dipole interaction, hydrogen bonding, hydrophobic interactions, and so on [25]. Definitely, not all the reactions are taking place in the antibiotic–BC system because the interactions depend on the antibiotic chemical structure as well as the desorption and adsorption condition. The adsorption capacity and mechanism are influenced by major factors such as the nature of adsorbate, pore structure, physicochemical properties, and relative affinity. The efficient adsorption was carried out using the optimal selection of system conditions and components [26]. Although, the removal of antibiotics is very difficult because of the types of pollutants that fix in the environmental matrices. So, it is important to select the selective adsorbent matrix that targets the antibiotics. To create interest in these engineered BCs, some advantages are discussed here (Figure 2).

4 Biochar-based adsorbents of antibiotics

On account of BC adsorbents, there are numerous approaches for the properties control with the goal which shows the most ideal properties of adsorption. Contingent upon methodology and the "headway" of alteration and amalgamation techniques, the materials such as inferred will be ordinarily partitioned into engineered BCs and pristine BCs. The last mentioned can be divided into BC composites and altered BCs (Figure 3).

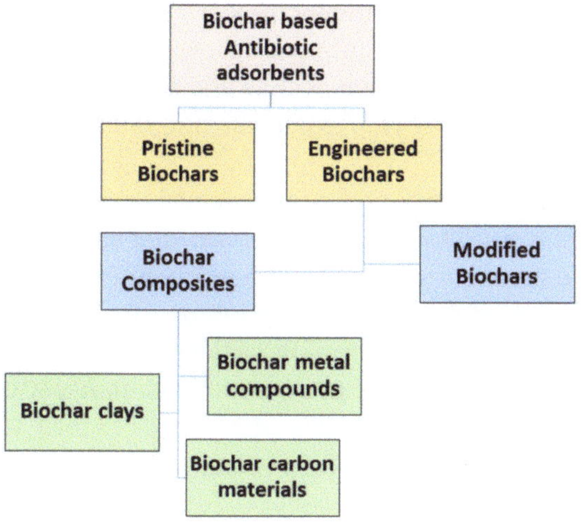

Figure 3: Classification of biochar-based materials.

4.1 Pristine BCs

Pristine BCs are delivered through moderate biomass pyrolysis. In connection to the adsorption of antibiotics, BC created attributes, specifically the parameters such as sorption (volume and measurement of pore size) just like its hydrophilicity, aromaticity, and thickness of oxygen surface functional gatherings or debris; this all will play a fundamental role. The BC properties that got to rely on the feeds and conditions for pyrolysis measure are rate of warming, temperature, type of carrier gas, and time of pyrolysis [27]. To date, numerous examines have directed on antibiotics agent's adsorption onto BCs delivered under changing properties and from various stocks.

4.1.1 Effects of pyrolysis conditions

The choice of pyrolysis conditions as time and temperature of heating rate, transporter gas, and pyrolysis, permits for the change of obtained yield and BCs properties. Moderate pyrolysis is described by the rate of heating as (<10 °C min^{-1}) for getting this BC as a principle item (more than 30% yield). Rather than quick pyrolysis, the heating rate will be extremely quick (>10 °C s^{-1}) which favors the bio-oil creation (more than half yield). Other than improving proficiency, higher warming rate influences the BCs properties. The surface zone and porosity will be diminished by the cell structure destruction. At a similar temperature, the residence time increment impacts the carbonization productivity measure diminishes the augments porosity and yield. Notwithstanding, a few reports delayed that the time can also diminish the surface area of BC which causes deformities and obliteration in the structure of BC. Without a doubt, the BC properties influencing key factors is pyrolysis temperature [28]. The result of the higher temperature cycle shows significant BC carbonization level like hydrogen and oxygen content decline and carbon content increment; this will helpful for the pore volume (V_p) expansion and the improvement of S_{BET}.

At high temperature, produced BCs are generally portrayed using higher aromaticity, hydrophobicity, and pH esteems, just like low polarity shows that expanding the temperature of pyrolysis increase the adsorption BCs limit toward different antibiotics. The kind of transporter gas utilized for the admittance of oxygen for other pyrolysis properties influencing the adsorption capacity of BC materials delivered. Li et al. researched the impact that affects conditions of pyrolysis, that is, transporter gas and temperature composition on the properties of BCs with regards to their TC adsorption and sulfadiazine adsorption. An expansion of temperature and the expansion of air toward nitrogen during the process of pyrolysis created more pores in the material. The mesopores extent in the produced BC especially porosity was extended. Hence, the vital boundary in the bigger mass or size particle adsorption like antibiotics is about to think it in the wording of pores size of sorbent to maintain a strategic distance from the size-avoidance impact [29].

4.1.2 Effects of feedstock type

Other parameters may affect a huge amount of adsorption of antibiotics with regards to BC utilization for the production of BC using the sort of feeds. Substance like hemicellulose and cellulose with that lignin are the first feedstock materials utilized to create the BC essentially to influence the yield of pyrolysis; they are also the natural composition and debris of the delivered materials. In synopsis, antibiotics adsorption proficiency to BCs adsorbent was subject to the physical properties and the structure of chemical substance which is medication molecule just like the conditions of adsorption cycle; whatever, the physicochemical properties of BC is primarily reliant. The properties of BC are prevalently dictated to the type of feeds utilized for pyrolysis and its temperature, transporter gas, and time. The further expansion of temperature improves the adsorption of BC to antibiotics by the oxygen present during the process of pyrolysis. This will be connected to BC surface graphitization to a higher degree and raise the fragment particles medication by hydrophobicity. Besides, the improvement of S_{BET} using the high temperature of pyrolysis also extents the pore size especially in the oxygen environment of BCs, in which the pore was loading up in a significant way using the dominated of medication. The stocks utilized for BC creation affects the cycle productivity, compound arrangement, hydrophobicity/extremity, and debris substance of the BC created [30]. The properties which were previously mentioned are the key factors for restricting antibiotics using polar particles or metals with complexation happening in debris. To get a specific and protective adsorbent of BC, the selection of paper feedstocks and analyzing the conditions for production are more important (Table 1).

4.2 Engineered biochars

4.2.1 Modified BCs

The adsorbents like BC pristine-based are showing adsorption which is unsuitable for different antibiotics agents. The absence of procaffinity livity among adsorbent and mdrugedication is generally connected which influenced by shows the impact of BC

Table 1: Pristine BC adsorption capacity on antibiotics.

Biochar feedstock	Pyrolysis conditions			Antibiotics	Adsorption conditions				Maximum adsorption capacity (Q_{max}) (mg g^{-1})	Reference
	Temperature (°C)	Time (h)	Atmosphere		Temperature (°C)	Time (h)	pH	Adsorbent dose (g L^{-1})		
Red pine	400	2	Aerobic	Sulfamethoxazole (SMX)	RT	72	6	10–15 mg L^{-1}	1.9 SMX	[49]
	500	2	Aerobic	Sulfa pyridine (SPY)	RT	72	6	10–15 mg L^{-1}	1.5 SPY	
Rice straw (swine manure)	400	2	Oxygen	Tetracycline TC	25	72	5	3	8.3	[30]
	600	2	Oxygen	Tetracycline TC	25	72	5	3	14.2	
	400	2	Oxygen	Tetracycline TC	25	72	5	3	6.5	
	600	2	Oxygen	Tetracycline TC	25	72	5	3	8.1	
Astragalus mongholicus	200	5	Oxygen	Ciprofloxacin (CIP)	25	12	–	2	5	[50]
	400	5	Oxygen	Ciprofloxacin (CIP)	25	12	–	2	4	
	500	5	Oxygen	Ciprofloxacin (CIP)	25	12	–	2	20	
	600	5	Oxygen	Ciprofloxacin (CIP)	25	12	–	2	25	
	700	5	Oxygen	Ciprofloxacin (CIP)	25	12	–	2	35	
	800	5	Oxygen	Ciprofloxacin (CIP)	25	12	–	2	38	

Bull manure	600	–	Nitrogen	Lincomycin (LIN)	25	48	10	1	0.30	[51]
Dairy manure	600	–	Nitrogen	Lincomycin (LIN)	25	48	10	1	0.37	
Poultry manure	600	–	Nitrogen	Lincomycin (LIN)	25	48	10	1	0.43	
Pinewood (*Pinus radiata*)	600	3	Nitrogen	Tetracycline (TC) Sulfadiazine (SD)	25	48	–	0.08–0.5	TC 5.53 SD 1.27	[29]
	600	3	Air:N_2 1:5	Tetracycline (TC) Sulfadiazine (SD)	25	48	–	0.08–0.5	TC 6.42 SD 2.20	
	700	3	Nitrogen	Tetracycline (TC) Sulfadiazine (SD)	25	48	–	0.08–0.5	TC 7.38 SD 2.84	
	700	3	Air:N_2 1:10	Tetracycline (TC) Sulfadiazine (SD)	25	48	–	0.08–0.5	TC 17.5 SD 9.32	
	700	3	Air:N_2 1:15	Tetracycline (TC) Sulfadiazine (SD)	25	48	–	0.08–0.5	TC 101 SD 123	
	800	3	Nitrogen	Tetracycline (TC) Sulfadiazine (SD)	25	48	–	0.08–0.5	TC 8.66 SD 4.94	
	800	3	Air:N_2 1:10	Tetracycline (TC) Sulfadiazine (SD)	25	48	–	0.08–0.5	TC 26.6 SD 11.4	
	800	3	Air:N_2 1:5	Tetracycline (TC) Sulfadiazine (SD)	25	48	–	0.08–0.5	TC 261 SD 163	
Alfalfa (*Medicago sativa L*)	500	0.5	Nitrogen	Tetracycline (TC)	25	5 days	5	0.1	372.31	[52]
Bermudagrass (*Cynodon dactylon*)	500	0.5	Nitrogen	Tetracycline (TC)	25	5 days	5	0.1	44.24	

(continued)

Table 1 (continued)

Biochar feedstock	Pyrolysis conditions			Antibiotics	Adsorption conditions				Maximum adsorption capacity (Q_{max}) (mg g^{-1})	Reference
	Temperature (°C)	Time (h)	Atmosphere		Temperature (°C)	Time (h)	pH	Adsorbent dose (g L^{-1})		
Waste *Auricularia auricula* dregs	300	2	Nitrogen	Tetracycline (TC)	RT	12	–	0.8–4	7.22	[53]
	500	2	Nitrogen	Tetracycline (TC)	RT	12	–	0.8–4	9.90	
	700	2	Nitrogen	Tetracycline (TC)	RT	12	–	0.8–4	11.90	
Rice straw	300	2	–	Tetracycline (TC)	25	24	–	3	4,136	[54]
	500	2	–	Tetracycline (TC)	25	24	–	3	8,756	
	700	2	–	Tetracycline (TC)	25	24	–	3	14,157	
Corn cob	500	24	Oxygen	Chlortetracycline (CTC)	25	24	5	10	12.39	[55]
Sugarcane bagasse	500	24	Oxygen	Chlortetracycline (CTC)	25	24	5	10	16.96	

polarity, aromaticity, with that the presence or absence of not required or required functional groups of the surface. In frameworks, the adsorption components mostly depend on pore fillings, due to the low porosity parameter the adsorption capacity is very low including the $S_{BET,}$ V_p, also additionally extremely thin pore width are related to the size of antibiotics molecules. For the improvement of BCs property, particularly the adsorption properties of various mixtures, including the antibiotics, BCs are exposed for changing the properties. The widest division of these alterations depends on synthetic and physical strategies [31].

4.2.1.1 Physical modification

BC's physical change strategies contain gas actuation such as microwave adjustment additionally known as microwave pyrolysis and ball processing. The first two strategies contain purported post-union techniques, wherein the recently blended BCs change completely. The third sort change contains heating through pyrolysis by utilizing the microwaves as the source, during this BC is produced as aftereffect for the activation of microwaves, the design and properties are altered. The structure of BC pores is improved according to adjustments of physical properties, especially on gas change. This will identify the deficient burning items expulsion, unblocking BC pore, and degasification during the interaction change. The strategies applying incorporate their minimal effort, immaculateness, and have the potential to control interaction [32].

In another research on adsorption of TC shows the heat treatment method which blends with nitrogen and air used for the BC adjustment. BC extracted from the pinus radiata dust heated at 700 °C under the climate of nitrogen at different rates of the stream. For the adjustment of S_{BET} and arrangement of mesopores, gas with the presence of air is utilized. For the quantity expansion of oxygen, BC utilitarian gatherings was used. Likewise, it seems that expanding air and gas combination which cause ascent synchronous in BC porosity and oxygen content percentage, will drove the enhancement of adsorption ability of TC. BC change will utilize the supply of air, the limit of sorption will be higher than BC adjusted without air over multiple times. It tends to be derived from the kind of material, the interaction among the adsorption of TC is identified with the pore-filling component, likewise connections including electrostatic interaction and hydrogen bonding [33].

4.2.1.2 Chemical modification

Alteration chemical methods include synthetic activity mixture, normally with reducing and oxidizing property, perfect BCs would have the capacity to undergo simple drying or are developed further for microwave treatment. In presynthesis method, the feeds are exposed for the chemical substance activity and subsequently pyrolyzed. Synthetic components like HCl, H_2O_2, and $KMnO_4$ are utilized ordinarily in both post- and presynthesis strategies. These kinds of reagent activity are prompt BCs oxidation surface molecules like aldehyde and hydroxyl and are also used for the carboxyl moieties development which is equipped for restricting cations, like the S_{BET} and microporosity improvement. The bases which act as reducer are KOH, and NaOH which normally leads to decrease in the hydroxyl group measure on surface of BC; this results in the reduced hydrophilicity and porosity of changed BC, these techniques considerably increase the expulsion of non-calcinated BC deposits. In contrast to physical changes, whose principal design is to build up the BCs porosity and adjust, the chemicals predominantly cause changes in the synthetic idea of BCs, particularly oxygen surface [34]. This sort of modification is of exceptional importance according to antibiotics adsorption. Nonetheless, the correlation with a physical process and usage of chemical techniques will be hazardous for the environment and the reagents are costly.

Liu et al. [35] explored the TC adsorption to the rice husk BC with H_2SO_4 (BC–A) or KOH (BC–B). In the examination with pristine BC, the TC adsorption limit expands toward both changed materials: corrosive modification shows 35% expansion in adsorption of TC, though KOH change ends up being more effective, giving a 250% expansion. These varieties were because of the independent property of adsorbents. BC exposed for the activation of base (BC–B) had two–three times high than the S_{BET} and V_p, increased carbon percentage content, least oxygen percentage content, and decrease the debris components comparatively with unblemished BC and altered corrosive (BC–A). This will increase the adsorption limit of TC subject to the porosity of BC. The surface oxygen substance containing functional groups was additionally discovered to be important because the adsorption of TC takes place.

Fan et al. [36] researched the adsorption of chloramphenicol (CAP) on bamboo-determined BC just like BC adjusted with H_2SO_4 (BC–A) and KOH (BC–B). BC–B shows the most noteworthy adsorption capacity toward CAP. This was compared with the change that happened in chemical arrangement and types of functional group in BC. BC–B had the most elevated content of oxygen just like the most noteworthy C=O percentage of surface molecules and structure of graphite is contrasted with the other two BCs (unblemished BC and BC–A).

Some research shows that elaborate pre-union adjustment of sewage sludge determined BC utilizing NaOH, build more than multiple times. NaOH adjustment likewise caused a more than the four-crease increase in S_{BET} and volume of mesopores, just like the eight-overlap increment in V_p estimation of mesopores of BC. In this situation, the upgraded adsorption of TC was related to increased porosity and with the pore filling system.

Jang et al. [37] contemplated the sorption of TC onto BCs created from *Medicago sativa* at a temperature of 300 °C. Three BCs were tried: BC pristine carbonized at 800 °C (BC-800), pristine BC (BC), and pristine BC exposed to the activity of NaOH and carbonized at 800 °C (BCB-800). The best outcomes of adsorption of TC were received from BCB-800 values decided for the BCB-800 which were multiple times more noteworthy than those for BC-800 and BC pristine, individually. Besides, BCB-800 shows higher S_{BET} than the pristine BC. BCB-800 will be portrayed to the least polar just like the high aromatic content and carbon content. The isotherms adsorption on the TC restricting energy with individual BCs demonstrated a unique restricting mechanism relying upon the BC type. On account of perfect BC and BC-800, the information shows that TC will bind through physisorption, while on account of BCB-800, it will bind through chemisorption.

TC adsorption will additionally be examined on the husk of rice inferred BC which adjust with methanol. The fermented liquor solution is utilized for esterification of the carboxyl group; in this manner, the properties of surface oxygen changing will take part in antibiotics binding. The modified BC was described using diminished carbonyl groups number, and the number of hydroxyl and ester group numbers will correlate with BC pristine. An increase in the S_{BET} of BC by 25% was recorded after the alteration of methanol, which brought the adsorption of TC increment limit by 46% compared to pristine BC; after 13 days, this shows the distinction of about 17%.

Hydrogen peroxide (H_2O_2) was additionally used for the adjustment of antibiotic BC adsorption. Utilizing the pyrolysis stalk, Tan et al. created three sorts of perfect BC at 300, 450, and 600 °C (BC-300, BC-450, and BC-600, individually). BC's adjustments were done by the postcombination method by utilizing 30% H_2O_2 (BCM-300, BCM-450, and BCM-600). Pristine BCs delivered at the most reduced temperature show high adsorption on TC. Curiously, BC-300 shows the most reduced boundaries of pores compared to the other tried BCs. The higher adsorption of TC on BC will be clarified by functional group presence on its residues and surface of natural non-carbonized that collaborate the TC. On account of changed BCs delivered at increased temperatures, the TC adsorption will expand by 30%, when arriving at maximum for BCM-600. The action of oxidative H_2O_2 emerges cause the oxygen number expansion taking part in the TC adsorption atoms through the hydrogen bond [38] (Table 2).

4.3 Biochar composites

The complex BC composites are utilized for antibiotic evacuation. As indicated by the definition of the International Union of Pure and Applied Chemistry, composite is known as multicomponent material involving numerous unique areas, while nanoparticle measurement is one of the stages, it is known as nanocomposite. In a ceaseless stage, different segments of BC were joined; most regularly, they are hydroxides and metal oxides, mud minerals, graphene oxide (GO), carbon materials carbon nanotubes (CNTs), graphene, and polymers. Commonly, the composites BC will be divided into two types, on account of the BCs modified. In this technique, the segments composite will consolidate into stocks before the pretreatment of pyrolysis, while in the subsequent technique, composite segments were added to BC which is completed after pyrolysis through the impregnation with instant nanoparticles, and oxides precipitation, furthermore, coatings. A benefit of this methodologies is that underlying the other matrices along with BC at the same time by which they consolidate the qualities of the underlying parts; subsequently, the material acquired displays new characteristics. On account of the composites of BC, the lattice of BC might be answerable for composite adsorption which is hydrophobic in nature related to high properties. At long last, the properties can pass through inferred composite. It ought to be considered that specific properties may decay. By changing the physicochemical properties with the growing new functionalities of composites of BC, this will be combined with synthesis process which is a simple and large number of particles available; this type of materials have tried numerous investigation from water for expulsion and adsorption of antibiotics in present years [39].

4.3.1 Bio-char clay composites

Due to their layer construction, high S_{BET}, and high particle trade limit, dirt minerals, it can eliminate different foreign substances including colors, substantial metals, and natural compounds. Also, they are available in many ways and minimal effort materials. Also, clay minerals are minuscule particles like nano-order, which upsets their application, in the substrate adsorption fixed-bed. The limit of utilizing earth fixed-bed minerals framework has lower flowability of water–dirt minerals, particularly if there should be an occurrence of the swellable. Along these lines, it appears to be prudent to join them with composites in different materials that have permeable design this will play out the capacity of the host like BCs or polymers.

Premarathna et al. [40] utilized two clay minerals, red earth clay (RE), and waste of municipal to create composites of BC through pretreatment technique. The composites which are delivered were named BC–MMT and BC–RE, individually; the pristine BC along with adsorption of TC using pH range about 3–9. Both mud minerals expansion caused almost two-overlay expansion in S_{BET} acquired composites, the proof which non-attendance phenomena of blocking the pore after incorporating permeable particles of minerals into BC. The adsorption of TC limit did not contrast essentially between pristine BC and the BC–RE. The comparable estimations of adsorption of TC limits on pristine BC and BC–RE proof for previously mentioned composite, the grid of BC is mostly mindful for adsorption, while the RE presence does not have huge effects on most amount and adsorption system. The MMT composite will display higher adsorption of TC limit compared to two other materials.

Ashiq et al. [41] researched the ciprofloxacin (CPX) adsorption with biocomposite recieved from the waste of municipal with bentonite mineral expansion. The

Table 2: Modified BC adsorption on antibiotics.

Biochar feedstock	Pyrolysis conditions			Modified sorbents	Antibiotics	Adsorption conditions				Maximum adsorption capacity (Q_{max}) (mg g^{-1})	Reference
	Temperature (°C)	Time (h)	Atmosphere			Temperature (°C)	Time (h)	pH	Adsorbent dose (g L^{-1})		
Bur-cucumber (*Sicyos angulatus* L)	700	2	Oxygen	Pristine BC	Sulfamethazine (SMT)	25	72	3	1	20.56	[31]
	700	2	Oxygen	Steam activated BC (5 mL min^{-1}, 45 min, 700 °C)	Sulfamethazine (SMT)	25	72	3	1	37.73	
Tea waste	700	2	Oxygen	Pristine BC	Sulfamethazine (SMT)	25	72	3	1	7.12	[56]
	700	2	Oxygen	Steam activated BC (5 mL min^{-1}, 45 min, 700 °C)	Sulfamethazine (SMT)	25	72	3	1	33.81	
Pinus radiata sawdust	400	2	Nitrogen	N2 200 mL min^{-1} + air flux mL min^{-1}: 0	Tetracycline (TC)	25	48	6.8	1	10.4	[33]
	400	2	Nitrogen	min: 30	Tetracycline (TC)	25	48	6.8	1	55.3	
	400	2	Nitrogen	min: 50	Tetracycline (TC)	25	48	6.8	1	90.4	
	400	2	Nitrogen	min: 70	Tetracycline (TC)	25	48	6.8	1	96.1	

(continued)

Table 2 (continued)

Biochar feedstock	Pyrolysis conditions			Modified sorbents	Antibiotics	Adsorption conditions				Maximum adsorption capacity (Q_{max}) (mg g^{-1})	Reference
	Temperature (°C)	Time (h)	Atmosphere			Temperature (°C)	Time (h)	pH	Adsorbent dose (g L^{-1})		
Rice straw (swine manure)	700	2	Oxygen	Pristine BC	Tetracycline (TC)	25	216	–	0.16–0.33	150.2	[30]
	700	2	Oxygen	Modified BC	Tetracycline (TC)	25	216	–	0.16–0.33	166.3	
	700	2	Oxygen	Pristine BC	Tetracycline (TC)	25	216	–	0.16–0.33	127.8	
	700	2	Oxygen	Modified BC (post-synthesis modif. BCs by immersion (24 h, 25 °C) in 14% H_3PO_4)	Tetracycline (TC)	25	216	–	0.16–0.33	159.7	
Rice husk	500–550	–	–	Pristine BC	Tetracycline (TC)	30	–	–	5	16.95	[35]
	500–550	–	–	H_2SO_4	Tetracycline (TC)	30	–	–	5	23.26	
	500–550	–	–	KOH (post-synthesis modif. by stirring (1 h, 60–70 °C) in (v/v) 10% H_2SO_4 or 3 mol L^{-1} KOH)	Tetracycline (TC)	30	–	–	5	58.82	

Bamboo	450	–	–	Pristine BC	Chloramphenicol (CAP)	25	0.5	–	8	0.6	[36]
	450	–	–	10% H_2SO_4	Chloramphenicol (CAP)	25	0.5	–	8	1.5	
	450	–	–	10% KOH (post-synthesis modif. by stirring (6 h, 60 °C)	Chloramphenicol (CAP)	25	0.5	–	8	2.5	
Treatment plant sludge	700	–	–	Pristine BC	Tetracycline (TC)	25	8 days	–	–	302	[57]
	700	–	–	Presynthesis modif. impregnation sludge in NaOH (50 g/100 g w/w)	Tetracycline (TC)	25	8 days	–	–	1,248	
Pinus taeda	300	–	Nitrogen	Pristine BC (post-synthesis modif. by mixing BC with 4 M NaOH solution for 2 h and carbonized (800 °C, 2 h))	Tetracycline (TC)	20	3 days	6	0.1	29.42	[37]
	300	–	Nitrogen	Pristine BC (Post-synthesis modif. by mixing BC with 4 M NaOH solution for 2 h and carbonized (800 °C, 2 h))	Sulfamethoxazole (SMX)	20	5 days	5	0.1	274.81	

(continued)

Table 2 (continued)

Biochar feedstock	Pyrolysis conditions			Modified sorbents	Antibiotics	Adsorption conditions				Maximum adsorption capacity (Q_{max}) (mg g^{-1})	Reference
	Temperature (°C)	Time (h)	Atmosphere			Temperature (°C)	Time (h)	pH	Adsorbent dose (g L^{-1})		
Alfalfa hays	300	–	Nitrogen	Pristine BC	Tetracycline (TC)	20	5 days	5	0.1	30.70	[58]
	300	–	Nitrogen	Pristine BC carbonized at 800 °C for 2 h	Tetracycline (TC)	20	5 days	5	0.1	55.62	

CPX adsorption increases 40% on composite contrasted with pristine BC. The porosity improvement in the adsorption will be related to the expansion of composite surface presence, and the CPX particle.

4.3.2 Biochar–carbon composites

Aside from the expansion of mineral clays, blending the carbon materials like graphene, GO, diminished GO, and CNTs with BC are utilized. The motivation behind this is that adding a carbon material with BC will raise the S_{BET} composites, mechanical strength, or resistance of heat. Furthermore, nanocarbons with functional groups are regularly used for surface grafting, also utilized for the improvement of properties of adsorption will create the composite. Carbon–BC composites were discovered for the expulsion application of different sorts of toxins, for example, hefty metals, colors, pesticides, PAHs, and medications.

Huang et al. [42] examined the adsorption of sulfamethazine (SMT) onto BC from bamboo sawdust and its composite with the expansion of GO. The composite was acquired by pretreatment. It was described by more than twofold SMT adsorption limit comparative with BC pristine. Also, GO raises the S_{BET} and measure the oxygen groups practicals by 90% surface composite, this will reflect the adsorption of SMT. The appeared SMT will go through chemisorption, the best commitment of π–π EDA interaction. Adsorption will examine the arrangements of various ionic and pHs qualities. The adsorbed antibiotics will increase by reducing the strength of ions, the most elevated arrangements of SMT particle's pH will electrically unbiased.

Inyang et al. [43] led an examination for the adsorption of SMT BC composites with CNTs. To make BC composites, sugarcane bagasse (SB) and hickory chips (HC) are utilized for alternatives of two pretreatments: (1) with the suspension of CNTs, and (2) with a CNTs suspension containing a surfactant like sodium dodecyl benzenesulfonate; (SDBS). The expansion of nanotubes fundamentally expands the evacuation of SMT; this will accomplish the arrangement of composites within the surfactant sight. Simultaneously, the higher productivity of composites was described using the reduced pH, which responsible for an acid contribution that will take part in the SMT limiting. In both, the SB and CNTs suspension got from composites were also with the CNTs suspension with the SDBS addition, this research discovers that when compared with pristine BC–SB the S_{BET} will be increased.

4.3.3 Biochar–metal composites

Metals like FeO, AgO, and their mixtures, typically oxides like MgO, Al_2O_3, and hydroxides ($Mg(OH)_2$), are naturally used for the composite BC blends. BC helps for changing the oxides and hydroxides of metals which are intended to build the porosity, and change the properties of physicochemical of derived composite; specifically, the surface charge will change from negative to positive. This kind of composite will show the adsorption application of compounds from water which are charged negatively, especially oxyanions, they are unsafe for wellbeing with that the climate and shows the pristine BCs affinity which is weak. The result of composite BCs has a much better adsorption rate than pristine BCs that will attribute the higher porosity properties of composites and functional group presence equipped for particles complexing.

Yang et al. [44] considered the adsorption of TC on BC pristine determined from the slime sewage and to composite (BC–Fe) produced using mixtures of ion and sewage muck. Before the pretreatment pyrolysis, the feeds were impregnated with iron sulfate. X-beam photoelectron spectroscopy researched and affirmed that FeS and Fe_2O_3 present in the composite. The $Fe_2(SO_4)^3$ arranged composite portrayed will expand by higher porosity and hydrophilicity as contrasted by BC pristine. The outcome of the iron mixture present in a composite expansion of about 300% in the adsorption of TC contrast limit and unmodified BC. The uncovered investigation of hydrogen holding complexion metal may have the contribution of TC binding in BC–Fe material, instead of BC pristine.

Zhou et al. [45] determined the compared outcomes and conclude the adsorption of TC using the saw-dust-based composite BC with the zinc and chloride iron salts expansion. The iron and zinc materials consist composite shows the three-overlay higher adsorption limit which connects to TC than BC pristine. Besides, the delivered material exposing to feedstock activity of the combined two salts exhibits the increment four-overlay.

Arrangement of BC composite and nanoscale zero-valent ion (nZVI) was done by Deng et al. with the goal of SMT oxidative debasement within hydrogen peroxide sight. The explored research on SMT proficiency presence will lead to the non-appearance of H_2O_2 for three materials: perfect BC, the composite, and ZVI nanoparticles. Without the arrangement of H_2O_2, the SMT mostly used will be taken out from the pristine BC (31%) and composite (25%). On account of ZVI nanoparticles, expulsion will be 7% simple. The higher qualities of expulsion found in the first two materials SMT adsorption, that are BC pristine and composite BC. The most elevated parameters of adsorption were portrayed using BC pristine, which shows the SMT elevated proficiency expulsion. The portrayed lower porosity composite when compared to the BC pristine, the SMT expulsion is just connected to oxygen groups presence, which will remember those beginning metals and takes part in binding the antibiotics [46].

Potato stems and leaves composites are just like attractive manganese oxide (BC–FeMn) These were used as adsorbents for three agents of antibiotics which belong to the class of quinolone. The biomass of plants including the leaves and stems of potato was impregnated in the $FeCl_3.6H_2O$ solution and therefore the impregnated and dried before pyrolysis with $KMnO_4$. The contrasting outcomes of BC were distinctly made from leaves and stems of potato. In the pristine BC correlation, the composites of BC show an extremely high limit of adsorption about 20–63% for all tried antibiotics. The order accompanying the expansion of adsorption: NOR < CIP < ENR. The varieties of modification that impact the adsorption of antibiotics individually because of the various properties and constructions of the molecules. The proposal of the creators will increase the porosity and the extra oxygen presence on the surface of the adsorbents raise the adsorption rate of the composite when compared to BC pristine. The multitude of antibiotics adsorption raised in pH diminishing and arrangement of ionic strength. It will show that after the three cycles of reuse, BC–FeMn kept up practically 100% of the original adsorption limit [48].

The antibiotics expulsion by utilizing the water composites of BCs with metal expansion and its mixture could completely adsorb not just only through adsorption, however, it is connected toward the synergist properties of composites, which may lead to degradation of antibiotics.

In synopsis, the utilization of designed BCs is intended to improve the productivity of expulsion of fluid antibiotic deposits. This depends on both adsorbate and adsorbent physicochemical properties of antibiotics also sorption conditions. The actual adjustment ordinarily prompts an improvement of the BC permeable design to raise the aromaticity and hydrophobicity. Consequently, it may be effectively utilized BC-drug, from where the adsorption depends on the pore filling measure, hydrophobic connections. Nonetheless, the sorption property is based on connections including polar practicals, for example, the electrostatic associations, hydrogen holding, typically subject to pH, the actual change is not adequate. In this manner, it is prudent to synthetically change the BCs, driving generally to build the polarity that changes inside surface function and BC pHpzc esteem, also expand porosity of BCs which empowers interaction of the abovementioned BC-antibiotic.

On account of composites of BC, the extra segment presence may lead to the change in properties or potential and give new ones, which impacts antibiotic cycles with that method of antibiotics expulsion and adsorption. The expansion of the BC matrices depends on clay types which can expand the pores, surface charge changes, and give polar oxygen groups or cations. Henceforth, the adsorption of BC increases the interaction involves in the polar groups. Also, the drug intercalation and exchange of ions take place.

The CNTs blended with the matrix of BC normally bring an expansion in the porosity, due to the presence of oxygen molecules. As an outcome, among the pathways, the pore filling and interaction of the oxygen functional groups play a major contribution to the adsorption of antibiotics on these short composites.

Metal mixtures which present in the metal composite of biochar are responsible for surface charge changing and give mainly oxygen functional groups. The hydrogen holding, complexation of metals, and electrostatic interaction are distinguished. Besides, a few metal mixtures show the activity of photocatalytic/reactant, so the BC is used as an antibiotics adsorbent as well as a transporter for catalyst. Within the sight of an activator like free radicals or light-producing oxidant, then the antibiotics can be degraded by chemically using the catalyst [48] (Table 3).

Table 3: Biochar composites adsorption on antibiotics.

Biochar feedstock	Pyrolysis conditions			Sorbents	Antibiotics	Adsorption conditions				Maximum adsorption capacity (Q_{max}) (mg g^{-1})	Reference
	Temperature (°C)	Time (h)	Atmosphere			Temperature (°C)	Time (h)	pH	Adsorbent dose(g L^{-1})		
Municipal solid waste	500	0.5	–	Pristine BC	Tetracycline (TC)	–	12	3–9	2	3.9	[40]
	500	0.5	–	BC–MMT	Tetracycline (TC)	–	12	3–9	2	4.2	
	500	0.5	–	BC–RE	Tetracycline (TC)	–	12	3–9	2	77.9	
Wheat straw	400	6	–	Pristine BC	Norfloxacin (NOR)	25	48	3–11	1.25	10.6	[59]
	400	6	–	BC–MMT	Norfloxacin (NOR)	25	48	3–11	1.25	25.5	
Municipal solid waste	450	0.5	–	Pristine BC	Ciprofloxacin (CPX)	25	12	6–7	1	114.0	[41]
	450	0.5	–	BC–BT	Ciprofloxacin (CPX)	25	12	6–7	1	190.0	
Bamboo sawdust	600	1	Nitrogen	Pristine BC	Sulfamethazine (SMT)	25	48	3–9	1	3.0	[42]
	600	1	Nitrogen	BC–GO	Sulfamethazine (SMT)	25	48	3–9	1	6.5	
Treatment plant sludge	750	2	Oxygen	Pristine BC	Tetracycline (TC)	25	–	6	2	13.4	[44]
	750	2	Oxygen	BC–Fe	Tetracycline (TC)	25	–	6	2	40.8	

Sawdust	600	2	Nitrogen	Pristine BC	Tetracycline (TC)	25	–	6	1	18.0	[45]
	600	2	Nitrogen	BC–Zn	Tetracycline (TC)	25	–	6	1	60.0	
	600	2	Nitrogen	BC–Fe	Tetracycline (TC)	25	–	6	1	62.0	
Corn husk	300	1	Nitrogen	IP–BC Fe	Tetracycline (TC)	30	24	–	0.8	102.0	[60]
	300	1	Nitrogen	PI–BC Fe	Levofloxacin (LEV)	30	24	–	0.8	149.1	
Potato stems and leaves	500	0.5	Oxygen	Pristine BC	Norfloxacin (NOR)	35	24	3	2	5.8	[47]
	500	0.5	Oxygen	BC–Mn Fe	Ciprofloxacin (CIP)	35	24	3	2	6.9	
	500	0.5	Oxygen	BC–Mn Fe	Enrofloxacin (ENR)	35	24	3	2	5.8	
Bagasse	475	2	Oxygen	Pristine BC	Tetracycline (TC)	25	24	4	0.67	62.0	[61]
	475	2	Oxygen	BC–Mg/Al	Tetracycline (TC)	25	24	4	0.67	125.0	
Bamboo	550	1	Ar	Pristine BC	Sulfamethoxazole (SMX)	RT	24	3–9	0.2	128.2	[62]
	550	1	Ar	BC–CZF	Sulfamethoxazole (SMX)	RT	24	3–9	0.2	212.8	
Cornstalk	700	2	Nitrogen	Pristine BC	Sulfamethazine (SMT)	20	12	–	30	30.8%	[63]

5 Conclusion

Antibiotics contamination in the natural environment gives potential impacts on the life and health of organisms. Antibiotics drug consumption will increase year by year, which creates the entire problem and raise the rate of contamination. For this reason, many materials and methods are designed to remove these drug residues from the natural environment. The outcome displays that BC adsorbent, that is, BC pristine, is chemically and physically changed under various conditions, but composite BC results in effective antibiotics removal, which is the major substance used to treat antibiotic residues. The action mode of this substance is in the form of adsorption or amalgamation of adsorption with catalyst oxidation, this leads to the antibiotic's degradation. The process carried out here depends on the specific type and efficiency of more factors mostly depends on physicochemical and composition of antibiotics and adsorbent also the application condition. The adsorption process mechanism involved here will determine all the factors. The antibiotics or adsorbates which is having an aromatic structure with that functional group presence will have the ability to accept or donate electrons with the surface of BC. The functional group's presence has the capacity to promote the dissociation and formation of environmental interactions with BC. The binding of antibiotics on the BC adsorbent surface is based on the functional group present and pH of BC. Due to this fact, the adsorption surface of the specific antibiotic particles is particular for the BC materials. This review will demonstrate the mechanism and relationships which is commonly involved here. So that, it is most important for further research which has the aim to know about different antibiotics binding mechanism. On this basis, it is possible to match the adsorbent with antibiotics and enhance the performance of created materials which is tailored. Besides, the adsorption study carried out till were restricted only for the lab. The designed research is optimized for the conditions of adsorption using samples obtained from the environment also carried out the ecotoxicological and economic analysis, which is most important for this research area. The demonstration by performing the antibiotics contaminants removal from the real contamination of the world also needs for the hour. From this, it is understandable that the BC increase its importance toward the economy and gives the possibilities of application.

References

[1] Gould K, Antibiotics: from prehistory to the present day, J. Antimicrob. Chemother., 2016, 71, 572–575, doi: https://doi.org/10.1093/jac/dkv484.
[2] Stockwell VO, Duffy B, Use of antibiotics in plant agriculture, Rev. Sci. Tech. OIE., 2012, 31, 199–210, doi: https://doi.org/10.20506/rst.31.1.2104.
[3] Hu X, Zhou Q, Luo Y, Occurrence and source analysis of typical veterinary antibiotics in manure, soil, vegetables and groundwater from organic vegetable bases, northern China, Environ. Pollut., 2010, 158, 2992–2998, doi: https://doi.org/ 10.1016/j.envpol.2010.05.023.
[4] Grenni P, Ancona V, Barra Caracciolo A, Ecological effects of antibiotics on natural ecosystems: a review, Microchem. J., 2018, 136, 25–39, doi: https://doi.org/ 10.1016/j.microc.2017.02.006.
[5] Asghar HMA, Hussain SN, Brown NW, Roberts EPL, Comparative adsorption–regeneration performance for newly developed carbonaceous adsorbent, J. Ind. Eng. Chem., 2019, 69, 90–98, doi: https://doi.org/10.1016/j.jiec.2018.09.012.
[6] Homem V, Santos L, Degradation and removal methods of antibiotics from aqueous matrices – a review, J. Environ. Manage., 2011, 92, 2304–2347, doi: https://doi.org/10.1016/j.jenvman.2011.05.023.
[7] Lehmann J, Joseph S, Biochar for Environmental Management: Science and Technology, Earthscan, London, 2009.
[8] Walsh C, Molecular mechanisms that confer antibacterial drug resistance, Nature, 2000, 406 (6797), 775–781, doi: 10.1038/35021219.
[9] Critically important antimicrobials for human medicine 6th. Revision 2018. Ranking of medically important antimicrobials for risk management of antimicrobial resistance due to non-human use, World Health Organization, 2019. https://apps.who.int/iris/handle/10665/255027.
[10] Klein EY, Van Boeckel TP, Martinez EM, et al., Global increase and geographic convergence in antibiotic consumption between 2000 and 2015, Proc. Natl. Acad. Sci., 2018, 115, E3463–E3470, doi: https://doi.org/10.1073/pnas.1717295115.
[11] Hao H, Cheng G, Iqbal Z, Ai X, et al., Benefits and risks of antimicrobial use in food-producing animals, Front. Microbiol, 2014, 5, doi: https://doi.org/10.3389/fmicb.2014.00288.
[12] Lulijwa R, Rupia EJ, Alfaro AC, Antibiotic use in aquaculture, policies and regulation, health and environmental risks: a review of the top 15 major producers, Rev. Aquac., 2019. doi: https://doi.org/10.1111/raq.12344.
[13] Kasprzyk-Hordern B, Initiatives for addressing antimicrobial resistance in the environment: current situation and challenges: white paper, Wellcome Trust (2018) https://wellcome.ac.uk/sites/default/files/antimicrobial-resistance-environment-report.pd
[14] Gelband H, Miller-Petrie M, Pant S, et al., The state of the world's antibiotics 2015, Wound Heal. South. Afr., 2015, 8, doi: https://hdl.handle.net/10520/EJC180082.
[15] Kim K-R, Owens G, Kwon S-I, et al., Occurrence and environmental fate of veterinary antibiotics in the terrestrial environment, Water Air. Soil Pollut., 2011, 214, 163–174, doi: https://doi.org/10.1007/s11270-010-0412-2.
[16] Li WC, Occurrence, sources, and fate of pharmaceuticals in aquatic environment and soil, Environ. Pollut., 2014, 187, 193–201, doi: https://doi.org/10.1016/j.envpol.2014.01.015.
[17] Ben Y, Fu C, Hu M, Liu L, Wong MH, Zheng C, Human health risk assessment of antibiotic resistance associated with antibiotic residues in the environment: a review, Environ. Res., 2019, 169, 483–493, doi: https://doi.org/10.1016/j.envres.2018.11.040.
[18] Jechalke S, Heuer H, Siemens J, Amelung W, Smalla K, Fate and effects of veterinary antibiotics in soil, Trends Microbiol., 2014, 22, 536–545, doi: https://doi.org/10.1016/j.tim.2014.05.005.

[19] Kummerer K, Significance of antibiotics in the environment, J. Antimicrob. Chemother., 2003, 52, 5–7, doi: https://doi.org/10.1093/jac/dkg293.

[20] Wu P, Ata-Ul-Karim ST, Singh BP, et al., A scientometric review of biochar research in the past 20 years (1998–2018), Biochar, 2019, 1, 23–43, doi: https://doi.org/10.1007/s42773-019-00002-9.

[21] Inyang MI, Gao B, Yao Y, et al., A review of biochar as a low-cost adsorbent for aqueous heavy metal removal, Crit. Rev. Environ. Sci. Technol., 2016, 46, 406–433, doi: https://doi.org/10.1080/10643389.2015.1096880.

[22] Premarathna KSD, Rajapaksha AU, Sarkar B, et al., Biochar-based engineered composites for sorptive decontamination of water: a review, Chem. Eng. J., 2019, 372, 536–550, doi: https://doi.org/10.1016/j.cej.2019.04.097.

[23] Homem V, Santos L, Degradation and removal methods of antibiotics from aqueous matrices – a review, J. Environ. Manage., 2011, 92, 2304–2347, doi: https://doi.org/10.1016/j.jenvman.2011.05.023.

[24] Peiris C, Gunatilake SR, Mlsna TE, Mohan D, Vithanage M, Biochar based removal of antibiotic sulfonamides and tetracyclines in aquatic environments: a critical review, Bioresour. Technol., 2017, 246, 150–159, doi: https://doi.org/101016/j.biortech.2017.07.150.

[25] Ahmed MB, Zhou JL, Ngo HH, Guo W, Adsorptive removal of antibiotics from water and wastewater: progress and challenges, Sci. Total Environ, 2015, 532, 112–126, doi: https://doi.org/10.1016/j.scitotenv.2015.05.130.

[26] Ahmad M, Rajapaksha AU, Lim JE, et al., Biochar as a sorbent for contaminant management in soil and water: a review, Chemosphere, 2014, 99, 19–33, doi: https://doi.org/101016/j.chemosphere.2013.10.071.

[27] Yang X, Zhang S, Ju M, Liu L, Preparation and modification of biochar materials and their application in soil remediation, Appl. Sci., 2019, 9(1365), doi: https://doi.org/10.3390/app9071365.

[28] Stefaniuk M, Oleszczuk P, Characterization of biochars produced from residues from biogas production, J. Anal. Appl. Pyrolysis, 2015, 115, 157–165, doi: https://doi.org/10.1016/j.jaap.2015.07.011.

[29] Li C, Zhu X, He H, et al., Adsorption of two antibiotics on biochar prepared in air-containing atmosphere: influence of biochar porosity and molecular size of antibiotics, J. Mol. Liq., 2019, 274, 353–361, doi: https://doi.org/10.1016/j.molliq.2018.10.142.

[30] Chen T, Luo L, Deng S, et al., Sorption of tetracycline on H3PO4 modified biochar derived from rice straw and swine manure, Bioresour. Technol., 2018, 267, 431–437, doi: https://doi.org/10.1016/j.biortech.2018.07.074.

[31] Rajapaksha AU, Vithanage M, Ahmad M, et al., Enhanced sulfamethazine removal by steam-activated invasive plant-derived biochar, J. Hazard. Mater., 2015, 290, 43–50, doi: https://doi.org/10.1016/jjhazmat.2015.02.046.

[32] Wang B, Gao B, Fang J, Recent advances in engineered biochar productions and applications, Crit. Rev. Environ. Sci. Technol., 2017, 47, 2158–2207, doi: https://doi.org/10.1080/10643389.2017.1418580.

[33] Zhu X, Li C, Li J, et al., Thermal treatment of biochar in the air/ nitrogen atmosphere for developed mesoporosity and enhanced adsorption to tetracycline, Bioresour. Technol., 2018, 263, 475–482, doi: https://doi.org/10.1016/j.biortech.2018.05.041.

[34] Sizmur T, Fresno T, Akgül G, Frost H, Moreno-Jiménez E, Biochar modification to enhance sorption of inorganics from water, Bioresour. Technol, 2017, 246, 34–47, doi: https://doi.org/10.1016/j.biortech.2017.07.082.

[35] Liu P, Liu W-J, Jiang H, et al., Modification of bio-char derived from fast pyrolysis of biomass and its application in removal of tetracycline from aqueous solution, Bioresour. Technol., 2012, 121, 235–240, doi: https:// doi.org/10.1016/j.biortech.2012.06.085.

[36] Fan Y, Wang B, Yuan S, et al., Adsorptive removal of chloramphenicol from wastewater by NaOH modified bamboo charcoal, Bioresour. Technol., 2010, 101, 7661–7664, doi: https://doi.org/10.1016/j.biortech.2010.04.046.
[37] Jang HM, Yoo S, Choi Y-K, Park S, Kan E, Adsorption isotherm, kinetic modeling and mechanism of tetracycline on Pinus taeda-derived activated biochar, Bioresour. Technol., 2018, 259, 24–31, doi: https://doi.org/10.1016/j.biortech.201803.013.
[38] Tan Z, Zhang X, Wang L, et al., Sorption of tetracycline on H 2 O 2 -modified biochar derived from rape stalk, Environ. Pollut. Bioavailab., 2019, 31, 198–207, doi: https://doi.org/10.1080/26395940.2019.1607779.
[39] Dawoud MM, Saleh HM, Introductory chapter: background on composite materials, In: El-Din H, Saleh M, Koller M (Eds.), Characteristic Some Composite Materials, IntechOpen, 2019, https://www.intechopen.com/books/characterizations-ofsome-composite-materials/introductory-chapter-background-on-composite-materials accessed October 2, 2019.
[40] Premarathna KSD, Rajapaksha AU, Adassoriya N, et al., Clay-biochar composites for sorptive removal of tetracycline antibiotic in aqueous media, J. Environ. Manage., 2019, 238, 315–322, doi: https://doi.org/10.1016/j.jenvman.2019.02.069.
[41] Ashiq A, Adassooriya NM, Sarkar B, et al., Municipal solid waste biochar-bentonite composite for the removal of antibiotic ciprofloxacin from aqueous media, J. Environ. Manage., 2019, 236, 428–435, doi: https://doi.org/10.1016/j.jenvman.2019.02.006.
[42] Huang D, Wang X, Zhang C, Zeng G, et al., Sorptive removal of ionizable antibiotic sulfamethazine from aqueous solution by graphene oxide-coated biochar nanocomposites: influencing factors and mechanism, Chemosphere, 2017, 186, 414–421, doi: https://doi.org/10.1016/j.chemosphere.2017.07.154.
[43] Inyang M, Gao B, Zimmerman A, Zhou Y, Cao X, Sorption and cosorption of lead and sulfapyridine on carbon nanotube-modified biochars, Environ. Sci. Pollut. Res., 2015, 22, 1868–1876, doi: https://doi.org/10.1007/s11356-014-2740-z.
[44] Yang X, Xu G, Yu H, Zhang Z, Preparation of ferric-activated sludge-based adsorbent from biological sludge for tetracycline removal, Bioresour. Technol, 2016, 211, 566–573, doi: https://doi.org/10.1016/j.biortech.2016.03.140.
[45] Zhou Y, Liu X, Xiang Y, Wang P, et al., Modification of biochar derived from sawdust and its application in removal of tetracycline and copper from aqueous solution: adsorption mechanism and modelling, Bioresour. Technol., 2017, 245, 266–273, doi: https://doi.org/10.1016/j.biortech.2017.08.178.
[46] Deng J, Dong H, Zhang C, Jiang Z, et al., Nanoscale zero-valent iron/biochar composite as an activator for Fenton-like removal of sulfamethazine, Sep. Purif. Technol., 2018, 202, 130–137, doi: https://doi.org/10.1016/j.seppur.2018.03.048.
[47] Li R, Wang Z, Zhao X, Li X, Xie X, Magnetic biochar-based manganese oxide composite for enhanced fluoroquinolone antibiotic removal from water, Environ. Sci. Pollut. Res., 2018, 25, 31136–31148, doi: https://doi.org/10.1007/s11356-018-3064-1.
[48] Krasucka P, Pan B, Sik Ok Y, et al., Engineered biochar – a sustainable solution for the removal of antibiotics from water, Chem. Eng. J., 2020, 126926, 1–27, doi: 10.1016/j.cej.2020.126926.
[49] Xie M, Chen W, Xu Z, Zheng S, Zhu D, Adsorption of sulfonamides to demineralized pine wood biochars prepared under different thermochemical conditions, Environ. Pollut., 2014, 186, 187–194, doi: https://doi.org/10.1016/j.envpol.2013.11.022.
[50] Shang JG, Kong XR, He LL, Li WH, Liao QJH, Low-cost biochar derived from herbal residue: characterization and application for ciprofloxacin adsorption, Int. J. Environ. Sci. Technol., 2016, 13, 2449–2458, doi: https://doi.org/10.1007/s13762-016-1075-3.

[51] Liu C-H, Chuang Y-H, Li H, et al., Sorption of lincomycin by manure-derived biochars from water, J. Environ. Qual., 2016, 45(519), doi: https://doi.org/10.2134/jeq2015.06.0320.
[52] Jang HM, Kan E, A novel hay-derived biochar for removal of tetracyclines in water, Bioresour. Technol., 2019, 274, 162–172, doi: https://doi.org/10.1016/j.biortech.2018.11.081.
[53] Dai Y, Li J, Shan D, Adsorption of tetracycline in aqueous solution by biochar derived from waste Auricularia auricula dregs, Chemosphere, 2020, 238, 124432, doi: https://doi.org/10.1016/j.chemosphere.2019.124432.
[54] Wang H, Chu Y, Fang C, et al., Sorption of tetracycline on biochar derived from rice straw under different temperatures, PLOS ONE, 2017, 12, e0182776, doi: https://doi.org/10.1371/journal.pone.0182776.
[55] Zhang L, Tong L, Zhu P, Huang P, Tan Z, et al., Adsorption of chlortetracycline onto biochar derived from corn cob and sugarcane bagasse, Water Sci. Technol., 2018, doi: https://doi.org/10.2166/wst.2018.407.
[56] Rajapaksha AU, Vithanage M, Zhang M, et al., Pyrolysis condition affected sulfamethazine sorption by tea waste biochars, Bioresour. Technol., 2014, 166, 303–308, doi: https://doi.org/10.1016/j.biortech.2014.05.029.
[57] Gómez-Pacheco CV, Rivera-Utrilla J, Sánchez-Polo M, López-Peñalver JJ, Optimization of the preparation process of biological sludge adsorbents for application in water treatment, J. Hazard. Mater., 2012, 217–218, 76–84, doi: https://doi.org/10.1016/j.jhazmat.2012.02.067.
[58] Jang HM, Kan E, Engineered biochar from agricultural waste for removal of tetracycline in water, Bioresour. Technol., 2019, 284, 437–447, doi: https://doi.org/ 10.1016/j.biortech.2019.03.131.
[59] Zhang J, Lu M, Wan J, Sun Y, Lan H, Deng X, Effects of pH, dissolved humic acid and Cu 2+ on the adsorption of norfloxacin on montmorillonite-biochar composite derived from wheat straw, Biochem Eng. J., 2018, 130, 104–112, doi: https://doi.org/10.1016/j.bej.2017.11.018.
[60] Chen Y, Shi J, Du Q, Zhang H, Cui Y, Antibiotic removal by agricultural waste biochars with different forms of iron oxide, RSC Adv., 2019, 9, 14143–14153, doi: https://doi.org/10.1039/C9RA01271K.
[61] Tan X, Liu S, Liu Y, Gu Y, et al., One pot synthesis of carbon supported calcined-Mg/Al layered double hydroxides for antibiotic removal by slow pyrolysis of biomass waste, Sci. Rep., 2016, 6, doi: https://doi.org/10.1038/srep39691.
[62] Heo J, Yoon Y, Lee G, Kim Y, Han J, Park CM, Enhanced adsorption of bisphenol A and sulfamethoxazole by a novel magnetic CuZnFe2O4–biochar composite, Bioresour. Technol., 2019, 281, 179–187, doi: https://doi.org/10.1016/j.biortech.2019.02.091.
[63] Deng J, Dong H, Zhang C, Jiang Z, Cheng Y, Hou K, Zhang L, Fan C, Nanoscale zero-valent iron/biochar composite as an activator for Fenton-like removal of sulfamethazine, Sep. Purif. Technol., 2018, 202, 130–137, doi: https://doi.org/10.1016/j.seppur.2018.03.048.

Ashwini A. Waoo, Charu Vyas

Biochar for the remediation of contaminated soil

1 Introduction

The contamination of soil with heavy metals compromises and threatens nature, resulting in adverse effects on agricultural products and human health. The continual deterioration of land quality must be controlled by economic and interesting remediation techniques. The long-term stability of biochar is a key focus of modern soil environmental management and remediation technologies. It is attainable to apply biochar as a new carbon-rich material that is green, economically feasible, and environmentally friendly. Metals in contaminated soils are immobilized by this method and should be considered for use in soil remediation plans. The success rate of remediation of metal-contaminated soils using biochar is analyzed based on research. There are several perspectives on the current research status and progress in this field, for example, its preparation technologies and performance characteristics, remediation mechanisms and its effects, metal–biochar interactions in soils, several affecting factors, risks associated with biochar amendment, and its uses in soil remediation. The efficiency of biochar depends on many factors, including the type of biomass, residence time, heating rate, temperature, and pyrolysis. A need still exists for research to develop and improve practical methods and useful techniques for the preparation and application of diverse biochars that target specific heavy metals. The long-term effects of biochar applications on soil remediation, soil organisms, and plant growth must be explored by expanding the research scale from laboratory studies to large-scale and long-term studies to ensure that biochar can be produced safely and used responsibly.

The entire biosphere is toxic if heavy metal contamination exceeds the recommended limit [1]. Heavy metals are naturally occurring in soil and water in the environment, as well as anthropogenic sources, such as human activity [2]. Human activities generate wastes, and these wastes are collected, stored, and disposed of properly to avoid hazardous effects on the environment and public health [3].

This chapter is mainly concerned with the area of heavy metal contamination, bioaccumulation, bioavailability, biochar, immobilization, and soil remediation techniques.

Waste disposal, mining, smelting of metalliferous ores, unintentional and systemic spills, and sewage sludge application to agricultural soils are all chargeable with

Ashwini A. Waoo, AKS University, Satna, (M.P.), India, e-mail: ashwiniwaoo@gmail.com
Charu Vyas, AKS University, Satna, (M.P.), India

https://doi.org/10.1515/9783110734003-002

introducing contaminants to otherwise uncontaminated locations as dust and leachate. For some heavy metals in the water, there is a maximum recommended concentration limit regulated by the Comprehensive Environmental Response Compensation and Liability Act (CERCLA), USA, which is 0.01 mg L^{-1} for Ar, 0.05 mg L^{-1} for Cd, 0.01 mg L^{-1} for Cr, 0.015 mg L^{-1} for Pb, 0.002 mg L^{-1} for Hg, and 0.05 mg L^{-1} for Ag, respectively. According to the Indian heavy metals standards, the standard norm for soil is 3–6, 135–270, 75–150, 250–500, and 300–600 mg kg^{-1} for Cd, Cu, Ni, Pb, and Zn [4].

In nature, heavy metals occur in insoluble forms, such as in mineral structures, or in precipitated or complex forms that are not readily accessible to plants and have severe negative effects on human health, and other living organisms. Volcanic eruptions, comets, erosion, and the weathering of minerals are examples of natural processes which result in the release of heavy metals in the biosphere.

All living forms are adversely affected by heavy metals from anthropogenic sources with high bioavailability and their soluble reactive forms. The sources of anthropogenic pollution include alloy production, atmospheric deposition, coating, battery production, improper storage of industrial waste, explosive manufacturing, leather tanning, phosphate fertilizer, photographic materials, pesticides, steel industries, electroplating industries, mining, printing dyes, smelting, textiles dyes, wood preservation, sewage irrigation, and biosolids such as composts, livestock manures, and sewage sludge [5].

Food webs accumulate metal ions due to a variety of factors, including sources of heavy metals, soil properties, soil concentrations, degree of uptake by plants, and animal absorption [6]. Accumulation and the build-up of heavy metals concentration in the environment by man result in accumulation and deposition of heavy metals, posing a threat to all living things including humans, plants, animals, and aquatic organisms, above permissible levels [7].

These heavy metals' hazardous effects create environmental and public health concerns, which can cause disturbances to human and aquatic biota, and consequently raise the need for remediating heavy-metal-polluted environments. Consequently, it is imperative to eliminate heavy metal contamination for preventing the risk of the environment and uptake.

1.1 Effects of heavy metals on life forms and removal techniques of heavy metals

Heavy metals can affect a variety of biochemical, physiological, and metabolic processes in living organisms; they may be acting as micronutrients, stabilization of molecules, osmotic pressure regulators, co-factors for some enzymes, and all of them are very toxic in excessive amounts despite not having any known biological function [8].

Addis et al. [9] determined the heavy metal in soils used for the cultivation of *Allium sativum* in East Gojjam Zone, Ethiopia. The soils were not heavily contaminated or polluted by toxic heavy metals (Cd and Pb). Therefore, the soils studied were not dangerous for the cultivation of garlic or any other agricultural purpose. However, precautions should be taken going forward.

Heavy metals can be removed from polluted environments using a variety of techniques. Chemical oxidation and reduction reactions, evaporative recovery, adsorption processes, electrochemical techniques, chemical precipitation, ion exchange, and reverse osmosis are conventional methods for removing and recovering heavy metals from solution.

Nevertheless, these strategies are expensive, sometimes impractical, and are not specific to metal-binding properties. In addition, they destroy large amounts of soil structure and fertility and are highly dependent on specific conditions, such as contamination, soil properties, and site conditions. Metal ion removal is also characterized by unpredictable results due to its high reagent requirements, toxic waste generation, and other disadvantages. Metal concentrations in solutions below 100 mg L^{-1} render most of these strategies ineffective [11]. As a result of heavy metal salts' high solubility in solution, separation by physical and chemical means is also challenging.

The evaluation of alternative methodologies needs to be based on their reasonableness, appropriateness, and relevance to the local neighborhood conditions, and the established permissible limits must be followed. In consequence, microorganisms and plants are a potential answer to heavy metal pollution considering that they incorporate economic and sustainable remediation advancements technologies to rectify and re-establish the natural condition of the soil. Due to this, there is a need to develop and apply bioremediation techniques that can more widely be applied to heavy metal contamination [2].

1.2 Mechanisms of heavy metal uptake by microbes

The organism that accumulates heavy metals must be resistant to high concentrations of metals, and they must possess improved transformational abilities, transforming the harmful synthetic chemicals into harmless compounds that alleviate the poisonous effects of the heavy metal, and retaining the metal in various forms simultaneously [12–13].

Heavy metals can be absorbed in numerous ways. The cellular biomass can trap heavy metal ions that then adsorb onto the binding sites of the cell wall [14]. Passive uptake is the name given to this process. It does not involve metabolism. Metal ions are bound to the surface of the cell wall and extracellular material during this process. Adsorption of metals depends on both kinetic equilibrium and the composition of the metal at the surface of the cell.

Other mechanisms involve precipitation electrostatic interaction, ion exchange, surface complexation, and redox process [15]. These processes reach equilibrium very quickly. Both living and non-living biomass can be used for biosorption. Passive uptake can occur through surface complexation on the cell wall.

Through the cell's metabolic cycle, heavy metal ions also pass across the membrane and into the cytoplasm. The process is known as bioaccumulation or active uptake, or intracellular uptake. A living cell accumulates substances through a variety of chemical, physical, and biological mechanisms. Biosorption has a limited and ill-defined role in intracellular and extracellular processes [16].

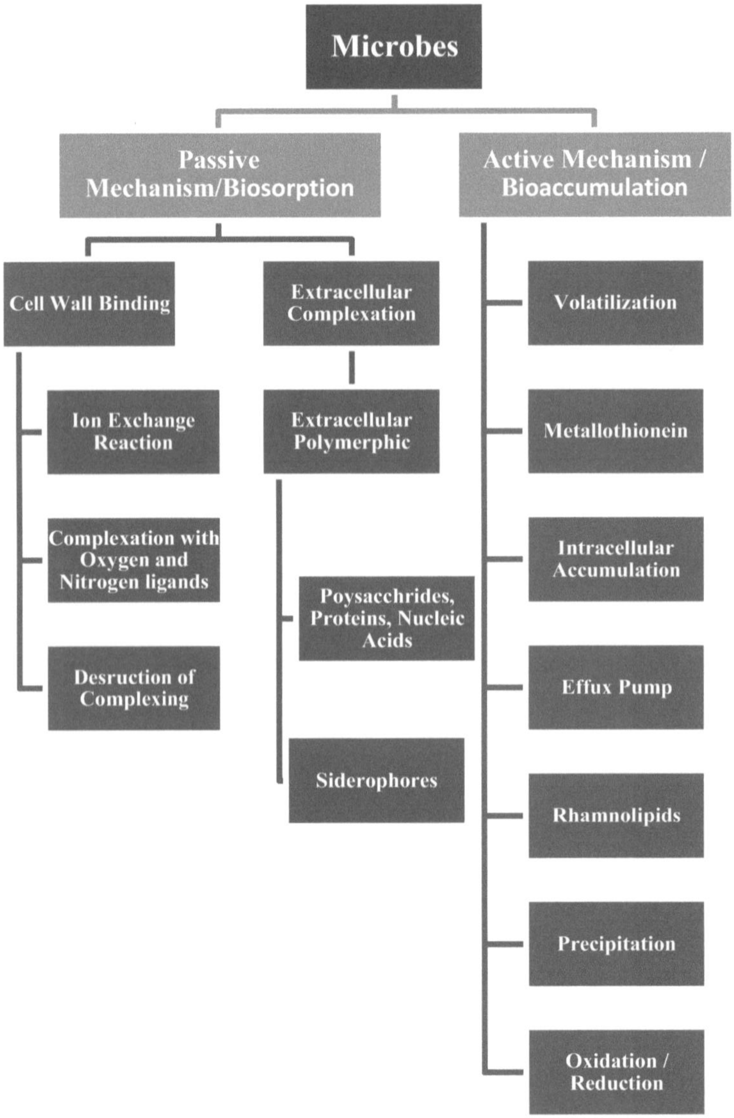

Figure 1: Mechanism of microbial metal tolerance.

2 Bioremediation

Recent decades have seen a remarkable development in the field of bioremediation. Biological restoration and rehabilitation of historically contaminated sites is the essence of bioremediation, as is cleaning up recent contaminated areas, whether by accident or incident, as a result of manufacturing, storing, and transporting natural or inorganic compounds. The use of bioremediation for heavy metal removal from contaminated wastewaters and soils is an environmentally friendly, cost-effective method. Using the activities of algae, fungi, bacteria, or plants, this can be accomplished by reducing and/or recovering heavy metal pollutants into less toxic forms [17].

The waste is degraded or converted into some other forms by bacteria. Regardless of whether it is a degraded or converted product, it is very important to ensure that the give-up product is not more poisonous and difficult than the original compound. Organisms are crucial for the degradation of xenobiotics and for maintaining a constant concentration of chemicals within the surrounding environment [18]. The biodegradation process varies greatly, but carbon dioxide is often the final product.

A large number of bioremediation advances have been discovered through microbiology, molecular biology, biochemistry, analytical chemistry, chemical, and environmental engineering, among others. As a result of these various fields, each with its methodological approach, bioremediation progress has improved dramatically in the last decade.

By using their enzymatic pathways (specific degradative enzymes), bacteria act as biocatalysts and encourage biochemical responses that break down the given pollutant. Microorganisms are acting against pollutants only when they have access to a range of material compounds to assist them to generate strength and nutrients (nutrients) for constructing greater cells via metabolic pathways via respiration, cometabolism, fermentation, and so on.

Many factors determine the effectiveness of bioremediation, including the chemical nature of pollutants, the concentration of pollutants, and the type of environment (physicochemical) in which they are found [19].

Bacteria and pollutants cannot come into direct contact; therefore, degradation occurs more slowly. Nature does not consistently spread microorganisms and contamination. A complex framework is required to control and improve bioremediation procedures on account of many factors [20].

Development and activity of microorganisms are affected by factors such as soil structure, redox potential, pH, temperature, oxygen concentration, moisture content, metal ions, toxic compounds, solubility in water, soil ion concentration, lack of skilled personnel, and physicochemical availability of pollutant (type, solubility, chemical structure, toxicity, and contaminant concentration). The above-mentioned factors influence the kinetics of degradation [21].

While biodegradation can occur under a wide range of pH values, 6.5 to 8.5 is commonly considered a benchmark for maximum aquatic and terrestrial biodegradation.

Moreover, moisture impacts the type and concentration of soluble materials that are accessible to contaminants, pH, and the osmotic pressure of aquatic and terrestrial environments [22]. Several factors play a role in the biodegradation process, including:
- Compound stereochemistry
- Compound toxicity,
- The compound concentration,
- Microbial strain performance,
- Situations of degradation,
- Retention time
- Other compounds and their concentration.

2.1 Bioremediation strategies

As part of bioremediation processes, various types of treatment technologies/methods and strategies or techniques can be used to remove heavy metals involving microorganisms. A bioremediation method can be categorized as bioattenuation, biostimulation, bioaugmentation, bioventing, bio-accumulation, biosorption, biopiles, biomineralization, and biotransformation. Purified biopolymers are bioprecipitated and absorbed by microbial cells. Based on the degree of saturation and aeration of an area, different strategies are employed.
1) In situ techniques (treatment at the contamination site)
2) Ex situ techniques (the contaminated area can be excavated and treated elsewhere)

2.1.1 In situ bioremediation

By introducing microbes and organic materials to the soil, in situ bioremediation occurs naturally and is eco-friendly [23]. This process does not involve the transportation of contaminants, excavating or removing soil or groundwater or sediments, and the materials are applied to the site with low cost and minimal disturbance. Using the nitrogen and carbon cycles to recycle nutrients was the focus of this strategy.

In situ bioremediation techniques rely on these natural geochemical cycles to degrade and recycle hazardous wastes and remediate contaminated soils. Microorganisms living on this contaminated soil degrade the contaminants, and they can only sustain themselves in that environment by using the hazardous substances for energy. Due to the complete distress of organic pollutants, no optional secondary streams are formed [24].

The depth of the soil at which a treatment can be effective limits the treatment. In many soils, effective oxygen diffusion for bioremediation extends only a few centimeters to about 30 cm into the soil, though depths closer to 60 cm and even more have been treated now and again in some cases [25]. In some instances, the immobilized toxic metal may be rereleased into water under special conditions, however, because it remains in soil and sediment. Biosparging, bioventing, and bioaugmentation are the most important treatments for lands.

2.1.2 Ex situ bioremediation

The ex situ bioremediation technique is faster, requires less time, easier to control, and can treat more pollutants and soil types than in situ methods. To facilitate microbial degradation, contaminated soil needs to be excavated or removed from the ground [26] or groundwater pumped. Bioreactors, land farming, composting, and biopiles are some of the most popular ex situ treatments.

Many contaminated sites have successfully used bioremediation of organic compounds; here, physical factors can limit the process. By injecting air into contaminated soil, biopiling enhances high-impact creosote degradation.

2.2 Principle of bioremediation

The most efficient bioremediation can be achieved when environmental conditions promote the growth and activity of microorganisms, so it is common for environmental factors to be manipulated to speed up the degradation process. Adding living things and fertilizers to the soil creates the process of bioremediation.

Bioremediation relies primarily on biodegradation. Organic toxic pollutants are completely emitted into naturally occurring compounds, such as water, carbon dioxide, and inorganic compounds that are safe for humans, plants, and aquatic life [27]. There have been several pathways for the biodegradation of a variety of organic compounds both with and without oxygen.

The term bioremediation refers to a process whereby natural wastes are biologically degraded to an innocuous state or tier below regulatory limits.

Microorganisms are ideal for killing contaminants since they have enzymes that enable them to consume environmental contaminants. Bioremediation aims to encourage their metabolism to degrade/detoxify substances that are harmful to nature and living things, by providing them with the finest levels of nutrients and chemical substances important for their metabolism.

Enzymes intervene in all metabolic responses. These include oxidoreductases, lyases, isomerases, hydrolases, transferases, and ligases. Under their affinity toward the pollutant, many enzymes can degrade it. The pollutants must be degraded enzymatically and converted to nonharmful forms by microorganisms as part of bioremediation.

2.3 The advantages of bioremediation

- The process is a natural one; it takes place over a short period, and it is a cost-effective means for reclaiming polluted materials, such as soil.
- Pollutants are degraded by microbes, which multiply until the pollutant is present. Biodegradative populations decline after the degradation of contaminants. Residues left behind from the remedy generally include water, carbon dioxide, and waste products produced by mobile cells.
- The procedure requires very few attempts and may often be performed on-site without causing much disruption to daily activities. Rather than transporting toxic waste off-site and creating potential hazards to human and the environment during transportation, this process can be conducted on-site.
- It can be applied cost-effectively as it loses less than the other conventional methods of cleaning up hazardous waste. A valuable method to treat oil-contaminated sites [28].
- Furthermore, it aids in the degradation of pollutants, and many toxic compounds can be converted into nontoxic, eco-friendly products. Additionally, the disposal of contaminated materials eliminates future hazards.
- A risky chemical compound is no longer used. Microbial growth is sped up and lives with the addition of compost. It is commonly used in gardens and lawns. When harmful chemicals are converted into water and harmless gases through bioremediation, they are destroyed [29].
- They are simple, easy to maintain, extensive, and inexpensive because of their natural function within the environment.
- Eco-friendly and sustainable.
- Contaminants are destroyed rather than transferred to different environments.
- Nonintrusive, allowing site usage to continue.
- Relatively easy implementation.
- Effectively clean up natural ecosystems from a range of contaminants while being environmentally friendly.

2.4 The disadvantages of bioremediation

- Biodegradable compounds are the only ones allowed. Every compound may not degrade completely and rapidly.
- Some people are concerned that the products of biodegradation are more chronic or toxic than the discernable compound.
- The biological world is often characterized by highly specialized processes. For fulfillment, important site factors include microbial populations that have the potential to produce metabolic products, optimum environmental conditions, and the right balance of vitamins (nutrients) and contaminants.

- Field operations cannot be extrapolated from the bench and pilot-scale studies.
- As a result of research and exploration, bioremediation advances will be made that are suitable for sites that have complex blends of pollutants that are not evenly distributed throughout the environment. Solid, liquid, and gaseous pollutants can be found in the environment.
- The process often takes longer than other options, such as excavation and incineration.
- Bioremediation performance standards remain uncertain due to regulatory uncertainty. Bioremediation is difficult to evaluate because there is no standard definition of smooth.
- Human power must be very skilled.

3 Role of microbes in bioremediation of heavy metals

The bioremediation process can break down or immobilize contaminants rapidly and inexpensively while protecting human health and the environment [30]. Moreover, because the technology relies on natural processes, it is considered more acceptable and eco-friendlier by the public. Multiple facilities are experimenting with exogenous, genetically modified microbes to improve bioremediation [31].

For a microbial bioremediation program to be effective and financially viable, it must consider hydrogeologic conditions, the pollutant, the microbiome, and others that may vary widely. A bioremediation cycle uses the contaminants as supplements or energy sources for the microorganisms [20]. By providing electron acceptors (oxygen), nutrients (nitrogen, phosphorus), or substrates such as methane, toluene, or phenol, microorganisms can catalyze bioremediation activity [32].

Kumar et al. [33] investigated the bioremediation of heavy metals from industrial effluents from the Haryana districts of Ambala, Karnal, and Yamunanagar using some fungal and bacterial strains. A few isolates of bacteria and fungi that can tolerate up to 100-ppm concentrations of heavy metals (Pb, Cd, and Cr) were tested for their ability to remove Pb, Cd, and Cr from liquid media at 50-ppm concentrations. *Aspergillus nidulans*, *Rhizopus arrhizus*, and *Trichoderma viride* were able to uptake 25.67 mg g^{-1} Pb, 13.15 mg g^{-1} Cd, and 2.55 mg g^{-1} Cr, respectively. Therefore, these microorganisms (fungi and bacteria) could serve as biosorbents for the elimination of high concentrations of metals from industrial effluents and wastewater.

Most microorganisms can be isolated from almost any ecological situation. In an anaerobic environment, with the presence of toxic compounds or on any waste stream, bacteria can adjust and grow at low temperatures, as well as under extreme heat, in desert conditions, with surplus oxygen in water, or with no oxygen. To function, one needs a source of vitality and a source of carbon. The adaptability of microbes and other biological systems makes them capable of degrading or remediating environmental threats [34].

Dwivedi et al. [35] demonstrated that heavy metals can be removed by fungi isolated from wastewater. Fungal isolates (*Aspergillus foetidus*, *Aspergillus awamori*, *Rhizopus sp.*, *Phanerochaete chrysosporium*, *Aspergillus flavus*, *Trichoderma viride*) that could tolerate up to 400 ppm concentration of Cd, Pb, Co, Cu, Cr, and Ni were further studied for removal of heavy metals from media at 50 ppm concentration. Pb, Ni, and Cr were reported to be eliminated by *Aspergillus flavus* and *Aspergillus niger*. Therefore, these fungi can remove heavy metals from wastewater and commercial effluents containing high levels of heavy metals as biosorbents.

Fazli et al. [36] studied the cadmium-tolerant fungi that can remove cadmium from cadmium-polluted sites in Zanjan province, Iran. The most tolerant fungi, *Aspergillus versicolor*, confirmed a tolerance index of 0.8 in 100 mg L^{-1} cadmium agar media and had been found to bioaccumulate 7 mg of cadmium per 1 g of. *Aspergillus versicolor* might be suitable for bioremediation processes. In large-scale heavy metal bioremediation, many of the properties of microorganisms have not been studied. To be successful, the organism of choice must be indigenous to the contaminated environment [37].

The improvement in understanding of the metabolism of microorganisms led to metal sequestration, improved microbial endurance rates, and altered metal adsorption mechanisms [38]. Adsorbate is the material aggregated at the interface, and adsorbent is the solid surface. Adsorption may lead to the production of a stable molecular phase, which can be represented by a surface complex at the interface. On the surfaces of microorganisms, functional groups such as –OH, –SH, and –COOH play a crucial role in the adsorption of metals [39].

According to some theories, a microbial cell acquires resistance to heavy metals through a variety of cellular processes, like excreting metal-chelating substances, when a transport system fails, which causes the metal ions to accumulate in the cell less. Metal ions are bound to intracellular molecules like metallothionein, vacuoles, or cells, as well as organelles like mitochondria resulting in alterations in the distribution of metal ions, which is the mechanism for resistance to heavy metals. Various types of microorganisms interact with metal ions, including metal siderophore, intracellular accumulation, extracellular polymerizations, vaporization of metals, extracellular mobilization, and immobilization of metal ions [10].

Metals are remedied by microbial activity in different ways. Metal bioavailability is determined by factors such as the concentration of pollutants, microbes, electron acceptors, pH, moisture content, nutrients, oxygen, soil structure, osmotic pressure, temperature, and water content. A soil's bioavailability depends on variables such as the buffering capacity, metal oxide content, cation exchange capacity, and organic matter. Therefore, heavy metal remediation is achieved by the expulsion of metal ions from the substratum.

Ecological conditions, history, and pretreatment must be established before selecting the most efficient biosorbent for a specific contaminant, from the readily available microorganisms. A certain metal might be of interest now and then, regardless of the levels of concentration achieved, or the interest might be to shorten contamination levels to fall within safe control limits. In addition, an enormous amount of metal may have to be recovered while achieving low equilibrium concentrations at the same time.

Agrawal et al. [40] studied physicochemical and microbiological factors influencing Ganga water in Kanpur, focusing on bioremediation of tannery effluent-contaminated soil. Except for one strain that could tolerate up to 100-ppm concentration of chromium, all four bacterial isolates could tolerate a 75-ppm concentration of chromium. Three strains had phosphate solubilizing potential ($3.14 > 2.50 > 2.37$) in addition to this. During testing, one strain had a high phosphate solubilizing potential, reducing the BOD content of the sewage. Bioremediation of wastewater and sewage using identified bacterial isolates was observed.

Microorganisms are resistant to heavy metals found in heavy-metal-contaminated sources. Microbes differ widely in their resistance and ability to eliminate heavy metals. The analysis and enumeration of autochthonous microbes (heavy metal tolerant microbes) from such contaminated niches (heavy-metal-contaminated source) are fundamental for the bioremediation of heavy metals [38, 41].

Hussain et al. [42] identified bacterial strains from drinking water sources in Kohat, Pakistan. They found a great deal of diversity among the bacteria. *Bacillus sp* and *Enterobacter sp* were the most prevalent bacteria, followed by *Pseudomonas sp*. Several coliform bacteria and fecal origin microorganisms (Enterobacteriaceae) were identified, which could pose a health hazard. Mulamattathil et al. [43] characterized the isolated bacteria using antibiotic resistance profiling and polymerase chain reaction (PCR) to identify *Pseudomonas* and *Aeromonas* species, and fecal indicator bacteria were identified. It was determined that various bacterial species, including *Aeromonas* and *Pseudomonas*, were present in the drinking water, and at times fecal and total coliforms. The bacteria were resistant to different classes of antibiotics.

Considering the level of microbial isolations in the examined sites (sampled sites), there is a correlation between the difference in richness of isolates and heavy metal contamination. The exposure of sediment and water to heavy metals for an extended period can produce considerable changes in the microbial community, that is, morphological, and physiological changes [44, 45], thus reducing their diversity, activity, and number. This causes the extinction of species that are sensitive to stress and increases the growth of species that are tolerant or resistant in that area.

Ezzouhri et al. [46] studied heavy metal resistance mechanisms on filamentous fungi from polluted sites in Tangier, Morocco, where the concentration of Cr, Pb, Zn, and Cu in the samples was higher than permissible limits (CCME, 1992). Almost all the isolates were resistant to Pb, Cr, Cu, and Zn, while only the fungus

Penicillium spp. was resistant to Cd. Eventually, it became possible to develop. *Aspergillus* and *Penicillium* isolates are the most tolerable of heavy metals and have shown strong growth, which makes them most suitable for future research regarding their ability to remove metals from contaminated wastewaters. The results obtained revealed that the response of isolates to heavy metals was related to the metals tested, the concentration of the metals in the medium, and the isolate tested. Therefore, both the strain tested and the location where it was isolated determined the level of resistance.

4 Remediation of biochar on heavy metal polluted soils

Pyrolysis is a controlled process by which organic material within agricultural and forestry wastes is converted into biochar. Despite its appearance, it is a form of carbon that can be stored safely. Plant materials, wood chips, leaves, or animal matter are burned in a container with very little oxygen in pyrolysis. There are virtually no contaminating fumes released during the burning process. As a result of this combustion, carbon is created in the form of biochar, which is stable and cannot be easily released into the atmosphere.

A wide range of soil heavy metals are absorbed into the food chain via crops and accumulate in organisms via diet, respiratory inhalation, or other exposure routes, all of which are harmful to human health. As HMs contamination degrades soils, it poses a threat to nature and urges action. To reduce the environmental risk of HMs and improve the quality and security of agricultural land by implementing different productive in situ and ex situ remediation methods. Controlling the continual deterioration of land quality requires certain economic and compelling remediation techniques.

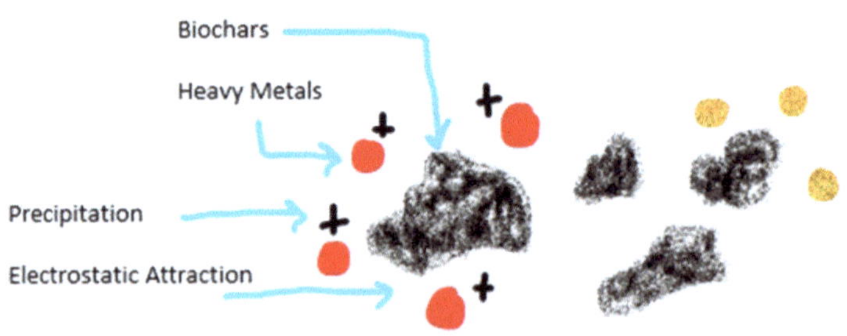

Figure 2: Mechanism of biochar in remediation of heavy-metal-contaminated sites.

Biochar's long-term stability is becoming a key characteristic of modern soil environmental management and remediation. The application of biochar as a novel carbon-rich material to immobilize metals in contaminated soils is receiving increasing attention as a green, economical, and environmentally friendly method. Figure 2 illustrates the mechanism of biochar remediation of metal-contaminated soils based on research achievements. Research status and progress in this area are analyzed from a few significant viewpoints, including preparation technologies, performance characteristics, remediation mechanisms, effects and affecting factors, metal–biochar interactions in soils, potential risks of biochar amendments, and use of biochar as an amendment. Biochar efficiency depends on a variety of factors, including biomass type, heating rate, residence time, and pyrolysis temperature.

Several research gaps remain in the development and improvement of practical methods and useful techniques for preparing and applying diverse biochars that target specific heavy metals. Research needs to be expanded from laboratory scale to large-scale and long-term studies to examine the long-term effects of biochar applications on soil remediation, soil organisms, and plant growth for ensuring the safe production and reasonable use of biochar in remediation technologies in the future.

In recent years, several methods such as physical remediation (vitrification, leaching, immobilization, and electrokinetics), chemical remediation (leaching), and biological remediation have been employed to achieve this goal. Additionally, these strategies pretty much have their limitations, such as complicated technique, low feasibility, inefficiency, high economic cost, short duration, and high secondary risk.

At present, the most promising in situ remediation technique is applying amendments to heavy-metal-contaminated soil. Composted organic matter, phosphate compounds, clay minerals, liming materials, metal oxides, coal fly ash, and biochar are often used in soil augmentation. In summary, HMs can be immobilized by complexation, reduction, adsorption, and precipitation reactions, which result in the redistribution of heavy metals from soil-liquid phases to solid phases and thus reduce their mobility and bioavailability.

A soil polluted with heavy metals affects agricultural products and human well-being. To control the crumbling and degradation of land quality, it is important to select economic and effective remediation methods. An overview of biochar use and application will be presented in this chapter, including preparation technologies, remediation strategies, and the processes and outcomes of remediation, as well as impacts on the bioavailability of heavy metals.

Plant- and animal-based biomass is thermochemically transformed into biochar, a carbon-neutral or carbon-negative product. In the field of environmental amendments and agricultural science, biochar is becoming a hot topic and exciting issue due to its incredible potential in soil improvement, pollution repair, contamination removal, and use of waste biomass assets.

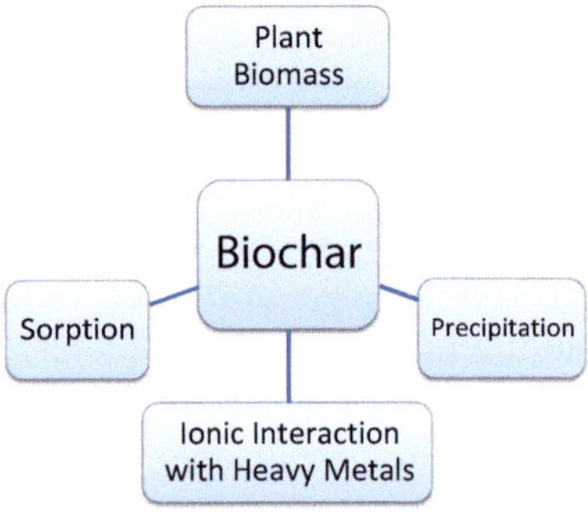

Figure 3: Interaction of biochar with heavy metals.

A remarkable adsorption effect of biochar on heavy metal pollution has been proven, with the mechanisms being mainly adsorption of cationic function, electrostatic adsorption, ion exchange, and precipitation of specific metal complexes. The use of biochar for assimilation of heavy metal pollution has a few benefits, including low-cost technology and high environmental stability. The animal excrement biochar has a higher priority than the sludge and plant residue biochars.

While biochar has a promising future in the remediation of heavy metal polluted soil, particularly animal fecal biochar, there are still a few issues with its hypothetical framework, theoretical system, and research directions.

Biochar is used for heavy metal remediation in laboratories and a small amount is used in large areas of contaminated sites. In addition, biochar cannot remove all types of heavy metal pollution. Biochar composition and properties, including ash and carbon content, aroma, and pH, are affected by the pyrolysis temperature, properties, and cost of the raw materials; therefore, we must develop optimum "specific biochars" and efficient biochar restorers for concrete heavy metal polluted soil in practical applications to maximize the potential of biochar innovation technology.

There is not a clear understanding of how biochar changes with time in soil and how it influences the environment over time. Furthermore, biochar itself may contain a limited quantity of heavy metals, which should be assessed for its potential negative impact before its application. Finally, the potential dangers from in situ remediation techniques of biochar make future research necessary to ensure the production and use of biochar safely and sustainably.

According to Figure 4, biochar has several advantages including increasing soil pH value and carbon content, reducing the accessible part of heavy metals, improving soil water retention capacity, increasing crop yield, and inhibiting heavy metal uptake and accumulation.

The key determinants of the properties of biochar are temperature, biomass type, pyrolysis rate, and residence time. Biochar and heavy metals interact primarily through reduction, complexation, cation exchange, precipitation, and electrostatic attraction. These mechanisms are influenced by pH, dissolved organic carbon content, and ash composition and content.

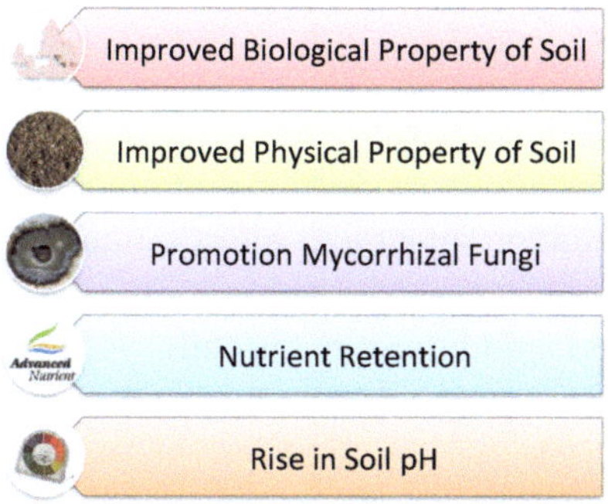

Figure 4: Advantages of biochar.

Plant- and animal-based biomass is thermochemically converted into carbon-neutral or negative biochar. It increases soil fertility in various ways. Therefore, increasing soil pH and organic carbon content improves soil water-holding capacity, reduces heavy metal availability, and increases agricultural crop production.

Biochar is a porous organic material containing an abundance of active functional groups and aromatic structures that has a pH value between neutral and alkaline, large surface area, high cation exchange capacity, and a negative surface charge. In the biochar-altered HM-contaminated soil, seed germination, crop yields, plant growth, and microbial activity and population have improved significantly.

Moreover, biochar production is considered a value-added method of managing an enormous number of organic wastes, which has certain advantages in terms of economic benefits and feasibility aspects. In the field of use of biochar in remediation technologies, there are still some potential threats and these may impede its further application.

Applying biochar for remediation of HM-contaminated soil can be green, economical, and environmentally friendly [47]. By pyrolysis, biochar is prepared using a variety of thermochemical techniques, and the quality of the biochar is greatly influenced by the feedstock materials, the pyrolysis temperature, and the residence time. HMs are immobilized by a variety of mechanisms, including complexation, cation exchange, electrostatic attraction, precipitation, and electrostatic attraction. Changing the DOC content, soil pH value, and other alkaline minerals content will reduce the heavy metal uptake and accumulation by plants, and these changes will promote plant growth, diminish HM bioavailability, and improve soil quality at the same time.

Environmental management and remediation are beginning to take interest in biochar's long-term stability and the ecological responses of pollutants with the toxicological parts. Due to its important effect on the restoration of metal-contaminated soil, biochar is gaining attention.

5 Affecting factors

Research achievements regarding biochar remediation of heavy-metal-contaminated soils over the past few years have been reviewed, including production and properties of biochar, remediation mechanisms, and remediation effect and affecting factors. Findings include:
- Biochar costs less and is highly efficient, green, and improves the soil.
- A biomass material and pyrolysis temperature influence the physicochemical properties of biochar.
- There are several strategies for increasing the adsorption of heavy metals on biochar through activation, magnetization, oxidation, and digestion, and bioavailability includes a conflict between the goals of immobilizing the metals or mobilizing them to enhance their availability.
- A solidifying agent can be added to biochar to enhance the strength of treated soil.
- Ion exchange, physical adsorption, electrostatic interaction, complexation, and precipitation are among the adsorption mechanisms by biochar.

Several areas for future research are suggested, including establishing a uniform classification criterion for biochar, exploring the effectiveness of biochar on the remediation of multi polluted sites, showing the interactions between biochar and heavy metals in soil, undertaking trials in large scales and long-term studies, and conducting solidification experiments to determine the mechanical properties of biochar for use in soil remediation.

As a carbon-rich material, biochar has been applied to soils to remove heavy metals. Obtained by thermally treating biomass in an oxygen-restricted compartment, biochar has a porous structure and is obtained as a solid product. As a carbon storage material, biochar possesses desirable physicochemical properties. Biochar has advantages such as high cation exchange capacity, low production cost, pH, porous structures, and surface functional groups that make it a suitable adsorbent for removing heavy metals from contaminated soil. Heavy metals in soil can form complexes with biochar, which reduce their bioavailability [48].

Reviews have been published regarding metal–biochar interactions in contaminated soils, risks associated with biochar amendments, and the use of biochar in soil remediation.
1) The effect of biochar remediation depends on its properties and its interactions with soil
2) The application of biochar could reduce the bioavailability of heavy metals in soils and their accumulation in plants
3) Biochar applications may pose health and environmental risks, such as the release of toxic substances into soils or the inhalation of biochar dust [49].

6 Future prospective

As far as cost and manageability go, bioremediation is a better option than other regular solutions. Bioremediation is not without its limitations, however. As chlorinated natural or very fragrant hydrocarbons are resistant to microbial attack, several microbes cannot break down toxic metals that are biodegradable.

Future studies must incorporate interdisciplinary approaches, such as engineering, nanotechnology, microbiology, geology, ecology, and chemistry, to ensure that bioremediation is effective under all adverse conditions. Researchers should study novel species in the area in the future. Many players are already involved in the bioremediation field, producing a variety of products to deal with a nutrient deficiency in fish farms, improving nutrient release in agricultural land [50], composting industrial waste, and removing toxic chemicals, and the list is growing. The development of new products to clean up our environment remains a strong focus of research.

7 Summary

Biochars have been specifically developed to target specific HMs, but there are still gaps in the production of practical methods for making and applying them. There needs to be a consideration in the future regarding the long-term effects and no risk of biochar applications for soil remediation.

Microbial interest is a very powerful method for fixing polluted environments through biodegradation. Biodegradation can be used for remediating, cleaning, managing, and recovering polluted environments. In most cases, genetically engineered microorganisms are not needed since a wide variety of naturally occurring microbes can be used instead.

Whenever ecological conditions allow microbial action and development, bioremediation can be compelling. Building up an understanding of microbial communities and their networks as well as their ability to respond to common habitats and toxins/pollutants, in addition to extending knowledge about their genetic traits, it is possible to develop our capacity for biodegrading degraded materials, and conducting field trials with these remarkable organisms for bioremediation procedures would provide cost-effective technologies and possibly open a door for a breakthrough. Diverse bioremediation techniques have been used globally, with varying degrees of success.

Fundamentally, the focal points outweigh the impediments, as evidenced by the growing number of places choosing to utilize this bioremediation innovation. There are different species explored from different sites, and they are powerful in controlling the environment.

References

[1] Dixit R, Malaviya D, Pandiyan K, Singh UB, Sahu A, Shukla R, Singh Rai BP, Sharma JP, Lade PK H, Bioremediation of heavy metals from soil and aquatic environment: An overview of principles and criteria of fundamental processes, Sustainability, 2015, 7, 2189–2212.

[2] Banik S, et al., Recent advancements and challenges in microbial bioremediation of heavy metals content, JSM Biotech. Biomedical Eng., 2014, 2(1), 1–9.

[3] Gupta N, Yadav KK, Kumar V, A review on current status of municipal solid waste management in India, J. Env. Sci., 2015, 37, 206–217.

[4] Nagajyoti P, Lee K, Sreekanth T, Heavy metals, occurrence, and toxicity for plants: A review, Environ. Chem. Lett., 2010, 8, 199–216.

[5] Fulekar M, Singh A, Bhaduri AM, Genetic engineering strategies for enhancing phytoremediation of heavy metals, Afr. J. Biotechnol., 8, 2009, 529–535.

[6] Bolan N, Kunhikrishnan A, Thangarajan R, Kumpiene J, Park J, Makino T, Kirkham MB, Scheckel K, Remediation of heavy metalloids contaminated soils – To mobilize or to immobilize? J. Hazard. Mater., 266, 2014, 141–166.

[7] D'Amore JJ, Al-Abed SR, Scheckel KG, Ryan JA Methods for speciation of metals in soils: A review, J. Environ. Qual., 34(2005), 1707–1745.

[8] Fashola M, Ngole-Jeme V, Babalola O, Heavy metal pollution from gold mines: Environmental effects and bacterial strategies for resistance, Int. J. Environ. Res. Public Health, 13, 2016, 1047.

[9] Addis W, Abebaw A, Determination of heavy metal concentration in soils used for cultivation of allium sativum L. (Garlic) in East Gojjam Zone, Amhara Region, Ethiopia, Cogent Chem., 3, 1419422, 2017.

[10] Siddiquee S, Rovina K, Azad S, Naher L, Suryani S, Heavy metal contaminants removal from wastewater using the potential filamentous fungi biomass: A review, J. Microb. Biochem. Technol., 7, 2015, 384–393.

[11] Ahluwalia SS, Goyal D, Microbial and plant derived biomass for removal of heavy metals from wastewater, Bioresour. Technol., 98, 2007, 2243–2257.

[12] Rajendran P, Muthukrishnan J, Gunasekaran P, Microbes in heavy metal remediation, Indian J. Exp. Bio., 41, 2003, 935–944.

[13] Mosa KA, Saadoun I, Kumar K, Helmy M, Dhankher OP, Potential biotechnological strategies for the cleanup of heavy metals and metalloids, Front. Plant Sci., 7, 2016, 1–14.

[14] Malik A, Metal bioremediation through growing cells, Environ. Int., 30, 2004, 261–278.

[15] Yang T, Chen ML, Wang JH, Genetic and chemical modification of cells for selective separation and analysis of heavy metals of biological or environmental significance, TrAC Trends Anal. Chem., 66, 2015, 90–102.

[16] Fomina M, Gadd GM, Biosorption: Current perspectives on concept, definition, and application, Bioresour. Technol., 160, 2014, 3–14.

[17] Akcil A, Erust C, Ozdemiroglu S, Fonti V, Beolchini F, A review of approaches and techniques used in aquatic contaminated sediments: Metal removal and stabilization by chemical and biotechnological processes, J. Clean. Prod., 86, 2015, 24–36.

[18] Misal SA, Lingojwar DP, Shinde RM, Gawai KR, Purification and characterization of azoreductase from alkaliphilic strains Bacillus badius, Process Biochem., 46(6), 2011, 264–269.

[19] El Fantroussi S, Agathos SN, Is bioaugmentation a feasible strategy for pollutant removal and site remediation? Curr. Opin. Microbiol., 8, 2005, 268–275.

[20] Tang CY, Criddle QS, Fu CS, Leckie JO, Effect of flux (transmembrane pressure) and membranes properties on fouling and rejection of reverse osmosis and nanofiltration membranes treating perfluoro octane sulfonate containing waste water, J. Environ. Sci. Technol., 41, 2007.

[21] Adams GO, Fufeyin PT, Okoro SE. Bioremediation, Biostimulation and, Bioaugmentation: A review. Int. J. Environ. Bioremed. Biodegrad., 3(1), 2015, 3–28.

[22] Cases I, de Lorenzo V, Genetically modified organisms for the environment: Stories of success and failure and what we have learned from them, Int. Microbiol., 8, 2005, 213–222.

[23] Litchfield CD, In situ bioremediation: Bases and practices, In: Biotechnology of Industrial and Hazardous Waste, Levin MA, Gealt MA, eds., McGraw-Hill, USA, 1993, 167–195.

[24] Dott W, Feidieker D, Steiof M, Becker PM, Kaompfer P, Comparison of ex situ and in situ techniques for bioremediation of hydrocarbon polluted soils, Int. Biodeterior. Biodeg., 1995, 35, 301–316.

[25] Boopathy R, Factors limiting bioremediation technologies, Bioresour. Technol., 74, 2000, 63–67.

[26] Paniagua-Michel J, Garcia O, Ex-situ bioremediation of shrimp culture effluent using constructed microbial mats, Aquac. Eng., 28, 2003, 131–139.

[27] Jain PK, Bajpai V, Biotechnology of bioremediation – A review, Int. J. Environ. Sci., 3(1), 2012, 535–549.

[28] Montagnolli RN, Matos Lopes PR, Bidoia ED, Assessing Bacillus subtilis biosurfactant effects on the biodegradation of petroleum products, Environ. Monit. Assess., 187(4116), 2015, 1–17.

[29] Sharma S, Bioremediation: features, strategies, and applications, Asian J. Pharm. Life Sci., 2 (2), 2012, 202–213.

[30] Gadd GM, Bioremedial potential of microbial mechanisms of metal mobilization and immobilization, Curr. Opin. Biotech., 11, 2000, 271–279.

[31] Hassan BA, Venkateshwaran AA, Fredrickson JK, Daly MJ, Engineering Deinococcus geothermalis for bioremediation of high temperature radioactive waste environments, Appl. Environ. Microbiol., 69, 2003, 4575–4582.

[32] Baldwin BR, Peacock AD, Park M, Ogles DM, Istok JD, McKinley JP, Multilevel samplers as microcosms to assess microbial response to bio stimulation, Groundwater, 46, 2008, 295–304.

[33] Kumar R, Singh P, Dhir B, Sharma AK, Mehta D, Potential of some fungal and bacterial species in bioremediation of heavy metals, J. Nucl. Phys. Mater. Sci. Radiat. Appl., 1(2), 2014, 213–223.

[34] Stabili L, Schirosi R, Licciano M, Mola E, Gianyrande A, Bioremediation of bacteria in aquaculture waste using the polychaete Sabella spallanzanii, New Biotechnol., 2010.

[35] Dwivedi S, Mishra A, Saini D, Removal of heavy metals in liquid media through fungi isolated from waste water, Int. J. Sci. Res. (IJSR), 1(3), 2012.

[36] Fazli MM, Soleimani N, Mehrasbi M, Darabian S, Mohammadi J, Ramazani A, Highly cadmium tolerant fungi: Their tolerance and removal potential, J. Environ. Health Sci. Eng., 13(19), 2015, 9.

[37] Sharma PK, Balkwill DL, Frenkel A, Vairavamurthy MA, A New Klebsiella Planticola strain (Cd-1) grows anaerobically at high cadmium concentrations and precipitates cadmium sulfide, Appl. Environ. Microbiol., 66, 2000, 3083–3087.

[38] Gavrilesca M, Removal of heavy metals from the environment by biosorption, Eng. Life Sci., 4 (3), 2004, 219–232.

[39] Gadd GM, Biosorption: Critical review of scientific rationale, environmental importance, and significance for pollution treatment, J. Chem. Technol. Biotechnol., 84, 2009, 13–28.

[40] Agrawal S, Bhatt A, Rai SK, Agrawal PK, Physiochemical and microbiological analysis of Ganga Water at Kanpur with special reference to bioremediation, Octa J. Environ. Res. Int. Peer-Rev. J., Oct. Jour. Env. Res., 3(4), 2015, 290–301.

[41] Pagnanelli F, Petrangeli MP, Toro L, Trifoni M, Veglio F, Biosorption of metal ions on Anthrobacter sp.: Biomass characterization and biosorption modelling, Environ. Sci. Technol., 34(13), 2000, 2773–2778.

[42] Hussain T, Roohi A, Munir S, Khan J, Anees M, Ahmed I, Hermann VE, Kim KY, Biochemical characterization and identification of bacterial strains isolated from drinking water sources of Kohat, Pakistan, Afr. J. Microbiol. Res., 7(16), 2013, 1579–1590.

[43] Mulamattathil SG, Bezuidenhout C, Mbewe M, Ateba CN Isolation of environmental bacteria from surface and drinking water in Mafikeng, South Africa, and characterization using their antibiotic resistance profiles, J. Pathog., Volume 2014, Article ID 371208, 2014, 11, pages.

[44] Vadkertiova R, Slavikova E, Metal tolerance of yeasts isolated from water, soil and plant environments, J. Basic Microbiol., 46, 2006, 145–152.

[45] Verma T, Srinath T, Gadpayle RU, Ramteke PW, Hans RK, Garg SK, Chromate tolerant bacteria isolated from tannery effluent, Bioresour. Technol., 78, 2001, 31–35.

[46] Ezzouhri L, Castro E, Moya M, Espinola F, Lairini K, Heavy metal tolerance of filamentous fungi isolated from polluted sites in Tangier, Morocco, Afr. J. Microbiol. Res., 3(2), 2009, 035–048.

[47] Sun W, Zhang S, Su C, Impact of biochar on the bioremediation and phytoremediation of heavy metal(loid)s in soil, Adv. Bioremediat. Phytoremediat., 2018, 149–168.

[48] Lyu H, Gong Y, Gurav R, Tang J, Potential application of biochar for bioremediation of contaminated systems, In: Biochar Application, Elsevier, 2016, 221–246.

[49] Puga AP, Abreu CA, Melo LC, Beesley L, Biochar application to a contaminated soil reduces the availability and plant uptake of zinc, lead and cadmium, J. Environ. Manage., 2015, 159, 86–93.

[50] Karppinen EM, Stewart KJ, Farrell RE, Siciliano SD, Petroleum hydrocarbon remediation in frozen soil using a meat and bonemeal biochar plus fertilizer, Chemosphere, 173, 2017, 330–339.

Authors profile

Dr. Ashwini A. Waoo

Associate Professor, Department of Biotechnology
Faculty of Life Sciences and Technology, AKS University, SATNA, M.P., India.

E-mail-ashwiniwaoo@gmail.com Contact No: 9981721676

Fifteen years of academic and research experience, including doctorate, GATE, CSIR-NET Life Science, M. Phil (Biotechnology), and M.Sc. (Biotechnology) along with IIT-Bombay and SWAYAM certifications. More than 25 research papers were published in international journals. Honored with Women Researcher Award, "1st International Scientist Awards in Engineering, Science and Medicine," organized by VDGood Association in Chennai, India. Acquired young scientist's scholarship from MP Council of Science and Technology (MPCST). Research contribution includes UGC major research project and Supervision of six Ph.D. students along with more than 100 dissertations at UG and PG level of students; also contributed in many international journals as a reviewer. Recently, was recommended as Bentham Ambassador to India for Bentham publications journals. Academic Coordinator of PG Biotechnology Courses, Coordinator of Academic Committee, Member of Training, and placement cell were the responsibilities that were undertaken.

Charu Vyas

Charu Vyas is a Ph.D research scholar in the Department of Biotechnology of AKS University, Satna, (M.P.), India. She had done M.Sc. (Biotechnology). Her areas of interest are plant tissue culture, biochemistry, immunology, genetic engineering, and cell and molecular biology.

Charu Vyas participated in various workshops, national and international seminars, and conferences. She had worked at ACBR, Delhi University, and in BCIL, (under BCIL-BITP, sponsored by the Department of Biotechnology, Ministry of Science and Technology, Govt. of India.) as a research trainee for a year. She has participated in the “Bioinformatics Workshop on Genomics, Proteomics, Drug Design & High-Performance Computing” at the Supercomputing Facility for Bioinformatics & Computational Biology (SCFBIO), IIT Delhi.

Shikha Kumari, Poonam Kataria, Geeta Dhania

Wastewater treatment using biochar technology

Abstract: Biochar is a high-carbon compound obtained from biomass in the absence of oxygen. Biochar shows unbelievable potential for the wastewater treatment and soil remediation. The application of biochar is rising due to available renewable feedstock and simple methods of preparation. The removal of toxic metals and pollutants depends on surface area, pore size, surface functional group, size of molecules to be removed, and adsorbent property. Biochar is a cost-effective and also eco-friendly method for treatment of industrial wastewater, agricultural wastewater, and sewage wastewater. This chapter includes the application of biochar in the removal of various toxic metals and other pollutants from wastewater.

Keywords: biochar, pyrolysis, wastewater treatment, activated sludge treatment

1 Introduction

The constant release of a large number of organic and inorganic contaminants such as dye, pesticides, pharmaceuticals, waste from industries, and municipal waste into water bodies is deteriorating the quality of water resources. To improve the water quality and eliminate the contaminants from water a number of technologies like ion exchange method, phytoremediation, and biological and chemical treatment are used, but these technologies require proper maintenance. These technologies are costly and are found to be ineffective in the complete removal of contaminants from water [1, 2]. Biochar is a good alternative to such expensive technologies.

Biochar enhances fertility of acidic soil and also improves the plant health resulting in increased productivity. It is also being used for the treatment of different kinds of wastewater say agricultural wastewater, sewage wastewater, and also industrial wastewater. Biochar possesses the ability to adsorb toxic heavy metals, organic as well as inorganic pollutants from wastewater. Modified biochar having large surface area and pore size effectively adsorbs nitrogen and phosphorous from industrial effluents [3].

2 Production of biochar

Biochar is a stable, solid, fine-grained and carbon-rich compound obtained from non-edible biomass. Besides agricultural biomass, biochar may also be obtained from municipal solid waste and sewage sludge [4, 5]. The conversion of biomass

https://doi.org/10.1515/9783110734003-003

into biochar involves various processes like pretreatment (physical, chemical, and biological), thermal processes (pyrolysis, gasification, and carbonization) followed by posttreatment which include ball milling and magnetization (see Figure 1). Biochar can be used to replace activated carbon due to its adsorption properties. The adsorption properties of biochar are affected by a number of factors including surface area, pore size, and ion exchange capacity. The technologies adopted, kind of feedstock used, and the methods employed for the preparation of biochar show a significant effect on the physical and chemical properties of biochar.

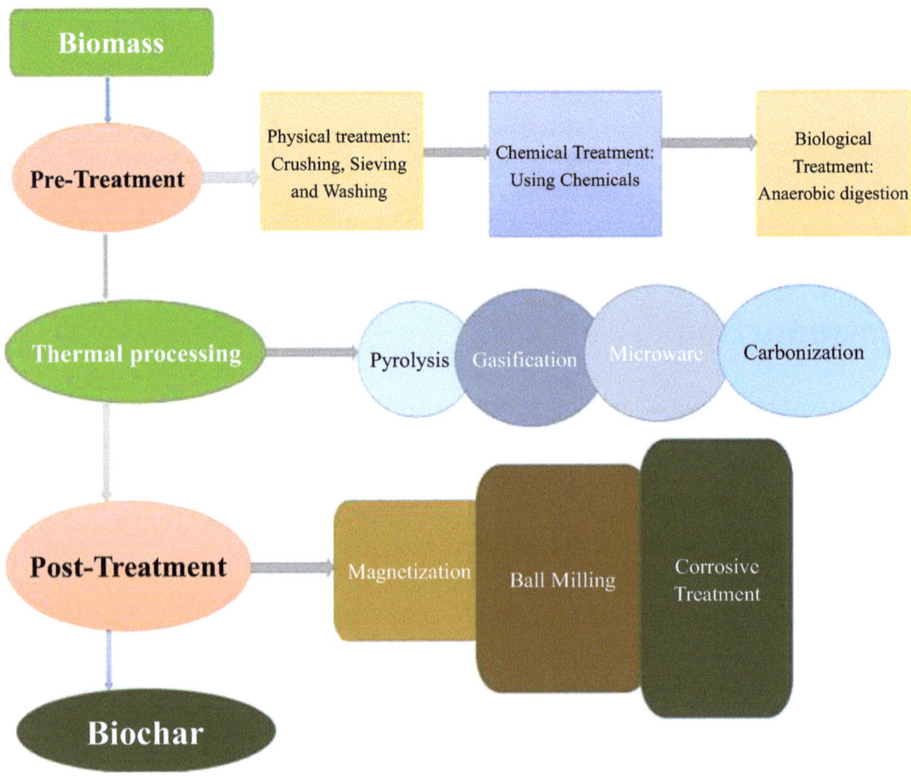

Figure 1: Technologies for production of Biochar.

2.1 Pyrolysis

Pyrolysis is a thermal process used for converting organic waste into carbonaceous things in the lack of oxygen and presence of high temperature (see Figure 2). The agro-residue used for biochar production mainly comprises lignin, hemicellulose, fats, and cellulose; these components are broken down using heat to improve the

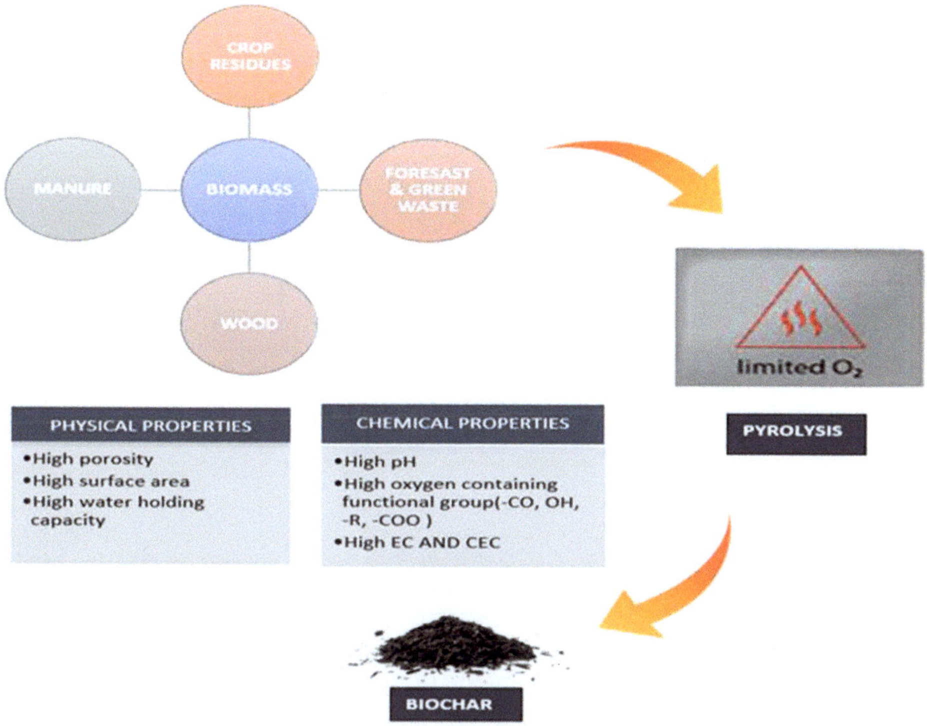

Figure 2: Production of biochar from biomass and its properties.

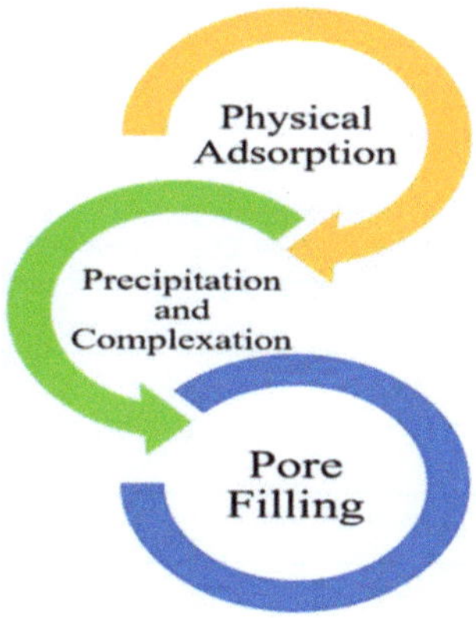

Figure 3: Mechanism of Adsorption.

carbon content of the feedstock. Depending on the time taken for the reaction and temperature, pyrolysis may be rapid, flash, or slow.

The pyrolysis temperature, heating rate, and reactor conditions directly affect the biochar production. Slow pyrolysis is usually carried out at slow heating rate

Figure 4: Application of Biochar.

and a temperature of about 350–500 °C. The biochar obtained as a result of slow pyrolysis possess small pore size, low surface area, and low carbon content. The heating rate and temperature are relatively higher during fast pyrolysis. The biochar obtained through fast pyrolysis is suitable for use in wastewater treatment due to greater hydrophobicity, rich carbon content, larger pore size, high surface area, and less ash content. The pH of the biochar increases with increase in temperature. This kind of biochar is highly effective in the removal of pollutants from water [6–9].

Tables 1 and 2 show the change in properties of biochar based on the raw material and pyrolysis method.

Table 1: Change in properties of biochar based on the raw materials.

Source	pH	Surface area ($m^2\ g^{-1}$)	CEC ($mmol\ kg^{-1}$)
Corn	9.27	107.2	607
Wheat/barley	8.8	26.65	103
Rice straw	9.17	42.15	212
Hazelnut shells	7.86	467.5	83.8
Bagasse	7.59	113.6	115
Food waste	9.09	0.803	81
Other green waste (grass, leaves, etc.)	8.72	119.8	290
Softwoods	7.48	194.2	145

Table 1 (continued)

Source	pH	Surface area ($m^2 g^{-1}$)	CEC (mmol kg^{-1})
Hardwoods	7.94	171.3	138
Paper mill waste	9.13	10.08	52
Poultry manure	9.8	50.35	538
Dairy manure	9.45	33.38	342
Sewage sludge	6.9	102.1	23.6

Table 2: Change in properties of biochar based on the pyrolysis method.

Pyrolysis temperature and type	pH	Surface area ($m^2 g^{-1}$)	CEC (mmol kg^{-1})
Fast, 300–499 °C	8.33	44.74	28.8
Fast, 500–699 °C	7.7	40.99	No data
Fast, 700–900 °C	10.1	178.2	No data
Slow, <300 °C	5.01	1.686	327
Slow, 300–499 °C	7.81	81.32	268
Slow, 500–699 °C	9.09	180.5	218
Slow, 700–900 °C	10.1	189.8	41.5

3 Biochar adsorption mechanism

The mechanism of adsorption relies on the properties of pollutant and also the adsorbent surface. The surface area of adsorbent must be free from any kind of adsorbed material, thus providing free surface area for the adsorption of different kinds of pollutants present in the wastewater. The most commonly used adsorbent is activated charcoal. Biochar resembles activated carbon in many properties [10]. There are mainly three stages involved in the adsorption of a pollutant. The first and foremost stage is physical adsorption which involves the settling down of the pollutant on the surface area of biochar. The second stage includes complexation and precipitation; here, the pollutant gets deposited on the adsorbents surface. The third stage is pore filling where adsorbate is reduced into the pore of the adsorbent. Electrostatic attraction, pore filling, π–π electron–donor acceptor interaction, hydrogen bonding, complexes adsorption, and hydrophobic interactions aid in the adsorption of adsorbent. For the removal of heavy metals, multiple mechanisms may affect the different stages [11–13].

4 Biochar application in wastewater treatment

Biochar has been used for the treatment of wastewater from various industrial discharges.

4.1 Industrial wastewater

Biochar is a method to eliminate many pollutants from industrial wastewater. Mining and smelting chemical industries are the main sources of the industrial wastewater and it mainly comprise of organic pollutants and heavy metals. Biochar along with chitosan is used as an adsorbent and it proved effective for heavy metal adsorption. Biochar made from sugarcane bagasse was used for treatment of effluent from battery manufacturing industry. The adsorption capacity was found to be around 12.7 mg g^{-1} [14].

Glue residue biochar modified with $ZnCl^{-}$, was used as an adsorbent for chromium and the maximum sorption capacity was about 325.5 mb g^{-1}. Dyes are also a major part of industrial wastewater. Bamboo cane was used to produce nanoporous biochar, effective in the sorption of Lanasyn orange and Lanasyn gray; dyes used in wool carpet industries. In another study, pecan nutshell biochar was used for removal of Reactive Red 141 from wastewater [15, 16].

4.2 Sewage wastewater

Ammonium was successfully removed from sewage wastewater with the help of biochar obtained from digested sludge. The ammonium removal capacity was found to be highest in the biochar produced at 450 °C [17]. Biochar obtained from waste sludge was used as a catalyst for ozonation of wastewater; considerably higher removal of total organic carbon was observed. The presence of metallic oxides and functional carbon groups in the biochar enhanced the formation of hydroxyl radical and mineralized petroleum contaminants.

It is possible to treat sewage wastewater with the help of biochar obtained from municipal biowaste. Biochar possesses high porous surface area which makes it able to act as biofilter for treatment of sewage wastewater. A 90%, 89%, 64%, and 78% reduction in the concentration of COD, TSS, TKN, and TP, respectively, was observed after passing the wastewater through biochar filter [18, 19].

4.3 Agricultural wastewater

The major contaminants in agricultural wastewater consist of heavy metals and pesticides. Rice straw biochar showed a significant reduction in the concentration of atrazine; a common pesticide used in agricultural activities. Removal of sulfamethazine was carried out by using steam-activated biochar. Biochar is also successfully utilized for the removal of carbaryl, diuron, and glyphosate from agricultural wastewater. Zero valent iron magnetic paper mill sludge biochar was effective for the removal of pentachlorophenol from wastewater [20, 21].

4.4 Indicator organisms and pathogens

The urban storm wastewater consists of a number of contaminants, and this polluted water is released into water resources. The use of this water for the irrigation of fields results into microbial growth and contamination of vegetables with microbes. Biochar application is found for removing pathogens as well as indicator organisms. Rice husk Biochar is used for removal of microbes from wastewater.

4.5 Organic pollutant removal

Organic pollutants can adversely affect human beings, animals, plants, and also pesticides, herbicides, and antibiotics. Presence of organic pollutants decreases dissolved oxygen in water and also damages the water system and human health. Research studies have shown that biochar effectively removes the organic contaminants from wastewater. The mechanism for adsorption of organic contaminants includes hydrophobic effect pore filling, hydrogen bonds, and electrostatic interaction. Polarity, aromaticity index, oxygen functional groups, and the electron donor–acceptor interactions also affect the process of adsorption.

Antibiotics, such as sulfonamides and tetracyclines, can also be removed with the use of biochar. Tetracycline was successfully eradicated from aqueous solution using iron and zinc-doped sawdust char. Bridge enhancement, site recognition, and site competition were observed to be the mechanism of adsorption. Digested bagasse biochar was used to remove sulfamethoxazole which is a kind of antibiotic widely used for animals and humans. A direct effect of the pH value on the adsorption process was noticed. Switchgrass biochar was used for the removal of metribuzin herbicide. There was no significant effect of change in temperature on the adsorption capacity of switchgrass biochar [22–25].

4.5.1 Nitrogen and phosphorus removal

The discharge of immense domestic sewage causes nutrient over enrichment in aquatic environments which contains nitrogen and phosphorus. In agriculture, excessive use of pesticides or chemical fertilizers increases the release of nitrogen and phosphorus into the aquatic systems. As a result of these activities, water eutrophication increases; it leads to degradation of aquatic biodiversity. Nitrogen and phosphorus are removed from industrial wastewater to minimize the effects of eutrophication with the help of adsorption using biochar. The biochar produced through fast pyrolysis and the modified biochar have large surface area and pore size; which provides free space for the adsorption of N and K. Presence of polar as well as nonpolar surface site and internal porosity enhance adsorption [26].

Biochar has the capability of adsorbing nitrogen and phosphorus from wastewater. Modification and posttreatment have a significant effect on the adsorption capacity of biochar. The adsorption capacity of maple wood biochar was observed to be less than the oxidized maple wood biochar. Pyrolysis temperature is another factor affecting the adsorption process. Biochar derived from pine sawdust at temperature 300 °C was observed to adsorb highest ammonium concentration from aqueous solution. A 73% removal rate of phosphorus was observed using digested sugar beet biochar. Magnesium oxide treated biochar obtained from wood waste was found to be good in the removal of phosphate as well as ammonium [27–29].

4.6 Heavy metals removal

Heavy metals one of the major contaminants released along with industrial wastewater. Heavy metals are toxic and carcinogenic and thus may affect the human health adversely. These pollutant has the ability to enter the food chain through a process called bioaccumulation. With industrialization and urbanization, the use and release of toxic heavy metals have also increased.

Research studies suggest that biochar can be used for adsorption of heavy metals. The presence of oxygen-containing functional groups (carboxyl group and phenolic group) on the surface of biochar initiate and enhance the adsorption process. Different mechanisms say electrostatic interactions, physical sorption, ion exchange precipitation and complexation are involved in the process of removal of heavy metals. Through electrostatic attractions, various biochar having negatively charged surface can absorb positively charged metals. Biochar can interact with numerous metals through specific ligands and functional groups to form complexes or precipitates of their solid mineral phases [30].

Rice straw biochar and coconut coir biochar were utilized for removal of aluminum and chromium, respectively. Copper and zinc were removed from the water using corn straw biochar. Biochar obtained from rice husk and dairy manure showed a significant reduction in the concentration of lead, copper, zinc and cadmium. Biochar modified using $Zn(NO_3)_2$ showed adsorption of arsenic up to 7.0 μg g^{-1}. Cadmium and lead were successfully eradicated from the wastewater using biochar obtained from sugarcane bagasse, sawdust, and chicken manure [31, 32].

5 Conclusion

Biochars have found numerous applications in resolving global problems mainly climate change and removal of contaminants from water and soil. Biochar improves the soil quality and nutrient recycling. This helps to reuse and recycle the waste which allows the material and product in use. It encompasses water, carbon, and nutrient cycles and directly impacts the water, energy, and food systems.

Biochar has been found to be an effective alternative for treatment of wastewater released as a result of agricultural activities and also for the industrial effluents. The removal of various heavy metals, nitrogen, and phosphorus has been observed by many researchers. The future research work will focus on establishing its effectiveness, given that its application in wastewater treatment and soil have already been established.

References

[1] Nharingo T, Moyo M, Application of Opuntiaficus indica in bioremediation of wastewaters. A critical review, J. Environ. Manage., 2016, 166, 55–72.

[2] Vunain E, Masoamphambe EF, Mpeketula PMG, Monjerezi M, Etale A, Evaluation of coagulating efficiency and water borne pathogens reduction capacity of Moringaoleifera seed powder for treatment of domestic wastewater from Zomba, Malawi, J. Environ. Chem. Eng., 2019, 7, 103–118.

[3] Xiang W, Zhang X, Chen J, Zou W, He F, Hu X, Daniel CW, Tsang YSO, Bin G, Biochar technology in wastewater treatment: a critical review, Chemosphere, 2019, 252.

[4] Enaime G, Ennaciri K, Ounas A, Bacaoui A, Seffen M, Selmi T, Yaacoubi A, Preparation and characterization of activated carbons from olive wastes by physical and chemical activation: application to indigo carmine adsorption, J. Mater. Environ. Sci., 2017, 8, 4125–4137.

[5] Selmi T, Sanchez-Sanchez A, Gadonneix P, Jagiello J, Seffen M, Sammouda H, Celzard A, Fierro V, Tetracycline removal with activated carbons produced by hydrothermal carbonisation of Agave americana fibres and mimosa tannin, Ind. Crops Prod., 2018, 115, 146–157.

[6] Suliman W, Harsh JB, Abu-Lail NI, Fortuna AM, Dallmeyer I, Garcia-Perez M, Influence of feedstock source and pyrolysis temperature on biochar bulk and surface properties, Biomass Bioenergy, 2016, 84, 37–48.

[7] Windeatt JH, Ross AB, Williams PT, Forster PM, Nahil MA, Singh S, Characteristics of biochars from crop residues: potential for carbon sequestration and soil amendment, J. Environ. Manage., 2014, 146, 189–197.

[8] Brewer CE, Hu YY, Schmidt-Rohr K, Loynachan TE, Laird DA, Brown RC, Extent of pyrolysis impacts on fast pyrolysis biochar properties, J. Environ. Qual., 2012, 41, 1115–1122.

[9] Gul S, Whalen JK, Thomas BW, Sachdeva V, Deng H, Physico-chemical properties and microbial responses in biochar-amended soils: mechanisms and future directions, Agric. Ecosyst. Environ., 2015, 206, 46–59.

[10] Ambaye GT, Vaccari M, Van Hullebusch DE, Amrane A, Rtimi S, Mechanism and adsorption capacities of biochar for the removal of organic and inorganic pollutants from industrial wastewater, Int. J. Environ. Sci. Technol., 2020.

[11] Rosales E, Meijide J, Pazos M, Sanroman MA, Challenges and recent advances in biochar as lowcostbiosorbent: from batch assays to continuous-flow systems, Bioresour. Technol., 2017, 246, 176–192.

[12] Ahmad M, Rajapaksha AU, Lim JE, Zhang M, Bolan N, Mohan D, Vithanage M, Lee SS, Ok YS, Biochar as a sorbent for contaminant management in soil and water: a review, Chemosphere, 2014, 99, 19–33.

[13] Lu H, Zhang W, Yang Y, Huang X, Wang S, Qiu R, Relative distribution of Pb^{2+} sorption mechanisms by sludge-derived biochar, Water Res., 2012, 46, 854–862.

[14] Hussain A, Maitra J, Khan KA, Development of biochar and chitosan blend for heavy metalsuptake from synthetic and industrial wastewater, Appl. Water Sci., 2017, 4525–4537.

[15] Pradhananga R, Adhikari L, Shrestha R, Adhikari M, Rajbhandari R, Ariga K, Shrestha L, Wool carpet dye adsorption nonnanoporous carbon materials derived from agro-product, J. Carbon Res., 2017, 3, 3020012.

[16] Zazycki MA, Godinho M, Perondi D, Foletto EL, Collazzo GC, Dotto GL, New biochar from pecan nut shells as an alternative adsorbent for removing reactivated 141 from aqueous solutions, J. Clean. Prod., 2018, 171, 57–65.

[17] Tang Y, Alam MS, Konhauser KO, Alessi DS, Xu S, Tian W, Liu Y, Influence of pyrolysis temperature on production of digested sludge biochar and its application for ammonium removal from municipal wastewater, J. Clean. Prod., 2019, 209, 927–936.

[18] Chen C, Yan X, Xu Y, Yoza BA, Wang X, Kou Y, Ye H, Wang Q, Li QX, Activated petroleum waste sludge biochar for efficient catalytic ozonation of refinery wastewater, Sci. Total Environ., 2019, 651, 2631–2640.

[19] Manyuchi MM, Mbohwaa C, Muzenda E, Potential to use municipal waste bio char in wastewater treatment for nutrients recovery, Phys. Chem. Earth, 2018, 107, 92–95.

[20] Rajapaksha AU, Vithanage M, Ahmad M, Seo DC, Cho JS, Lee SE, Sang SL, Yong SO, Enhanced sulfamethazine removal by steam-activated invasive plant-derived biochar, J. Hazard. Mater., 2015, 290, 43–50.

[21] Wei D, Li B, Huang H, Luo L, Zhang J, Yang Y, Guo J, Tang L, Zeng G, Zhou Y, Biochar-based functional materials in the purification of agricultural wastewater: fabrication, application and future research needs, Chemosphere, 2018, 197, 165.

[22] Essandoh M, Wolgemuth D, Pittman CU, Mohan D, Mlsna T, Adsorption of metribuzin from aqueous solution using magnetic and nonmagnetic sustainable low-cost biochar adsorbents, Environ. Sci. Pollut. Res., 2017, 24, 4577–4590.

[23] Yao Y, Zhang Y, Gao B, Chen RJ, Wu F, Removal of sulfamethoxazole (SMX) and sulfapyridine (SPY)fromaqueous solutions bybiochars derived from anaerobically digested bagasse, Environ. Sci. Pollut. Res., 2018, 25, 25659–25667.

[24] Peiris C, Gunatilake SR, Mlsna TE, Mohan D, Vithanage M, Biochar based removal of antibiotic sulfonamides and tetracyclines in aquatic environments: a critical review, Bioresour. Technol., 2017, 246, 150–159.

[25] Sun P, Li Y, Meng T, Zhang R, Song M, Ren J, Removal of sulfonamide antibiotics and human metabolite by biochar and biochar/H2O2 in synthetic urine, Water Res., 2018, 147, 91–100.

[26] Yin Q, Zhang B, Wang R, Zhao Z, Biochar as an adsorbent for inorganic nitrogen and phosphorus removal from water: a review, Environ. Sci. Pollut. Res., 2017, 24.

[27] Yang HI, Lou K, Rajapaksha AU, Ok YS, Anyia AO, Chang SX, Adsorption of ammonium in aqueous solutions by pine sawdust and wheat straw biochars, Environ. Sci. Pollut. Res., 2017, 1–10.

[28] Xu X, Zheng Y, Gao B, Cao X, N-doped biochar synthesized bya facile ballmilling method for enhanced sorption of CO_2 and reactive red, Chem. Eng. J., 2019, 365, 564–572.
[29] Yao Y, Gao B, Inyang M, Zimmerman AR, Cao X, Pullammanappallil P, Yang L, Biochar derived from anaerobically digested sugar beet tailings: characterization and phosphate removal potential, Bioresour. Technol., 2011a, 102, 6273–6278.
[30] Inyanga MI, Gaoa B, Yaoa Y, Xueb Y, Zimmermanc A, Mosad A, Pullammanappallila P, Oke YK, Caof X, A review of biochar as a low-cost adsorbent for aqueous heavy metal removal, Crit. Rev. Environ. Sci. Technol., 2016, 46.
[31] Van Vinh N, Zafar M, Behera S, Park HS, Arsenic (III) removal from aqueous solution by raw and zinc-loaded pine cone biochar: equilibrium, kinetics, and thermodynamics studies, Int. J. Environ. Sci. Technol., 2015, 12, 1283–1294.
[32] Higashikawa FS, Conz RF, Colzato M, Cerri CEP, Alleoni LRF, Effects of feedstock type and slow pyrolysis temperature in the production of biochars on the removal of cadmium and nickel from water, J. Clean. Prod., 2016, 137, 965–972.

K. Harinisri, Dr. B. Thamaraiselvi, Reshmi Gopalakrishnan,
M. Manovina

Application of biochar in wastewater treatment: a review

Abstract: Biowaste like agricultural waste consists of various active components which are rarely employed in environmental application. The massive generation of waste in various countries are not being managed and other ways like mass production of beneficial products are accomplished rather than ruining the atmosphere. Generation of carbon-rich material biochar from organic waste raised heed in the field of ecological treatment. The cost-efficient biochar has the finest absorbing properties which remove the contaminants like heavy metals, pollutants, organic substances from the wastewater. Because of the drastic need, the review focuses on biochar production, properties and application in wastewater treatment.

Keywords: agricultural waste, biochar, wastewater treatment, biowaste, waste management

1 Introduction

Population growth and rising urbanization elevated waste generation both in quantity and variety. Includes industrial waste, agricultural waste, domestic waste, and so on which are renewable and organic sources. Every year, about 350 million agricultural waste has been generated in India. The utilization and recycling of the waste with the help of various technologies have also been increased. Renewable sources are potential and alternative energy sources for the creation of by-products that can be in solid, liquid, or gaseous form.

Over the last 50 years, the production of agriculture been increased due to soil extension for cultivation, increasing growth population, and green revolution technology [1]. The sudden increase in world population also created substantial pressure on the environment, especially bleak outcomes on soil, water, and air resources which risks the ecosystem [2]. For the highest biomass production, agriculture plays a major role [3] and also supports the development of various valuable products like biofuels, enzymes animals feed, and so on. [4]. In many countries, the disposal of waste, especially agricultural biomass, becomes a major issue and causes various ecological issues.

Many countries and agricultural industries dispose of their agrowaste by burning which also causes air pollution and also leads to mislaying of nutrients like nitrogen, potassium, and so on in the soil (Figure 1). Generally, agricultural practices increase greenhouse gas emissions like carbon dioxide, methane, nitrous oxide,

https://doi.org/10.1515/9783110734003-004

Figure 1: Burning of agricultural waste.
Source: Wikimedia.

and so on which in turn causes global warming and climate change [5]. Long-term storage of waste in the cultivating field may cause various soil issues and environmental impacts. Though organic waste causes various impacts on the environment, it also consists of enriched organic sources like proteins, minerals, carbohydrates, lignocellulosic, and so on are utilized in the conversion of new products.

Recycling waste is an essential process for the reduction of modern waste. Various products like biofertilizers, biofuel, fermentation medium, organic acid production, biochar, and so on. Biochar created a curiosity among industries because of its physical and chemical properties. Biochar plays an important role in conserving sustainable agriculture [6]. The high carbon-containing substance has a high absorption capacity used as a source for the treatment of wastewater. Because of the resource management and natural water body contamination treatment, water has become an essential need for the upcoming generations. Treatment reduces the pollutants, contaminants and can be used for secondary activities. The review chapter majorly focuses on agricultural waste and the value of organic components present in the renewable wastes. The waste has been recycled into various by-products and their importance in numerous fields is explained. Biochar, a carbon substance, holds a major place in the treatment of wastewater and its effective application in the environment is elucidated. The critical objective is to discuss the importance of biochar in wastewater treatment with future perspectives that pave research hints for further effective innovations.

2 Biowaste and recycling

Biowastes are biodegradable materials of biological origin which includes various wastes like biosolids (sewage sludge), animal effluent, wood waste, green waste which consist of high concentrations of organic matter, and nutrients that can also be utilized in the production of various copious valuable products. Generation of waste is increasing by the year and so is the recycle of waste into many bio-based products. Based on data released in 2018 by the United Nations, about 11.2 billion tons of solid waste generated on Earth every year contributes to greenhouse gas emission. To reduce the impacts created, many countries set up recycling units for waste management and product production. Based on that criterion in 2018, Eunomia Research and Consulting teamed up with the European Environmental Bureau to release data on the top nations for recycling rates in the world (Chart 1).

Beyond these countries, many other countries and their citizens contribute to the waste management process. Swedish citizens separate their waste based on their properties into colored bags. The non-renewable sources are combusted and used as energy sources where renewable sources are converted by various methods for new products. The Japanese government encourages the separation of waste and citizens to separate their waste send their categorized waste into recycling units. Netherland recycled their plastics and reused them in the construction of entire roads. Canada people donate their discarding materials for recycling units and these are used for the construction of roads. Many other countries like Ukraine, Palestine, Kazakhstan, Serbia, and Botswana, Bermuda are the least ranked countries that do not recycle the accumulated waste on large scale.

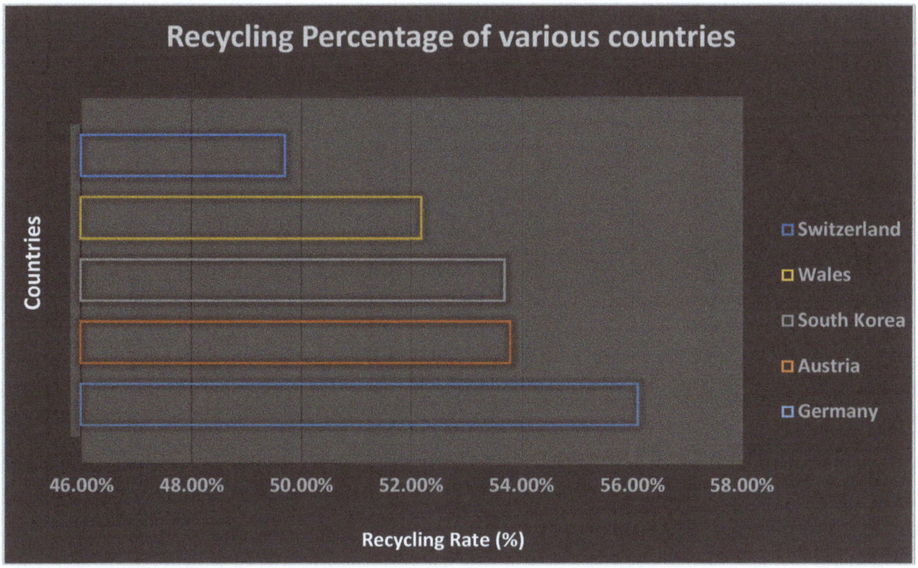

Chart 1: Top five countries recycling rate of waste in percentage (%).
Source: NS Packaging.

The waste also contains contaminants and pathogens that may cause an ill effect on the environment and human beings. Improper discarding of waste is both expensive and harmful to the environment. If waste is effectively applied to degraded lands, biowastes can generate both economic and environmental advantages. The numerous categories of biowaste include crops and residues, biodiesel production waste, forestry crops and residues, municipal solid and sewage waste, agro/food industry waste, fermentation processing waste, animal processing waste, and marine processing waste.

3 Agricultural waste

Agricultural waste is the kind of waste which is generated during agriculture processing. Generally, agrowaste is categorized as non-natural waste like packaging and non-packaging plastics, agrochemicals, animal health elements, and natural waste or organic waste like bagasse, straw, roots, and so on [7]. The basic classification of agrowaste types is illustrated in Figure 2.

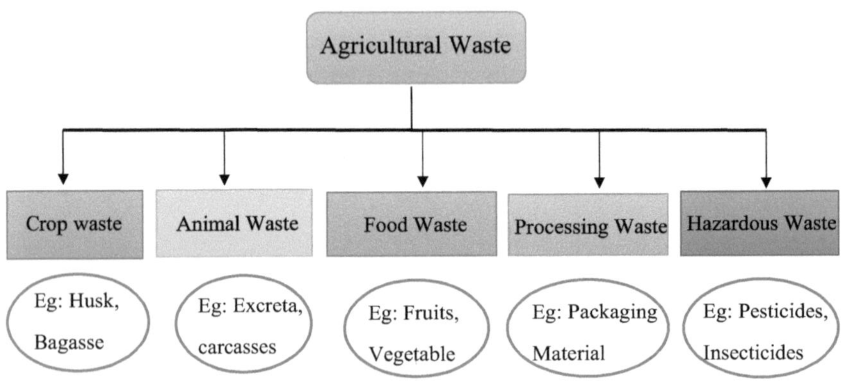

Figure 2: Classification of agricultural waste.

The crop waste is generally classified as field residues and process residues Figure 3. Field waste is the type of waste which is produced after harvesting. Seed pods, stalks, leaves, and so on are the waste generated after harvest. Most of the crop residues are directly used in the ground to increase irrigation and to reduce erosion or burnt [8]. After processing the crop, the waste generated like husks, bagasse, molasses, and so on is called process residues. These kinds of waste are being used as animal feed, fertilizer production, and other bioprocess units. Sugarcane bagasse, rice straw, and corn straw are the biomass utilized in large for the production of bioproducts like biofuel. Rice straw consists of energy sources like carbohydrates, different types of sugar like glucose, arabinose, and so on. Without complete knowledge on the sources of agrowastes, it has been burnt, messed in water resources, or stacked in the resource lands. These activities cause various pollution and nutrient discharge from the fertile soil [9].

Animal manure or livestock manure consists of liquid manure, solid manure, and wastewater. The untreated manure can cause land pollution and also leaches nutrients from the soil. Manures may contaminate the surface water and support greenhouse gas emissions [10]. Fruit and vegetable waste like mango, cabbage, and so on are a part of agrowaste which is majorly produced from agriculture and industry processing. This waste also includes market waste and processing waste which cause pollution on the land surface.

Industrially generated agrowaste is generated in large amounts and is utilized in various forms. These residues consist of insoluble chemical components like cellulose, hemicellulose, nitrogen, lignin, and so on, and soluble components like amino acids, sugars, and organic acids. These residues act as an economic source for the farmers for the improvement of the fertile land.

Figure 3: Types of agricultural waste.
Source: Pixabay, Pxhere, Flicker.

Waste management actions are implemented by various countries and include the collection of waste, transport, treatment, and disposal. The disposal of waste has been regulated and monitored. Management is generally divided into eight groups. The waste disposal methods include landfills, incineration or combustion, recovery and recycling, composting, energy recovery, biomedical waste disposal, and so on. Recovery and recycling in various countries were implemented with limited regulations and techniques were adopted for the reuse and recycle of waste, especially recycling of agrowaste into varied products. There are innumerable beneficial approaches that have been implemented by industries with the use of agrowastes.

3.1 Biofertilizers

In ancient days, manure was used as fertilizer in fertile lands [11]. Later, wood ash is also used as a biofertilizer which is a cost-effective product for the effective growth of the crops. Based on the need for nitrogen, potassium, and phosphorous sources, the waste has been used as an alternative source for synthetic fertilizers. The waste is also enriched with various micronutrients and trace elements for soil enrichment [12]. The fertilizer production can be done in various methods like composting, vermicomposting, anaerobic digestion, and thermal methods. The organisms in various forms can also be used as fertilizer for the enrichment of soil and plants. The organisms include cyanobacteria and arbuscular mycorrhiza fungi are widely used.

3.2 Anaerobic digestion

Manures are widely used in the production of methane and heating and drying gas generated by the anaerobic digestion process. Various types of gases are produced which are methane, carbon dioxide, hydrogen, nitrogen, and so on. About 50–70% of gas is produced during the microbial conversion of waste elements. The digested sludge does not have any stinky nature, and the nutrients and the organic sources are not utilized during the digestion process [13].

3.3 Heavy metal removal

Because of developed industrialization, the effluent consists of various harmful heavy metals which cause damage to the environment and living beings. The organic waste in the effluent is being utilized and degraded by the microbial actions whereas heavy metals are not degraded. In successive years, the use of agricultural waste has become effective in the removal of heavy metals. The heavy metals, such as copper, cadmium, mercury, zinc, chromium, and lead, are harmful and cause disease in prolonged stages. Wastes like rice husk, coconut husk, and so on are used as an alternative source for elimination with low economic budgets [13].

3.4 Solid-state fermentation

The solid support used in solid state, such as rice straw, husks, bran, seeds, sawdust, and so on, supports the fermentation. The sources used for fermentation should be easily available, be low in their cost, and not require water or any other soluble materials. These are the criteria taken into account while setting up the solid-state fermentation process. The different substrate selection is based on the product to be recovered and the process to be carried out. The detailed substrate utilization from agrowaste is given in Figure 4.

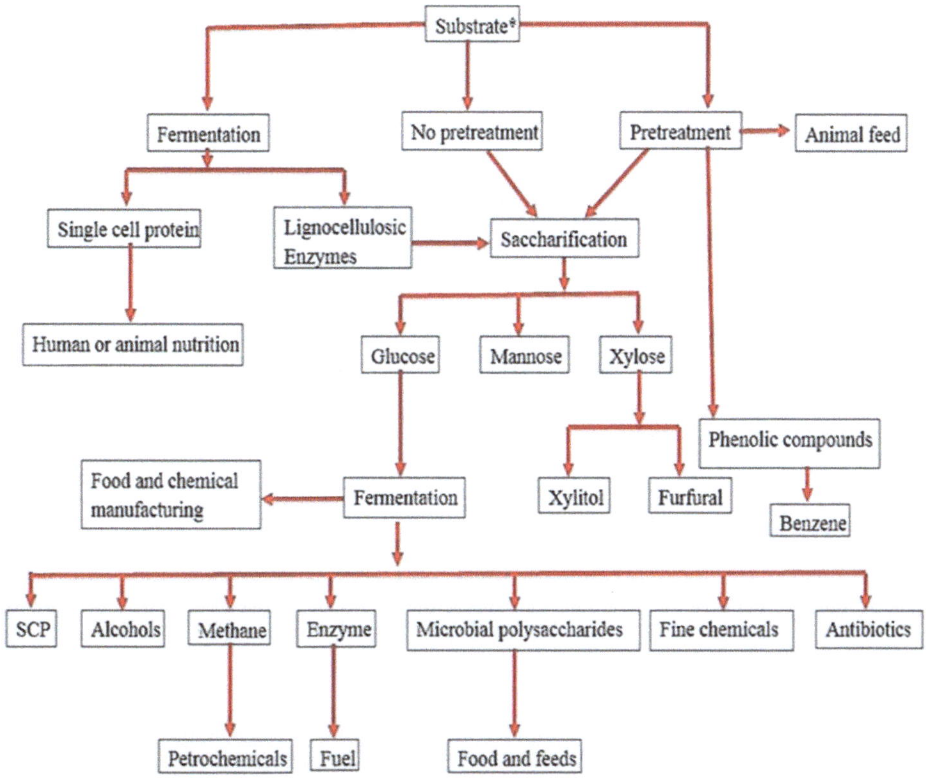

Figure 4: Schematic representation of different substrate application from biowaste.
Source: SpringerOpen.

3.5 Antibiotics

Antibiotic is a substance that inhibits the growth of microorganisms even at very low concentrations. Waste like corncobs, sawdust, and so on are used as raw material for the antibiotic production of oxytetracycline [14]. Groundnut shells and coconut oil cake produced maximum production of antibiotics when compared with other wastes. Agrowaste is the low-cost carbon source used in antibiotic production. Besides, various other specific compounds along with waste enhanced the antibiotic production obtained by the down streaming process.

3.6 Biofuel

Either by biochemical conversion or by thermochemical conversion, lignocellulosic waste is utilized and converted into biofuel products. In biochemical conversion, appropriate microorganism and enzymes are involved for the conversion of biogas, bioethanol, and biodiesel products. In the case of thermochemical conversion, heat gas and various types of chemicals are utilized for end-product production like bio-oil, biochar, and so on [10] Figure 5.

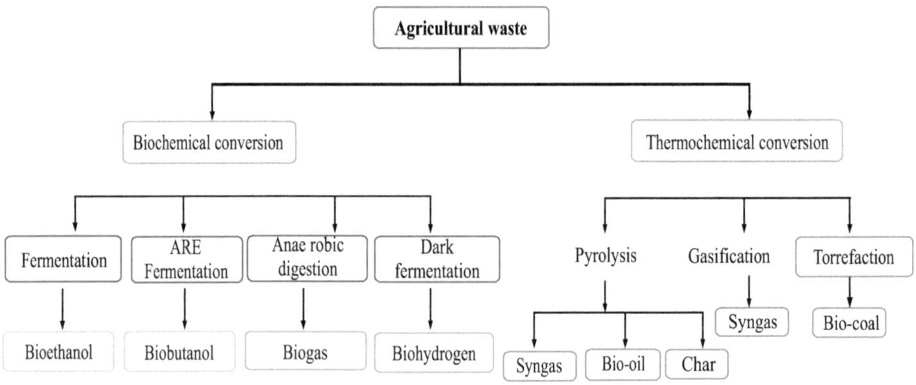

Figure 5: Conversion of agricultural waste to biofuel by various techniques.

3.7 Enzyme production

Various sources in the waste enhance the microbes to produce a wide variety of enzymes which have industrial applications. The sources and enzymes produced are detailed in Table 1.

Table 1: Agricultural waste and the enzymes produced.

Substrates	Enzymes
Papaya waste	α-Amylase
Groundnut oil cake	Lipase
Wheat bran and orange peel	Pectin methylesterase
Linseed oil cake	Lipase
Orange Peel	α-Amylase
Coconut oil cake	α-Amylase
Rice bran	α-Amylase
Corn bran	α-Amylase
Rice, wheat, black gram, and soybean bran	α-Amylase
Fruit peel waste	Invertase

3.8 Mushroom cultivation

Lignocellulosic waste is used as a raw material for the cultivation of protein-rich food mushroom. One of the major solution related to the agrowaste can be solved by using as raw material for the production of various by-products. By using coffee husk as a major source, oyster mushroom was cultivated [15]. Paddy straw, banana stalks, rice bran, wheat bran, and so on are the various sources used as raw material for the production of mushroom with increased nutrient contents.

3.9 Single-cell protein production

Cucumber peel and orange peels are used as a source of raw materials for the production of single-cell protein by fermentation method [16]. Cucumber peel has high productivity than orange peel. The product consists of high nutrient content while using waste as the raw source. Single-cell protein is produced by the use of suitable microorganisms which utilized the raw materials and produced the end products.

3.10 Animal feed

Crop residues consist of major nutrients like protein, fat, and starch. Animal feeds lack protein nourishment, and needs are solved by supplementing forage and grains. Rice bran, husk, corn stalks, bagasse, and banana stem are the types of waste used as animal feed [17].

4 Biochar

A solid product obtained from the biomass is subject to a higher temperature in the absence of oxygen. Various sources of biomass like rice husk, bagasse, paper products, and other agricultural waste are used for the production of char. About 50–70% of biochar is made of carbon. Production ranges from small scale with cook stoves to large-scale production with pyrolysis are done. Based on the types of production conditions it result in different types of biochar. Biochar used in soil fertility also results in improved crop yield. It also impacts soil pH, nutrient availability in soil, and soil organic matters while applying on the field. It can also impact climate change by storing and stabilizing the carbon and in the utilization of waste into various valuable products.

4.1 History of biochar

Charcoal-like carbon substances produced by the burning of biomass like agricultural waste and forestry waste are known as biochar [18]. The usage of biochar has a historic background that it was originated 2000 years ago in the Brazilian Amazon. Archaeological studies reveal that the ancient people employed biochar and organic waste in undernourished soils in the period of 400 BC. By creating a low-oxygen environment by burning and burying, the biomass was converted into biochar, and technique was called slash-and-char agriculture. It showed more carbon sequestration activity. The slash-and-burn method yields a larger amount of ash but has little sequestration activity. Even today, biochar utilized lands can be seen in the regions of Amazon and laid out affirmation of beneficial facts of biochar to the modern era [19].

4.2 Production of biochar

Biomass conversion leads to the development of various by-products, especially biochar, holds a unique place in soil conservation. A heterogeneous compound with aromatic carbon and mineral product involves several steps in production. The steps involved are illustrated in Figure 6.

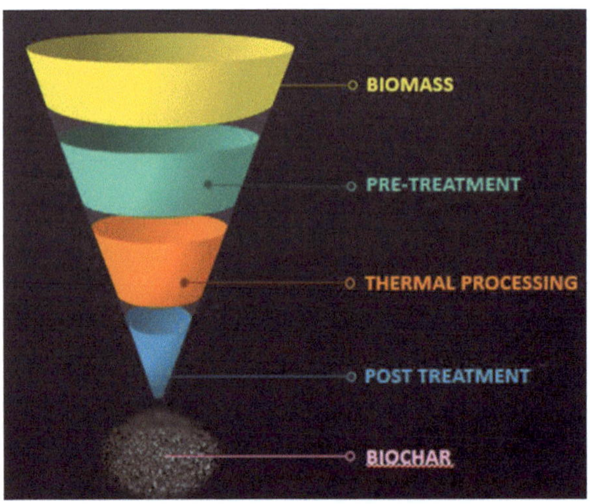

Figure 6: Steps involved in biochar production.

4.2.1 Pretreatment

Biomasses from various sources have been collected and subjected to the pretreatment process. Generally, the pretreatment process includes physical, chemical, and biological methods. The waste materials collected are dried in case of moisture samples to constant weight at a high temperature of around 105 °C and then ground into smaller pieces [20]. Chemical pretreatment includes the use of chemicals like ferric chloride, magnesium chloride, and so on are immersed in chemicals and treated [21]. It can also be treated with nanomaterials and nanocomposites. By using microorganisms, the waste is biologically treated by anaerobic digestion [22]. It may also be treated with heavy metals by the bioaccumulation method.

4.2.2 Thermal processing

Various sources for biochar materials are processed by thermochemical conversion techniques. The thermochemical techniques include pyrolysis, hydrothermal carbonization, gasification, and so on. The technique employed should be based on the biomass and optimum process parameters like heating rate, temperature, time, and so on.

Pyrolysis is a general technology employed for the production of biochar (Figure 7). The thermal breakdown of biomass to char, liquid (bio-oil), or gas (syngas, carbon dioxide, carbon monoxide) state in the limited supply or absence of oxygen produces a carbon residue. The lignocellulosic materials undergo depolymerization, fragmentation, and cross-linking at a specific temperature and produce various products [23]. Various reactors like bubbling fluidized bed, paddle kiln, and so on are used for char production. In general, the pyrolysis temperature works in the range of 250–900 °C. Based on the various parameters, like pressure, temperature pyrolysis process is classified into two types: fast pyrolysis and slow pyrolysis.

Fast pyrolysis majorly converts the solid components into liquid form like bio-oil. The temperature ranges from 500 to 1,000 °C. Slow pyrolysis maintains a temperature range of 300–700 °C for an extended period. It has a better yield while comparing the other methodologies. The basic methods of biomass pyrolyzing include external heating, internal heating, and combustion. A separate heat source is provided to the biomass container and products are obtained. Flame is set with air circulation is set up for internal heating and products are recovered. The gas inside the chamber is combusted and re-entered inside the chamber [24]. Temperature, time, and heating rate are the factors to be considered for the pyrolysis process.

Microwave-assisted pyrolysis is used to produce biochar, bio-oil, and biogas from biomass [25]. The time required is shorter than the conventional method; it has effective heat transfer, and low energy is required for the process. The factors that need to be considered during the processing of biomass are microwave power and irradiation power. If the microwave power increases, the yield is found to be reduced and may degrade if a high temperature is maintained.

Hydrothermal carbonization is a wet drying process with a temperature range of 120–260 °C. The structured carbon from this technology is known as hydrochar which is considered an effective method for the production of solids. The biomass is

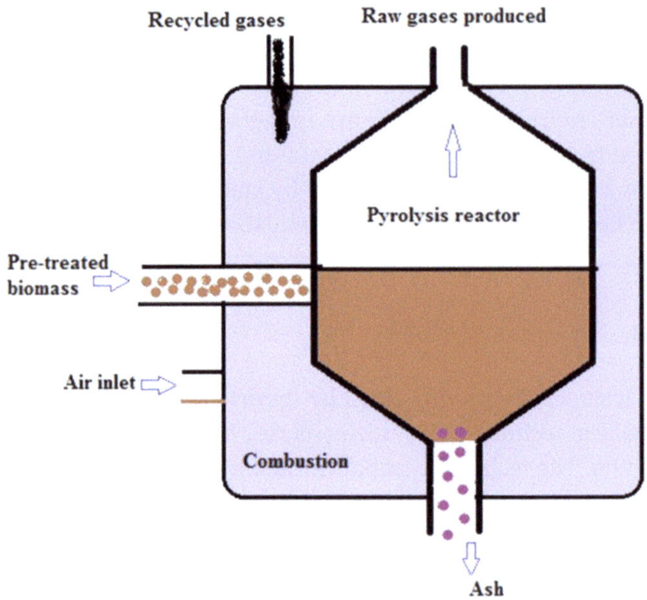

Figure 7: Mechanism of pyrolysis.

added with water in the reactor gradually the temperature is increased. At different temperature ranges, various products are obtained. At the temperature range below 250 °C, biochar is collected. The liquids are collected at the temperature of 250–400 °C and gases like syngas, methane, and carbon dioxide are collected above the temperature range of 400 °C [23]. It consists of acidic functional groups elevated in the adsorption capacity [26]. For the synthesis of nanostructured carbon, humus, and so on, hydrothermal carbonization method can be employed. About 75–80% moisture-containing source can be used for this technique (Table 2).

The conversion of biomass into gas fuels using gasification agents at a high temperature of above 800 °C is known as gasification [27]. The product from gasification consists of alkali salts, alkaline minerals, syngas, carbon monoxide, carbon dioxide, and methane are obtained. The alkaline products enhance the absorption of heavy metals from the wastewater. It also enhances soil productivity when applied to the agriculture fields. Gasification has various subprocesses like drying, combustion, torrefaction, and flash carbonization. The moisture content in the biomass is removed by the drying process. The gasification agents carry out the combustion process for the release of gas and water. The oxygen, moisture, and carbon dioxide present in the samples are removed by the torrefaction process at the temperature of 300 °C. Process of torrefaction can be classified as steam torrefaction, wet torrefaction, and oxidative torrefaction.

Table 2: Technologies employed in char production.

Thermal carbonization technologies	Key parameters	Temperature/ power range	Residence time	Desired product
Pyrolysis	temperature: heating rate: residence time	300–850 °C	1–3 h	Biochar
Microware-assisted pyrolysis	nicrowave power: microwave irradiation time	400–550 W	1–10 min	Biochar and biofuel
Hydrothermal carbonization	irradiation time temperature: residence time: pressure: water-to-biomass ratio	120–260 °C	1–16 h	Hydrochar
Gasification	temperature: particle size: residence time: pressure: gasification agent/ biomass ratio	>800 °C	10–20 s	Syngas

4.2.3 Posttreatment

The end product is either physically or chemically modified for enhancement in the application. Pore volume, functional compounds, and surface chemistry can be treated for efficiency [28]. Magnetic biochar is produced where magnetic iron oxides are added and easily recovered when applied in aqueous treatment [21]. The ball milling method is followed to produce the nano form of biochar. Acid, alkali, and oxidative treatment are the most commonly used techniques that majorly alter the surface of the biochar product.

4.2.4 Factors influencing the yield

Various factors like biomass sources, temperature, heating rate, and so on influence the formation of biochar during production. Biomass consist of composite materials generally divided as woody, includes tree and forest remnant, and non-woody biomass includes animal and agricultural wastes. Based on the biomass, the formation by pyrolysis can be varied. Moisture plays a major role in the production of biochar.

High moisture content slows down the process of char formation whereas low moisture containing biomass is preferred for effective production [23].

Biochar is produced in the atmosphere without oxygen supply and at raised temperature. Pyrolysis is carried out in different ranges and is named slow pyrolysis, moderate pyrolysis, and quick pyrolysis. The temperature has an impact on the structure of biochar, pores, functional groups, and so on. Prolonged time reaction at slow pyrolysis influence the yield of biochar.

Many pretreatment processes are also considered as the major factor for the production of solid chars. Basic treatment includes reduction of size which increases yield. Certain methods like nitrogen and metal doping, soaking, and steaming influence the properties of char. Baking is considered an effective method because it can increase carbon content and reduce the moisture and oxygen content present in the biomass samples.

4.3 Characterization of biochar

The complete characterization of produced biochar is important before application in various fields. The characterization is based on their physical, chemical, and surface properties. The analyzed physical properties include bulk density, particle density, pore volume, pore size, and density. The chemical characters including pH, biochar compositions, electrical conductivity, toxic compounds, carbon/hydrogen ratio, and cationic and anionic exchange capacity are evaluated. The structure elucidation is done by scanning electron microscopy, transmission electron microscopy, X-ray diffraction method, Fourier transform infrared spectroscopy, and so on [29].

The functional groups on the char include carboxylic group, hydroxyl group, amine group, amide group, and so on. These groups increase the adsorption activity of biochar while treating. If there is an elevation in pH and porosity, the presence of functional groups in char may be reduced. Apart from other physical characters, temperature also play a role in influencing the functional groups and their properties during char production. The functional group analysis can be studied using Fourier transform infrared spectroscopy and nuclear magnetic resonance instruments.

During pyrolysis, dehydration of the sample causes pores in the biochar. These pores enhance the sorption activity when treating with wastewater. The pore varies from micro to macro sizes. The small pore size in the biochar cannot remove the molecules of pesticides hence pore size matters for effective removal. The pore size can be characterized by scanning electron microscopic studies. Surface area may also differ based on treated biomass and untreated biomass. Activation during char production may increase the surface area which supports the enhancement of sorption activity [23].

Because of different methodologies and varied range of temperature, the scanning electron microscope analysis showed unique surface properties. The prolonged temperature during processing conveyed advancement in pore characteristics of char. Fourier transform infrared spectroscopy is used for the analysis of functional groups present on the char material. The structure analysis is done using the X-ray diffraction method using radiation as the source of analysis. The stability of the biochar is analyzed based on the carbons present in their structure through various experimental methods. Many experiments are carried out, and stability analysis is categorized into three ways. Indirect or direct analysis of carbon structures, analysis of stable carbon by thermochemical methods, and carbon mineralization modeling [30].

Various factors impact the physicochemical properties of biochar. The source of raw materials and the substrate influence the characteristics of the biochar. The production processes like mode of pyrolysis, temperature range, rate of heat, and time duration also dominate the properties of biochar. Not only in the properties, these factors influence the quality of the biochar. High carbon content, the aromatic group with oxygen and high pore size are effective in biochar structure and especially used in the removal of organic contaminants. If it is produced at low temperature, it possesses functional groups with oxygen with the least pores in it. They can be used for the removal of inorganic components. Stable molecular size, surface area, porosity, functional groups, and carbon are the major properties that are considered for the adsorption of pollutants from the wastewater [31].

5 Application of biochar in wastewater treatment

The basic mechanism of biochar in removing contaminants is sorption theory whereas biochar act as adsorbents. The atoms or molecules deposit on the surface of the adsorbent. The absorbent and adsorbate precipitate and form a complex structure. The pores in the adsorbent are concentrated. These processes occur in three phases: initial phase is called clean zone, the intermediate zone is called mass transfer zone, and the final zone is called the exhausted zone. The contaminants in the wastewater are captured by the biochar and treated. Basically, in industrial wastewater treatment,the major contaminants like heavy metals, dyes, pesticides, organic contaminants, and so on are being removed.

5.1 Heavy metals

Heavy metals are a serious contaminant in the water treatment process which slowly accumulates in the tissues and cause health issues in human even at low concentrations [32]. Biochar can be applied as an alternative approach to remove metal ion contaminants from water. Rice husks, olive pomace, and orange wastes are pyrolyzed and hydrothermalized to yield biochar which removes copper ions efficiently [33]. Another study includes the use of biochar produced from peanut, canola, and soybean straw attained by muffle furnace which had very good adsorption of copper ions when compared to commercially available activated carbon. Removal of copper ions, zinc ions, cadmium ions, and lead ions is achieved by pig and cow manure [34]. Lead ion, cadmium ion, and arsenic ion adsorption showed high potency in oakbark char. Besides, pine char also has an efficient role in fluoride removal.

The process where metals diffuse by binding with biochar is known as the surface adsorption mechanism. Based on the processing temperature, the formation of surface area and pores are dependent. Biochar which is prepared from pinewood at high temperature showed greater activity against uranium. Based on affinity, the metal interaction with the solid char differs.

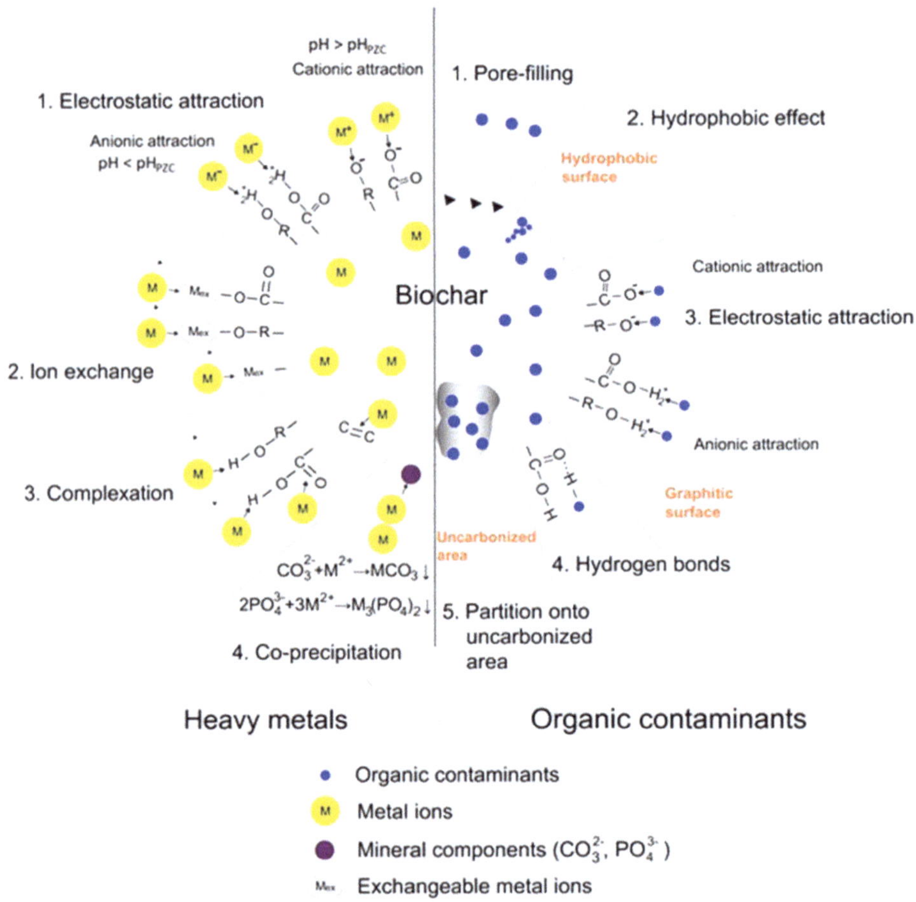

Figure 8: Sorption mechanism of heavy metals and organic contaminant removal using biochar [31].

Metal ion and charged biochar interact electrostatically and remove the toxic metals from the water. This mechanism is known as electrostatic interaction (Figure 8). Through various experiments, biochar from rice and wheat showed greater activity in the removal of lead especially with negatively charged biochar. The enhancement in electrostatic effect can be increased by temperature during the production of biochar by pyrolysis technique. The removal mechanism also depends on the pH of the treating environment.

Heavy metal removal depends on the pollutant size and functional groups present on the char material. Protons and ionized cation exchange on the surface of the biochar and the mechanism collectively called cation/ion exchange capacity. If the temperature increases, the exchange mechanism seems to be reduced in treating wastewater. The inorganic pollutants precipitate on the surface of the biochar. The plant-based biochar is especially made of cellulose; hemicellulose with alkaline property has enhanced precipitation mechanism.

Biochar produced at a lower temperature can bind generously with heavy metals. The oxygen groups on the surface of the solid product increase the complexation by surface oxidation. Plant-derived product char has greater activity in the removal than many other sources [31].

5.2 Organic contaminants

Biochar can remove the organic contaminants from the wastewater. The mechanism includes pore filling, hydrophobic effect, hydrogen bonds, and electrostatic interaction (Figure 8). Pore filling is an important mechanism for the sorption of organic contaminants. Removal of organic components can also be achieved by partition technique. In this mechanism, the organic pollutants interact with the non-carbonized region of biochar and lead to sorption. This adsorption depends on the characteristics of biochar. Solid product from manure has high adsorption capacity toward the atrazine pollutant through partitioning mechanism. Other sources like grass, wood based biochar revealed the efficiency on norflurazon pollutants through various experiments. In the case of higher organic contaminants, this medium of removal is highly efficient and more visible while treating.

Based on the properties like polarity, char type, and pore type, the organic pollutants interact. Many plant material (e.g., oak) are productive in the removal of catechol pollutant while comparing the other mechanisms. For a high potent mechanism, the char produced should contain very least amount of volatile matter and low availability to the organic pollutants.

Ionizable pollutants are adsorbed onto the positive side of the biochar by electrostatic interaction. Based on the pH and ionic properties, the solid char attracts the pollutants. The pH of the biochar remains positive at lower pH and carriers negative charge when pH in aqueous increases. Biochar prepared from bagasse in nanoparticle form can effectively remove the methylene blue pollutants.

The aromatic compounds in biochar present graphene-like structure that are interacted by electron donor and acceptor mechanism. The electron density on the biochar depends on the pyrolysis temperature. If the temperature of biochar is below 500 °C, it acts as an electron acceptor. If the temperature is above 500 °C, it acts as an electron donor. The neutral and hydrophobic components are interacted by hydrophobic interaction with less energy utilization. Some pollutants like benzoic acid and other acid components interact with biochar through the hydrophobic mechanism.

5.3 Phenol and pesticide removal

Nitrophenols and chlorophenols are major pollutants released from plastics, dyes, drugs, antioxidants, and pesticide industries. Higher phenol adsorbing capacity was found in rice husk and corncob biochar when prepared at constant temperature and varying residence time [35]. Pesticides, such as dibromochloropropane, are removed by using almond shell activated biochar, which showed maximum adsorption capacity [36]. Maximum adsorption of 1-naphthol and naphthalene was achieved by using orange peel biochar [37].

5.4 Color removal

The release of textile into the river streams causes a color change in the water which becomes a challenging approach to remove such contaminants. Rice straw biochar successfully removed the malachite green reported [38]. Besides, canola straw, peanut straw, and rice hulls adsorbed methyl violet from water [39].

Adsorption of contaminants is affected by the pH of the aqueous solution. Adsorption capacity is varied based on the surface charge dependent on pH. pH has a great impact on the treatment of industrial wastewater. At alkaline pH, the solid char effectively removes the textile dyes. The effectiveness of adsorption also depends on the quantity of the biochar. Char prepared from hardwood and corn straw productively remove toxic metals from the wastewater. A high amount of biochar can constructively associate with various contaminants like organic pollutants and toxic metals.

The biochar prepared from the cow dung is effectively used against the removal of phosphorous and nitrogen. The industrial wastewater consists of nitrogen and phosphorous contaminants, which leads to ecosystem eutrophication. The treated material can also be used as fertilizer for crop development.

6 Application in various field

6.1 Industrial wastewater treatment

Industrial wastewater majorly consists of heavy metals, organic pollutants, chemical industry waste, dyes, and so on. The appropriate ratio of chitosan and biochar copper, lead, arsenic, and cadmium are removed [40]. Bagasse can absorb lead and remove it from the wastewater. Biochar can also reabsorb nutrients like ammonium and phosphate from dairy wastewater.

6.2 Municipal wastewater treatment

Either biochar or filter can be used to treat municipal wastewater. Engineered biochar can be used as a biofertilizer for the slow release of phosphorous. Sludge-based biochar is used in the removal of ammonium compounds from municipal wastewater [41].

6.3 Agricultural wastewater treatment

Due to the increased population, the agriculture sector has also been increased in the production, usage of pesticides, and so on. The large-scale use of pesticides is absorbed in the fertile lands [42]. Many types of research showed biochar effectiveness in agricultural wastewater treatment. Rice straw biochar, corn straw, and soybean straw are effective against components of pesticides [43]. Heavy metals are removed by mechanisms like electrostatic interaction, ion exchange, and intermolecular interactions [42].

Various types of biomass differ in their nutrient sources; hence, they are not equally used before characterization. The quality of biomass is a major factor that needs to consider during biomass production. In case higher temperature is employed for the production of biochar, the total carbon content in the biochar gets reduced and is not employed. Cost-effectiveness is high when nanosized biochar is produced. Based on the pH, the usage should be devised.

In general, biochar increases soil fertility and enhances plant growth. The nutrients can be reabsorbed by using biochar and the cheapest source for the recovery processes. It enhances the microbial flora and resists various kinds of pathogens. It reduces the soil pH and prevents it from leaching. It can also be used as a biofertilizer which slowly releases nutrients to the soil. It can also retain the amount of water in the soil and helps in the retaining of groundwater level.

Because of their retention capacity, the plants and trees are limited by the water sources. If harmful chemicals are sorbed, it will contaminate water bodies and fertile soils. In humans, because of their fine size, it may cause respiratory tract infections.

7 Conclusion and future perspectives

Biochar is an economical carbon by-product produced from agricultural residues. This article reviews the significance of renewable resources, their organic components, and recycling to produce new products. It also epitomized the importance of agricultural waste sources and the application of residues in various fields. Further research findings are required for the identification of new adsorption mechanisms for the effective removal of contaminants from various sources. To analyze the positive impacts of biochar on the environment, various analysis procedures should be carried out. The use of various new technologies characterization studies can be widened for effective utilization. Biochar as a new trend and alternative treatment method requires more characterization studies and optimization studies. The core focus of biochar and their production processes like pretreatment, thermal processing, and posttreatment are summarized. The mechanism of biochar with wastewater treatment is primarily concentrated. Different types of application like industrial wastewater treatment, municipal wastewater treatment, and agricultural wastewater treatment are outlined.

The utmost conclusions of this chapter are:

- Biochar raw material resources and their source properties.
- The production process of biochar includes biomass collection, pretreatment of biomass, thermal processing, and conversion of biochar with the help of various technologies and posttreatment process.
- The removal of pollutants and containments from municipal waste, agricultural waste, and industrial waste are explained.
- The advantages, application, and importance of biochar in further wastewater treatment are elucidated.

References

[1] Food and Agriculture Organization of the United Nations (FAO).. Strategic Work of FAO for Sustainable Food and Agriculture 2017. http://www.fao.org/3/a-i6488e.pdf

[2] Food and Agriculture Organization of the United Nations (FAO). 2017. The State of Food and Agriculture. Leveraging Food Systems for Inclusive Rural Transformation. 2017. http://www.fao.org/3/a-i7658e.pdf

[3] European Commission. EIP-AGRI Workshop 'Opportunities for Agriculture and Forestry in the Circular Economy. 2015. https://ec.europa.eu/eip/agriculture/sites/default/files/eip-agri_ws_circular_economy_final_report_2015_en.pdf

[4] Sadh PK, Duhan S, Duhan JS, Agro-industrial wastes and their utilization using solid-state fermentation: a review, Bioresour. Bioprocess., 2018, **5**(1), doi: https://doi.org/10.1186/s40643-017-0187-z.

[5] International Atomic Energy Agency (IAEA). Greenhouse gas reduction. https://www.iaea.org/topics/greenhouse-gasreduction#:~:text=Agriculture%20produces%20a%20substantial%

20amount,global%20warming%20and%20climate%20change.&text=This%20does%20not%20yet%20take,of%20which%20is%20largely%20unrecognized.
[6] Parmar A, Nema PK, Agarwal T, Biochar production from agro-food industry residues: a sustainable approach for soil and environmental management, Curr. Sci., 2014, 107.
[7] Nagendran R, Chapter 24 – Agricultural Waste and Pollution, In: Waste, 2011, https://doi.org/10.1016/B978-0-12-381475-3.10024-5.
[8] Richards BK, Muck RE, Walter MF, Variation in line transect measurements of crop residue cover, J. Soil Water Conserv., 1984, 39, 1.
[9] Billa SF, Angwafo TE, Ngome AF, Agro-environmental characterization of biochar issued from crop wastes in the humid forest zone of Cameroon, Int. J. Recyl. Org. Waste Agric., 2019, 8, 1–13. doi: https://doi.org/10.1007/s40093-018-0223-9.
[10] Pattanaik L, Pattnaik F, Saxena DK, Naik SN, Biofuels from agricultural Wastes.
[11] Gary Hergert RN, A historical overview of fertilizer use, *Farm Journal's AGPRO*, 2015.
[12] Samoraj M, Tuhy Ł, Valorization of biomass into micronutrient fertilizers, *Waste Biomass Valorization*, 2019.
[13] Obi FO, Ugwuishiwu BO, Nwakaire JN, Agricultural Waste Concept, Generation, Utilization, and Management, Nigerian J. Tech. (NIJOTECH), 2016, doi: http://dx.doi.org/10.4314/njt.v35i4.34.
[14] Ifudu ND, Indigenous resources for antibiotic production, In: Expansion Today Nigeria, African University Press Ltd., 1986, pp 52–53.
[15] Murthy PS, Manonmani HK, Bioconversion of coffee industry wastes with white rot fungus *Pleurotus florida*, Res. J. Environ. Sci., 2008, 2.
[16] Mondal AK, Sengupta S, Bhowal J, Bhattacharya DK, Utilization of fruit wastes in producing single-cell protein, Intern. J. Sci. Environ. Technol., 2012, 1, 5.
[17] Leng RA, Choo BS, Arreaze C, Practical technologies to optimize feed utilization by ruminants, In: Legume Trees and Other Fodder trees as Protein Sources for Livestock, FAO, 1992, 145–120.
[18] Spears S. What is Biochar? Regeneration International. 2018. https://regenerationinternational.org/2018/05/16/what-is-biochar/
[19] King R. Biochar: a brief history and developing future. Mongabay.2013. https://news.mongabay.com/
[20] Zhang X, Gao B, Creamer AE, Cao C, Li Y, Adsorption of VOCs onto engineered carbon materials: a review, J. Hazard Mater., 2017, 338.
[21] Xiang W, Zhang X, Chen J, Zou W, Feng H, Xin H, Daniel CW, Tsang YSO, Gao B, Biochar technology in wastewater treatment: A critical review., Chemosphere, 2020.
[22] Wang B, Gao B, Fang J, Recent advances in engineered biochar productions and applications, Crit. Rev. Environ. Sci. Technol., 2017.
[23] Yaashikaa PR, Senthil Kumar P, Sunita Varjani A, Saravanan A, A critical review on the biochar production techniques, characterization, stability and applications for circular bio economy, Biotechnol. Rep., 2020, doi: https://doi.org/10.1016/j.btre.2020.e00570.
[24] Joseph S, Taylor P, Cowie A. Basic principles and practice of Biochar production and kiln design. https://biochar.international/guides/basic-principles-of-biochar-production/#introduction
[25] Dai L, et al., 2017, Production of bio-oil and biochar from soapstock via microwave-assisted co-catalytic fast pyrolysis, Bioresour. Technol., 2017.
[26] Kambo HS, Dutta A, A comparative review of biochar and hydrochar in terms of production, Physico-chemical properties and applications, Renew. Sustain. Energy Rev., 2015.
[27] You S, et al, A critical review on sustainable biochar system through gasification: energy and environmental applications, Bioresour. Technol., 2017.

[28] Van Vinh N, Zafar M, Behera S, Park H, 2015, Arsenic (III) removal from aqueous solution by raw and zinc-loaded pine cone biochar: equilibrium, kinetics, and thermodynamics studies, Int. J. Environ. Sci. Technol., 2015.

[29] Adilla N, Yusup S, A Mini Review of Biochar Synthesis, Characterization, and Related Standardization and Legislation, Applications of Biochar for Environmental Safety, 2020, doi: 10.5772/intechopen.92621.

[30] Leng L, Huang H, Li H, Li J, Zhou W, Biochar stability assessment methods: a review, Sci. Total Environ., 2019, 210–222.

[31] Ambaye TG, Vaccari M, Van Hullebusch ED, *et al.*, Mechanisms and adsorption capacities of biochar for the removal of organic and inorganic pollutants from industrial wastewater, Int. J. Environ. Sci. Technol., 2020, doi: https://doi.org/10.1007/s13762-020-03060-w.

[32] Davydova SL, Heavy metals as main pollutants of the next century, Crit. Rev. Anal. Chem., 1999, 28(4), 377–381.

[33] Pellera FM, et al, Adsorption of Cu(II) ions from aqueous solutions on biochars prepared from agricultural by-products, J. Environ. Manage., 2012, 96(1), 35–42.

[34] Kołodyn´ska D, et al., Kinetic and adsorptive characterization of biochar in metal ions removal, Chem. Eng. J., 2012, 197, 295–305.

[35] Liu WJ, Zeng FX, Jiang H, Zhang XS, Preparation of high adsorption capacity bio-chars from waste biomass, Bioresour. Technol., 2011.

[36] Klasson KT, Ledbetter CA, Uchimiya M, Lima IM, Activated biochar removes 100% dibromochloropropane from field well water, Environ. Chem.Lett., 2013, 11(3), 271–275.

[37] Chun Y, Sheng G, Chiou CT, Xing B, Compositions and sorptive properties of crop residue-derived chars, Environ. Sci. Technol., 2004.

[38] Hameed BH, et al, Kinetics and equilibrium studies of malachite green adsorption on rice straw-derived char, J. Hazard. Mater., 2008, 153, 701–708.

[39] Xu RK, et al., Adsorption of methyl violet from aqueous solutions by the biochars derived from crop residues, Bioresour.Technol., 2011.

[40] Hussain A, Maitra J, Khan KA, 2017, Development of biochar and chitosan blend for heavy metals uptake from synthetic and industrial wastewater, Appl. Water Sci., 2017.

[41] Tang Y, et al., Influence of pyrolysis temperature on production of digested sludge biochar and its application for ammonium removal from municipal wastewater, J. Clean. Prod., 2019.

[42] Wei D, et al., 2018, Biochar-based functional materials in the purification of agricultural wastewater: fabrication, application and future research needs, Chemosphere, 2018.

[43] Mandal A, Singh N, Optimization of atrazine and imidacloprid removal from water using biochars: designing single or multi-staged batch adsorption systems, Int. J. Hyg Environ. Health., 2017.

Helen La, J. Patrick A. Hettiaratchi

Role of biochar in the removal of organic and inorganic contaminants from waste gas streams

Abstract: Typical air treatment processes for environmental contaminants, such as combustion, are neither environmentally friendly nor cost effective. An alternative is sorption using biochar – a rich carbon solid residue originating from biomass that is typically generated under pyrolysis. Biochar has been shown to have excellent sorption capabilities for the removal of both inorganic and organic contaminants. Further, microorganisms can colonize on the biochar and biologically transform the adsorbed air pollutants into benign by-products. The sorption process is driven by mass transfer until equilibrium is attained and involves physical adsorption on the surface, partitioning, pore filling, and chemical and biological oxidation of contaminants. Biochar prepared at high temperatures can sorb organic contaminants due to enhanced pore structure, increased hydrophobicity, and carbonized elements. In contrast, biochar prepared at low temperatures (<500 °C) has more surface polar functional groups and is better equipped at removal of inorganic contaminants.

Keywords: adsorption, inorganic/organic pollutants, contaminant removal, biochar, waste gas

1 Introduction

Industrial, commercial, agricultural, animal husbandry, energy production processes, and landfills produce waste gas containing odorous, toxic, and environmentally adverse gaseous components. These may include acid rain precursors (e.g., NO_X and SO_X), greenhouse gases (GHGs), hydrogen sulfide (H_2S), ammonia (NH_3–N), mercury (Hg), and volatile organic compounds (VOCs). Exposure to H_2S can cause adverse respiratory problems at lower concentrations (<120 ppmv) and is fatal at higher concentrations (>500 ppmv) [1, 2]. Apart from being odorous, sulfurous gaseous components are toxic, corrosive, and have the potential to cause acid rain. When emitted into the atmosphere, NH_3–N can lead to the formation of haze-inducing aerosols, particulate matter, and the deposition of nitrogenous components in potentially vulnerable ecosystems [3]. According to the IPCC [4], animal husbandry is the main contributor to global methane (CH_4) and nitrous oxide (N_2O) emissions – each of which exert a global warming potential (GWP) of 28 and 265, respectively, over a 100-year time horizon. VOCs contribute to photochemical smog, ozone depletion, and global warming. For instance, the treatment of manure for soil amendment applications,

https://doi.org/10.1515/9783110734003-005

typically by composting, inadvertently releases GHGs (CO_2, CH_4, and N_2O) and produces $NH_3{-}N_{(g)}$ as a secondary pollution source – reducing the total carbon and nitrogen content. In fact, losses of $NH_3{-}N_{(g)}$ can account for up to 94% of the total nitrogen content [5]. Today, a large contribution to Hg emissions comes from coal combustion [6]. Mercury is a highly persistent environmental toxin. When released into the environment, it is highly mobile and readily exchanges between the soil, water, and atmosphere. Because Hg is nondegradable, and due to its ability to redistribute across various environmental media, it is highly persistent once present in the environment.

There have been a number of proposed technologies for the remediation of environmental contaminants from waste gas streams: adsorption with activated carbon, regenerative thermal oxidation, regenerative catalytical oxidation, incineration, hydrolysis, membrane separation, and biofiltration are some examples. Some of these technologies are uneconomical for treating low pollutant concentrations in waste gas streams (e.g., regenerative thermal oxidation), ineffective for treating high pollutant concentrations (e.g., activated carbon adsorption), may require frequent catalyst replacements (e.g., regenerative catalytical oxidation), entail high energy consumption (e.g., thermal oxidation), or may be dangerous to operate (e.g., incineration of explosive gases).

A cost-effective alternative to the above treatment technologies is the use of biochar for the treatment of low-concentration waste gas constituents. Traditionally, activated carbon has been the most commonly used sorbent due to its ideal sorption properties (e.g., large pore volume and high specific surface area); however, they are economically impractical – especially for large-scale applications, consist of a narrow reaction temperature range, and exhibit poor regeneration. As a sorbent, biochar has wide applications for the removal of both inorganic and organic contaminants. Biochar can be subjected to chemical modification to produce activated carbon, using steam or chemicals, for the purpose of enhancing the surface area for the increased sorption of contaminants. Although unmodified biochar has a lower surface area than activated carbon, they compensate by possessing surface-active oxygen-containing functional groups [7]. This chapter will provide an overview of the application of biochar and its use in waste gas treatment.

2 Biochar

Biochar is a relatively inexpensive carbonaceous material and a cost-effective solid adsorbent. Essentially any carbonaceous material, including synthetic, geologic (e.g., coal), biomass, and solid waste, can be used as a carbon source to produce char. As biomass is a renewable feedstock, when utilized to produce biochar, the process can be performed sustainably – especially when utilizing resources that are readily available and do not compete with other uses, such as agricultural residues and algal blooms (e.g., manure, coconut shell, and orange peel). Biochar formed from the by-products of other processes (e.g., biofuel production) can not only reduce solid waste handling but also is, in and of itself, a waste treatment method and allows for waste recovery, resource recycling, and the mitigation of GHG emissions by offsetting the waste treatment and handling of these biogenic sources. Further, biochar is increasingly applied as a soil amendment to boost soil fertility. For instance, due to its alkaline pH, the application of biochar can neutralize the pH of soil that has acidified as a result of the repeated application of nitrogen-based fertilizers and the deposition of NO_X and SO_X – by-products of fossil fuel combustion. Since biomass combustion is considered carbon neutral, the adsorption of carbon-based contaminants using biochar can be thought of as being "carbon negative" or a "carbon sink."

The production of biochar occurs through thermochemical conversion processes, including slow and fast pyrolysis, gasification, torrefaction, and hydrothermal carbonization, under oxygen-limited and high-temperature conditions and using carbonaceous biomass feedstocks. Of these methods, pyrolysis is the most common [7]. The thermochemical conversion process alters the physicochemical structure of the biomass and condenses the carbon atoms into stable arrangements that are resistant to microbiological and thermal breakdown [8]. The sorption of gaseous components on biochar is determined by their physicochemical properties such as pore size distribution, specific surface area, surface functional groups, and carbonized fractions, which are dictated by pyrolysis conditions and feedstock [2]. In general, the carbonization of biochar that are produced under low temperatures (e.g., 100–400 °C) are incomplete and will result in greater yields and a chemical structure that consists of a diverse array of organic functional groups [9]. In contrast, at higher temperatures (e.g., 400–800 °C), the resulting biochar will consist of a greater proportion of aromatics, with a reduction of aliphatics, oxygen, and hydrogen contents and an enhanced presence of pore structures. Surface area, pore volume, pH, and carbon content of the resulting biochar also increases with treatment temperature [9, 10].

Pyrolysis of the different biochemical components in the biomass (e.g., fat, starch, cellulose, hemicellulose, and lignin) generates biochar, bio-oil, and non-condensable gases (or syngas), such as CO_2, CH_4, H_2, and CO. Slow pyrolysis generates a greater biochar yield while fast pyrolysis produces more bio-oil [7]. By selectively removing elemental components such as C, H, and O through the formation of non-condensable gases, the resulting biochar will display varying degrees of O/C and H/C ratios [11]. These properties affect the aromaticity, polarity, and biodegradability of the biochar, which will have a direct impact on the targeted removal of organic or inorganic pollutants in a waste gas stream. In other words, biochar can be made amenable to target inorganic or organic pollutants. Further, biochar has been shown to be reusable even after repeated sorption–desorption cycles, which is economically attractive for the treatment of chemical constituents found in waste gas. Nonetheless, it is important to note that different biomass feedstocks and pyrolysis temperatures will produce a biochar consisting of different physicochemical properties, which will have a direct influence on the sorption of waste gas constituents. These properties can include biochar yield, surface area, pore volume, cation exchange capacity (CEC), ash content, pH, nitrogen content, carbon content, and volatile matter. Currently, there is increasing research on modifying biochar to enhance their adsorption properties via chemical, physical (e.g., CO_2, steam, or air at 600–1,200 °C), and biological modification [12].

There are three pathways involved in the removal of a sorbate: (i) diffusion of the sorbate from the waste gas stream onto the surface of the sorbent (i.e., external mass transfer); (ii) diffusion of the sorbate into the micropores of the sorbent (i.e., intraparticle diffusion); and (iii) oxidation of the sorbate onto the active sites of the sorbent (i.e., chemisorption). The rate-controlling step of removal will depend on the slowest pathway and can be determined by undertaking a kinetic study. If the rate of reaction follows the pseudo first-order or second-order model, the rate-controlling step is controlled by pathway (i) and (iii), respectively. To determine whether pathway (ii) is the rate-controlling step, the Weber–Morris model can be used.

Below is a description of the characteristics of biochar that can be used to target organic and inorganic pollutants. Table 1 summarizes studies of targeted gaseous constituents from waste gas that have been treated with biochar.

2.1 Organic contaminants

As compared to lower pyrolysis temperatures (e.g., <500 °C), biochar produced under higher pyrolysis temperatures (>500 °C) displays lower O/C and H/C ratios, a concurrent increase in the degree of aromaticity, and a decrease in polarity resulting from the removal of hydrogen and oxygen atoms via dehydration reactions [13]. This implies that the hydrophobicity of biochar increases with higher pyrolysis temperatures; however, the degree to which these changes occur varies between feedstock. For instance, Wang et al. [11] attributed the higher surface polarity in swine manure-derived biochar, in comparison to maize straw biochar, to the higher degree of ash content and surficial calcium phosphates, although both were prepared under similar conditions. Nevertheless, with increasing treatment temperatures and regardless of feedstock, oxygen-containing functional groups have a tendency to decrease and thereby lower the overall bulk polarity of the biochar, making them more amenable to the sorption of organic contaminants.

Sorption of organic constituents onto biochar is believed to include two dominant mechanisms: partitioning into the noncarbonized organic matter fraction and adsorption onto the carbonized fraction. With increasing treatment temperatures, biochar

Table 1: Selected results on waste gas treatment using biochar. Maximum reported adsorption results under specified conditions are displayed wherever possible.

Contaminant	Feedstock	Pyrolysis temperature	Treatment	Adsorption capacity	Reference
H_2S	80% woodchips + 20% anaerobic digester residue	600 °C	n/a	**C_o = 1,020 ± 20 ppmv** **MC = 85.5 ± 2.3%** **Sterilized biochar:** 98%	Kanjanarong et al. [43]
a. Benzene b. *m*-Xylene c. *o*-Xylene d. *p*-Xylene	Corn stalk	600 °C	Ball milling with NH_4OH and H_2O_2	**NH_4OH:** a. 116.70 mg g^{-1} b. 133.05 mg g^{-1} c. 126.13 mg g^{-1} d. 133.01 mg g^{-1}	Zhang et al. [18]
m-Xylene *o*-Xylene *p*-Xylene	a. Oxford hardwood b. Wheat straw c. Corn straw d. Rice straw e. Rape straw	300 °C 400 °C 500 °C	n/a	**At 500 °C:** a. 36.82 mg g^{-1} b. 50.87 mg g^{-1} c. 25.88 mg g^{-1} d. n/a e. 7.22 mg g^{-1}	Rajabi et al. [16]

(continued)

Table 1 (continued)

Contaminant	Feedstock	Pyrolysis temperature	Treatment	Adsorption capacity	Reference
a. Benzene b. Toluene c. Methyl chloride d. Xylene e. Chloroform f. Carbon tetrachloride	Bamboo Sugarcane Neem	350 °C 450 °C 550 °C	n/a	**At 550 °C and Neem:** a. 54.6 mg g^{-1} b. 65.5 mg g^{-1} c. 39.61 mg g^{-1} d. 60.2 mg g^{-1} e. 30.81 mg g^{-1} f. 40.99 mg g^{-1}	Kumar et al. [15]
a. Acetone b. Cyclohexane c. Toluene	Bamboo Sugarcane bagasse Brazilian pepper wood Sugar beet tailings Hickory wood	300 °C 450 °C 600 °C	n/a	**Sugarcane:** a. 91.16 mg g^{-1} (300 °C) b. 69.3 mg g^{-1} (600 °C) c. 62.91 mg g^{-1} (300 °C)	Zhang et al. [13]
Simulated landfill gas: a. H_2S b. Toluene c. Acetone d. Chlorobenzene	Municipal solid waste (as-is and modified with sewage sludge)	300 °C 350 °C 400 °C 450 °C 500 °C	Sewage sludge	**Waste char (400 °C):** a. 317.12 mg g^{-1} b. 441.06 mg g^{-1} c. 179.24 mg g^{-1} d. 425.22 mg g^{-1}	Qin et al. [20]
H_2S	Macroalgae: a. Enteromorpha b. Sargassum	400 °C 600 °C 800 °C	n/a	**At 800 °C:** a. 582 mg kg^{-1} b. 5,800 mg kg^{-1}	Han et al. [2]

NH_3–N	Oak (Hydrochar and biochar)	**Hydrochar:** a. 250 °C **Biochar:** b. 450 °C c. 650 °C	KOH	**C_o = 43 mg NH_3:** a. 18.8 mg g^{-1} b. 2.9 mg g^{-1} c. 0.7 mg g^{-1} **C_o = 450 mg NH_3:** a. 28.5 mg g^{-1} b. 4.3 mg g^{-1} c. 8.1 mg g^{-1}	Takaya et al. [3]
CO_2	Microalgae powders: a. *Chlorella* b. *Spirulina*	800 °C	Urea and KOH	a. 3.44 mmol g^{-1} b. 3.09 mmol g^{-1}	Shi et al. [39]
CO_2	Coffee grounds	400 °C (plus hydrothermal treatment at 160 °C)	a. 3-Aminopropyltrimethoxysilane + KOH b. Aniline + KOH c. Melamine + KOH	a. 2.04 mmol g^{-1} b. 2.22 mmol g^{-1} c. 2.67 mmol g^{-1}	Liu and Huang [47]
CO_2	Sunflowers: a. Receptable (SR) b. Stalk (SS) c. Stalk (pyrolysis at 1,000 °C)	500 °C 1,000 °C (no activation, stalk only, (c))	KOH	a. 4.09 mmol g b. 4.08 mmol g c. 4.52 mmol g	Sun et al. [38]
CO_2	Marine biomass: a. *Sargassum* b. *Enteromorpha*	400 °C 600 °C 800 °C	KOH	**At 800 °C and 1:1 KOH/biomass (mass basis) and 12% CO_2:**a. 1.05 mmol g^{-1} b. 0.52 mmol g^{-1}	Ding and Liu [36]
H_2S	Marine biomass: a. *Sargassum* b. Enteromorpha	800 °C	H_2SO_4/H_2O_2	a. 7.28 mg kg^{-1} b. 8.5 mg g^{-1}	Chen et al. [44]

(continued)

Table 1 (continued)

Contaminant	Feedstock	Pyrolysis temperature	Treatment	Adsorption capacity	Reference
H_2S	a. Corn stoverd. b. Maple wood	500 °C	$FeCl_3$	**C_O = 1,000 ppm:** a. 8.2 mg g^{-1} b. 23.9 mg g^{-1}	Choudhury and Lansing [41]
Hg^0	*Sargassum*	400 °C 600 °C 800 °C	NH_4Br	**C_O = 60 µg m^{-3}** **NH_4Br = 5 wt%** a. 8% O_2 = 94.3% b. 600 ppm SO_2 = 91.58% c. 800 pt NO = 95.73% d. 1% H_2O (g) = 91.58%	Yang et al. [10]
Hg^0	Tea leaves	500 °C 700 °C (post-pyrolysis)	$Fe(NO_3)_3$	**C_O = 65 µm m^{-3}** **$Fe(NO_3)_3$ = 0.46 mol L^{-1}** **T = 180 °C** **10% H_2, 10% CO, 400 ppm H_2S, N_2 (balance)** 94.81% ≥98% (N_2 + 400 ppm H_2S)	Altaf et al. [49]
Hg^0	Wood:tire	500 °C 600 °C 700 °C 800 °C 900 °C	n/a	**C_o = 90 µm m^{-3}** **50:50 blend** **800 °C**	Xu et al. [50]

transitions from an amorphous mass of carbon to graphene-like arrangements that give rise to nanopores of less than 2 nm in diameter [9]. In fact, surface area, as evidenced by CO_2 gas adsorption analyses, has been shown to be positively correlated to aromaticity and negatively correlated to aliphaticity, which are dominant under high and low treatment temperatures, respectively [11]. That is, hydrogen bonding with aliphatic hydrocarbons is the dominant sorption mechanism at lower treatment temperatures while π–π electron–donor–acceptor interactions, a non-covalent bonding force, dominate at higher temperatures. This section provides a summary of sorption studies on organic compounds that are divided into two groups: non-methane volatile organic compounds (NMVOCs) and CH_4.

2.1.1 Non-methane volatile organic compounds

Chronic exposure to NMVOCs can pose serious health risks and these compounds are known to be carcinogenic. NMVOCs originate from natural and anthropogenic sources, including biomass combustion, fossil fuel extraction and processing, landfills, and industrial processing. NMVOCs are prevalent in our daily lives and even contribute to indoor air pollution through off-gassing of objects made from particle board (e.g., furniture, flooring, and walls) [14]. The dominant removal mechanism of NMVOCs using biochar prepared at lower temperatures is related to the natural organic matter (NOM) content, also referred to as the noncarbonized organic matter fraction, which decreases as pyrolysis temperature increases. At higher pyrolysis temperatures, the increased surface area plays a larger role in the separation and removal of NMVOCs [15]. For example, Rajabi et al. [16] observed increasingly improved sorption capacity for xylene with pyrolysis temperature, from 300 °C to 500 °C. Although, as the reaction temperature was increased from 25 °C to 45 °C, Rajabi et al. [16] observed that the sorption of xylene onto wheat straw biochar decreased substantially more with biochar pyrolyzed at 500 °C (e.g., reduction of up to 91%) than at 300 °C (e.g., reduction of up to 71%). Since physical adsorption is the dominant removal mechanism for biochar prepared at higher pyrolysis temperatures, as the reaction temperature was raised, desorption of xylene occurred, and a larger reduction of xylene removal was therefore observed for the biochar pyrolyzed at 500 °C. On the other hand, partitioning of xylene still occurred in both biochar regardless of the reaction temperature. However, not all feedstocks will result in greater sorption when prepared at higher pyrolysis temperatures as this will depend on the physicochemical properties of the biochar, such as the specific surface area, pore volume, and so on. For instance, Zhang et al. [13] observed higher acetone adsorption on bamboo and sugarcane bagasse biochar that were produced at 300 °C versus at 600 °C, and this was attributed to the VOCs partitioning into the NOM fraction of the biochar. The opposite effect was observed for biochar prepared from Brazilian pepperwood, sugar beet tailings, and hickory wood, where acetone was removed via adsorption reactions as a result of the increased surface areas developed under higher temperatures.

The choice in feedstock used to generate biochar will also have a direct impact on the physicochemical properties of the resulting adsorbent. Kumar et al. [15] prepared three biochar from bamboo, sugarcane, and neem in which the latter exhibited the highest sorption capacity for VOCs under the same conditions. Neem was observed to have higher surface area and pore size. In general, feedstock consisting of higher lignin content, or wood-based biomass, tend to produce biochar with higher surface area, porosity, and electrical conductivity [7]. An increase in surface area with pyrolysis temperature is due to the decomposition of cellulose and hemicellulose into lower molecular weight components below 450 °C, the transformation of lignin into aromatic hydrocarbons, and the formation of biochar [8]. This is followed by the transformation of cellulose carbon into amorphous char [9, 17]. Currently, there is increasing research on modifying biochar predominantly with physical or chemical methods to raise the sorption capacity of the sorbent, especially with inexpensive and simple techniques. One such approach is ball milling. Ball milling is a physical process where chemical bonds are broken, in which the energy of moving balls is used to break down the particle size into ultrafine particles [12]. The process is meant to increase the surface area, pore volume, negative zeta potential, and oxygen functional groups – both internally and externally. Zhang et al. [18] used ball milling with ammonium hydroxide (NH_4OH) and hydrogen peroxide (H_2O_2) to synthesize nitrogen-doped and oxygen-doped biochar. Similarly, Lyu et al. [19] were able to increase the surface area of sugarcane bagasse biochar prepared at a low pyrolysis temperature (300 °C) from 0 to 10.8 $m^2\ g^{-1}$ and at high temperature (600 °C) from 0 to 359 $m^2\ g^{-1}$ after ball milling. Ball milling biochar increased the sorption capacity of phenyl compounds, relative to untreated biochar, but the sorption capacity was even higher with chemical treatment, especially with NH_4OH. Further, ball milling and chemical treatment increased the acidity (i.e., a proxy for polarity) and volatile organic matter content (i.e., a proxy for oxygen-containing functional groups), especially for oxygen-doped biochar, which may suggest that they would be better equipped to sorb inorganic compounds. These effects were least improved in nitrogen-doped biochar and, correspondingly, they also displayed the highest removal capacity for phenyl compounds (i.e., benzene, *m*-, *p*-, and *o*-xylene).

Few studies using biochar for the separation and removal of NMVOCs have considered biological oxidation. However, when the treatment of waste gas can be coupled between physisorption, chemisorption, and biological oxidation, a continuous operation may be possible since desorption of the sorbent may be extended to longer intervals. Qin et al. [20] compared the sorption capacity of waste-derived biochar and waste-derived biochar modified with sewage sludge for simulated landfill gas comprising of H_2S, toluene, acetone, and chlorobenzene. When sewage sludge was used as a natural inoculum of microorganisms, the sorption capacity for the simulated landfill gas was reduced due to some degree of pore blocking from the sewage sludge. However, the VOCs and H_2S were thought to have been simultaneously biodegraded by the microorganisms present in the sewage sludge, such as methane oxidizing bacteria (type I and type II), sulfur-oxidizing bacteria, sulfur-reducing bacteria, heterotrophic bacteria, and nonmethane oxidizing bacteria – although DNA sequencing was not performed to confirm the bacterial strains that were present within the sample.

2.1.2 Methane

Methane (CH_4) is an important GHG that is released under anaerobic conditions through the activities of methanogens. Anthropogenic CH_4 emissions from the ongoing growth and expansion of animal husbandry, agricultural practices, landfills, and fossil fuel extraction are responsible for the growing CH_4 concentrations in the atmosphere [21]. A promising technique for the reduction, separation, and removal of CH_4 from these sources is the application of biochar. For the purposes of CH_4 mitigation, biochar has been applied to paddy cultivation and aerobic composting of animal manure, as a floating cover for manure storage, in landfill cover soils, and as a feed additive for ruminants. As an amendment, biochar can potentially alter the physical and chemical properties of a medium to promote microbial oxidation of CH_4 by way of water retention and improving aeration. In landfill biocovers and biofiltration systems comprised of soil or compost, biochar amendments can help to prevent desiccation fissures near the surface, which can inadvertently create preferential pathways that lead to the subsequent release of CH_4 through advective flux [22]. Compared to soil/compost, the larger particle sizes of biochar add to the air-filled porosity and promote methane oxidation, which is especially important for traditional landfill covers as they are typically heavily compacted. Compacted clay is widely used as a final cover in landfills due to its low cost and simplicity in design; however, it is subject to environmental factors that may cause it to lose its structural integrity over time, resulting in shifts and cracks that lead to the release of CH_4. Recent interest has diverted toward the use of biochar as a final cover layer in place of traditional materials. Li et al. [23] studied the relationship between biochar content and the degree of compaction on the air permeability coefficient of a biochar-clay mixture. The air permeability coefficient was observed to decrease as the biochar content increased (from 0% to 20%, mass basis) in a biochar-clay mixture at a dry clay density of 1.42 g cm^{-3}. The same trend was observed at dry clay densities of 1.56 and 1.65 g cm^{-3} with the exception that it began to increase as the biochar content reached 20% w/w. This was attributed to biochar filling the clay pores such that at 20% w/w; porous nature of the biochar began to play a more evident role on promoting air permeability through the medium. Aside from the physical benefits, there are also biological benefits to biochar amendments in soil media, adding another route of CH_4 uptake. Cui et al. [24] demonstrated that soils amended with biochar experienced a lower degree of microbial mass carbon reduction as compared to unamended soils with increasing temperature from 5 °C to 25 °C. Interestingly, it was observed that biochar amendments suppressed soil CH_4 uptake regardless of temperature while soil N_2O production was significantly reduced. Cui et al. [24] attributed this observation to the consumption of organic compounds other than CH_4 as a growth substrate by facultative methanotrophs. Although CH_4 was not necessarily reduced, this demonstrated that biochar amendments are still beneficial in mitigating the even more powerful GHG, N_2O.

Along similar lines, biochar has been applied as a cover for manure storage. The sorption of CH_4 using a thin cover of floating biochar over manure digestate was studied by di Perta et al. [25]. Emissions of CH_4 were reportedly higher initially and declined over time, but the overall removal efficiency was only 33% in comparison to the higher 78% removal efficiency observed for NH_3–N, which was studied simultaneously. Based on these results, it was not clear whether there was an interactive effect between sorption properties of NH_3–N and CH_4 on the woodchip-based biochar used in the study.

Biochar amendments can lower the level of CH_4 emissions during aerobic composting [26] by promoting air exchange through the media and encouraging the activities of methane oxidizing bacteria. Aerobic composting may lead to CH_4 emissions that account for 1–2% [27, 28] of total organic carbon. Although this amount may seem minor, the GWP of CH_4 is 28 times that of CO_2 over a 100-year horizon and, unlike the CO_2 emitted from compost [21], CH_4 emissions are not considered to be carbon neutral. This may occur when the organic matter is degraded faster than the rate at which oxygen can be supplied to the aerobic microorganisms responsible for the degradation, thereby creating an anaerobic zone. Studies have shown that when biochar is used in composting and animal operations, it can effectively reduce the emissions of nitrogenous and sulfuric constituents, GHGs, and VOCs [29–31]. In addition to being a sorbent, biochar has been shown to be an effective bulking agent in aerobic composting. In composting operations, the organic matter degradation process can lead to a significant removal of carbon and nitrogen content through the volatilization of NH_3–N and GHGs such as CH_4 and N_2O, adversely affecting the fertilization value of the resulting compost. The addition of biochar in composting can combat nutrient losses by promoting air diffusion. For instance, when applied as a bulking agent, biochar has been demonstrated to reduce the emissions of CH_4, NH_3–N, and N_2O by promoting the diffusion of O_2, increasing aeration, and thereby providing a suitable habitat for aerobic microorganisms [26, 31]. Further, a higher degree of organic matter decomposition has been observed when compost was supplemented with biochar [31]. Chen et al. [31] observed a reduction in CH_4 emissions in biochar amended compost of up to 26.1% using cornstalk biochar in comparison to a control. A mixing ratio of 1:0.3:0.14 (w/w) of layer manure, sawdust, and biochar was used. The addition of biochar as a bulking agent increased aeration and minimized the opportunities for anaerobic conditions to develop, which is especially important in the initial stages of composting where there is typically an abundance of nutrients and organic substrate to promote microbial growth. Higher rates of microbial growth and aerobic degradation can deplete oxygen levels in the compost and, thereby, promote anaerobic conditions. Guo et al. [32] found that the combination of a bamboo charcoal biochar and bamboo vinegar mixture (i.e., an aqueous liquid generated during pyrolysis) produced the highest level of CO_2 emissions, bacterial abundances, and organic matter degradation, relative to the control, particularly during the thermophilic stage of composting, indicating that there was greater organic matter decomposition due to higher levels of microbial degradation activities than if each amendment were used alone. The *mcrA* gene in methanogens was suppressed best by the vinegar mixture, effectively limiting the amount of CH_4 emitted.

Rice paddy cultivation represents an expanding man-made flux of CH_4. On an annual basis, up to 40 T_g CH_4 are released into the atmosphere – representing 10% of all anthropogenic sources [21]. Sequestration of soil carbon is a common management practice of both returning organic carbon into soils and mitigating GHG emissions. Producing biochar from biomass and returning it to the soil is a form of soil carbon sequestration. In a continuous water flooding rice cultivation study, Qi et al. [33] concluded that the addition of biochar to rice paddy soil reduced the abundance of methanogenic archaea, relative to the control, which effectively mitigated CH_4 emissions. Methanotrophic bacteria increased with increasing levels of biochar amendment, but their presence was complicated by the urea-based fertilizer that was used as the methanotrophs were inhibited at an application concentration of 200 mg N kg^{-1}. On the contrary, Cao et al. [34] observed an increase in CH_4 emissions following the addition of straw-based biochar and attributed this phenomenon to the stimulation of methanogenic archaea from carbon supplied in the plant rhizosphere as a result of increased plant photosynthesis. However, urea and biochar application increased CO_2 uptake and lowered N_2O emissions relative to urea alone.

2.2 Inorganic contaminants

Common inorganic waste gas constituents of targeted interest include NH_3–N, H_2S, and CO_2 – all of which may be emitted from manures, composts, landfills, and industrial activities producing flue gas or waste gas [30]. In general, biochar prepared under lower pyrolysis temperatures (e.g., <400 °C) are partially carbonized and will display lower surface area, higher O/C and H/C ratios, and smaller pore sizes [9, 35]. The higher O/C ratios indicate higher oxygen contents, which are favorable for the formation of oxygenated functional groups. If the removal process is controlled by chemical sorption (or chemisorption), then the availability of functional groups to participate in the sorption process will be important. These oxygenated functional groups can be beneficial for the removal of inorganic contaminants. Enhancing the degree of oxygen functional groups can reduce the hydrophobicity of biochar and help in targeting polar waste gas constituents. In particular, oxygenated functional groups are responsible for the surface acidity (e.g., carboxyl) or basicity (e.g., ketone, carbonyl, alcohol, and ether) of the biochar and attraction to hydrophilic waste gas chemical constituents. For instance, Chun et al. [35] examined the Fourier transform infrared (FTIR) spectra of wheat residue biochar prepared at lower pyrolytic temperatures and observed a higher degree of acidic functional groups (e.g., carboxyl and carbonyl groups) than for biochar prepared at higher treatment temperatures. Similarly, Ding and Liu [36] prepared a biochar that exhibited a greater intensity of carbonyl and carboxyl functional groups – which were conducive to CO_2 sorption. However, when the process is controlled by physical sorption (or physisorption), the pore structure will play a principal role in dictating removal efficiency [37]. In either case, a lower pyrolytic temperature may not provide a char with sufficient specific surface area, pore volume, or the surface functional groups to sorb the inorganic components of waste gas constituents.

Other than elevating the pyrolysis temperature to enhance the pore structure of biochar by releasing more volatiles, chemical modification can increase the microporosity, total pore volume, and specific surface area. For example, chemical activation by KOH has been shown to result in a micro- and nano-porous structural morphology that is advantageous for gas storage [38]. Generally, adsorption capacity is correlated to specific surface area. Shi et al. [39] enhanced the surface area of two microalgae, *Chlorella* and *Spirulina*, by 84.5 and 71 times, respectively, using urea and KOH at one and two times the mass ratio, respectively, of biochar. Not only was pore volume enhanced, there was also a greater abundance of micropores as evidenced by a smaller average pore diameter. Sun et al. [38] found that the surface area was higher upon chemical activation with KOH than by pyrolyzing at a treatment temperature as high as 1,000 °C. Without chemical activation, the biochar will lack sufficient microporosity to promote physical sorption reactions. The microporosity can be defined by the micropore ratio:

$$V_{\text{micro}}/V_{\text{total}} \tag{1}$$

where V_{micro} is the micropore volume and V_{total} is the total pore volume. The higher the micropore ratio, the greater the microporosity offered by the biochar to participate in adsorption. In tandem, the diameter of the micropores will provide an indication of the strength of the potential interaction between the sorbate and sorbent. A smaller diameter may increase the sorption capacity between the sorbate and the pore walls of the biochar.

Reagents that have been used to modify surface active sites include KOH, urea, H_3PO_4, HNO_3, H_2SO_4, H_2O_2, K_2CO_3, NaOH, and $ZnCl_2$. Chemical modification can modify the surface functional groups of biochar for the enhanced and targeted removal of inorganic contaminants. Depending on the reagents used, the pH of the resulting biochar can be tuned for the removal of waste gas constituents. For example, activation by KOH can increase the basicity and lead to the preferential removal of acidic chemical constituents such as CO_2. This was seen in the seaweed biochar (i.e., *Sargassum* and *Enteromorpha*) prepared by Ding and Liu [36], who observed approximately three quarters of the surface functional groups belonging to ketones, carbonyls, ethers, and alcohols – increasing in basicity – following activation with KOH and a pyrolysis temperature of 800 °C.

2.2.1 Hydrogen sulfide

Hydrogen sulfide (H_2S) is an odorous and toxic gas that is produced in energy production processes, waste processing, and in the natural environment. Due to the risk they pose on human health and on industrial processing equipment, their efficient removal and separation from waste gas is of great importance. The separation and removal of H_2S can be performed through physicochemical methods (e.g., adsorption and oxidation) or by biological methods (biofiltration, biotrickling filters, etc.). Due to the high costs associated with operation and maintenance, many of these methods are uneconomical. Biochar has been used as a filter bed medium in biofiltration and as an adsorbent in external scrubbing systems. However, most studies that utilize biochar typically apply it purely as a sorbent and not as a growth medium for microorganisms for H_2S separation and removal.

A benefit in using biofiltration for the treatment of waste gas constituents, such as H_2S, is that microorganisms can utilize it for food and energy and, thereby, permanently separate and eliminate the chemical compound through biological oxidation into sulfate (SO_4^{2-}). This is in addition to the filter bed adsorbing H_2S when an adsorbent such as biochar is used – albeit, more continuous removal could be achieved through the biological oxidation route since sorption may be limited by the availability of active sites. Biochar, with its high water-holding capacity and pore structure, provides microorganisms with a preferable environment for attached growth as well as serve as an effective sorbent for the retention of the sorbate. Water is required periodically to wash out SO_4^{2-} and prevent it from accumulating, thereby preventing the inhibitory effects of SO_4^{2-} on the sulfur-oxidizing bacteria. The accumulation of SO_4^{2-} will drop the pH of the system and inhibit the growth of sulfur-oxidizing chemolithotrophs. Das et al. [40] used a compost–biochar mixture as the filter bed packing medium in a biofilter to treat H_2S. Up to 32.7 g m^{-3} h^{-1} of H_2S was removed, corresponding to a 100% removal efficiency when a compost–biochar mixture was used (3:1, by volume). However, accumulation of SO_4^{2-} inhibited the sulfur-oxidizing bacteria and the removal efficiency subsequently dropped.

The mechanism of removal of H_2S using biochar is assumed to be similar to that of activated carbon as a sorbent [41]: (i) surface adsorption; (ii) dissolution into the water film; (iii) dissociation of H_2S into H^+ and HS^- within the water film; (iv) oxidation of H_2S to S^0 and SO_2; and (v) oxidation of SO_2 to H_2SO_4 catalyzed by water and metallic species.

For adsorption, the acid-alkali property of the biochar may play an important role in the separation of H_2S since H_2S is an acidic gas, therefore, a biochar exhibiting an alkaline pH is preferable for the successful removal of H_2S by pH neutralization. Biochar prepared under higher treatment temperatures typically tend to exhibit higher alkalinity and pH due to increases in carbonates, organic anions, and ash content, which correspond to higher alkaline constituents such as alkaline salts [42]. Kanjanarong et al. [43] utilized a 4:1 wood chip and anaerobic digester residue prepared at 600 °C through pyrolysis for the removal of H_2S. Because the initial pH of the biochar (7.98) was higher than the first dissociation constant of H_2S ($pK_a = 7.2$), the pH of the biochar promoted its subsequent dissociation. However, pH may not always be a determining factor on H_2S removal. Iron trichloride ($FeCl_3$) has also been used to modify the surface of biochar to increase the selectivity toward H_2S, as iron has a high affinity for the acidic gas. A recent study by Choudhury and Lansing [41] observed higher sorption of H_2S with maple biochar (pH = 9.1) and activated carbon (pH = 8.3) than corn stover biochar even though the latter had a higher pH (i.e., pH = 10.2). This was attributed to direct oxidation of gaseous H_2S by reactive iron oxide (Fe_3O_4) within the water film into elemental sulfur (S^0) and SO_4^{2-} [41]. Han et al. [2] observed increasing pH values on two macroalgae-based biochar with increasing calcination temperatures from 400 to 800 °C, owing to oxygen-containing functional groups including C=O, C–O, COO, and O–H. These alkaline functional groups contributed to the removal of H_2S through oxidation reactions resulting in SO_4^{2-} and S^0, which were observed through XRD analysis. Other than increasing the treatment temperature of pyrolysis, the biochar surface can be chemically modified to increase oxygen-containing functional groups. The oxidants H_2O_2 and H_2SO_4 have been used individually or in tandem; however, a greater level of specific surface area may be developed when they are employed in tandem. When both H_2O_2 and H_2SO_4 have been used as modifying reagents, they were able to promote the development of pore structure [44]. The increased pore structure supports adsorption of the H_2S molecules followed by oxidation into $CaSO_4$ through oxidation-reduction or substitution reactions [44]. Although a higher pyrolysis temperature can increase the surface area of the biochar and positively affect physical adsorption, chemical sorption may play a more determining role in the removal of H_2S [41]. The addition of water to the biochar can either promote or hinder the removal of inorganic contaminants [2, 43]. In the presence of water, H_2S is ionized to HS^-, which is more readily oxidized than gaseous H_2S. However, when too much water is added, the pores of the sorbent are blocked and can hinder adsorption [44]. Generally, it is expected that as reaction temperature is increased the breakthrough time will subsequently decrease due to desorption of the sorbate; however, with H_2S, as the reaction temperature increases, sorption of H_2S increases, promoting desulfurization. This phenomenon may be a result of an intensification of the rate of chemisorption between H_2S with chemically active sites.

On the contrary, some studies have indicated physisorption is more important than chemisorption for H_2S removal. Han et al. [2] found that the rate controlling step for H_2S sorption onto two types of macroalgae could be characterized by the pseudo first-order adsorption kinetics, or that the external mass transfer of the H_2S from the gas phase and onto the surface of the biochar, dominated the adsorption process. Similar results were obtained by Chen et al. [44] who experimented on seaweed biochar (*Sargassum* and *Enteromorpha*) modified with H_2O_2/H_2SO_4 and both biochar followed pseudo-first-order kinetics. According to Chen et al. [44], this was further supported by Gibbs free energy quantifications based on their experimental data (i.e., –20 kJ mol^{-1}) where values were within the range to support physical adsorption (i.e., –20 kJ mol^{-1} to 0 kJ mol^{-1}) and chemical adsorption (i.e., –400 to –80 kJ mol^{-1}) – indicating both processes were responsible for desulfurization – with the former having a dominant role in the process.

2.2.2 Ammonia and nitrous oxide

Ammonia (NH_3–N) is primarily emitted from agricultural activities and animal husbandry such as through manure storage and manure application. Aside from their negative environmental impacts to air quality, the release of NH_3–N from manure lowers their value as a fertilizer through the loss of nitrogen during storage (e.g., up to 40% of TKN [45]) and composting. As NH_3–N is the primary form of nitrogen that is lost in the process [26], there is an incentive to minimize their volatilization. Among the methods of NH_3–N emissions reductions are adjustments to processing conditions, inoculation with microorganisms, and the use of bulking agents, such as biochar. Not only can biochar reduce the emissions of NH_3–N, but it also has the potential to sorb ammonium nitrogen (NH_4^+ –N) which adds to the fertilization value of the manure or compost. In fact, biochar is applied as a soil amendment to retain water and nutrients to boost soil fertility, pH, CEC, and microbial activity. A rise in the CEC can help to limit nutrient leaching in soil systems. Biochar may limit the production of N_2O from soils by limiting nitrogen cycling through the adsorption of inorganic nitrogen. Specifically, it can suppress autotrophic nitrification and denitrification; decrease the production of N_2O through the action of ammonium oxidizing bacteria; and improve soil aeration, thereby preventing denitrification processes from taking place, which occurs under anaerobic conditions [24].

Common methods that are used to minimize NH_3–N emissions are the use of permeable (e.g., straw and geotextiles) or impermeable (concrete, steel) cover systems and acidification of the animal slurry with sulfuric acid. However, permeable covers typically have a shorter lifetime; impermeable covers are more costly to implement and to operate; and acidification leads to the potential exposure of sulfur leaching, foaming, acid handling and can contribute to N_2O emissions (a potent

GHG). More recently, attention has focused on the use of biochar as a cost-effective NH_3–N mitigation effort.

Biochar has been successfully applied in the sorption of NH_3–N. Typically, both NH_3–N and NH_4^+ –N are studied in unison in sorption studies as their transformation and the magnitude of NH_3–N emissions are linked to their equilibrium reaction with one another. Emissions of NH_3–N are greatly impacted by the pH of the system [46] and are primarily released between a pH of 7 and 10, minimally below a pH of 7, and virtually not at all below a pH of 4.5. Therefore, lowering the pH of the system has the potential advantage of diminishing the release of NH_3–N.

di Perta et al. [25] saw a 78% removal efficiency of NH_3–N from a manure digestate using a thin biochar cover at bench-scale, relative to an uncovered control. Chen et al. [31] observed a reduction in NH_3–N emissions during the composting of layer manure amended with five different biochar (i.e., cornstalk, wood, bamboo, manure, and coir) that were statistically different from the control, with the cornstalk biochar exhibiting the highest removal efficiency. Precursors to NH_3–N, including NH_4^+ –N, urea, and uric acid may have successfully undergone sorption by the biochar, especially to the acidic surface functional groups, and therefore minimized the formation and subsequent NH_3–N volatilization – especially under the high pH and temperatures that were reached during the composting process. In composting, achieving high temperatures of >50 °C during the thermophilic stage is required to achieve maturity. The production of NH_3–N and the decomposition of organic acids can contribute to a rise in pH; however, biochar-amended composts can counteract this rise in pH through the sorption of NH_3–N and NH_4^+ –N while also minimizing the volatization of NH_3–N by promoting an aerobic environment for the nitrifying bacteria to transform NH_3–N to NO_3^- –N [31]. In the study by Chen et al. [31], a cumulative effect of making biochar amendments to the compost was the retention of total nitrogen by up to 29.4% that would have otherwise been lost. Further, higher NO_3^- –N contents were observed in biochar amended composts, demonstrating that cornstalk and woody biochar established an environment that promoted the conversion of NH_3–N to NO_3^- –N, an aerobic process, retaining nitrogen in the compost through the activities of nitrifying bacteria. In another study, Covali et al. [46] observed a short-term effectiveness of minimizing NH_3–N emissions by utilizing a reed canary biochar cover (torrefied at 300 °C) over a biogas digestate. Biochar is generally hydrophobic, allowing them to be used as an effective cover system. This property is attributed to the large presence of aliphatic surface functional groups. Further, in comparison to the control and to the biochar mixed into the digestate, the biochar cover was observed to limit changes in pH by minimizing the exposure of the surface of the digestate to the ambient air where NH_4^+ may be converted to NH_3–N through the release of hydroxyl molecules. However, the biochar cover may have encouraged the release of N_2O due to the development of anaerobic conditions within the digestate [46].

The surface functionality of biochar can be chemically modified to enhance the sorption of NH_3–N. Covali et al. [46] noted that the specific surface area and surface functional groups of the biochar were increased when the biochar were acidified with H_2SO_4, H_3PO_4, or H_2O_2, which also beneficially lowered the pH and alkalinity of the biochar. Interestingly, biochar treated with H_2SO_4 and H_3PO_4 were able to suppress NH_3–N emissions better than when just the acids were added to the biogas digestate; however, because the alkalinity of biochar could not be entirely removed with acidification, the suppression could be reversed as the pH of the system may increase over time. Similarly, Takaya et al. [3] compared hydrochar and biochar prepared at two pyrolysis temperatures for the sorption of NH_3–N. Hydrochar, prepared by hydrothermal carbonization, is generated at temperatures <200 °C and under moderate pressures. Due to the lower temperatures used in their preparation, the resulting char exhibited a greater diversity of functional groups. Takaya et al. [3] observed a higher degree of NH_3–N sorption to oak hydrochar produced by hydrothermal carbonization at 250 °C than by biochar produced by pyrolysis at 450 and 650 °C (Table 1). While the hydrochar was able to sorb 45%, or 18.8 mg g^{-1}, on the other hand, the two biochars underwent sorption of <10%, or <3 mg g^{-1}, from a total NH_3–N headspace of 43 mg. In fact, the as-received oak was able to sorb more NH_3–N (>30%) than the two biochar. However, as the headspace concentration increased to 450 mg, the removal efficiencies for all adsorbents dropped. Nevertheless, the hydrochar persistently exhibited greater sorption capacity than the as-received oak while the two biochars experienced the least sorption of NH_3–N. To test whether surface functionality would impact the sorption of NH_3–N, the hydrochar and the biochar were chemically modified with H_2SO_4, H_3PO_4, H_2O_2, and KOH. Following treatment, the hydrochar benefited from H_2O_2 and KOH treatment while the biochar pyrolyzed at 450 °C experienced a three-fold increase in sorption capacity for NH_3–N from all chemical treatments. A four-fold increase in NH_3–N sorption was observed for biochar pyrolyzed at 650 °C. An explanation was offered for each chemical used: an increase in sulfate adsorbent sites are believed to enhance NH_3–N sorption as a result of H_2SO_4 modification; labile carbon and lignocellulosic materials were chemically transformed into ketones through H_3PO_4 modification; and peroxide and KOH treatment increased both surface hydroxyl and carbonyl groups.

Not many have simultaneously studied the microbiological and sorption effects of biochar on the treatment of waste gas. In a recent study by Chen et al. [26], a chicken manure-integrated microbial consortium was tested as a co-amendment with biochar for the mitigation of NH_3–N and CH_4 from a chicken manure compost. Emissions of NH_3–N were higher when the microbial consortium was used alone – although slightly lower than the control where no amendments were made. Higher emission reductions were only observed with increasing biochar quantities. From a microbiological perspective, the mitigation of air pollutants, such as NH_3–N and CH_4, through the use of biochar as a bulking agent is believed to be due to their ability to foster improved air exchange through the medium, allowing for microbiological growth and contaminant degradation.

Although composting is an effective method for the treatment of livestock manure, emissions of N_2O can occur during composting because of incomplete nitrification and denitrification resulting from the nonhomogeneous distribution of O_2 within the compost, even when biochar is used as a bulking agent. Both nitrification at the biochar surface, where O_2 is more plentiful, and denitrification within the compost can lead to the formation of N_2O. Chen et al. [26] observed a gradual increase in N_2O emissions as the temperature of the compost dropped. The higher pH and temperatures observed initially during the thermophilic stage of composting effectively inhibited nitrifying bacteria from generating N_2O. As the temperature of the compost dropped, a concurrent increase in NO_3^-–N concentration led to an increase in the formation of N_2O. Increasing the percentage of biochar amendments to the compost lowered N_2O emissions, relative to controls. Chen et al. [26] concluded that biochar amendments may foster denitrification reactions where NO_3^-–N is reduced to $N_{2(g)}$ directly.

2.2.3 Carbon dioxide

Carbon dioxide is a GHG that is principally produced from the combustion of fossil fuels and industrial processing. To combat the effects of climate change, the capture of CO_2 for storage or utilization has been a heavily studied topic in recent years. Numerous methods have been proposed for the separation of CO_2 from waste gas (e.g., amine absorption, electrocatalysis, membrane separation, biological fixation, and adsorption). Of these methods, adsorption has several advantages over the other CO_2 separation methods including ease of use and low energy requirements. Both carbonaceous (e.g., activated carbon and biomass-based carbon) and noncarbonaceous (e.g., silica, zeolite, and metal oxides) adsorbents are available but biomass-based carbons, or biochar, are particularly attractive due to their low cost, wide-ranging selection, and availability.

The removal of CO_2 using biochar can be controlled by either physical or chemical sorption; however, more can be removed when both physical and chemical sorption is capitalized simultaneously. In a study comparing microwave pyrolysis versus conventional pyrolysis using wheat straw biochar, Huang et al. [37] observed that CO_2 adsorption generally increased with pyrolysis temperature. The FTIR spectra between the raw feedstock and the two chars were not significantly different from one another, indicating that physical adsorption was the dominant CO_2 removal process [37]. Surface modification of the biochar may have increased the sorption capacity for CO_2, as discussed below. Most CO_2 sorption studies have utilized chemical activation to enhance the properties of biochar. This is because CO_2 is a weak Lewis acid and is not sufficiently removed by untreated, or raw, biochar. The most common activating reagent used for CO_2 sorption is KOH. Activation by KOH is believed to result from the following reaction:

$$6KOH + 2C \rightarrow 2K + 3H_2 + 2K_2CO_3 \tag{2}$$

A caveat with the above reaction is that chemical activation with KOH occurs at 400 °C [36]. With increasing temperature, metal complexes can become gasified and, thereby, support further pore formation. Upon further temperature increases (e.g., 800 °C), K_2CO_3 and K_2O will react with carbon to form carbon monoxide (CO) as seen in equation [3] and [4]. The release of CO from the surface of the biochar will promote further pore production [36].

$$2K_2CO_3 + 2C \rightarrow 2K + CO \tag{3}$$

$$K_2O + C \rightarrow 2K + CO \tag{4}$$

Shi et al. [39] observed minimal changes to the surface functional groups of their microalgae-based biochar (*Chlorella* and *Spirulina*) following chemical modifications with urea and KOH. However, the adsorption was improved following chemical activation with KOH alone. The removal mechanism was attributed to physical adsorption as the KOH caused an increase in specific surface area (84.5× for *Chlorella* and 71× for *Spirulina*) and pore volume (28× for *Chlorella* and 23× for *Spirulina*). Liu and Huang [47] generated a hydrochar that possessed thermally stable N species, including pyridinic and pyrrolic nitrogen, following chemical treatment with ammoxidizing reagents and KOH activation. Aside from being responsible for a more thermally stable adsorbent, when melamine was used as the ammoxidizing reagent, the activation by KOH was responsible for the enhanced pore structure, which in turn resulted in an enhancement in adsorption capacities by up to 19× with ammoxidation-alone, as compared to 3× when no chemical treatment was used. According to Liu and Huang [47], because the heat of adsorption for analine and KOH-activated hydrochar was >40 kJ mol^{-1} (i.e., 49.9 kJ mol^{-1}), chemical sorption played a role in the removal of CO_2 as compared to the other chars treated with 3-aminopropyltrimethoxysilane and aniline, which displayed a heat of adsorption of <40 kJ mol^{-1} and were, therefore, principally controlled by physical adsorption. Although KOH is a promising activating reagent for CO_2 sorption, excessive activation can reverse the effects detrimentally and lead to the structural failure of the carbon wall [36]. Ding and Liu [36] found that the specific surface area and total pore volume following activation with KOH decreased with increasing KOH-biomass weight ratio from 1:1 to 4:1, with the latter exhibiting values similar to that of the untreated biochar.

2.2.4 Mercury

Mercury (Hg) is released into the environment from both natural and anthropogenic sources, with the latter accounting for 90% of all emissions [48]. Upon its release, Hg will readily distribute across all forms of environmental media. As a heavy metal, Hg is toxic and will bioaccumulate. In response to the global threat of Hg proliferation and its impact on the environment and human health, the Minimata Convention was formed to address Hg usage and controls across its entire life cycle. Currently, there are 128 signatories and 131 ratified parties who have signed into the convention [48]. Emissions of Hg can occur in the elemental form (Hg^0), ionic form (Hg^{2+}), or as particulates. Of these three forms, Hg^0 is relatively difficult to separate and remove from a waste gas stream owing to its volatility and low solubility in water. Untreated biochar has limited affinity to Hg^0, but surface modification with halides, sulfides, and metal oxides may improve their sorption capacities. Yang et al. [10] used NH_4Br to modify a seaweed biochar (i.e., *Sargassum*) and found that doping biochar decreased the surface area and pore volume; however, the removal efficiency increased substantially from <40% (i.e., untreated biochar prepared at a pyrolysis temperature of 800 °C) to >90% and >98% at a doping concentration of 5 and 9 wt%, respectively. This indicated that chemisorption played a more significant role than physisorption in the removal of Hg^0. Several factors were tested to determine whether they would have a positive impact on the removal efficiency of Hg^0, including reaction temperature, O_2 levels, SO_2 levels, NO levels, and water vapor – all of which were found to support Hg^0 removal. Quantitative estimates of the activation energy (i.e., −45.86 kJ mol^{-1}) and enthalpy (i.e., +18.31 kJ mol^{-1}) supported chemisorption as the mechanism of Hg^0 removal, and, therefore, the system benefited from higher reaction temperatures to drive the reaction forward. In contrast, a system that is controlled by physisorption would experience increasing levels of desorption as reaction temperature were to increase. Yang postulated that Hg^0 is removed through multiple routes at the surface of the biochar: (i) reaction with O_2 to form HgO, (ii) oxidation by carbonyl functional groups, (iii) oxidation by SO_3, derived from the oxidation of SO_2, to form $HgSO_4$, (iv) reaction between HgO and NO_2/NO^+, formed from the oxidation of NO, to form $Hg(NO_3)_2$ and HgO, (v) reaction with OH radicals, formed from the decomposition of water vapor, to form HgO, (vi) reaction with alkylated bromide to form $HgBr_2$, and (vii) reaction with active waste gas constituents. A caveat is that excess SO_2 and water vapor can reduce the oxidation of Hg^0 by competing for both O_2 and active sites.

Altaf et al. [49] prepared a magnetized biochar made of tea leaves with $Fe(NO_3)_3$ in order to separate the sorbent and sorbate from coal fly ash for the removal of Hg^0 in the presence of syngas. This is especially important for Hg^0 as the disposal of fly ash may lead to the secondary release of Hg back into the environment. Chemically modifying the biochar with $Fe(NO_3)_3$ successfully increased the specific surface area, pore volume, C=O bonds, and overall adsorption with an optimal doping concentration of 0.46 mol L^{-1}, where a removal efficiency of 90.6% was achieved as compared to 30% with the untreated biochar. The proposed mechanism of Hg^0 removal was as follows: (i) Hg^0 was adsorbed to the biochar surface; (ii) adsorbed Hg was oxidized by lattice oxygen on the biochar surface to HgO; and (iii) $H_2S_{(g)}$ adsorbed to the surface of the biochar was oxidized and subsequently reacted with HgO to form HgS. Fe_3O_4 served as an important molecule for the chemisorption and oxidation of Hg^0, essentially catalyzing the reaction. Because chemisorption was responsible for the separation and removal of Hg^0, increasing the reaction temperature to 180 °C promoted the oxidation of Hg^0 with iron oxides; however, above this temperature, at 240 and 300 °C, was counterproductive as desorption of Hg^0 from the biochar surface occurred. Further, in the presence of H_2S, which is commonly found in syngas, removal of Hg^0 was enhanced from 81.22% to ≥98% when the concentration of H_2S was 50 and 400 ppm, respectively. When H_2S was oxidized by surface O_2, the resulting oxidized sulfur species reacted with Hg^0 and accelerated their removal. Additionally, H_2S had a counteracting effect of reversing the loss of occupied surface-active sites that were chemically reduced by CO and H_2 – molecules found in syngas. Even water molecules, which occupied surface active sites in a pure N_2 environment (removal efficiency = 49%), were no longer inhibitory in the presence of syngas and removal efficiencies of 90% could be obtained.

Xu et al. [50] did not use a chemical modifier but were able to harness the sulfuric constituents found in wood and tire scrap. A 50/50 blend achieved higher removal efficiency than either wood or tire scrap alone, even though char from tire scrap displayed a higher BET surface area. This suggested that a synergy resulted from the blend where the removal of Hg^0 was controlled by chemisorption. For instance, 100% wood chars exhibited less active functional groups and physisorption was likely responsible for Hg^0 removal since the removal efficiency dropped with increasing reaction temperature. Higher levels of Hg^0 removal in the blended char were attributed to the organic sulfur and S^0 that were formed with increasing pyrolysis temperatures, up to 800 °C before decomposing at 900 °C, which were not present to a significant degree in either 100% wood or tire scrap chars. Both S^0 and organic sulfur served as active sites for the oxidation of Hg^0, as confirmed by XPS analysis. Increasing the reaction temperature to 140 °C led to higher removal efficiency by accelerating the chemical oxidation reactions, reaching 89.9% before declining to 30% at 220 °C due to desorption of Hg^0 from the surface of the char. Because chemisorption was responsible for Hg^0 removal in the blended char, chemical additions in the forms of dioxygen, nitric oxide, hydrochloric acid, and sulfur dioxide all helped to increase the removal efficiency.

2.3 Sorption and desorption

Up to this point, the focus of this chapter has been on the sorption effects of biochar for waste gas constituents. However, it is important to also consider the regeneration performance of the biochar for cyclic operation. Only some studies have considered the sorption capacity of biochar following regeneration. Ease of regeneration, frequency of adsorbent replacement, and costs are key factors when considering the long-term renewability of biochar. A number of feedstocks have shown good regeneration performance. For instance, Shi et al. [39] observed a drop in sorption potential of only 10% following ten cycles of regeneration using a microalgae-based biochar. Similarly, seaweed biomass has shown only a 13–15% decline in performance following ten cycles [36]. Rice straw biochar experienced an 8–10% reduction in sorption capacity following eight regeneration cycles [37]. Hydrochar prepared from coffee grounds experienced only a 6% drop following ten cycles [47]. Using an iron-doped tea leaf-based biochar, Altaf et al. [49] was able to demonstrate good regeneration performance as total removal efficiency of Hg^0 remained >90% after six consecutive adsorption/desorption cycles.

On the other hand, Han et al. [2] found that increasing the calcination temperature of biochar can improve the physical properties of the biochar such that they are amenable to adsorption, including increasing the surface area and pore size – both of which can lead to longer breakthrough time, but after two times of thermal regeneration, there was a reduction in sorption capacity for H_2S by 40–60%. Chen et al. [44] observed a drop in performance to 40% and 50% for H_2S by their chemically modified seaweed biochar following the second desorption cycle. The drop in performance was believed to be caused by a breakdown in pore structure leading to lower active sorption sites. Zhang et al. [18] has suggested that a reduction in sorption capacity may be due to incomplete desorption for compounds that have strong affinity toward the sorbent.

The isoteric heat of adsorption can be determined to estimate the strength of the interaction between the sorbate and the sorbent. A lower value (e.g., <40 kJ mol^{-1} for CO_2 [47]) suggests physical adsorption may be the dominant interaction while the reverse suggests chemical adsorption has taken place. Since the interaction between the sorbate and sorbent are weaker if the process is dominated by physical adsorption than by chemical adsorption, then the energy associated with the isoteric heat of adsorption would correspond accordingly. Further, a lower heat of adsorption would indicate that the regeneration of the biochar can be performed at relative ease.

3 Conclusion

This chapter described the individual removal of a compound or element in waste gas stream using biochar as a sorbent; however, in a waste gas stream, several constituents will likely be present simultaneously that may interfere negatively or positively with their separation and removal. For instance, SO_2 may interfere with the separation of CO_2 by competing for sorption sites [36]. Likewise, desulfurization of H_2S may be negatively impacted by the presence of SO_2, NO, and CO_2 due to competition for active sites [44]. In some instances, a small amount of humidity (i.e., $H_2O_{(g)}$) can enhance the performance of the system (e.g., ionizing H_2S into HS^- ions, which are in turn oxidized more readily). Few studies have thus far pursued the potential interactions between multiple chemical constituents and their effect on the separation and removal of each constituent. Future studies should consider implementing factorial experimental design to examine the effects of multiple factors within a complete trial to identify the main contributing factors and possible interactions between two or more physicochemical components (e.g., pyrolysis temperature and modifying agent), elements, compounds, and media. The results of the analysis could then be interpreted through a quantitative regression model for field implementation.

References

[1] Xia G, Zhou X, Hu J, Sun Z, Yao J, Chen D, Wang J, Simultaneous removal of carbon disulfide and hydrogen sulfide from viscose fibre waste gas with a biotrickling filter in pilot scale, J. Clean. Prod., 2019, 230, 21–28.
[2] Han X, Chen H, Liu Y, Pan J, Study on removal of gaseous hydrogen sulfide based on macroalgae biochars, J. Nat. Gas Sci. Eng., 2020, 73, 103068.
[3] Takaya CA, Parmar KR, Fletcher LA, Ross AB, Biomass-derived carbonaceous adsorbents for trapping ammonia, Agriculture, 2019, 9, 16.
[4] IPCC., Climate change 2014: Synthesis report contribution of working groups i, ii and iii to the fifth assessment report of the intergovernmental panel on climate change, Pachauri RK, Meyer LA, eds Geneva, Switzerland, 2014, 151.
[5] De Guardia A, Petiot C, Rogeau D, Druilhe C, Influence of aeration rate on nitrogen dynamics during composting, Waste Manage., 2008, 28, 575–587.
[6] Giang A, Stokes LC, Streets DG, Corbitt ES, Selin NE, Impacts of the minamata convention on mercury emissions and global deposition from coal-fired power generation in Asia, Environ. Sci. Technol., 2015, 49, 5326–5335.
[7] Qian K, Kumar A, Zhang H, Bellmer D, Huhnke R, Recent advances in utilization of biochar, Renewable Sustainable Energy Rev., 2015, 42, 1055–1064.
[8] Lee Y, Park J, Ryu C, Gang KS, Yang W, Park Y-K, Jung J, Hyun S, Comparison of biochar properties from biomass residues produced by slow pyrolysis at 500°C, Bioresour. Technol., 2013, 148, 196–201.

[9] Keiluweit M, Nico PS, Johnson MG, Kleber M, Dynamic Molecular Structure of Plant Biomass-Derived Black Carbon (Biochar), Environ. Sci. Technol., 2010, 44, 1247–1253.
[10] Yang W, Liu Z, Xu W, Liu Y, Removal of elemental mercury from flue gas using sargassum chars modified by NH4Br reagent, Fuel, 2018, 214, 196–206.
[11] Wang Z, Han L, Sun K, Jin J, Ro KS, Libra JA, Liu X, Xing B, Sorption of four hydrophobic organic contaminants by biochars derived from maize straw, wood dust and swine manure at different pyrolytic temperatures, Chemosphere, 2016, 144, 285–291.
[12] Xiang W, Zhang X, Chen J, Zou W, He F, Hu X, Tsang DCW, Ok YS, Gao B, Biochar technology in wastewater treatment: A critical review, Chemosphere, 2020, 252, 126539.
[13] Zhang X, Gao B, Zheng Y, Hu X, Creamer AE, Annable MD, Li Y, Biochar for volatile organic compound (VOC) removal: Sorption performance and governing mechanisms, Bioresour. Technol., 2017, 245, 606–614.
[14] Jeon J, Kim H-I, Park JH, Wi S, Kim S, Evaluation of thermal properties and acetaldehyde adsorption performance of sustainable composites using waste wood and biochar, Environ. Res., 2021, 196, 110910.
[15] Kumar A, Singh E, Khapre A, Bordoloi N, Kumar S, Sorption of volatile organic compounds on non-activated biochar, Bioresour. Technol., 2020, 297, 122469.
[16] Rajabi H, Mosleh MH, Mandal P, Lea-Langton A, Sedighi M, Sorption behaviour of xylene isomers on biochar from a range of feedstock, Chemosphere, 2021, 268, 129310.
[17] Zhao L, Cao X, Mašek O, Zimmerman A, Heterogeneity of biochar properties as a function of feedstock sources and production temperatures, J. Hazard. Mater., 2013, 256–257, 1–9.
[18] Zhang X, Miao X, Xiang W, Zhang J, Cao C, Wang H, Hu X, Gao B, Ball milling biochar with ammonia hydroxide or hydrogen peroxide enhances its adsorption of phenyl volatile organic compounds (VOCs), J. Hazard. Mater., 2021, 403, 123540.
[19] Lyu H, Gao B, He F, Zimmerman AR, Ding C, Huang H, Tang J, Effects of ball milling on the physicochemical and sorptive properties of biochar: Experimental observations and governing mechanisms, Environ. Pollut., 2018, 233, 54–63.
[20] Qin L, Xu Z, Liu L, Lu H, Wan Y, Xue Q, In-situ biodegradation of volatile organic compounds in landfill by sewage sludge modified waste-char, Waste Manage., 2020, 105, 317–327.
[21] IPCC, Climate change 2013: The physical science basis. Contribution of working group i to the fifth assessment report of the intergovernmental panel on climate change, In: Stocker TF, Qin D, Plattner G-K, Tignor M, Allen SK, Boschung J, Nauels A, Xia Y, Bex V, Midgley PM, eds Cambridge University Press, Cambridge, United Kingdom and New York, NY, USA, 2013, 1535.
[22] Reddy KR, Yargicoglu EN, Chetri JK, Effects of Biochar on Methane Oxidation and Properties of Landfill Cover Soil: Long-Term Column Incubation Tests, J. Environ. Eng., 2021, 147, 04020144.
[23] Li M-Y, Sun W-J, Wang Y-J, Sun D-A, Tan Y-Z, Air permeability of biochar-amended clay cover, Arabian J. Geosci., 2021, 14, 732.
[24] Cui J, Glatzel S, Bruckman VJ, Wang B, Lai DYF, Long-term effects of biochar application on greenhouse gas production and microbial community in temperate forest soils under increasing temperature, Sci. Total Environ., 2021, 767, 145021.
[25] Di Perta ES, Giudicianni P, Mautone A, Caro S, Cervelli E, Ragucci R, Pindozzi S Is the biochar an effective floating cover for manure storage to reduce ammonia emissions, adsorbing nitrogen at the same time?: 2020 IEEE International Workshop on Metrology for Agriculture and Forestry (MetroAgriFor). 2020.
[26] Chen H, Awasthi SK, Liu T, Duan Y, Ren X, Zhang Z, Pandey A, Awasthi MK, Effects of microbial culture and chicken manure biochar on compost maturity and greenhouse gas emissions during chicken manure composting, J. Hazard. Mater., 2020, 389, 121908.

[27] Wolter M, Prayitno S, Schuchardt F, Greenhouse gas emission during storage of pig manure on a pilot scale, Bioresour. Technol., 2004, 95, 235–244.
[28] Yang B, Ma Y, Xiong Z, Effects of different composting strategies on methane, nitrous oxide, and carbon dioxide emissions and nutrient loss during small-scale anaerobic composting, Environ. Sci. Pollut. Res., 2019, 26, 446–455.
[29] Sánchez-Monedero MA, Sánchez-García M, Alburquerque JA, Cayuela ML, Biochar reduces volatile organic compounds generated during chicken manure composting, Bioresour. Technol., 2019, 288, 121584.
[30] Meiirkhanuly Z, Koziel JA, Chen B, Białowiec A, Lee M, Wi J, Banik C, Brown RC, Bakshi S, Mitigation of gaseous emissions from swine manure with the surficial application of biochars, Atmosphere, 2020, 11, 1179.
[31] Chen W, Liao X, Wu Y, Liang JB, Mi J, Huang J, Zhang H, Wu Y, Qiao Z, Li X, Wang Y, Effects of different types of biochar on methane and ammonia mitigation during layer manure composting, Waste Manage., 2017, 61, 506–515.
[32] Guo H, Gu J, Wang X, Song Z, Yu J, Lei L, Microbial mechanisms related to the effects of bamboo charcoal and bamboo vinegar on the degradation of organic matter and methane emissions during composting, Environ. Pollut., 2021, 272, 116013.
[33] Qi L, Ma Z, Chang SX, Zhou P, Huang R, Wang Y, Wang Z, Gao M, Biochar decreases methanogenic archaea abundance and methane emissions in a flooded paddy soil, Sci. Total Environ., 2021, 752, 141958.
[34] Cao Y, Shan Y, Wu P, Zhang P, Zhang Z, Zhao F, Zhu T, Mitigating the global warming potential of rice paddy fields by straw and straw-derived biochar amendments, Geoderma, 2021, 396, 115081.
[35] Chun Y, Sheng G, Chiou CT, Xing B, Compositions and sorptive properties of crop residue-derived chars, Environ. Sci. Technol., 2004, 38, 4649–4655.
[36] Ding S, Liu Y, Adsorption of CO2 from flue gas by novel seaweed-based KOH-activated porous biochars, Fuel, 2020, 260, 116382.
[37] Huang Y-F, Chiueh P-T, Shih C-H, Lo S-L, Sun L, Zhong Y, Qiu C, Microwave pyrolysis of rice straw to produce biochar as an adsorbent for CO2 capture, Energy, 2015, 84, 75–82.
[38] Sun H, Yang B, Li A, Biomass derived porous carbon for efficient capture of carbon dioxide, organic contaminants and volatile iodine with exceptionally high uptake, Chem. Eng. J., 2019, 372, 65–73.
[39] Shi S, Ochedi FO, Yu J, Liu Y, Porous Biochars Derived from Microalgae Pyrolysis for CO2 Adsorption, Energy & Fuels, 2021.
[40] Das J, Rene ER, Dupont C, Dufourny A, Blin J, Van Hullebusch ED, Performance of a compost and biochar packed biofilter for gas-phase hydrogen sulfide removal, Bioresour. Technol., 2019, 273, 581–591.
[41] Choudhury A, Lansing S, Adsorption of hydrogen sulfide in biogas using a novel iron-impregnated biochar scrubbing system, J. Environ. Chem. Eng., 2021, 9, 104837.
[42] Yuan J-H, Xu R-K, Zhang H, The forms of alkalis in the biochar produced from crop residues at different temperatures, Bioresour. Technol., 2011, 102, 3488–3497.
[43] Kanjanarong J, Giri BS, Jaisi DP, Oliveira FR, Boonsawang P, Chaiprapat S, Singh RS, Balakrishna A, Khanal SK, Removal of hydrogen sulfide generated during anaerobic treatment of sulfate-laden wastewater using biochar: Evaluation of efficiency and mechanisms, Bioresour. Technol., 2017, 234, 115–121.
[44] Chen H, Han X, Liu Y, Gaseous hydrogen sulfide removal using macroalgae biochars modified synergistically by H2SO4/H2O2, Chem. Eng. Tech., 2021, 44, 698–709.
[45] Perazzolo F, Mattachini G, Tambone F, Calcante A, Provolo G, Nutrient losses from cattle co-digestate slurry during storage, J. Agric. Eng., 2016, 47, 94–99.

[46] Covali P, Raave H, Escuer-Gatius J, Kaasik A, Tõnutare T, Astover A, The Effect of Untreated and Acidified Biochar on NH3-N Emissions from Slurry Digestate, Sustainability, 2021, 13, 837.
[47] Liu S-H, Huang -Y-Y, Valorization of coffee grounds to biochar-derived adsorbents for CO2 adsorption, J. Clean. Prod., 2018, 175, 354–360.
[48] UN Environment Programme. Minimata Convention on Mercury. 2021. https://www.mercury convention.org/)
[49] Altaf AR, Teng H, Zheng M, Ashraf I, Arsalan M, Rehman AU, Gang L, Pengjie W, Yongqiang R, Xiaoyu L, One-step synthesis of renewable magnetic tea-biochar derived from waste tea leaves for the removal of Hg0 from coal-syngas, J. Environ. Chem. Eng., 2021, 9, 105313.
[50] Xu Y, Luo G, Zhang Q, Li Z, Zhang S, Cui W, Cost-effective sulfurized sorbents derived from one-step pyrolysis of wood and scrap tire for elemental mercury removal from flue gas, Fuel, 2021, 285, 119221.

Palas Samanta, Sukhendu Dey, Jinho Jung, Apurba Ratan Ghosh

Mechanism of metal sorption by biochar

Abstract: Biochar has received potential interest as a cost-efficient and environment-friendly sorbent material to remediate metal contamination recently. However, the mechanism of metal sorption capacity by agriculture-based biochar is still lacking. Accordingly, this review chapter demonstrates metal sorption mechanisms by biochar including physical binding, ion exchange, membrane filtration, complexation, precipitation, sorption, and electrostatic interactions and elucidate the role of corresponding biochar characteristics namely biochar preparation methods, surface area or charge, porosity, medium pH, presence of functional groups, mineral components, and pyrolysis temperature. Additionally, this review chapter addresses different techniques, namely steam or acid/base activation, composite impregnation with carbon-rich materials, minerals, organic compounds, and so on to improve metal sorption capacity either through functional improvement or providing efficient surface attributes to biochar. Furthermore, this chapter describes different mathematical models to check the metal sorption efficiency by biochar. Finally, this chapter highlights the future prospects of biochar-associated metal sorption in large-scale field application cost-effectively.

Keywords: biochar, metal sorption, functional groups, pyrolysis, complexation

1 Introduction

Heavy metals, which are ubiquitously present in natural environment, are considered as environmental pollutants and caused adverse impacts on public health because of nonbiodegradability and toxicity [1]. They are discharged into the natural environment from diverse natural and anthropogenic origins including smelting and mining activities, fertilizer and pesticide production units, and different manufacturing industries. As a result, the environment is deteriorating adversely and imposed toxic impacts on each environmental segment. Additionally, the bioaccumulation property of heavy metals due to their nondegradable nature has accelerated the adverse impacts [2].

Palas Samanta, Department of Environmental Science, Sukanta Mahavidyalaya, University of North Bengal, West Bengal, India, e-mail: samanta.palas2010@gmail.com
Sukhendu Dey, Apurba Ratan Ghosh, Department of Environmental Science, The University of Burdwan, West Bengal, India
Jinho Jung, Division of Environmental Science and Ecological Engineering, Korea University, Seoul, Republic of Korea

https://doi.org/10.1515/9783110734003-006

Consequently, different treatment techniques (biological and/or physicochemical) are very much needed to remediate metal from contaminated environment. In this regard, coagulation-flocculation, chemical-mediated precipitation, reverse osmosis (RO), ion exchange, adsorption, ozonation, membrane filtration, electrocoagulation, electrochemical treatment technologies, and biosorption are widely used techniques to immobilize/remediate the metal level [1, 2]. However, due to high operational cost and low efficiency their applications are restricted for removal of metals from contaminated environment. Biosorption method among these techniques is very convenient and emerged as very cost-effective method as biosorbent materials are environment-friendly and readily available in natural environment. Recently, "biochar" derived from biomaterials is considered as a main alternative agent for metal removal.

Biochar, a fine-grained carbon-rich porous substance, is generally synthesized by thermally induced decomposition of biological materials in oxygen-limited environment at < 900 °C temperature, called pyrolysis [3]. Highly porous structure and high cation exchange capacity due to the presence of hydroxyl, phenolic, and carboxyl, make it an excellent sorbent substrate for carbon sequestration (CS), soil fertilization, contaminants removal (organic/inorganic) including toxic metals, and even used as biofuel [3]. Accordingly, biochars have been produced from a diverse biomass specifically agricultural waste, sewage sludge, dairy manure, bunches of empty fruits, manures (human, poultry, swine, goat, etc.), and wood waste [4]. However, the biochar potentiality to remediate metal-contamination level is dependent upon biomass type, pyrolytic conditions (heating rate, temperature and residence time), and biochar modification techniques as well. Herein, we consider the novel concept of using agriculture waste-derived biochar (AWDBC) for metal sorption from different contaminated environment. Till now, few scientists demonstrated that AWDBC have higher metal sorption capacity compared to the existing remediation techniques [5, 6]. However, the use of agricultural wastes as biochar in treating different pollutants, particularly heavy metals, has gained increasing attention although there is a lack of available information as cost-efficient and environment-friendly sorbent materials.

In this review, we are focused to: (1) delineate different agriculture waste materials and biochar production methods in order to enhance metal sorption understanding, (2) describe different characteristics of biochar, (3) discuss the metal sorption mechanisms on biochar, (4) describe different factors influencing metal sorption mechanism, (5) unravel the advances on biochar modification/activation techniques, and (6) explore future research directions.

2 Source materials and biochar preparation methods

Biochars can be produced from diverse biomasses that have diverse physicochemical characteristics, which includes forest and agriculture-based materials, industrial wastes and its by-products, domestic-origin solid wastes, and nonconventional waste substances (such as waste tires, papers, newspaper scraps, food wastes, plastics, bones, and invasive plants). Recently, the use of agriculture-based waste materials for biochar production has gained focal interest by scientific community worldwide due to abundant availability and cost-effectivity [1, 7]. It has been demonstrated that around 500 million tons of agricultural-based wastes are generated yearly worldwide [1]. These waste materials are generally originated from harvesting and processing of crops like rice, corn, wheat, oil palm, sugarcane, peanut, and sorghum including yard, wood, or farm or domestic animal wastes (manure and litter such as bedding materials, spilled feed, and feathers) [1].

Different thermochemical processes are used for the preparation of biochar, including biofuel, syngas and other bio-based products. Table 1 represents different conventional techniques for biochar preparation from a variety of biomass materials. Conventionally, three techniques that are used for biochar production are summarized in Figure 1.

Table 1: The characteristic features of different conventional biochar preparation techniques.

Production process	Temperature (°C)	Heating rate	Reaction (residence) time	Biochar or solid (%)	Liquid or bio-oil (%)	Gas or syngas (%)
Pyrolysis						
Slow	<700	Slow (0.01–2 °C s^{-1})	Few hours to days	35	30	35
Fast	<1,000	Fast (>2 °C s^{-1})	Seconds (< 2 s)	10	70	20
Flash	775–1,025	Faster	Seconds (< 2 s)	10–15	70–80	5–20
Gasification	700–1,500	Faster	Seconds	10	5	85
Carbonization						
hydrothermal	<350	Slow	Min. to hours		50–80	
Flash	<700	Faster	Seconds	50	50	

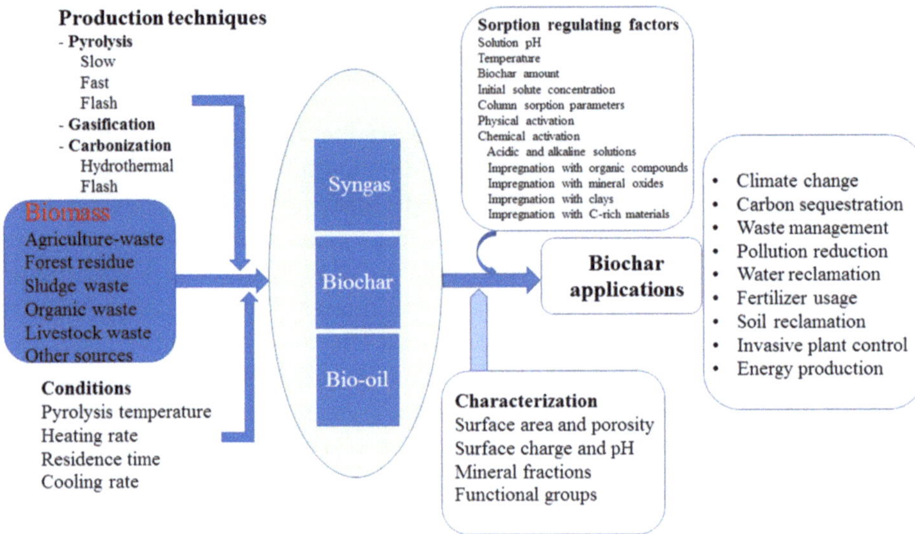

Figure 1: Biochar production techniques from different biomass and their applications including regulating factors.

2.1 Pyrolysis

Pyrolysis technique is dominant as well as the most preferred method for biochar preparation. In pyrolysis technique, the waste materials are decomposed at higher temperature under limited oxygen condition to produce biochar. Generally, three pyrolysis techniques are used based on temperature and reaction time for biochar preparation, namely slow, fast, and flash pyrolysis [8, 9].

Slow pyrolysis process is generally employed at low temperature (300–700 °C) condition under longer residence time (few hours to days) and the process produced equal amount of solid, liquid, and gaseous end-products. Additionally, it needs high pressure and low heating rates (0.01–2 °C s^{-1}), which is supplied from sources like external heaters and hot-gas recirculation process. Further, diverse reactors are employed for biochar production, for instance, agitated drum sand rotating kilns with screw pyrolyzers. This process produced high amount of large-sized biochar with higher ash and lignin contents because of increased cracking process, which suppresses the generation of liquid proportion [1]. The fast pyrolysis, on the other hand, is conducted at very short residence time (<2 s) and fast heating rates (>2 °C s^{-1}) under higher temperature (500–1,000 °C). This process produces a low proportion of biochar (12%) and gases (noncondensable, 13%) but generates high amount of bio-oil (75%). Fast pyrolysis derived biochar contains hard aromatic ingredients, but the process conducted under lower temperature produced large volume of partly pyrolyzed biochar materials [1]. Flash pyrolysis is generally employed at very short residence time (<2 s) but requires higher temperature (775–1,025 °C) and rapid heating rates. Flash pyrolysis process produces lesser amount of biochar (10–15%) and noncondensable gases (5–20%) but generates higher amount of liquids as bio-oil (70–80%). Comparatively, the solid end-product production in slow pyrolysis is higher (about 35%) than other two pyrolysis techniques, and accordingly slow pyrolysis among three pyrolysis techniques is considered as a dominant biochar production technique [1, 9].

2.2 Hydrothermal carbonization

Hydrothermal carbonization (HTC) is another important biochar preparation technique. This method is characterized by high efficiency and widely applied for a range of materials. This process is generally employed at high residence times (minutes to hours), at low temperature (<350 °C), and slow heating rates and produces higher amount of biochar (50–80%). HTC method requires less energy as it does not require biomass drying like slow pyrolysis. Additionally, the process doesn't produce any harmful materials/end-products. Further, HTC-derived biochars are more stable and suitable for wide range of contaminant removal from contaminated environment [9]. As a result, the hydrochar from HTC method can enhance biogas production and soil remediation. However, the major drawback of this method is that it requires high-pressure concomitant to high temperature and higher reactor cost due to time-consuming process. Accordingly, further research is needed for coupling HTC with anaerobic digestion process to treat pollutants for large-scale applications [9]. In addition to this, recently *flash carbonization technique* is used to produce biochar. *Flash carbonization technique* uses moderate temperature and heating rate to produce equal proportion of biochar (50%) and syngas (50%).

2.3 Gasification

Gasification technique is characterized by less residence times (seconds to minutes), high temperatures (700–1,500 °C), and rapid heating rates. In gasification process, the feedstock materials are converted into CO (carbon monoxide), CO_2 (carbon dioxide), H_2 (hydrogen), and CH_4 (methane) under carefully managed oxygen-level environment. The process produced only 10% solid biochar, 5% liquid (also called "tar" and contains aromatic hydrocarbons), and 85% gaseous mixture. The gaseous mixture mainly contains syngas having CO, H_2, CO_2, CH_4, C_2H_6 (ethane), C_2H_2 (acetylene), and N_2 (nitrogen) in air-rich oxidizing conditions [1]. Hu et al. [7] demonstrated that 823 °C temperature is the optimal biochar production temperature under gasification technique. This temperature enhances the biochar characteristics such as biochar's surface area, cation exchange capability, porosity, and carbon content.

3 Biochar's properties influencing metal sorption

Biochar properties depend upon the type of materials used, type of thermal conversion technology used, thermal conversion temperature, residence time, heating process, and so on (see Section 2 for details). Biochar properties (physical/chemical) significantly influenced their performances of metal sorption. Therefore, before elucidating the biochar's mechanisms of metal sorption, it is very essential to understand the biochar properties including porosity, surface area, biochar's surface charge, pH, role of functional units, and mineral fractions.

3.1 Surface area and porosity

Biochar's surface area and porosity are prime physical attributes which directly regulates metal sorption capability to reduce the contamination level. During biomass conversion (pyrolysis or HTC), micropores are produced within biochar structure due to water loss during dehydration step [1, 2]. Biochar pore size is highly variable, which includes macropores (>50 nm), micropores (<2 nm) and nanopores (<0.9 nm). The pore size plays a crucial role during sorption of metals, for example, small-pore-sized biochars are not able to capture big-sized sorbate, regardless of their surface polarity or charges [1]. Further, the biochar porosity and surface area increased proportionally with pyrolysis temperature, that is, elevated pyrolysis temperature increased the generation of small pore sized biochar and subsequently increases the surface area [2]. For instance, the biochar's surface area and porosity increased from 25.4 to 67.6 $m^2\ g^{-1}$ and 0.056 to 0.099 $cm^3\ g^{-1}$, respectively, with increasing temperature from 500 to 900 °C [10]. However, in some cases, the biochar's surface area and porosity reduced due to elevation of pyrolysis temperature. Jin et al. [11] recorded reduced surface area at 600 °C compared with 550 °C (5.99 vs 8.45 $m^2\ g^{-1}$) and opined that this reduction was resulted either from biochar's porous structure destruction or by tar production. Apart from this, biochar porosity and surface area depend on type of feedstock materials used. For instance, the biosolid and manure-derived biochar's surface area and porosity (5.4–94.2 $m^2\ g^{-1}$ and 0.053–0.068 $cm^3\ g^{-1}$) at temperature 500–700 °C are much smaller than that of the plant-derived biochar (112–642 $m^2\ g^{-1}$ and 0.076–1.90 $cm^3\ g^{-1}$) such as wheat, oak wood, corn stover, and pine needle. Generally, lignin-rich biomass (e.g., coconut shell and bamboo) produced predominantly macroporous-sized biochar, while cellulose-rich biomass (e.g., husks) produced microporous-sized biochar predominantly [2].

3.2 Surface charge and pH

Biochar's surface charge plays a crucial role during metal sorption. Generally, solution pH influenced the surface properties of sorbent as well as speciation and finally, the degree of metal sorption. The pH also regulates the adsorbent's protonation degree, thus controlling the functional group's specific charge, and finally, adsorbent's sorption performances. Generally, the cationic metal ions' uptake capability in lower pH conditions is very low because of strong competition from H^+ ions. On the other hand, the cationic metal ions' uptake capability increased several folds during pH surges as H^+ ionic concentrations decline in the medium. The biochar's pH_{PZC} (point of zero charge) denotes the pH at which biochar's surface net charge is zero. Generally, the biochar binds to metal cations when it is negatively charged ($pH > pH_{PZC}$) and binds to metal anions when biochar is positively charged ($pH < pH_{PZC}$). Chen et al. [10] demonstrated that biosolid derive biochar showed elevated pH_{PZC} value from 8.58 to 10.2 due to conversion temperature enhancement from 500 to 900 °C. Yuan et al. [12] observed that corn, canola, peanut straw and soybean derived biochars produced at 300 and 500 °C were negatively charged at pH 3–7, while those produced at 700 °C were less negatively charged, suggesting that pyrolysis temperature increment enhances pH_{PZC}. They opined that this is generally due to presence of lower number of functional units (negative charged) on biochar's surface, namely –OH, $-COO^-$, and –COH.

Likely, biochar pH varies significantly with biomass type and pyrolysis temperature. Generally, biochar is alkaline in nature but in some exceptions, it is acidic in nature depending on feedstock. Nguyen et al. [13] demonstrated that biochar produced at 350 and 600 °C from oak wood was acidic (4.84–4.91). In addition, biochars produced at low temperature (200–400 °C) from wheat straw, corn stover and wastewater sludge were also acidic, pH ranges from 4.87 to 6.11 [13, 14]. Generally, the pH of biochar increased proportionally with temperature elevation (pyrolysis). Similar positive relationships between biochar pH and pyrolysis temperature for biochars yielded from oak wood, wheat, corn, biosolids, manure, maize residues, and *Conocarpus* wastes have been reported by several authors [11, 13, 14]. Generally, enhanced pyrolysis temperature yields higher amount of ash, which seems to be contributing factor for higher pH [11]. In addition to this, temperature elevation enhanced base cations and carbonates levels in biochar, which ultimately contributed to higher pH [12]. Moreover, this elevated pH value at higher temperature could be due to loss of acidic functional units like –COOH. For instance, in biochar produced from *Conocarpus* wastes at elevated temperature the amount of basic functional units enhanced to 3.55 mmol g^{-1} from 0.15 mmol g^{-1}, while acidic functional units reduced to 0.22 mmol g^{-1} from 4.17 mmol g^{-1}, the findings are consistent with biochar's pH elevation to 12.4 from 7.37 [15].

3.3 Mineral fractions

Apart from functional units, biochar contains different mineral components, including sodium (Na), potassium (K), calcium (Ca), magnesium (Mg), and phosphorus (P) as oxides, phosphates, and carbonate complexes [16]. Mineral fractions also played a vital role for metal sorption either through precipitation or exchange with metals, accordingly regulating their application in diverse directions. Still, they are given less importance compared with carbon because biomass materials are predominantly plant-based that contain fewer mineral proportions [1]. The presence of minerals within biochar generally acts as a direct source of plant growth attributes and the removal of anionic contaminants including OH^-, CO_3^{2-}, SO_4^{2-}, and PO_4^{3-} helps to reduce heavy metal levels. However, the biochar produced without deashing contains higher level of toxic heavy metals and this could adversely affect the human health and have the potentiality to lower the sorption potentiality for organic substances [9].

For biochar that contains lower amount of mineral fractions, for instance residue-derived biochar, the sorption of heavy metals occurred predominantly by surface complexation process [1]. This complexation is generally employed through oxygen-rich functional units like –OH and –COOH [1]. On the contrary, for biochar having high mineral fractions produced from sludge and manure, the heavy metals sorption occurred predominantly by precipitation technique between proportions of minerals and heavy metals, and removed 90% heavy metals [15]. Biochar enriched with sulfur ingredients removes toxic metals from contaminated water by complex formation. On the other hand, sulfur–carbon-rich substance-derived biochars are reducing metal (Pb, Cd, Cu) levels in a selective way through tight attaching of metal sulfides (sparingly soluble) on their surface [17]. Generally, sorption of heavy metals onto carbon–sulfur material-based biochar is dependent upon biochar's S-containing thiol unit.

Likely, biochar mineral fraction is primarily dependent on the biomass type and pyrolysis temperature. Biomass type is generally regulating mineral fractions in biochar. For example, P level (0.03–0.06%) in oak wood–derived biochar is lower than biosolid-derived biochar (1.82–3.60%), poultry litter and swine manure biochar (1.6–5.9%) [10, 14]. Besides biomass type, the concentration of different mineral fractions (K, Mg, Ca, and P) is increased proportionally with temperature elevation of conversion techniques (mainly pyrolysis) for biochar produced from manure, biosolids, oak wood, *Conocarpus* wastes, and so on [10, 14, 15]. Generally, the biochar production is lower at temperature elevation, thereby at elevated temperature mineral fraction in biochar is enriched. In addition to this, water-soluble mineral fractions (K, Mg, Ca, and P) enhanced when biomass are converted at 200 °C but after that water-soluble mineral fraction production decreased. This phenomenon is basically because of either enhanced crystallization as evidenced by whitlockite [$(Ca, Mg)_3(PO_4)_2$] formation (less soluble) or its incorporation into silicon structure at 500 °C [2].

3.4 Functional groups/units

Functional units namely hydroxyl, amino, or carboxylic units play vital roles during sorption of metals. Similar to mineral fractions, temperature (pyrolytic) and biomass type are prime determinants to control the presence of functional units on biochar surface. Generally, the number (or amount) of functional units on biochar surface is reduced with elevation of pyrolytic temperature. This is probably due to atomic ratios (O/C, H/C, and N/C) reduction, resulting from higher degree of carbonization. The pyrolysis of lignocellulosic materials at elevated temperature recorded almost complete loss of functional units. Yuan et al. [12] recorded decreased peak intensity corresponding to hydroxyl (–OH) and carboxylic (–COOH) groups of canola/peanut straw/corn/soybean-derived biochars due to temperature increment from 300 to 700 °C. Loss of Fourier transform infrared spectroscopy (FTIR) spectral features were also noticed for biochars (manure-derived) produced at 700 °C compared with 350 °C [2]. Generally, the biochars produced at low temperature (100–200 °C) did not show any FTIR spectra differences, indicated that low temperature has no impact on functional units. In addition to this, dehydration of ligneous and cellulosic components of biomass started at 300 °C (at wavenumber, 3,500–3,200 cm^{-1}) whereas hemicellulose transformation started at 400 °C (multiple peaks at 1,600–700 cm^{-1}). Moreover, an elevated condensation degree was noticed at charring temperature 500 °C (intensity loss at 1,650–1,500 cm^{-1} relative to 885–752 cm^{-1}) [2].

On the other hand, Li et al. [18] produced biochars (straw and bran) through dehydroxylation/dehydrogenation followed by aromatization technique at 100–800 °C and observed the yield of functional groups on respective biochars using two-dimensional ^{13}C nuclear magnetic resonance (NMR) spectroscopy. Generally, anomeric O–C–O carbons and O-alkylated groups are formed during initial stage with increasing pyrolysis temperature prior to formation of aromatic structures. More specifically, the biochars formed at lower temperature (<300 °C) produced aliphatic O-alkylated carbons predominantly, and beyond that temperature (>300 °C) the aliphatic O-alkylated carbon property of biochar lost, and on time course aromatic property of structures is dominant [18]. For example, Zhang et al. [14] reported reduced O-alkyl carbon group (20–54% to 6.9–13%) in biochars derived from lignosulfonate and wheat straw due to temperature elevation to 600 °C from 200 °C.

4 Mechanisms of metal sorption by biochar

Different conventional technologies including their advantages and disadvantages have been employed for removal of toxic metals from contaminated environment, in particular aqueous medium (Table 2). The predominant mechanisms governing the metal sorption over conventional techniques by biochar from solution phase are surface sorption, cation/ion exchange, electrostatic interactions (chemisorption), complexation, and precipitation (Figure 2). However, the role of each mechanism is totally different to remove metals from aqueous solution, that is, metal-specific. Here, we reviewed the different sorption mechanisms for metals as a whole in the following sections:

Table 2: Different conventional techniques of metal removal, including their advantages and disadvantages.

Techniques	Advantages	Disadvantages
Membrane filtration	– Highly selective separation – Need small space – Less pressure required	– Huge operation cost – Process is very complex – Lower permeate flux output
Coagulation and flocculation	– Solve turbidity problems – More sludge formation – Sludge has de-watering features	– Greater sludge volume – Not able to remove metals from wastewater
Electrodialysis	– Highly selective separation	– Huge operation cost very – Need high energy – Fouling of membrane
Flotation	– Higher specificity for metals – Great removal efficiency – Higher rates of overflow – Short detention time – Highly cost-effectiveness	– Huge installation cost – Maintenance cost very high – Huge operation cost
Photocatalysis	– Efficient for metals and organics – Byproducts are nontoxic	– Restricted applications – Need high duration time
Chemical precipitation	– Investment cost very low – Easy operation	– Form sludge – Limits for low metal concentrations – Extra cost for sludge dumping

4.1 Surface sorption

Surface sorption, also called physical sorption, is a physical sorption technique that involves the diffusion movement of metal ions via formation of chemical bonds through pores/cavities of sorbents. The biochar's (sorbent) surface area and pore volume depend upon the carbonization temperature as well as biomass type. Generally, biochars produced from organic wastes, namely, plants and animals possess higher surface area/pore volume with temperature elevation. For example, biochars produced from pine wood and switch grass at 300 and 700 °C have the capability to remove Cu and uranium (U) by sorption technique [19]. In particular, the biochar developed at 700 °C (high temperature) entirely removes Cu and uranium in

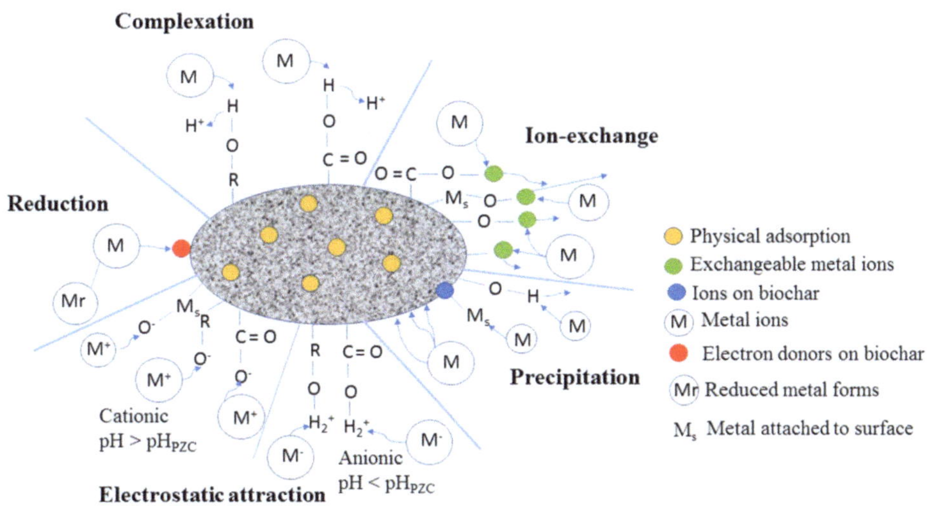

Figure 2: Metal sorption mechanisms by biochar (modified after Li et al. [2], Ambaye et al. [9] and Shakoor et al. [1]).

comparison to biochar produced at lower temperature, that is, 300 °C [19]. This finding indicated that high carbonization (i.e., enhancing pyrolytic temperature) enhances biochar's surface area and pore volume. Choy and McKay [20] reported that animal bone biochar is highly efficient to sorb Zn (33.03 mg g^{-1}), Cu (45.04 mg g^{-1}), and Cd (53.6 mg g^{-1}) by biochar's surface pores. In another study, Wang et al. [17] noticed that hickory wood–derived biochar treated with $KMnO_4$ is highly efficient to adsorb different quantities of toxic metals like Cu (34.2 mg g^{-1}), Pb (153.1 mg g^{-1}), and Cd (28.1 mg g^{-1}) from the aqueous solution, mainly due to affinity differences.

4.2 Cation/ion exchange

The positively charged ion (protons) exchange on biochar's surface with ionized cations from dissolved salts (particularly, heavy metals) is the prime principle of cation/ion exchange technique, and the process is considered as potent mechanism of metal sorption. The adsorption capacity of this method primarily depends on the size of metals and biochar's surface characteristics, particularly the presence of functional groups [9]. Additionally, the metal sorption capacity of ion exchange mechanism depends on pyrolysis temperature proportionately. However, on the contrary, Ambaye et al. [9] recorded lower cation exchange capacity with increasing pyrolysis temperature, particularly greater than 350 °C. Generally, biochar's higher ion exchange capability indicates higher metal adsorption, thereby higher metal removal potentiality. Moreover, metal sorption through cation/ion exchange technique is governed by substitution of ionized toxic metal ions with biochar's positive charge species (functional units) in aqueous phase.

Qian et al. [21] observed that the removal of Cr(III) by colloids biochar was significantly enhanced with temperature elevation from 100 to 400 °C. They also demonstrated that sorption of Cr(III) attributed to the biochar-derived cationic presence (also including Mg and Ca) in solution matrix, suggesting the ion exchange was a prime sorption process. Trakal et al. [22] investigated the efficiency of Cd immobilization by agriculture waste-derived biochars (grape husks, grape stalk, plum stones, wheat straws, and nut shells) and postulated that grape stalk biochar sorbed more Cd (0.45 mmol kg^{-1}) than plum stone–derived biochar (0.04 mmol kg^{-1}), indicating ion exchange technique as dominant mechanism for Cd sorption. In another study, Lu et al. [23] demonstrated that biosolid–biochar assisted lead sorption is predominantly governed by ion exchange technique, and Mg and Ca were prime attributors for presence of cations in solution medium (40–52% and pH 2–5), while K and Na were the least contributors (4.8–8.5%). Zhang et al. [14] reported that cationic concentration (Ca, Mg, K, and Na) generated in solution matrix from water hyacinth–derived biochar was equivalent to sorbed Cd level on biochar, revealing that ion exchange technique was a dominant Cd sorption process. In another study, Trakal et al. [24] documented Cd and Pb removal performances using agriculture waste-derived biochars (grape husks, grape stalk, plum stones, wheat straws, and nut shells). The results revealed that wastes that have iron oxides showed higher Pb and Cd removal efficiency, which means that presence of iron act as a prime regulator to enhance the biochar's cation/ion exchange capacity.

4.3 Complexation

The process is characterized by the arrangement of outer- or inner-sphere multiatom complex formation through interactions between metals and the biochar's ligands. This mechanism is highly suitable for removing transition metals. Transition metals possess partially filled d-orbitals, and due to that these metals have strong attraction for ligands [1]. Generally, biochar produced at low temperature holds oxygen-rich functional units like carboxyl, phenolic, lactonic on their surface [2, 9]. This oxygen content enhanced biochar's surface oxidation capacity as well as generation of more carboxyl groups, which ultimately lead to higher metal complexation due to higher ligand–metal interactions [4]. Generally, plants-based biochars are highly efficient to diverse heavy metals including Cu, Ni, Cd, and Pb for removing them through metal complex formation with phenolic and carboxylic units compared with animal-based (dairy manure and poultry litter) biochar [25]. In particular, animal-based biochars (dairy manure and chicken litter) possess phosphate ligands and bind with metals largely with Pb [25]. They also postulated that plant-based biochars displayed enhanced complexation and cation/ion exchange capacities. Qian et al. [21] documented that colloidal biochars displayed higher performances in Cr removal than residue-originated biochars, probably due to the presence of large amount of oxygen-rich units on colloidal biochar's surface, which also indicated the dominance of complexation technique during removal of Cr(III) from contaminated water.

Trakal et al. [22] documented that there was a minor spectral shift of carboxyl groups of unloaded- and Cd-loaded biochar during Cd sorption under FITR spectroscopy, indicating that complexation mechanism (between Cd and carboxyl groups) displayed little role in Cd sorption. On the other hand, Harvey et al. [4] demonstrated that honey mosquito- or loblolly pine- or cordgrass-based biochars performs Cd complexation by negatively charged graphene-like structures of biochars (mainly functional groups), and complexation was the dominant mechanism for Cd removal among other techniques. Kołodynska and Bak [26] documented that hydroxyl and carboxyl units played pivotal roles during Cu sorption as revealed by the before and after FTIR analysis. Dong et al. [27] documented that during Hg complexation by Brazilian pepper-derived biochar, the functional units play a dominant role. They also demonstrated that enhancing pyrolytic temperature (300–600 °C) significantly lowered sorption of Hg metal (24.2–15.1 mg g^{-1} @ pH 7) while biochar derived at 300 and 450 °C displayed elevated Hg sorption (23–31% and 77–69%, respectively) in the presence of phenolic, hydroxyl, or carboxylic groups. On the contrary, significantly higher Hg sorption (91%) was noticed when biochar fabricated with graphite-like aromatic structure at 600 °C, and only 9% sorption of Hg metal was recorded during complexation in presence of phenolic/ hydroxyl units. They also postulated that this reduction during sorption of Hg metals by pepper biochar is basically due to loss of functional units with increasing pyrolytic temperature. Xu et al. [28] also reported complexation process as prime mechanism during sorption of Hg from contaminated water by bagasse or hickory chips-derived biochars using X-ray photoelectron spectroscopy (XPS) spectra analysis. They also noticed that bagasse-derived biochar was more capable to bind with Hg complexes than hickory chips-derived biochars.

4.4 Electrostatic interactions (chemisorption)

Electrostatic interaction, also called chemisorption, is another prime sorption technique which integrates adsorption of ionizable compounds to biochar's positive-charged units through electrostatic interaction/reaction. The capability of this mechanism depends on aqueous solution's pH and ion strength, pyrolysis temperature, and biochar's PZC [9]. Mukherjee et al. [29] documented the pH effect during electrostatic reaction between heavy metals and biochar surface. The biochar surface generally displayed positive charge at low pH and negative charge at higher pH. This phenomenon indicates that solution pH played a vital role in controlling surface charge of biochar. Keiluweit and Kleber [30] described that elevated pyrolysis temperature plays a pivotal role for efficient sorption of metal ions by graphene-like biochar during electrostatic reactions. Accordingly, Qiu et al. [31] recorded higher Pb sorption efficiency by wheat and rice straw biochar because of strong electrostatic attractions between the metal's (Pb) negative charge and biochar's negative charge units. Efficient chromium sorption was also reported by electrostatic reactions between cationic biochar surfaces and anionic Cr (VI) at pH 2 [32].

4.5 Precipitation

The precipitation technique for metal sorption is generally characterized by generating solids on biochar surface. Precipitation is very suitable for removal of inorganic pollutant, particularly heavy metals by biochar. Generally, it involves mineral precipitate formation within solution matrix or onto the biochar's surface (sorbent material). Shakoor et al. [1] reported that metals like Zn, Ni, Cu, and Pb possess moderate ionization potentials (2.5–9.5) and due to that they produce more precipitates on biochar surface than other metals. Cao and Harris [33] demonstrated that biochars produced from plant-based materials through pyrolysis at elevated temperatures (>300 °C) have alkaline property and during metal sorption from aqueous solution form precipitates of metal ions. Recently, Puga et al. [34] prepared biochar from sugarcane and straw dust, and observed that prepared biochar enhanced the precipitation of Zn and Cd from the aqueous solution. They also opined that the performance of surface precipitation for metal removal by biochar depends on diverse factors like pyrolytic temperature; hence, further research is needed to determine the optimized pyrolysis temperature.

5 Factors affecting contaminants' sorption

5.1 Solution pH

It is a prime determinant regulating metal sorption and the success of metal sorption generally depends on it. Precisely, solution pH regulates surface properties of biochar (mainly functional groups), ionization of biochar, speciation, and the degree of metal sorption as ionization of biochar are pH-dependent [35]. The deprotonation of functional units generally occurred when pH of solution increased, thereby enhanced biochar's sorption capacity toward cationic metals. However, decrease in pH enhanced electrostatic repulsion forces between functional groups and metal ions within solution. Consequently, the produced cations from biochar competes with each other, resulting in reduced sorption capability of biochar for metal ions [23].

Chen et al. [36] documented increased adsorption capacity of Zn, Cu/Pb cations by corn and hardwood straw biochars with increasing pH from 2.0 to 5.0. They also pointed out decreased adsorption capacity beyond pH 5.0 due to hydroxide complex formation [23]. Zhang et al. [35], contrarily, demonstrated that decreased pH level of the medium enhanced anionic adsorption capacity of Cr(VI) ions. They postulated that enhanced sorption is because of electrostatic reactions between biochar's positive charge and chromate ion's negative charges at low pH. They also proposed that elevated pH level reduced the Cr adsorption by biochar due to OH^- competition with Cr(VI) for binding biochar's active/charged sites. Therefore, it could be demonstrated that Cr(VI) decontamination is totally regulated by pH as suggested by Shang et al. [37]. Liu et al. [38] demonstrated that sorption of Pb(II) by corn stalk–derived biochar enhanced surprisingly due to solution pH increment from pH 4 to 6. They also postulated that acidic conditions, that is, presence of excess H^+ inhibits sorption of Pb ions. Recently, Senthilkumar et al. [39] described that As(V) decontamination by *Ulva reticulata* generated biochar significantly elevated to 93% from 55% due to pH surges to 4 from 2, but beyond that pH value biochar's adsorption potentiality decreased. They also postulated at acidic (pH 4–6) circumstances As(V) oxyanion occurs mostly as $H_2AsO_4^-$ and $HAsO_4^{2-}$. Accordingly, the comparatively higher sorption of As(V) at lower pH value by *Ulva reticulate*–based biochar was probably because of biochar's highly protonated binding sites (positively charged) leading to excess H^+ ions saturation, consequently increasing sorption of As(V) using electrostatic reactions. Apart from these, solution pH also controls the adsorption capacity of organic contaminants originated from industrial activities [40].

5.2 Temperature

We already explained how temperature influenced the metal sorption capacity by biochar preparation temperature. Temperature influenced the adsorbent's sorption efficiency and reaction's kinetics rate. This indicates that temperature helps to differentiate among different sorption techniques based on the increase or decrease of adsorption capacity by biochar. More specifically, based on temperature changes, the sorption mechanism designated as endothermic when adsorption capacity increases proportionately with elevated temperature; whereas in exothermic sorption process sorption efficiency decreases with temperature elevation [41]. Accordingly, a number of scientific findings have documented that temperature plays a dominant role during biochar-induced heavy metal sorption [38].

5.3 Biochar (sorbent) amount

Biochar amount also affected the adsorption capacity, that is, metal sorption capacity. Generally, the percent metal remediation is directly dependent upon adsorbent dosage. More precisely, the adsorbent dose directly enhances biochar's removal efficiency of metals. For instance, Chen et al. [36] documented that heavy metal (Cu, Cd, Zn, and Pb) removal potentially from industrial effluents by corn and hardwood straw biochar was significantly enhanced with increasing biochar concentration (0.5–5 g L^{-1}). This is particularly due to enhanced number of biochar's functional units and surface area [40]. Likely, Sun et al. [42] demonstrated that increasing swine manure-derived biochar dose to 8 g L^{-1} from 1 g L^{-1} led to enhanced active functional sites to adsorb contaminants. Recently, Wang et al. [43] also demonstrated that enhancing biochar amount significantly influenced organic pollutants adsorption as well as potentially toxic elements. On the other hand, the sorption efficiency reduced simultaneously with increasing adsorbent dosage [41], mainly due to the interaction types between sorbate (metal ions) and sorbent. More specifically, at higher biochar amount, the sorbate in solution matrix is very less in quantity in comparison to existing exchangeable functional units on biochar surface, which ultimately reduced metal uptakes [41].

5.4 Initial solute concentration (ISC)

Initial solute (metal) concentration is another vital parameter that directly influenced the metal removal capacity by the biochar. Generally, the sorption capacity is enhanced with increasing initial solute level. Generally, at ISC the ratio of initial solute moles to biochar surface was very low, and consequently, adsorption became independent of ISC. However, at higher solute level, the sorbent's accessible functional units become fewer compared with solute moles and hence, removal efficiency of metals is adversely affected by ISC. For instance, Senthilkumar et al. [39] documented that during As(V) adsorption by *Ulva reticulate*–generated biochar, the As(V) uptake capacity significantly enhanced from 4.65 to 7.40 mg g^{-1} with increasing initial As(V) concentration to 25 mg L^{-1} from 10 mg L^{-1} but percent removal capacity decreased from 93.0 to 59.2%.

5.5 Column sorption parameters

Although a dearth of data is available on column sorption compared with batch sorption to remove metal ions from aqueous solutions, it was revealed that column sorption is primarily dependent upon parameters like bed depth, flow rate, and influent metal concentration [41]. Additionally, although batch sorption process has several advantages over column sorption, batch sorption technique does not provide real-time scale-up applications during industrial wastewater treatment. Packed columns are effective practices for elution processes/cyclic adsorption, as they utilized concentration differences very effectively, which is considered a prime driving force for metal sorption. Accordingly, column process permits more efficient utilization of adsorbent, resulting in fast and high rate of yielding and easy scale up. Further, column process allows to treat the large amount of contaminated water continuously using small amount of sorbent. In addition to this, sorbent regeneration and subsequent reuse can also be performed by using appropriate elutant [41]. For instance, during continuous column adsorption to remediate Co(II) and Ni(II) ions by *Tectona grandis* leaves-derived biochar, Vilvanathan and Shanthakumar [44] demonstrated that column exhaustion time is enhanced if bed depth increasing and exhaustion time is reduced if metal ion concentrations and flow rate are decreased. Likely, Senthilkumar et al. [39] used *U. reticulate*–derived biochar to remove As(V) from aqueous solutions through column (2 cm diameter and 35 cm depth) bed loaded with initial As(V) concentration 25 mg L^{-1} at flow rate 0.3 L h^{-1}. They observed exhaustion and breakthrough times at 13 and 3.25 h, respectively, and total 3.9 L As(V) solution was decontaminated but column adsorption capability and percent removal of As(V) was 8.12 mg g^{-1} and 59.5%, respectively.

5.6 Biochar modification

Although biochar has good sorption capability, still its sorption capacity can be improved through different techniques called biochar activation, particularly aimed to enhance its surface morphology like surface area, PZC, porosity, and predominantly functional units (Table 3). Generally, two types of activation techniques are employed to further improve its sorption efficiency, and detailed information is described in the following sections:

Table 3: Impacts of biochar modification on metal sorption capacity (+ denotes positive influence).

Modification type	Biomass type/pretreatment	Pyrolysis type	Metal/ sorption type	References
Organic molecules composites	*Eichhornia crassipes*/Fe and chitosan	Post	Cr/+	Zhang et al. [5]
Activated by base	Municipal sewage sludge/2 M KOH	Post	AsV/+	Jin et al. [11]
Biochar composites – minerals	Pine wood/Mn oxide	Post	Pb/+	Wang et al. [17]
Composites – metal modify	Different feedstocks/Fe	Post	Pb/+	Trakal et al. [24]
Composites – metal modify	Fe/different feedstocks	Post	Cd/+	Trakal et al. [24]

Table 3 (continued)

Modification type	Biomass type/pretreatment	Pyrolysis type	Metal/ sorption type	References
Physical activation (PA)	Poultry manure/steam activation	Post	Zn/+	Lima and Marshall [45]
PA	Poultry manure/steam activation	Post	Cu/+	Lima and Marshall [45]
PA	Pine sawdust/steam activation	Post	Cu/+	Lou et al. [46]
Chemical activation (CA)	Cactus cladodes/HNO_3	Post	Cu/+	Hadjittofi et al. [47]
CA	Fruit peel/KOH	Post	Cu/+	Hamid et al. [48]
CA	Ipomoea biomass/KOH	Post	Cd/+	Goswami et al. [49]
Composites – organic molecules	Commercial sawdust/amino-modification by nitrification and reduction	Post	Cu/+	Yang and Jiang [50]
Composites – organic molecules	Rice husk/polyethylenimine-methanol	Post	Cr/+	Ma et al. [51]
Composites – organic molecules	Hyacinth/graphene oxide	Pre	Cr/+	Shang et al. [52]
Composites – carbonaceous materials	Wheat straw/graphene oxide	Pre	Hg/+	Tang et al. [53]

5.6.1 Physical activation

It is also called steam activation and has gained wide attention as an alteration technique, generally employed to enhance biochar's surface porosity for metal sorption and subsequently reduces impurities associated with incomplete combustion. In physical activation, a steam (activated with high temperature) is passed across biochar's pore to enhance its surface area. For instance, Lima and Marshall [45] prepared activated poultry manure-derived biochar at 700 °C following steam activation in different time periods and water flow rates at 800 °C. They observed that prolonged duration and enhanced flow rates improved Cd, Cu, and zinc sorption by activated biochar. Contrarily, Shim et al. [3] recorded no significant changes in Cu sorption capacity by *Miscanthus sacchariflorus*–derived biochar (@ 500 °C/ slow pyrolysis) and *Miscanthus*–derived stream activated biochar at 800 °C. However, they noticed that biochar's surface area and porosity were almost doubled due to steam activation but biochar's functional units decreased dramatically. Likely, Lou et al. [46] recorded similar enhanced surface area of biochar (pine sawdust) activated by steam activation but no significant differences in functional units on biochar's surface were recorded. Additionally, they opined that steam activation did not play any significant role in phosphate sorption potential by activated biochar due to repulsion force between phosphate ions and biochar's negatively charged functional groups. These findings concluded that steam activation technique is more fruitful and productive when pretreatment step/second modification (helps to increase surface functional groups) is employed as only surface area increases in steam activation.

5.6.2 Chemical activation

Chemical activation is another most important modification technique to enhance biochar's sorption capacity. In this technique, already prepared biochar undergoes either acidic or alkaline treatment to improve the number of O-rich functional units on biochar surface. Apart from this, biochar can be modified by a variety of means such as impregnation with organic compounds, mineral oxides, clays, and coating with carbon-rich materials. The details are described further.

5.6.2.1 Acidic and alkaline solutions

The treatment of biochar by acidic solution improved the quantity of O-rich functional units (mainly negatively charged units –OH and –COOH) on biochar surface as well as enhanced biochar potentiality to attract more cationic metals during sorption process. Acidic treatment increased the amount of carboxylic groups on biochar surface as well as porosity, which led to enhance biochar's surface area [1]. For instance, Hadjittofi et al. [47] observed that nitric acid modification of cactus fibers-derived biochar improved Cu sorption and removal efficiency mainly due to presence of –COOH units on biochar surface in higher quantity. They also opined that sorption of Cu was totally pH regulated, improved (higher) sorption efficiency at higher pH (pH 6.5 vs 3) mainly due to precipitation (chemisorption). Qian et al. [54] also recorded higher quantity of O-rich functional groups on activated (sulfuric and nitric acid mixture) rice straw-derived biochar as evidenced by higher O/C ratios in different posttreatment materials. Li et al. [55] observed almost double Pb absorption and removal capacities (46 vs 99%) of nitric acid modified bagasse biochar than normal bagasse biochar.

Alkali activation is generally governed by different metal hydroxides such as KOH and NaOH. It improves sorption capacity of toxic metals by enhancing porosity, surface area, and providing more O-rich functional units onto biochar's surface. During alkali activation, biochar's surface functional units provide exchangeable sites (positive charged) for sorption of cationic metals present in contaminated water [1]. For example, Liu et al. [56] observed that KOH-exposed biochar showed higher surface area, porosity, and high C/O composition, resulting in higher sorption of tetracycline, inorganic pollutants including metals. Hamid et al. [48] demonstrated that biochar treated with KOH solution sorbed more Cu ions from contaminated water than without KOH-treated biochar, mainly by chemical sorption process. Goswami et al. [49] recorded that biochar combined and treated with KOH at 350–550 °C temperature resulted in the opening of biopolymer rings and molecules and subsequently enhanced sorption capacity of metals by biochar. Likely, Li et al. [57] in their study recorded 92% greater sorption efficiency of Cd by NaOH-treated rape straw biochar compared with untreated rape straw biochar.

5.6.2.2 Biochar impregnation with organic compounds

Organic compound that possesses amino groups has great potential to modify biochar's surface morphology through complexation between amino groups and aqueous metal ions. The major objective of biochar impregnation is improving biochar's surface characteristics, thereby they act as high supporting base for metal sorption. Generally, this modification is performed either by chemical reactions or direct biochar mixing with polymers (e.g., chitosan) which possess amino groups in higher quantity [1]. For example, Yang and Jiang [50] documented that organic compound–treated saw dust biochar using sulfuric acid (H_2SO_4) and nitric acid (HNO_3) mixture followed by sodium dithionite ($Na_2S_2O_4$) reduction holds more amino acid groups on modified biochar and showed five times higher Cu sorption capacity than untreated saw dust biochar, attributed mainly by amino groups complexation with Cu ions. Ma et al. [51] observed that alkali/acid activation of modified rice husk–derived biochar with polyethylenimine showed substantially improved Cr(VI) sorption mainly due to the presence of excess amino groups on modified rice husk biochar. Deng et al. [58] produced novel pristine feedstock biochar by pyromellitic dianhydride (PMDA) and chitosan modification for sorption of Cu, Pb, and Cd ions from contaminated water. They observed through XPS and FTIR analyses that biochar modification improves the quantity of N–C=O units which play a crucial role during sorption of Cu and Pb sorption but other N-rich units sorbed Cd ions mainly by precipitation process.

5.6.2.3 Biochar impregnation with mineral oxides

Minerals that have different properties along with positive charge serve as promising sorbent to eliminate oxyanions from contaminated water when impregnated with high-surface-area biochars. In this regard, hematite, magnetite, goethite, magnetite, $FeCl_3$, zerovalent Fe, $MgCl_2$, calcium oxide, FeO, $Fe(NO_3)_3$, and hydrous Mn oxide were employed as excellent impregnation/loading agents. The mineral impregnation with biochar is governed simply by biochar or its feedstock soaking, resulting in oxyanions binding homogeneously with biochar's functional units. The soaked biochar (including its feedstock) is then pyrolyzed at 50–300 °C to remove the chlorides and nitrates as Cl_2 and NO_2 gases, respectively, and finally to transform them into metal oxides. For example, Han et al. [59] reported that peanut hull–derived biochar impregnated with γ-Fe_2O_3 removes 1–2 times higher Cr(VI) than unimpregnated biochar. Shang et al. [37] observed that herb-residue originated biochar with zerovalent iron (nZVI) impregnation have potentiality to remove 98% Cr from water (waste/contaminated water). Duan et al. [60] documented that magnetic Fe-impregnated cotton stalk biochar originated from iron sludge showed higher Cr sorption capacity (67.44 mg g^{-1}) compared with unloaded biochar (40.91 mg g^{-1}), mainly due to holding improved surface area (129.2 m^2 g^{-1}) and porosity (0.1711 cm^3 g^{-1}). Lyu et al. [61] prepared a composite-based biochar by nanoscale iron sulfide (FeS) impregnation for sorption of Cr(VI) ions from waste or contaminated water and results showed enhanced Cr(VI) sorption (130.5 mg g^{-1}) in comparison to untreated biochar (25.4 mg g^{-1}).

5.6.2.4 Biochar impregnation with clays

The basic purpose of impregnation of biochar with clay composites is enhancing biochar's surface features, thereby promoting higher metal sorption from contaminated water. The preparation method of clay composites–based biochars is totally different with regard to traditional methods (like chemical activation) as completely new functional units (–COOH/–OH) are attached on biochar surface. Graphene oxides, among C-based substances, are employed widely for new composites synthesis, which have high metal-binding capacity. Generally, binding of –COOH/–OH units with metals was performed by chemical oxidation process [1]. The enlarged surface area of biochars due to higher porosity enables them to stabilize C-based nanomaterials and, thus, widen their remediation applications [1]. For instance, Shang et al. [52] documented that biochar composite produced from hyacinth feedstock modified by graphene oxide significantly improved Cr(VI) removal potentiality from aqueous medium in comparison to unmodified biochar.

5.6.2.5 Biochar impregnation with carbon-rich materials

Graphene oxides, among carbon-rich materials, are widely used as precursor material during biochar impregnation because of their high binding capacity with diverse metals including Cd, Zn, Co, Cu, and Pb [1]. Generally, graphene oxide is commonly used to fabricate graphene for enhancing biochar's sorption capacity. Particularly during chemical reactions, graphene binds with graphene oxide's O-rich functional units (–COOH/–OH) through chemical oxidation techniques [53]. Generally, the biochar materials are first soaked in graphene oxide suspension solution and then undergo pyrolysis process to prepare the composites (graphene oxide-biochar), which improves biochar's surface area and the amount of O-rich functional units mainly –COOH/–OH compared with untreated biochar [52, 53].

6 Analysis of interaction mechanism

6.1 Mathematical modeling approaches

Mathematical modeling is a very common approach for interaction measurement. Generally, different equilibrium models based on sorption isotherms are employed to analyze reactions performance based on two-, three-, and four-parameter isothermal models. The most common among them during metal sorption includes Freundlich, Langmuir, Sips, and Toth models [1]. Langmuir model, among these, is the most widely used model which is expressed as follows:

$$Q = \frac{Q_{\mathrm{max}} b_{\mathrm{L}} C_e}{1 + b_{\mathrm{L}} C_e}$$

where, Q denotes sorption capability (mg g^{-1}); C_e denotes equilibrium level (mg L^{-1}); Q_{max} denotes adsorbent's maximum toxicant uptake capacity (mg g^{-1}) and b_{L} denotes equilibrium coefficient (L mg^{-1}).

The Freundlich model, generally empirically derived equation, describes that metal sorption is done on multilayer heterogeneous surface but sorption (metal) sites are not homogeneous and energetically equivalent. Freundlich model is calculated using the following equation:

$$Q = K_{\mathrm{F}} C_{\mathrm{e}}^{1/n_{\mathrm{F}}}$$

where, n_{F} is model exponent and K_{F} denotes model coefficient (L g^{-1})$^{1/nF}$.

The Sips model demonstrates that adsorbent's reaction sites possess diverse strengths and only one sorbate molecule binds with each site. The model is represented as follows:

$$Q_{\mathrm{e}} = \frac{K_{\mathrm{S}} C_{\mathrm{e}}^{\beta_{\mathrm{S}}}}{1 + a_{\mathrm{S}} C_{\mathrm{e}}^{\beta_{\mathrm{S}}}}$$

where, a_{S} and K_{s} denote the model coefficient (L mg^{-1})$^{\beta\mathrm{S}}$, β_{S} denotes the model exponent.

Toth's model considers three parameters and generally uses quasi-Gaussian energy distribution model to demonstrate metal–adsorbent isotherms. Toth model is represented as follows:

$$Q = \frac{Q_{\mathrm{max}} b_{\mathrm{T}} C_{\mathrm{e}}}{\left[1 + (b_{\mathrm{T}} C_{\mathrm{e}})^{1/n\mathrm{T}}\right]^{n_{\mathrm{T}}}}$$

Where, b_T denotes model constant (L mg^{-1}) and n_{T} denotes the model exponent.

Apart from these, different kinetic models are used to know the rate of reaction and process design on practical background. The commonly used sorption kinetic models are described below. The intraparticle diffusion is measured following the equation described by Weber and Morris through kinetic data fitting to an intraparticle diffusion plot:

$$Q_t = k_i t^{1/2}$$

where, Q_t denotes sorption capability at time t (mg g^{-1}); k_i denotes diffusion constant (intraparticle).

The pseudo-first-order model is another most important kinetic model that describes that solute uptake rate at particular time is directly proportional to the differences between solid uptake amount with time and saturation concentration. On the other hand, pseudo-second-order model is employed to calculate adsorption potentiality over experimentation time. The models are expressed as follows:

$$Q_t = Q_e(1 - \exp(-k_1 t))$$

Where, Q_e denotes equilibrium uptake (mg g^{-1}) and k_1 denotes pseudo-first-order constant (1/min).

$$Q_t = \frac{Q_e^2 K_2 t}{1 + Q_e K_2 t}$$

where, K_2 denotes pseudo-second-order constant.

6.2 Advanced spectroscopic analysis

Generally, reaction mechanisms are predominated by complexation adsorption, ion exchange, oxidation–reduction, and precipitation. Different techniques in this regard are employed to characterize reaction mechanism, namely FTIR, XPS, and X-ray absorption fine structure (XAFS) spectroscopy. Changes in peak intensity (red/blue peak shift) help to obtain semiqualitative information by FTIR. The XPS gives valuable information about chemical composition, oxidation state, water content, and biochar's functional unit's bonding information [7, 62]. On the other hand, XAFS, which is a qualitative and semiquantitative technique, provides information about reduction/oxidation status, bond distances, coordination number, and even microstructure at molecular level. For example, Kołodynska and Bak [26] documented that hydroxyl and carboxyl groups have a direct role during sorption of Cu ions onto biochar using before and after FTIR analysis during reaction mechanism. Wu et al. [62] determined the interaction of PB species with Pb-loaded biochars by XAFS and found that $Pb_3(PO_4)_2$, Pb-montmorillonite, $PbSO_4$, $Pb(C_2H_3O_3)_2$, and Pb-Al_2O_3 contributed to Pb binding.

6.3 Computational modeling approaches

Computational modeling is now a widely recognized aid to determine quantitative molecular-scale interactions where experiments cannot directly probe. It provides information such as structure, dynamics, energetics, and mechanisms as well at atomic or molecular scale level. Density functional theory (DFT) and molecular dynamic (MD) methods are two commonly used techniques in this regard [7]. Apart from these, techniques such as density of states (DOS), differential charge densities, electrostatic potential (ESP), frontier molecular orbital (FMO), and NCI noncovalent interaction index (NCI) analysis have also been widely employed to explore molecular-scale mechanism [7, 63].

7 Future perspectives

Biochar has great potential for metal sorption as well as treating wastewater and soil remediation. Additionally, studies on these aspects are on initial stage, particularly using agricultural-based biochar materials. Moreover, most of the studies are lab-based, dealing mainly single metal sorption, and few studies have addressed mixed metal sorption capacity by biochar. Till now, not a single study investigated metal removal capacity by biochar from contaminated wastewater at field level. Apart from these, the production cost of biochar is still the prime constraint in field/industry scale applications. Therefore, considering these facts, the future research on biochar-based metal sorption might focus on the following aspects:

- More efforts should be given on biochar's functionality and porosity enhancement through adopting techniques like sulfonation, amination, surface oxidation, and pores' structure modification during biochar preparation.
- To find out the optimum dose as well as suitable biochar preparation method for rapid removal of contaminants in order to aim at large-scale implementation and minimize biochar production and utilization costs.
- To develop more advanced methods for enhancing the biochar's adsorption and removal capacity.
- Economical biochar regeneration and standard operating methods are very much needed for its possible implementation on a large scale in field.
- To enhance public awareness about the biochar's capability in removing contaminants from the environment.

As very limited studies are available on these topics, further research is needed on these topics and their associated relationship among biomass materials, biochar processing parameters, and biochar regeneration to reduce environmental risks as described in Figure 3.

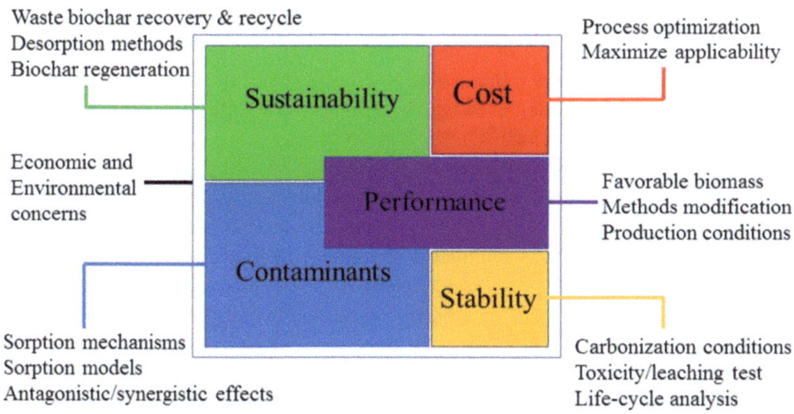

Figure 3: Proposed future research directions for biochar application in metal removal (modified after Ambaye et al. [9]).

8 Conclusions

This review chapter addressed that biochar has broad potential for metal sorption from contaminated environment. Biochar is a new advanced adsorbing material due to its higher efficiency and eco-friendly nature, mainly because of possessing large surface area or porosity, and existence of charged functional units. This review highlighted information about different types of biomasses used for biochar including agriculture-based sources, different production methods followed by biochar characterization, factors controlling biochar's sorption potentiality, their applications in metal sorption (namely surface sorption, ion exchange, electrostatic reactions (chemisorption), complexation, and precipitation) and interaction mechanism analysis. Still, further detailed study is warranted for large-scale biochar production cost-effectively to minimize biochar production and utilization costs in order to maintain environmental and economic feasibility. Finally, public awareness about the use of agriculture materials to produce biochar for environmental clean-up needs to be enhanced for its effective utilization; this requires urgent cooperation between environmentalists and researchers.

References

[1] Shakoor MB, Ali S, Rizwan M, Abbas F, Bibi I, Riaz M, Khalil U, Niazi NK, Khalil J, A review of biochar-based sorbents for separation of heavy metals from water, Inter. J. Phytoremediation, 2019, 10.1080/15226514.2019.1647405.

[2] Li H, Dong X, Da Silva EB, de Oliveira LM, Chen Y, Ma LQ, Mechanisms of metal sorption by biochars: Biochar characteristics and modifications, Chemosphere, 2017, 178, 466–478.

[3] Shim T, Yoo J, Ryu C, Park Y-K, Jung J, 2015, Effect of steam activation of biochar produced from a giant *Miscanthus* on copper sorption and toxicity, Bioresour. Technol., 2015, 197, 85–90.

[4] Harvey OR, Herbert BE, Rhue RD, Kuo LJ, Metal interactions at the biochar-water interface: Energetics and structure-sorption relationships elucidated by flow adsorption microcalorimetry, Environ. Sci. Technol., 2011, 45(13), 5550–5556.

[5] Zhang S, Dong Q, Zhang L, Xiong Y, High quality syngas production from microwave pyrolysis of rice husk with char-supported metallic catalysts, Bioresour. Technol., 2015, 191, 17–23.

[6] Khalil U, Shakoor MB, Ali S, Rizwan M, Tea waste as a potential biowaste for removal of hexavalent chromium from wastewater: Equilibrium and kinetic studies, Arab. J. Geosci., 2018, 11(19), 573. 10.1007/s12517-018-3932-5.

[7] Hu Y, Ai Y, Jin J, Hayat T, Alsaedi A, Zhuang L, Wang X, Efficient elimination of organic and inorganic pollutants by biochar and biochar-based materials, Biochar, 2020, https://doi.org/10.1007/s42773-020-00044-4.

[8] Sun Y, Gao B, Yao Y, Fang J, Zhang M, Zhou Y, Chen H, Yang L, Effects of feedstock type, production method, and pyrolysis temperature on biochar and hydrochar properties, Chem. Eng. J., 2014, 240, 574–578.

[9] Ambaye TG, Vaccari M, van Hullebusch ED, Amrane A, Rtimi S, Mechanisms and adsorption capacities of biochar for the removal of organic and inorganic pollutants from industrial wastewater, Int. J. Environ. Sci. Technol., 2020, https://doi.org/10.1007/s13762-020-03060-w.

[10] Chen Z, Xiao X, Chen B, Zhu L, Quantifcation of chemical states, dissociation constants, and contents of oxygen-containing groups on the surface of biochars produced at different temperatures, Environ. Sci. Technol., 2014, 49, 309–317.

[11] Jin JW, Li YA, Zhang JY, Wu SC, Cao YC, Liang P, Zhang J, Wong MH, Wang MY, Shan SD, Christie P, Influence of pyrolysis temperature on properties and environmental safety of heavy metals in biochars derived from municipal sewage sludge, J. Hazard. Mater., 2016, 320, 417–426.

[12] Yuan JH, Xu RK, Zhang H, The forms of alkalis in the biochar produced from crop residues at different temperatures, Bioresour. Technol., 2011, 102, 3488–3497.

[13] Nguyen BT, Lehmann J, Hockaday WC, Joseph S, Masiello CA, Temperature sensitivity of black carbon decomposition and oxidation, Environ. Sci. Technol., 2010, 44, 3324–3331.

[14] Zhang J, Liu J, Liu RL, Effects of pyrolysis temperature and heating time on biochar obtained from the pyrolysis of straw and lignosulfonate, Bioresour. Technol., 2015, 176, 288–291.

[15] Al-Wabel M, Al-Omran A, El-Naggar AH, Nadeem M, Usman ARA, Pyrolysis temperature induced changes in characteristics and chemical composition of biochar produced from conocarpus wastes, Bioresour. Technol., 2013, 131, 374–379.

[16] Xu X, Zhao Y, Sima J, Zhao L, Masek O, Cao X, Indispensable role of biochar-inherent mineral constituents in its environmental applications: A review, Bioresour. Technol., 2017, 2017(241), 887–899.

[17] Wang Z, Liu G, Zheng H, Li F, Ngo HH, Guo W, Liu C, Chen L, Xing B, Investigating the mechanisms of biochar's removal of lead from solution, Bioresour. Technol., 2015, 177, 308–317.

[18] Li XM, Shen QR, Zhang DQ, Mei XL, Ran W, Xu YC, Yu GH, Functional groups determine biochar properties (pH and EC) as studied by two-dimensional ^{13}C NMR correlation spectroscopy, PLoS One, 2013, 8, e65949.
[19] Kumar S, Loganathan VA, Gupta RB, Barnett MO, An assessment of U (VI) removal from groundwater using biochar produced from hydrothermal carbonization, J. Environ. Manage., 2011, 92(10), 2504–2512.
[20] Choy KK, McKay G, Sorption of cadmium, copper, and zinc ions onto bone char using Crank diffusion model, Chemosphere, 2005, 60(8), 1141–1150.
[21] Qian LB, Zhang WY, Yan JC, Han L, Gao WG, Liu RQ, Chen MF, Effective removal of heavy metal by biochar colloids under different pyrolysis temperatures, Bioresour. Technol., 2016, 206, 217–224.
[22] Trakal L, Bingol D, Pohorelý M, Hruska M, Komarek M, Geochemical and spectroscopic investigations of Cd and Pb sorption mechanisms on contrasting biochars: Engineering implications, Bioresour. Technol., 2014, 171, 442–451.
[23] Lu H, Zhang W, Yang Y, Huang X, Wang S, Qiu R, Relative distribution of Pb2þ sorption mechanisms by sludge-derived biochar, Water Res., 2012, 46, 854–862.
[24] Trakal L, Veselska V, Safarık I, Vıtkova M, Cıhalova S, Komarek M, Lead and cadmium sorption mechanisms on magnetically modified biochars, Bioresour. Technol., 2016, 203, 318–324.
[25] Cao XD, Ma LQ, Gao B, Harris W, Dairy-manure derived biochar effectively sorbs lead and atrazine, Environ. Sci. Technol., 2009, 2009(43), 3285–3291.
[26] Kołodynska D, Bak J, Use of three types of magnetic biochar in the removal of copper (II) ions from wastewaters, Sep. Sci. Technol., 2018, 53(7), 1045–1057.
[27] Dong X, Ma LQ, Zhu Y, Li Y, Gu B, Mechanistic investigation of mercury sorption by Brazilian pepper biochars of different pyrolytic temperatures based on X-ray photoelectron spectroscopy and flow calorimetry, Environ. Sci. Technol., 2013, 47(21), 12156–12164.
[28] Xu X, Schierz A, Xu N, Cao X, Comparison of the characteristics and mechanisms of Hg(II) sorption by biochars and activated carbon, J. Colloid Interface Sci., 2016, 463, 55–60.
[29] Mukherjee A, Zimmerman A, Harris W, Surface chemistry variations among a series of laboratory-produced biochars, Geoderma, 2011, 163(3–4), 247–255.
[30] Keiluweit M, Kleber M, Molecular-level interactions in soils and sediments: The role of aromatic p-systems, Environ. Sci. Technol., 2009, 43(10), 3421–3429.
[31] Qiu Y, Cheng H, Xu C, Sheng GD, Surface characteristics of crop-residue-derived black carbon and lead (II) adsorption, Water Res., 2008, 42(3), 567–574.
[32] Dong X, Ma LQ, Li Y, Characteristics and mechanisms of hexavalent chromium removal by biochar from sugar beet tailing, J. Hazard. Mater., 2011, 190(1–3), 909–915.
[33] Cao X, Harris W, Properties of dairy-manure-derived biochar pertinent to its potential use in remediation, Bioresour. Technol., 2010, 101, 5222–5228.
[34] Puga AP, Abreu CA, Melo LCA, Beesley L, Biochar application to a contaminated soil reduces the availability and plant uptake of zinc, lead and cadmium, J. Environ. Manag., 2016, 159, 86–93.
[35] Zhang X, Wang H, He L, Lu K, Sarmah A, Li J, Bolan NS, Pei J, Huang H, Using biochar for remediation of soils contaminated with heavy metals and organic pollutants, Environ. Sci. Pollut. Res., 2013, 20, 8472–8483.
[36] Chen XC, Chen GC, Chen LG, Chen YX, Lehmann J, McBride MB, Hay AG, Sorption of copper and zinc by biochars produced from pyrolysis of hardwood and corn straw in aqueous solution, Bioresour. Technol., 2011, 102, 8877–8884.

[37] Shang J, Zong M, Yu Y, Kong X, Du Q, Liao Q, Removal of chromium (VI) from water using nanoscale zerovalent iron particles supported on herb-residue biochar, J. Environ. Manage., 2017, 197, 331–337.
[38] Liu L, Huang Y, Zhang S, Gong Y, Su Y, Cao J, et al., Adsorption characteristics and mechanism of Pb(II) by agricultural waste-derived biochars produced from a pilot-scale pyrolysis system, Waste Manag., 2019, 100, 287–295.
[39] Senthilkumar R, Reddy Prasad DM, Govindarajan L, Saravanakumar K, Naveen Prasad BS, Synthesis of green marine algal-based biochar for remediation of arsenic(V) from contaminated waters in batch and column mode of operation, Int. J. Phytoremediation, 2020, 22, 279–286.
[40] Tsai WT, Chen HR, Adsorption kinetics of herbicide paraquat in aqueous solution onto a low-cost adsorbent, swine-manure derived biochar, Int. J. Environ. Sci. Technol., 2013, 10(6), 1349–1356.
[41] Senthilkumar R, Reddy Prasad DM, Sorption of Heavy Metals Onto Biochar, In: Applications of Biochar for Environmental Safety, http://dx.doi.org/10.5772/intechopen.92346.
[42] Sun L, Wan S, Luo W, Biochars prepared from anaerobic digestion residue, palm bark, and eucalyptus for adsorption of cationic methylene blue dye: Characterization, equilibrium, and kinetic studies, Bioresour. Technol., 2013, 140, 406–413.
[43] Wang P, Liu X, Wu X, Xu J, Dong F, Zheng Y, Evaluation of biochars in reducing the bioavailability of flubendiamide in water/sediment using passive sampling with polyoxymethylene, J. Hazard. Mater., 2018, 344, 1000–1006.
[44] Vilvanathan S, Shanthakumar S, Column adsorption studies on nickel and cobalt removal from aqueous solution using native and biochar form of *Tectona grandis*, Environ. Progress Sustain. Energy, 2017, 36, 1030–1038.
[45] Lima IM, Marshall WE, Adsorption of selected environmentally important metals by poultry manure-based granular activated carbons, J. Chem. Technol. Biotechnol., 2005, 80, 1054–1061.
[46] Lou K, Rajapaksha AU, Ok YS, Chang SX, Pyrolysis temperature and steam activation effects on sorption of phosphate on pine sawdust biochars in aqueous solutions, Chem. Speciat. Bioavail., 2016, 28, 42–50.
[47] Hadjittofi L, Prodromou M, Pashalidis I, Activated biochar derived from cactus fibres–preparation, characterization and application on Cu (II) removal from aqueous solutions, Bioresour. Technol., 2014, 159, 460–464.
[48] Hamid SBA, Chowdhury ZZ, Zain SM, Base catalytic approach: A promising technique for the activation of biochar for equilibrium sorption studies of copper, Cu (II) ions in single solute system, Materials, 2014, 7(4), 2815–2832.
[49] Goswami R, Shim J, Deka S, Kumari D, Kataki R, Kumar M, Characterization of cadmium removal from aqueous solution by biochar produced from *Ipomoea fistulosa* at different pyrolytic temperatures, Ecol. Eng., 2016, 97, 444–451.
[50] Yang GX, Jiang H, Amino modification of biochar for enhanced adsorption of copper ions from synthetic wastewater, Water Res., 2014, 48, 396–405.
[51] Ma Y, Liu WJ, Zhang N, Li YS, Jiang H, Sheng GP, Polyethylenimine modified biochar adsorbent for hexavalent chromium removal from the aqueous solution, Bioresour. Technol., 2014, 169, 403–408.
[52] Shang MR, Liu YG, Liu SB, Zeng GM, Tan XF, Jiang LH, Huang XX, Ding Y, Guo YM, Wang SF, A novel graphene oxide coated biochar composite: Synthesis, characterization and application for Cr (VI) removal, RSC Adv., 2016, 6(88), 85202–85212.
[53] Tang J, Lv H, Gong Y, Huang Y, Preparation and characterization of a novel graphene/biochar composite for aqueous phenanthrene and mercury removal, Bioresour. Technol., 2015, 196, 355–363.

[54] Qian K, Kumar A, Patil K, Bellmer D, Wang D, Yuan W, Huhnke RL, Effects of biomass feedstocks and gasification conditions on the physiochemical properties of char, Energies, 2013, 6(8), 3972–3986.
[55] Li K, Jiang Y, Wang X, Bai D, Li H, Zheng Z, Effect of nitric acid modification on the lead(II) adsorption of mesoporous biochars with different mesopore size distributions, Clean Technol. Environ. Policy, 2016, 18(3), 797–805.
[56] Liu P, Liu WJ, Jiang H, Chen JJ, Li WW, Yu HQ, Modification of bio-char derived from fast pyrolysis of biomass and its application in removal of tetracycline from aqueous solution, Bioresour. Technol., 2012, 121, 235–240.
[57] Li B, Yang L, Wang CQ, Zhang QP, Liu QC, Li YD, Xiao R, Adsorption of Cd (II) from aqueous solutions by rape straw biochar derived from different modification processes, Chemosphere, 2017, 175, 332–340.
[58] Deng J, Liu Y, Liu S, Zeng G, Tan X, Huang B, Tang X, Wang S, Hua Q, Yan Z, Competitive adsorption of Pb (II), Cd (II) and Cu (II) onto chitosan-pyromellitic dianhydride modified biochar, J. Colloid Interface Sci., 2017, 506, 355–364.
[59] Han Y, Cao X, Ouyang X, Sohi SP, Chen J, Adsorption kinetics of magnetic biochar derived from peanut hull on removal of Cr (VI) from aqueous solution: Effects of production conditions and particle size, Chemosphere, 2016, 145, 336–341.
[60] Duan S, Ma W, Pan Y, Meng F, Yu S, Wu L, Synthesis of magnetic biochar from iron sludge for the enhancement of Cr (VI) removal from solution, J. Taiwan Inst. Chem. Eng., 2017, 80, 835–841.
[61] Lyu H, Tang J, Huang Y, Gai L, Zeng EY, Liber K, Gong Y, Removal of hexavalent chromium from aqueous solutions by a novel biochar supported nanoscale iron sulfide composite, Chem. Eng. J., 2017, 322, 516–524.
[62] Wu W, Li J, Lan T, Muller K, Niazi NK, Chen X, Xu S, Zhen L, Chu Y, Li J, Yuan G, Wang H, Unravelling sorption of lead in aqueous solutions by chemically modified biochar derived from coconut fiber: A microscopic and spectroscopic investigation, Sci. Total Environ., 2017, 576, 766–774.
[63] Wei D, Zhao C, Khan A, Sun L, Ji Y, Ai Y, Wang X, Sorption mechanism and dynamic behavior of graphene oxide as an effective adsorbent for the removal of chlorophenol based environmental-hormones: A DFT and MD simulation study, Chem. Eng. J., 2019, 370, 121964.

Piyush Gupta, Namrata Gupta, Subhakanta Dash

Biochar for sustainable environmental management

Abstract: The concern for sustainable environmental management is continuously growing across the globe. The quality of water, air, and soil has been degraded due to several natural and human activities. This environmental degradation can be managed by employing biochar as a cheaper substitute for commercially available cost-effective products. Thermochemical conversion of biomass wastes as a feedstock results in production of carbonaceous biochar which is having tremendous potential for environmental cleanup. The wide availability of biomass makes it a low-cost resource for producing biochar having pollutant remediating characteristics. This chapter provides an insight on diverse applications of biochar in several fields ranging from agriculture, carbon sequestration, construction, and battery industries to pollution remediation.

Keywords: environment, biomass, thermochemical conversion, biochar, pollution, pollutants, agriculture

1 Introduction

In recent years, researchers around the world have thoroughly studied biochar and the wide variety of potential applications. Biochar is a carbonized fine-grained material, as described by the International Biochar Initiative (IBI), characterized by high levels of organic carbon (OC) and low susceptibility to degrade from biomass pyrolysis and biodegradable waste. It is a stabilized, recalcitrant OC compound made from organic vegetables that is produced when biomass is heated in range between 300 and 1,000 °C at lower oxygen levels. Biochar is prepared via thermal treatments (mostly slow pyrolysis) of biomass feedstocks, including agricultural residues such as rice straw, wheat straw, maize stoves, and rice husk, maize stoves, which are usually laid in fields after harvest and used in many areas for fodder, waste (or burning), municipal solid waste, manure, wood chips, and so on. Because of its unique properties, it is increasingly being recognized as a solution to a variety of current issues, such as farm productivity and the improvement of contaminated habitats, primarily through elimination of CO_2 from environment and the improvement of soil functions in terms of sorptive capabilities, surface area, and porosity. Although other applications may be equally significant in environmental remediation, these early applications mainly focused on biochar as a land modification in agriculture (for treatment of groundwater and soil, stormwater filter media).

https://doi.org/10.1515/9783110734003-007

Biochar has recently been recognized for its multifunctional properties of carbon sequestration, contaminant immobilization, greenhouse gas reductions, soil fertilization, and water filtration [1–3]. The heterogeneous complex composition of biochars provides a better platform for the removal of pollutants (Figure 1).

The chemical composition in biochars depends upon the feedstock type and pyrolytic conditions for instance, residence time, heating rate, temperature, and type of reactor used. Biochars consist mainly of carbon. The organic portion of biochar has high carbon content and, depending on the form of feedstock, the inorganic portion includes mostly minerals such as calcium, magnesium, potassium, and inorganic carbonates (carbonate ions). The pyrolysis temperature used in biochar development does not significantly form graphite and aromatic rings in the biochar are not perfectly formed in stacked and aligned sheets as they are graphite. During biochar processing, more irregular carbon arrangements are made and contain O and H [4]. In certain cases the formation of minerals depends on the type of feedstock. Biochars are not generally fully carbonized and exhibit carbonized and non-carbonized phases both. Additionally, soil biochar tends to an increment in water infiltration, a retention of soil water, an ion exchange capability and a retention of nutrients [5, 6].

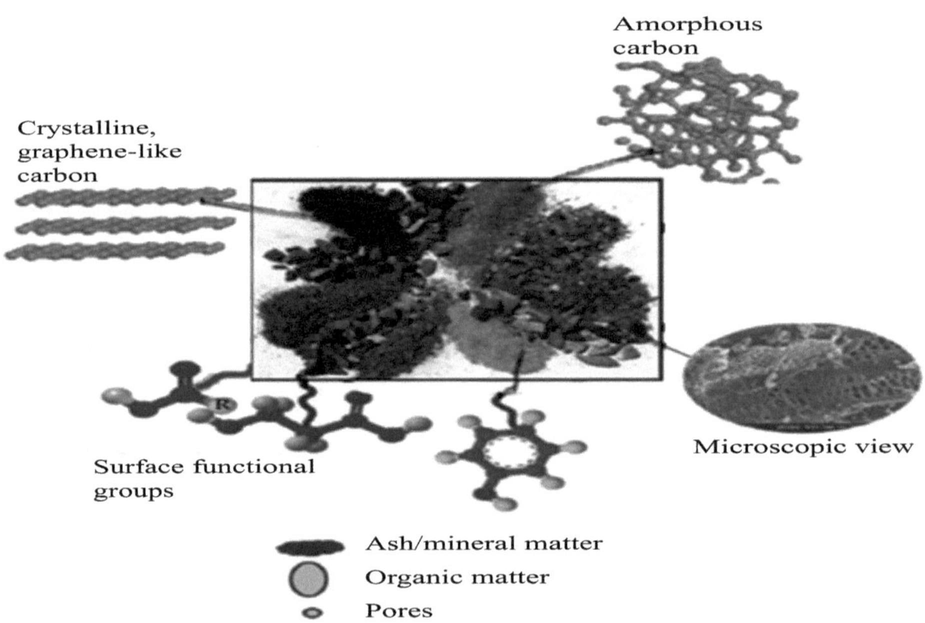

Figure 1: Heterogeneous complex compositional contents of biochar.

This chapter aims to present an overview on potential applications of biochar in various fields, some of which have to be explored, thanks to its highly variable and adaptable surface chemical framework.

2 History and emergence of biochar

The origin of biochar is in pre-Colombian times, when ancient American cultures of the area of Brazil Amazon rendered dark ground soils and terra preta, by slash-and-char. The soils are high in carbon content, up to 150 g C kg^{-1} of soil, compared with the environment (20–30 g C kg^{-1} of soil). Even hundreds of years after these soils were abandoned; this increasing carbon content is still seen to demonstrate that the organic matter is persistent in those dark earth soils. The cumulative carbon in these soils therefore tends to be greater than in neighboring soils. The persistence of carbon, soil fertility, and susceptibility of biochars has been the subject of considerable research. In 2000, Glaser et al. [7] proposed a new concept that recognizes black carbon (residues of incomplete combustion–biochar) as a possible cause and main ingredient in sustainable and fertile soils. The organic matter has shown a remarkable resemblance to biochar, leading scientists to this explanation, with its high carbon content and fertility. It is also observed by Smith in 1980 [8].

3 Biochar: an inexpensive precious resource

Biochar can be used in the extraction of renewable energy in an eco-friendly way from agricultural land. Typically, the biochar quality depends upon a variety of parameters: pyrolytic conditions, raw material used for carbonization, and biochar used for soil. The modification of the soil's biochar has proved useful in improvement of soil quality, retaining nutrients, and thereby increasing growth of plants [9]. The pH of soil, electrical conductivity (EC), phosphorus available (P), total nitrogen (TN), biocarbon (C), and cation-exchange capacities (CEC) have also increased because biochar is enriched with nutrients and organic matter. Earlier on, Verheijen et al. [10] reported deadliness, transport, and availability of many heavy metals in soil affected by the application of biochar as their ability for soil absorption increased. Availability of plant nutrients and possibility to use micro-organisms as a medium in biochar with its large surface areas has been recognized as the main causes for enhancing soil properties and absorption of plant nutrients in soils modified with biochar. Chan et al. [11] observed the use of biochar in soil and it was reported that biochar reduces soil core tensile strength and risk of soil compaction. Advantages of inoculating rhizobacteria in the soil are noticed, but adding biochar may also give soil some more nutrients and thus benefit agricultural crop yields.

Saxena et al. rendered "biochar and Bacillus species as the best combination for growth and yield of French beans" [12].

Adding biochar to soil can be helpful for both improving soil quality and stimulating plant development, and thus developing a sustainable agriculture system. The biochar modification has considered many applications and positive results as an efficient method to remove contamination from the soil without affecting natural environment and to receive higher crop yield [13]. Biochar shows positive impact on plant-growth and soil quality which indicates that biochar may be used for fixing nutrient shortcomings in the best possible way and an ideal method for improving nutrient cycles on farms.

4 Processing of biochar

The biochar contains the mineral elements, that is, carbon, sulfur, oxygen, hydrogen, nitrogen, and ash. Biomass is decomposed thermally under controlled-oxygen environments during pyrolysis. Biochar is a black-colored, porous, fine seeded, lightweight material having large surface, which attributes its implications in soil. It is used in soil to improve its consistency in order to take fundamental questions related to nature of soil degradation. The type of raw material used as biomass and the processing parameters define the characteristics of biochar.

4.1 Biomass as a feedstock in biochar production

Biochar is generally made using modern pyrolysis technology from residual biomass. There is a broad spectrum of organic materials which can be employed as a feedstock for producing biochar. There has been substantial use of agricultural waste (bark, straw, husks, seed, skins, bagasse, scarves, maize cobs, maize stalks, wood chips, manure, compost, wood residues, forest waste, green waste, cassava rhizome, etc.) and industrial waste, that is, bagasse and distillery's grain of urban/ municipal waste, thereby managing the waste by processing and use. The development of high-nutrient biochar was stated to rely on the use of raw material type and pyrolytic conditions. Feedstocks currently in commercial use include tree barks, wood chips, crop residues, grasses, and organic waste viz., sugarcane mill waste, grain from distillers, litter of chicken, milk manure, sewage sludge, pulp sludge. The biomass for biochar is primarily composed of hemicellulose, cellulose, and lignin. Cellulose was found to be the principal component of most of the plant biomasses, though lignin is also essential in woody biomass.

4.2 Biochar generation

Biochar can be achieved by means of pyrolysis using different biomass feedstocks. Pyrolysis is one of the mostly used thermal processes for the purpose. During pyrolysis, thermal breaking of lignin, hemicellulose, cellulose, and fat is perpetrated within feedstock to increase the carbon content in oxygen-free conditions, by removing by-products of oil and gases, noncarbon elements like oxygen and hydrogen [14]. The dry biomass waste is cut into smaller pieces (<3 cm in size) before use. The smaller pieces of feedstock are heated at temperatures between 350 and 700 °C (662–1,292 °F) either without oxygen or with low oxygen. Pyrolysis may be classified on the basis of temperature and heating time. Fast pyrolysis occurs at >500 °C and normally takes place in seconds (heating speeds as high as 1,000 °C min^{-1}). This condition maximizes bio-oil production. Slow pyrolysis typically consumes more time for feedstock to be completely pyrolyzed (heating speeds ~ 100 °C min^{-1}) from 30 min to a few hours and simultaneously gives more yields of biochar at 250–500 °C. Pyrolysis conditions and related distribution of by–products are summarized in Table 1. Moderate temperatures and longer residence times during slow pyrolysis are utilized in absence of oxygen for higher biochar outputs (30%), following increasing cracking reactions and reducing condensable fluids. In comparison, fast pyrolysis generates low mass yield of biochar (12%) and promotes biomass bio–oil formation (up to 75%) Figure 2.

Table 1: Thermal conversion processes and distribution of products.

Method	Temperature range	Heating rates	Residence period	By-products (%)		
				Biochar	Bio-oil	Syngas
Slow pyrolysis	350–980 °C	Slow (<10 °C min^{-1})	1 h	27–37	19–29	25–41
Fast pyrolysis	400–1,000 °C	Fast (~1,000 °C s^{-1})	<2 s	12	75	13
Hydrothermal Carbonization	180–300 °C	Slow	1–16 h	50–80	5–20	2–5
Gasification	700–1,500 °C	Moderate to fast	10–20 s	10	5	85
Torrefaction	200–300 °C	Slow (<10 °C min^{-1})	~10–60 min	80	0	20

In all pyrolysis reactions, high temperatures lead to lower char yields. It has been noticed that low temperature (<550 °C) biochar has an amorphous carbon structure with a lower aromaticity as compared to high temperature biochar. Pyrolysis has a considerable effect on biochar quality and its potential value usage for farming in context of carbon sequestration or agricultural production.

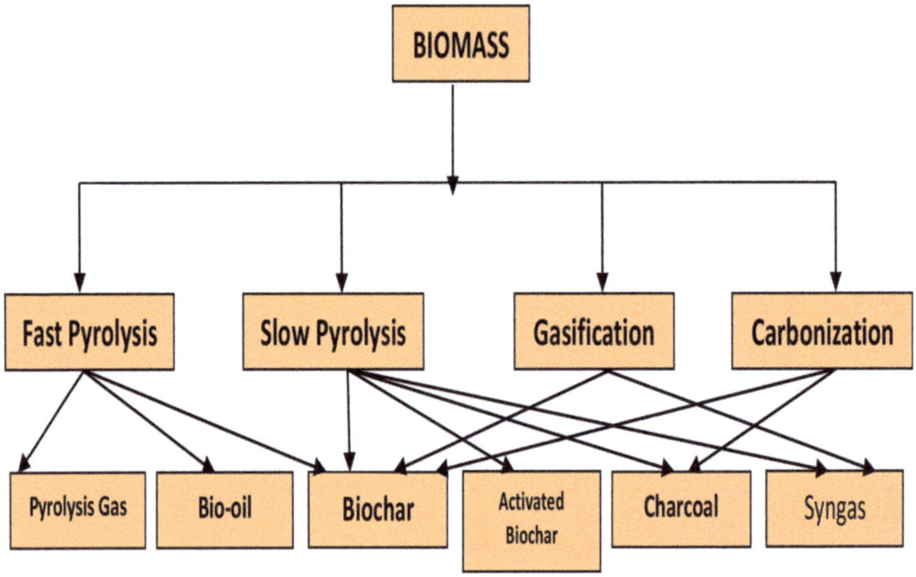

Figure 2: Production methods of biochar using biomass as feedstock.

The pyrolysis process can be shown as follows:

Biomass $\Rightarrow$ Biochar + liquid or oil (tars, water, etc.) + volatile gases (CO_2, CO, H_2)

5 Applications of biochar

Biochar is a significant source of carbon which can persists in the soil for thousands of years. It is produced for an addition to soil for carbon security and improvement of soil quality. Pyrolysis conditions and items utilized can enormously influence biochar properties. The physical characteristics of biochar enhance its functionality for environmental management [15]. Some of the applications of biochar in various dimensions of environment for its management are given further.

5.1 Biochar in amendment of soil

Food sustainability, climate change, reduced soil fertility, and profitability are drawing the focus of individuals to search new techniques or agricultural systems. The aim behind soil modification is to reduce risk factors of pollutant transmission to local waters and receptor species. Organic products, such as biochar, may be a good choice for this since they come from a biological source and can be applied directly to soils that do not need much pretreatment. Biochar modification is superior to other organic matter for two reasons: first, it has a high resistance to decay, allowing it to reside in soil for longer time periods, imparting long-term advantages to the soil; and second, it has the ability to retain nutrients. The addition of biochar to the soil boosts up the soil's stability by increasing pH, cation exchange, moisture retention, and microbial flora [16]. The availability of phosphorus, total nitrogen concentrations, and basic cations were reportedly increased after application of biochar to soil. Many soils get benefited from alkaline pH and biochar minerals (ash, which includes nitrogen, phosphorus, potassium, and trace elements). The modified soil became less acidic when the pH of the biochar was higher. As similar to other properties, pH of biochar is also determined by the type of feedstock used, processing temperature, and production time. Biochar may also be used for suppressing greenhouse gaseous emissions in soil. Zhang et al. observed the reduction in methane and nitrous oxide discharges by using agricultural soils [17]. Spokas et al. observed reduction in carbon dioxide production upon addition of biochar between 2% and 60% (w/w), and also a decrease in nitrous oxide levels [18]. According to several studies, biochar is capable to activate the soil microflora, resulting for the enhanced carbon accumulation in the soil. Biochar can provide home for actinomycetes, bacteria, and fungi in addition to adsorb organic substances, nutrients, and gases. Fast pyrolysis of biomass results in biochar with small pore sizes, less microorganisms, and more liquid and gaseous components. Biochar is capable to improve water conservation after application in soil and this may have an effect on soil microbial communities. Biochar offers an ideal environment for a diverse population of soil microorganisms, but the relationship between biochar and soil microorganisms is a complicated phenomenon.

5.2 Biochar in carbon sequestration

The sequestration of carbon is a carbon storage method in soil to remove greenhouse gas, that is, carbon dioxide from the atmosphere. The applicability of biochar for sequestering carbon in soil is receiving attention of researchers from various fields in recent years to introduce effective climate agricultural practices. Soil is having the capacity to hold and store three times more than the carbon present in air. During photosynthesis, green plants consume carbon dioxide which is a form of atmospheric carbon. It gives carbon back to soil in the form of plant residues, meanwhile a huge amount of carbon returns to environment as a result of decomposition process (in presence of microbes) of plant biomass which release carbon dioxide. An increase in soil carbon occurs when the carbon input rate exceeds the rate of microbial decomposition.

In the majority of cases, biochar makes up the majority of carbon (70–80%), and this fact has been used to its advantage: about four times the percentage of biomass than plant matter (approximately 40%) can (usually) as an amendment (40%).

Additionally, about 60% of this OC is more easily decomposed. Biochar is carbon negative and thus removes carbon from atmosphere. The pyrolytic stability of biochar can be improved by raising the biochar's temperature. Unsurprisingly, at 500 °C, biochar made at higher temperatures is less recalcitrant than lower ones. It considered to be more resistant to microbial decomposition, which means it can mitigate the impact of pollution on the atmosphere from agriculture. The temperature at which biochar decomposes appears to lose its glycerinability and becomes gaseous, relative to carbon dioxide content, is useful for bio sequestration in the soil. Using the methods of soil carbon sequestration and a primary production, it is the case that one must seek to maximize both yield and biochar stability. It is expected that biochar would have less effect on soils that are high in carbon compared to soils that are lower in carbon. Biochar can be used as a means of increasing the efficiency of the carbon sequestration process in rice paddies. Approximately 1 billion pounds of OC may be contained in biochar spread on land, or that the potential for soil organic carbon (SOC) [19].

5.3 Biochar in pesticides treatment

Adsorption starts immediately when pesticides are sprayed into the soil. The biochar's adsorption capability is the main factor which impacts environmental, desorption, or release behavior of pesticides, which must be monitored due to its link with bioavailability and effectiveness of pesticides. Reversible pesticide adsorption on biochar in amended or modified soil samples has been reported in various studies which involves following steps:

i. macro-pore network deformation due to sorbent swelling
ii. weakening of binding forces between biochar components and pesticides

Biochar inclusion improves the capacity to adsorb and decreases mobility. Carbon amendments are commonly found to significantly reduce pesticide loss due to the adsorption of pesticides into the micropores and/or permeability of the biochar particles in the soil matrix. Due to being microporous and carbon rich in nature, the biochar is a material of great interest for observing its influence on pesticide degradation with time. Pesticide biodegradation may be increased due to microbial stimulation in biochar–amended soil. The degradation of atrazine has been reportedly enhanced from 0.1% to 1% by adding wheat char in soil. It may be due to an increase in nutrients mainly phosphorus supplied by biochar, and thus stimulated microorganisms activity for effective biodegradation. The degradation of atrazine in clay soil modified with hardwood biochar produced at 450–500 °C tends to stimulate soil microflora by nutrients complemented by biochar. Low uptake of pesticides by plants grown in soil amended with biochar could be assigned to porosity of biochar which may increase sequestration of pesticides due to higher adsorption sites for pesticide molecules than unamended soil organic matter.

The amendment of soil using biochar lessens the bioavailability of pesticides to earthworms through:

a. enhanced adsorption capacities of biochar particles and
b. reduced food intake by earthworms.

Generally, earthworms are capable to sorb organic chemicals through skin present in soil or ingest soil particles directly depending upon the burrowing-feeding habits of earthworms. A decrease in pesticide bioavailability for earthworm species like *E. foetida* and *M. guillelmi* may be attributed to active adsorption sites on biochar surface which are responsible for reducing pesticide concentrations in soil [20].

5.4 Biochar in construction industry

Using biochar in building materials will reduce much of the greenhouse emissions and pollution caused by waste plastics, concrete, and organic waste. By sequestering the carbon from organic waste materials, biochar reduces the methane and carbon dioxide emissions released from landfill gas. If all of organic waste from landfills were to be converted to biochar, the reduction in landfill gas would amount to a potential 11% global reduction in human-sourced methane emissions. Furthermore, the addition of biochar in concrete bricks would reduce the amount of used cement, and thus, reducing the carbon dioxide released in the process. The current brick available for testing is composed of three parts biochar and one part portland cement. If these composite bricks are scaled globally, there is potential for a 6% decrease in the original 8% global carbon dioxide emissions produced from cement. Biochar can be mixed with waste plastic to fabricate brick to reduce the plastic waste problems. The biochar combined with clay and sand in 30–50% ratio and biochar-mud mixture comprising 50% of biochar, 30% of sand, and 20% of clay can be applied for plastering purpose. Biochar can also be added with lime and cement for plastering and exterior coatings. This plaster caters good insulation, moisture storage capacity and reduces the scaling. For achieving an optimal indoor climate, mud is recommended for the base material. The fabrication of biochar bricks by using lime, mud, and cement, as a binding material make these bricks a feasible material. Biochar has also been used for manufacturing concrete panel, plaster, biochar lime brick, and tile adhesive. Due to prominent properties viz. high chemical stability, low thermal conductivity, and low flammability it is considered to be the best construction material. Nowadays, biochar is treated as a feasible and cheaper mortar admixture and alternative modifier in bitumen for road construction. The addition of biochar in bitumen-based pavement material increases the durability while in asphalt, properties viz. resistance, fatigue, and ductility are enhanced. The char particles derived from carbonization of hazelnut and peanut shell affects the electromagnetic interference shielding performance and mechanical properties of concrete. Bamboo-derived char in 0.08% addition by weight of cement increases toughness and flexural strength of cement based composite materials. An inclusion of 2% biochar by mass of cement enhanced 40% strength of the concrete specimens [21–29].

5.5 Biochar in removal of pollutants

Biochar has a greater capability to combat with contaminants (organic, gaseous, and inorganic) due to wide accessibility of feedstock, low-cost and favorable physicochemical surface characteristics. However, research gaps and ambiguities associated with the use of biochar as an adsorbent are also identified [30].

5.5.1 Biochar in removal of organic pollutants

The adsorption characteristics and structural properties for treating malachite green and congo red were studied using biochar derived from litchi peel via carbonization and followed by activation. It was concluded that litchi peel based biochar comprised a higher surface area and pore volume of 1,006 $m^2\ g^{-1}$ and 0.588 $cm^3\ g^{-1}$, respectively, and adsorption capacities for congo red and malachite green were 404.4 and 2,468 mg g^{-1}, respectively which were assigned to pore filling, electrostatic interactions, hydrogen bonding, and π–π interactions. Regeneration and recycling of litchi peel derived biochar were also carried out for assuring the non-compliance of secondary environmental pollution [31].

Biochar provides a cheaper solution in comparison to other carbonaceous materials with very favorable achievement [32, 33] according to several studies performed on biochar particles. Jayawardhana et al. [34] employed biochar obtained after pyrolyzation of municipal solid waste in extraction of alkylated benzenes. The investigations were carried out by Kang et al. [35] for adsorption of phenanthrene on biochar-derived material. Lingamdinne et al. [36] studied the removal of nitro romatics by using rice husk and biochar. Mandal et al. [37] produced biochar from cucumber, tea, and mixed hardwood biomass waste for studying the adsorption of 2, 4-dichlorophenoxy acetic acid. Zhu et al. [38] subjected biochar surface to amination and oxidation which showed the improved absorption of 2,4-dichlorophenoxyacetic acid. Al Ameri et al. [39] achieved peat-derived biochar as a good bio-sorbent material for removing crude oil spills from synthetic seawater. He et al. [40] removed methylene blue from water using biochar particles. Hou et al. [41] discussed the use of bamboo shoot shells derived hydrochar for removal of rhodamine B. Zazycki et al. [42] observed the efficient removal of reactive Red 141 from aqueous medium using economical pecan nutshell-based biochar as an adsorbent. According to Heo et al. [43], bisphenol A and sulfamethoxazole can be removed by applying a $CuZnFe_2O_4$-tailored biochar composite. Kim et al. [44] gave an idea of ultrafiltration-activated biochar hybrid system for adsorbing 17 α-ethinyl estradiol, ibuprofen, and carbamazepine (47 wt%). Li et al. [45] analyzed cassava dregs derived biochars for the adsorption of ciprofloxacin. Jang et al. [46] examined sodium hydroxide-activated biochar to eliminate tetracycline traces. Mandal et al. [47] applied an agricultural waste stream-derived biochar in a multi-staged batch adsorption method for eliminating imidacloprid and atrazine from water medium.

5.5.2 Biochar in removal of gaseous pollutants

Das et al. [48] studied that biochar packed bio filter can be used for gas-phase hydrogen sulphide removal. Braghiroli et al. [49] studied the elimination of sulfur dioxide developed from anthropogenic sources. Shao et al. [50] demonstrated the improved adsorption capacity of biochar activated using carbon dioxide. Huang et al. [51] emphasized the production of biochar via microwave co-torrefaction of Leucaena wood and sewage sludge which observed 53 mg g^{-1} CO_2 uptake. Chiag et al. [52] described the CO_2 adsorption capacities of different biochar with surface microstructures and residual functionalities of carbon-rich substances. Zhang et al. [53] investigated the influence of nitrogen residual functionalities for the realization of a nitrogen-rich rice husk biochar appropriate for absorbing CO_2 at a rate of 59 mg g^{-1}. The pyrolysis of rice husk using microwave irradiation technique and modified with post-pyrolysis treatment methods [54, 55], basically exhibit very prominent CO_2 uptake values.

5.5.3 Biochar in removal of inorganic pollutants

The persistence of metal species in water domain make it polluted. It is crucial to manage this issue worldwide for the sake of human health. Huggins et al. [56] gave a comparative analysis of granular-activated carbon and granular wood-derived biochar used for studying wastewater stream employing batch and column systems. The results indicated that biochar-based filtering material decreases chemical oxygen demand (COD) and concentrations of phosphate, ammonium, arsenic, cadmium, chromium, lead, zinc, copper ions more effectively than activated carbon. Aran et al. [57] performed the copper adsorption studies from different feedstock-derived biochar materials. Banerjee et al. [58] fabricated a zirconium-caged steam-activated biochar to remove chromium ions (94 wt%) under flux rate of 4 mL min^{-1}. Zhang et al. [59] analyzed the cadmium and lead sorption using *Phyllostachys pubescens* biochar produced under a low-oxygen pyrolysis atmosphere. Trazzi et al. [60] performed adsorption and desorption studies for phosphate removal using biochar. The use of biochar in desalinization process represents a relevant application of water treatment methods.

5.6 Biochar in batteries production

Dai et al. [61] developed biochar-based hierarchical porous hollow carbon nanospheres which were operated as anode for Li-ion battery with a good discharge capacity of 1,169 mAh g^{-1}. Luna-Lama et al. [62] reported low porosity biochar from spent coffee grounds pyrolyzed at 800 °C which showed low performances (360 mAh g^{-1}) when used in batteries. Zhang et al. [63] observed 600 mAh g^{-1} capacity of microporous biochar. As per observations of Benitez et al. [64], specific capacity of lithium–sulfur battery can be achieved 915 mAh g^{-1} by using microporous biochar as a cathodic material. Chen et al. [65] cited that nitrogen doping may increase hierarchical porous biochar activity up to 550 mAh g^{-1} produced from pyrolization of pomegranate rind residues at 700 °C. Pan et al. [66] fabricated a silk-derived biochar with $ZnCo_2O_4$ nanocubes to generate a flexible performing anode material. Li et al. [67] decorated a pomelo pericarp biochar using Fe_3O_4 nanoparticles which showed good capacity in range of 635 mAh g^{-1}. Salimi et al. [68] produced an electrode material following the combination of Fe_3O_4 nanoparticle-tailoring process and algae-pyrolization which exhibited discharge capacity of 740 mAh g^{-1}. Saavedra Rios et al. [69] used a range of biochars from different biomass sources, as hard carbon anodes in sodium-ion battery applications.

6 Conclusion

The issue of land erosion brought about by the steadily expanding populace required an effective farming strategy. Biochar has been proposed as a method of cleaning up contaminated soil, upgrading soil fertility by reducing acidity, and expanding nutrient accessibility. Adding biochar to soil can likewise be perhaps the most proficient approaches to battle biotic pressure in the soil and improve crop yields. It can possibly improve soil quality, microbial bounty, natural nitrogen, and thus plant development.

Biochar is economical, renewable, and feasible in comparison to activated carbon. It is essential to search solutions to desorb and recover waste biochar to isolate the biochar loaded with different types of contaminants. Biochar laden with nitrogen (N) and phosphorus (P) can be used as a slow-release fertilizer in agriculture practices and environmental management methods. The biochar loaded with copper (Cu) or zinc (Zn) may be utilized as a micro-nutrient fertilizer also. However, release of any harmful components from biochar must be observed to avoid its sorption by crops which may lead to biomagnification. This chapter summarized the broad applications of biochar for managing all the dimensions of environment. It is better to activate or modify surface area, porosity, surface sorption sites of biochar to avail all its possible benefits. Future studies must be explored for commercial applications of biochar on a large scale which may considerably help society in maintaining the purity, nature regularity, and aesthetic aspects of environment.

References

[1] Ahmad M, Lee SS, Dou X, Mohan D, Sung JK, Yang JE, Ok YS, Effects of pyrolysis temperature on soybean stover–and peanut shell–derived biochar properties and TCE adsorption in water, Bioresour. Technol., 2012, 118, 536–544.

[2] Awad YM, Blagodatskaya E, Ok YS, Kuzyakov Y, Effects of polyacrylamide, biopolymer, and biochar on decomposition of soil organic matter and plant residues as determined by ^{14}C and enzyme activities, Eur. J. Soil Biol., 2012, 48, 1–10.

[3] Bolan NS, Kunhikrishnan A, Choppala GK, Thangarajan R, Chung JW, Stabilization of carbon in composts and biochars in relation to carbon sequestration and soil fertility, Sci. Total Environ., 2012, 424, 264–270.

[4] Joseph S, Socio–economic assessment and implementation of small–scale biochar projects, In: Lehmann J, Joseph S (Eds), Biochar for Environmental Management: Science and Technology, Earthscan, Sterling, VA, USA, 2009, 359–374.

[5] Cheng Y, Cai ZC, Chang SX, Wang J, Zhang JB, Wheat straw and its biochar have contrasting effects on inorganic N retention and N_2O production in a cultivated Black Chernozem, Biol. Fertil. Soils, 2012, 48(8), 941–946.

[6] Laird D, Fleming P, Wang B, Horton R, Karlen D, Biochar impact on nutrient leaching from a Midwestern agricultural soil, Geoderma, 2010, 158(3–4), 436–442.

[7] Glaser B, Balashov E, Haumaier L, Guggenberger G, Zech W, Black carbon in density fractions of anthropogenic soils of the Brazilian Amazon region, Org. Geochem., 2000, 31(7–8), 669–678.

[8] Smith NJH, Anthrosols and human carrying capacity in Amazonia, Ann. Am. Assoc. Geogr., 1980, 70, 553–566.

[9] Bonanomi G, Ippolito F, Cesarano G, Nanni B, Lombardi N, Rita A, Saracino A, Scala F, Biochar as plant growth promoter: better off alone or mixed with organic amendments, Front Plant Sci., 2017, 8, 1570.

[10] Verheijen F, Jeffery S, Bastos AC, Van Der Velde M, Diafas I Biochar application to soils–a critical scientific review of effects on soil properties, processes and functions. Luxemburg: EUR 24099, EN Office for the Official Publications of the European Communities, 2009, 149.

[11] Chan K, Van Zwieten L, Meszaros I, Downie A, Joseph S, Using poultry litter biochars as soil amendments, Aust. J. Soil. Res., 2008, 46(5), 437–444.

[12] Saxena J, Rana G, Pandey M, Impact of addition of biochar along with Bacillus sp, on growth and yield of French beans, Sci. Hortic., 2013, 162, 351–356.

[13] Placek A, Grobelak A, Kacprzak M, Improving the phytoremediation of heavy metals contaminated soil by use of sewage sludge, Int. J. Phytoremed., 2016, 18(6), 605–618.

[14] Zhu L, Lei H, Zhang Y, Zhang X, Bu Q, Wei Y, A review of biochar derived from pyrolysis and its application in biofuel production, SF J. Mater. Chem. Eng., 2018, 1(1), 1007.

[15] Ajema L, Effects of biochar application on beneficial soil organism, Int. J. Res. Stud. Sci. Eng. Technol., 2018, 5(5), 9–18.

[16] Mensah AK, Frimpong KA, Biochar and/or compost applications improve soil properties, growth, and yield of maize grown in acidic rainforest and coastal savannah soils in Ghana, Int. J. Agron, 2018, 1–8.

[17] Zhang A, Cui L, Pan G, Li L, Hussain Q, Zhang X, Zheng J, Crowley D, Effect of biochar amendment on yield and methane and nitrous oxide emissions from a rice paddy from Tai Lake plain, China, Agric. Ecosyst. Environ., 2010, 139(4), 469–475.

[18] Spokas KA, Koskinen WC, Baker JM, Reicosky DC, Impacts of woodchip biochar additions on greenhouse gas production and sorption/degradation of two herbicides in a minnesota soil, Chemosphere, 2009, 77, 574–81, 19.

[19] Purakayastha TJ, Kumari S, Sasmal S, Pathak H, Biochar carbon sequestration in soil – A myth or reality, Int. J. Bio–resource Stress Mgmt, 2015, 6(5), 623–630.
[20] Khorram MS, Zhang Q, Lin Q, Zheng Y, Fang H, Yu Y, Biochar: a review of its impact on pesticide behavior in soil environments and its potential applications, J. Environ. Sci., 2016, 44, 269–279.
[21] Schmidt HP, Novel uses of Biochar Proc USBI North American Biochar Symp, Center for Agriculture, Univ. of Massachusetts, Amherst, MA, 2013.
[22] Chebil S, Chaala A, Roy C, Use of softwood bark charcoal as a modifier for road bitumen, Fuel, 2000, 79(6), 671–683.
[23] Zhao MY, Enders A, Lehmann J, Short–and long–term flammability of biochars, Biomass Bioenergy., 2014, 69, 183–191.
[24] Zhao S, Huang B, Shu X, Ye P, Laboratory investigation of biochar–modified asphalt mixture, Transp. Res. Rec., 2014, 2445, 56–63.
[25] Walters RC, Fini EH, Abu–Lebdeh T, Enhancing asphalt rheological behavior and aging susceptibility using biochar and nanoclay, Am. J. Eng. Appl. Sci., 2014, 7(1), 66–76.
[26] Choi WC, Yun HD, Lee JY, Mechanical properties of mortar containing biochar from pyrolysis, J. Korea Inst. Struct. MaintInsp., 2012, 16(3), 67–74.
[27] Khushnood RA, Ahmad S, Restuccia L, Spoto C, Jagdale P, Tulliani JM, Ferro GA, Carbonized nano/microparticles for enhanced mechanical properties and electromagnetic interference shielding of cementitious materials, Front Struct. Civ. Eng., 2016, 10(2), 209–213.
[28] Ahmad S, Khushnood RA, Jagdale P, Tulliani JM, Ferro GA, High performance self–consolidating cementitious composites by using micro carbonized bamboo particles, Mater. Des., 2015, 76, 223–229.
[29] Low CY, Kua HW, Gupta S, Evaluation of using Biochar as Mortar Admixture, Dept of Building, National Univ of Singapore, Singapore, 2016.
[30] Tan X, Liu Y, Zeng G, Wang X, Hu X, Gu Y, Yang Z, Application of biochar for the removal of pollutants from aqueous solutions, Chemosphere, 2015, 125, 70–85.
[31] Jia WJ, Yang J, Feng P, Huang G, Xu C, Lin B, High–efficiency removal of dyes from wastewater by fully recycling litchi peel biochar, Chemosphere, 2020, 246, 125734.
[32] Palansooriya KN, Yang Y, Tsang YF, Sarkar B, Hou D, Cao X, Meers E, Rinklebe J, Kim KH, Ok YS, Occurrence of contaminants in drinking water sources and the potential of biochar for water quality improvement: a review, Crit. Rev. Environ. Sci. Technol., 2020, 1–63.
[33] Blanco–Canqui H, Biochar and water quality, J. Environ. Qual., 2019, 48, 2–15.
[34] Jayawardhana Y, Gunatilake SR, Mahatantila K, Ginige MP, Vithanage M, Sorptive removal of toluene and m–xylene by municipal solid waste biochar: simultaneous municipal solid waste management and remediation of volatile organic compounds, J. Environ. Manag., 2019, 238, 323–330.
[35] Kang S, Jung J, Choe JK, Ok YS, Choi Y, Effect of biochar particle size on hydrophobic organic compound sorption kinetics: applicability of using representative size, Sci. Total Environ., 2018, 619, 410–418.
[36] Lingamdinne LP, Roh H, Choi YL, Koduru JR, Yang JK, Chang YY, Influencing factors on sorption of TNT and RDX using rice husk biochar, J. Ind. Eng. Chem., 2015, 32, 178–186.
[37] Mandal S, Sarkar B, Igalavithana AD, Ok YS, Yang X, Lombi E, Bolan N, Mechanistic insights of 24–D sorption onto biochar: influence of feedstock materials and biochar properties, Bioresour. Technol., 2017, 246, 160–167.
[38] Zhu L, Zhao N, Tong L, Lv Y, Li G, Characterization and evaluation of surface modified materials based on porous biochar and its adsorption properties for 2,4-dichlorophenoxyacetic acid, Chemosphere, 2018, 210, 734–744.

[39] AlAmeri K, Giwa A, Yousef L, Alraeesi A, Taher H, Sorption and removal of crude oil spills from seawater using peat-derived biochar: an optimization study, J. Environ. Manag., 2019, 250, 109465.
[40] He J, Cui A, Deng S, Chen JP, Treatment of methylene blue containing wastewater by a cost-effective micro-scale biochar/polysulfone mixed matrix hollow fiber membrane: performance and mechanism studies, J. Colloid Interface Sci., 2018, 512, 190–197.
[41] Hou Y, Huang G, Li J, Yang Q, Huang S, Cai J, Hydrothermal conversion of bamboo shoot shell to biochar: preliminary studies of adsorption equilibrium and kinetics for rhodamine B removal, J. Anal. Appl. Pyrol., 2019, 143, 104694.
[42] Zazycki MA, Godinho M, Perondi D, Foletto EL, Collazzo GC, Dotto GL, New biochar from pecan nutshells as an alternative adsorbent for removing reactive red 141 from aqueous solutions, J. Clean. Prod., 2018, 171, 57–65.
[43] Heo J, Yoon Y, Lee G, Kim Y, Han J, Park CM, Enhanced adsorption of bisphenol A and sulfamethoxazole by a novel magnetic CuZnFe2O4–biochar composite, Bioresour. Technol., 2019, 281, 179–187.
[44] Kim S, Park CM, Jang A, Jang M, Hernández-Maldonado AJ, Yu M, Heo J, Yoon Y, Removal of selected pharmaceuticals in an ultrafiltration-activated biochar hybrid system, J. Membr. Sci., 2019, 570, 77–84.
[45] Li F, Feng D, Deng H, Yu H, Ge C, Effects of biochars prepared from cassava dregs on sorption behavior of ciprofloxacin, Proc. Environ. Sci., 2016, 31, 795–803.
[46] Jang HM, Kan E, Engineered biochar from agricultural waste for removal of tetracycline in water, Bioresour. Technol., 2019, 284, 437–447.
[47] Mandal A, Singh N, Optimization of atrazine and imidacloprid removal from water using biochars: designing single or multi-staged batch adsorption systems, Int. J. Hyg. Environ. Health, 2017, 220, 637–645.
[48] Das J, Rene ER, Dupont C, Dufourny A, Blin J, Van Hullebusch ED, Performance of a compost and biochar packed biofilter for gas-phase hydrogen sulfide removal, Bioresour. Technol., 2019, 273, 581–591.
[49] Braghiroli FL, Bouafif H, Koubaa A, Enhanced SO_2 adsorption and desorption on chemically and physically activated biochar made from wood residues, Ind. Crop Prod., 2019, 138, 111456.
[50] Shao J, Zhang J, Zhang X, Feng Y, Zhang H, Zhang S, Chen H, Enhance SO_2 adsorption performance of biochar modified by CO_2 activation and amine impregnation, Fuel, 2018, 224, 138–146.
[51] Huang YF, Chiueh PT, Lo SL, CO_2 adsorption on biochar from co-torrefaction of sewage sludge and leucaena wood using microwave heating, Energy Proc., 2019, 158, 4435–4440.
[52] Chiang YC, Juang RS, Surface modifications of carbonaceous materials for carbon dioxide adsorption: a review, J. Taiwan Inst. Chem. Eng., 2017, 71, 214–234.
[53] Zhang X, Zhang S, Yang H, Shao J, Chen Y, Feng Y, Wang X, Chen H, Effects of hydrofluoric acid pre-deashing of rice husk on physicochemical properties and CO_2 adsorption performance of nitrogen-enriched biochar, Energy, 2015, 91, 903–910.
[54] Huang YF, Chiueh PT, Shih CH, Lo SL, Sun L, Zhong Y, Qiu C, Microwave pyrolysis of rice straw to produce biochar as an adsorbent for CO2 capture, Energy, 2015, 84, 75–82.
[55] Shen Y, Fu Y, KOH-activated rice husk char via CO_2 pyrolysis for phenol adsorption Mater Today Energy, 2018, 9, 397–405.
[56] Huggins TM, Haeger A, Bi_nger JC, Ren ZJ, Granular biochar compared with activated carbon for wastewater treatment and resource recovery, Water Res., 2016, 94, 225–232.
[57] Arán D, Antelo J, Fiol S, Macías F, Influence of feedstock on the copper removal capacity of waste-derived biochars, Bioresour. Technol., 2016, 212, 199–206.

[58] Banerjee S, Joshi SR, Mandal T, Halder G, Application of zirconium caged activated biochar alginate beads towards deionization of Cr (VI) laden water in a fixed bed column reactor, J. Environ. Chem. Eng., 2018, 6, 4018–4029.
[59] Zhang C, Shan B, Tang W, Zhu Y, Comparison of cadmium and lead sorption by Phyllostachys pubescens biochar produced under a low–oxygen pyrolysis atmosphere, Bioresour. Technol., 2017, 238, 352–360.
[60] Trazzi PA, Leahy JJ, Hayes MHB, Kwapinski W, Adsorption and desorption of phosphate on biochars, J. Environ. Chem. Eng., 2016, 4, 37–46.
[61] Dai XH, Fan HX, Zhang JJ, Yuan SJ, Sewage sludge–derived porous hollow carbon nanospheres as high–performance anode material for lithium-ion batteries, Electrochim. Acta, 2019, 319, 277–285.
[62] Luna–Lama F, Rodríguez–Padrón D, Puente–Santiago AR, Muñoz–Batista MJ, Caballero A, Balu AM, Romero AA, Luque R, Non–porous carbonaceous materials derived from coffee waste grounds as highly sustainable anodes for lithium–ion batteries, J. Clean. Prod., 2019, 207, 411–417.
[63] Zhang H, Yu F, Kang W, Shen Q, Encapsulating selenium into macro–/micro–porous biochar–based framework for high–performance lithium–selenium batteries, Carbon, 2015, 95, 354–363.
[64] Benítez A, González–Tejero M, Caballero Á, Morales J, Almond shell as a microporous carbon source for sustainable cathodes in lithium–sulfur batteries, Materials, 2018, 11, 1428.
[65] Chen X, Du G, Zhang M, Kalam A, Su Q, Ding S, Xu B, Nitrogen–doped hierarchical porous carbon derived from low–cost biomass pomegranate residues for high performance lithium–sulfur batteries, J. Electroanal. Chem., 2019, 848, 113316.
[66] Pan P, Hu Y, Wu K, Cheng Z, Shen Z, Jiang L, Mao J, Ni C, Ge Y, Wang Z, Growth of ZnCo2O4 nanocubes on flexible biochar substrate derived from natural silk waste fabric for lithium–ion battery anode, J. Alloy Compd., 2020, 814, 152306.
[67] Li T, Bai X, Qi Y–X, Lun N, Yj B, Fe_3O_4 nanoparticles decorated on the biochar derived from pomelo pericarp as excellent anode materials for Li–ion batteries, Electrochim. Acta, 2016, 222, 1562–1568.
[68] Salimi P, Norouzi O, Pourhoseini SEM, Bartocci P, Tavasoli A, Di Maria F, Pirbazari SM, Bidini G, Fantozzi F, Magnetic biochar obtained through catalytic pyrolysis of macroalgae: a promising anode material for Li–ion batteries, Renew. Energy, 2019, 140, 704–714.
[69] Saavedra Rios C, Simone V, Simonin L, Martinet S, Dupont C, Biochars from various biomass types as precursors for hard carbon anodes in sodium–ion batteries, Biomass Bioenergy., 2018, 117, 32–37.

Poulomi Ghosh, Saprativ P. Das

Unraveling a dynamic ameliorant of heavy metal–polluted soil: biochar

Abstract: Urbanization acceleration along with advancement of industrial production in the recent years has led to a dramatic augmentation in heavy metal pollutants, harming utmost the ecological environment. There are global concerns of heavy alloy contamination in soils, thereby a greater requirement for soil indemnification. The accumulation along with leaching of heavy ores and metalloids, namely, arsenic, lead, cadmium, and mercury in soil is an issue of growing concern for their lethal, carcinogenic, and mutagenic effects corroborated with multiple organ failure in the living organisms. Contaminated soil leads to multiple physiochemical and biochemical changes ensuing in decreased growth of plants, affecting the yield, and thus, food unavailability. Biochar has been widely reported as an immobilizing agent for the heavy metals in contaminated soils aiding in soil remediation. This chapter furnishes an insight on the native properties of biochar, making it as a choice of adsorbent toward heavy metal remediation from soil.

Keywords: heavy alloy, soil remediation, biochar, immobilization, carcinogenic, lethal

1 Introduction

Soil is the ultimate abode for substantial or heavy metals (HMs) regardless of their common or anthropogenic sources. Mineral asset misusing and refining, metal electroplating, paint and covering handling, electronic gear fabricating, farmland sewage inundating and pesticides, and compound manure dissemination are the essential anthropogenic calisthenics exasperating HM pollution in soil [1, 2]. An excess of 30,000 tons of chromium and 800,000 tons of lead have been delivered into the climate all around the world in the past 50 years, the greater part of which at last aggregates in soils. It is accounted that roughly one-sixth of the absolute agrarian land territory in China and around 600,000 hectares of earthy colored field destinations in America have been debased by HMs [3]. As indicated by the post effects of a public soil review in 2014, the over-standard paces of cadmium, mercury, arsenic, copper, lead, chromium, zinc, and nickel in soils of Asian countries are 7.0%, 1.6%, 2.7%, 2.1%, 1.5%, 1.1%, 0.9% and 4.8%, individually [4]. Soil HMs enter the food chain primarily through farming harvests and at last gather in living beings through eating regimen, respiratory inward breath, skin contact, and other openness pathways, which straightforwardly or by implication rooting genuine negative consequences for

https://doi.org/10.1515/9783110734003-008

human welfare [5]. Soil decadence and decrease of agricultural crop yield in land caused by HM contamination brings a crucial requirement for the utilization of different productive in situ and ex situ remediation technique to minimize the ecological risk of HMs and maximize the quality and reliability of agricultural land.

Over the years, actual remediation (washing, warm desorption, hardening, and visitor land strategies), synthetic remediation (vitrification, draining, immobilization, and electrokinetic techniques), also, natural remediation (microorganisms and plants) approaches have been applied to accomplish this goal [6]. Eventually, these strategies pretty much exist with particular impediments, such as muddled method, shortcomings, helpless achievability, brief length, high monetary expenses, and heightened optional hazards. As of now, applying revisions to HM-tainted soil is viewed as one of the most encouraging in situ remediation methods [7]. The regular escalations of soil usage incorporate phosphate compounds, liming materials, dirt minerals, coal fly debris, natural fertilizers, metal oxides, and also biochar. In aggregate, the immobilization interaction of HMs can be accomplished principally through adsorption, complexation, decrease, and precipitation responses causing the rearrangement of HMs from soil fluid stages to strong stages, to decrease their versatility and bioavailability [8]. Biochar is a sort of carbon-rich, permeable substance with plentiful dynamic natural practical gatherings coupled with carbon fragrant designs with an unbiased to basic pH esteem, moderately high cation trade limit, enormous explicit surface region, and negative surface charge. Literature reveals that seed germination, plant development, crop yields, and microbial action have been altogether expanded in the HM-tainted soil corrected with biochars. Impacts of biochar on the immobilization/activation for various types of HMs have been affirmed in many pot examinations and field preliminaries [9]. Moreover, the creation interaction of biochar is viewed as a proficient administration strategy to discard an enormous number of natural squanders, exhibiting certain focal points in financial advantages with possible perspectives. Amazingly, it cannot be totally overlooked that there are still some expected dangers in the field utilization of biochar, and these potential dangers may hamper its further application. The associated portions of biochar are evaluated on the basis of (a) arrangement advancements, execution attributes furthermore impacting factors; (b) association components with various types of HMs; (c) consequences for plant development and HM bioavailability; and (d) expected dangers in field designing applications. The advancement and utilization of biochar are described in later sections.

2 Biochar and attributes

Biochar is a kind of baffling, stable, highly aromatic and high carbon content solid material produced by deserted biomass under oxygen-deficient conditions, high temperature, and slow pyrolysis (usually <700 °C) [10]. Biochar is a single co-product from the thermochemical processing of biomass used as a dirt remover or potential carbon sequestration specialist [10]. Biochar production method is presented in Figure 1.

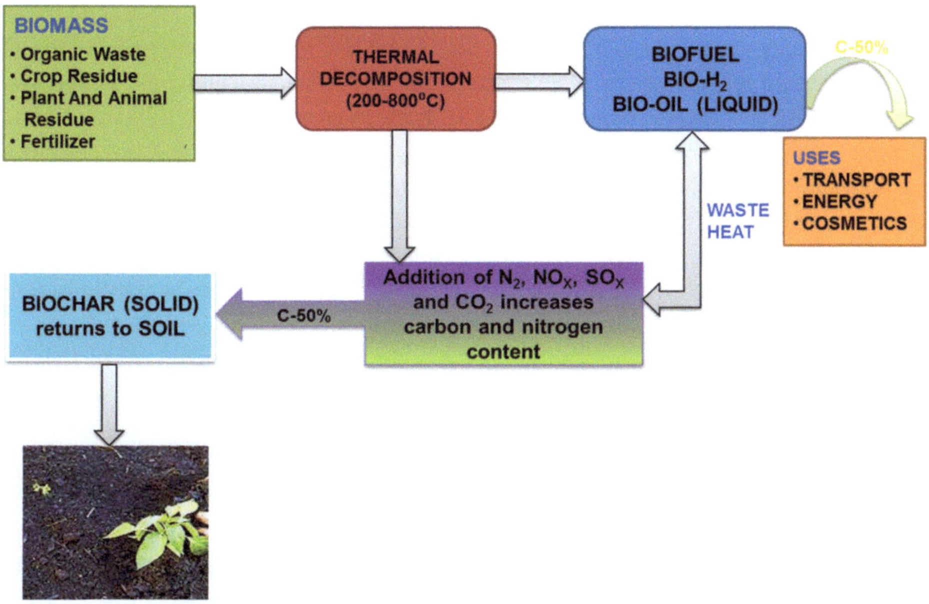

Figure 1: Biochar production method.

According to the source of biomass materials, biochar can be divided into three categories: plant debris biochar, animal excrement biochar, and sludge biochar.

2.1 Ash and mineral elements

Biochar predominantly comprised aromatized hydrocarbons and single carbon or carbon with graphite structures, with a carbon substance of over 60%, and different components, such as H, O, N and S. The synthetic creation and pyrolysis temperature of crude materials have an enormous effect on the properties of biochar. With the expanding temperature, the unpredictable segments of biochar decline; likewise the sections of hydrogen and oxygen with the step-by-step increment of fragments of carbon and debris [10]. In the pyrolysis interaction, huge quantities of mineral components like K, Na, Ca, Mg and supplement components, namely, N and P rich in organic waste and muck do not diminish with temperature, so the debris substance of animal waste and muck biochar is clearly higher than that of plant buildup biochar. Accordingly, the cation trade limit and the measure of replaceable particles in animal waste and muck biochars are moderately higher, while the carbon substance of them is clearly lower than that of plant buildup biochar [11].

2.2 Aromatic properties (H/C and O/C content)

The aromaticity of biochar can be demonstrated by H/C and O/C content [11]. Pyrolysis temperature influences the biochar aromaticity. The translucent construction and pore design of biochar change drastically with the expanding pyrolysis temperature. During the dewatering interaction and also polymerization, the lignin and cellulose in plant buildup biochar break down into more modest atoms with H/C and O/C content diminishing. Be that as it may, there are no lignocellulosic compounds in animal fecal and muck biochar, and the carbon content is altogether lower than that of plant buildups. Consequently, the aromaticity of plant buildup biochar is clearly higher than that of the two. The fragrant construction of biochar can be utilized as a π-electron giver or an acceptor, framing poison π bond with contaminations, further affecting the adsorption impact of biochar on contaminations [12].

2.3 Surface structure and functional groups

The pyrolysis temperature and crude materials likewise influence the surface design and morphology of biochar. As the pyrolysis temperature rises, the fat alkyl and ester bunch are annihilated and the sweet-smelling lignin atoms are presented to the surface, making the surface zone bigger [13]. The carbon substance of animal fecal and slime biochar is lower, H/C and O/C content are higher, and the level of aromatization is lower than that of plant buildup biochar, so their surface regions are more modest than that of plant biochar. In expansion, the practical design of biochar likewise changes clearly. The biochar delivered under high pyrolysis temperature has a solid aromaticity and its adsorption of substantial metals mostly connects with the particular surface region and π-bond. Contrastingly, biochar delivered under low temperature incorporates more oxygenic practical gatherings, and its adsorption of bulk metals is credited essentially to the complexation with utilitarian congregations [13].

2.4 pH of biochar

Most biochar is caustic and the alkalinity augments with the ascension of pyrolysis temperature. The higher pH of biochar might be related with countless soluble salts, alkaloid metals (Na, K, Ca and Mg), and furthermore, $CaCO_3$ [14]. Diverse crude materials project biochar with different pH, in which animal waste fertilizer has a clearly higher pH than that of plant buildup. The basic biochar produced using high pyrolysis temperature can improve the acidic soil, likewise, advancing the arrangement of metal hydroxide precipitation and adsorption of weighty metals [15].

3 Metal absorption mechanism by biochar

Till date, researchers chiefly centered on the remediation of substantial metals, such as As, Pb, Cd, Zn, Cr, Cu and Hg. In contrast to natural toxins, HMs are challenging to be biodegraded, in turn, intensifying the trouble of repayment. To that end, biochar has wide sources, permeable design, bigger surface region, and plentiful surface utilitarian assemblies, which can satisfactorily fix a specific measure of HM poisons [16]. Presently, the adsorption component of biochar on substantial metal contamination has not been resolved with great consideration of fundamental components and different mechanisms.

3.1 Ion exchange and adsorption of cationic π-function

The trade adsorption of biochar surface is one of the significant purposes behind the decrease of HM maneuvers. The greater the quantity of cation trade, the more grounded the maintenance of HMs [17]. The idea of particle trade is electrostatic association between negative charge bunches on the biochar surface and positive charges in the dirt. This sort of response with lower adsorption energy has a place with vague adsorption and has clear reversibility. The cationic π-function relies upon the aromatization of biochar. More the existence of the π-form sweet-smelling structure, further noteworthy is the negative charge in π-orbital changes; accordingly, the capacity of losing electrons of capacity bunches surges with the adsorption impact turning out to be more critical [18].

3.2 Co-precipitation

Biochar can effectively reduce the activities of HMs by adsorption and dissolution–precipitation of mineral constituents. The addition of biochar can upturn the pH of soil [19] and the reaction of heavy minerals ions with OH, PO_4^{3-}, CO_3^{2-} can form hydroxide, carbonate, or phosphate condensation, adequately solidifying the HM pollutants.

3.3 Complexation

This complexation is significant for the obsession of HM particles with solid proclivity. An enormous number of studies have revealed that the responses of substantial metal particles with oxygenic functional assemblies, namely hydroxyl gathering (-OH) carboxyl gathering (-COOH), and amino gathering (-NH_2) on biochar surface generate extraordinary commitments to the adsorption of HM particles [14].

3.4 Electrostatic absorption

The bigger surface zone and higher surface energy are useful for biochars to emphatically ingest the HM contaminations and eliminate them from the dirt.

Table 1: Remediation of different heavy metals by biochar.

Heavy metals	Type of biochar	Absorption technique	Inferences	References
Pb^{2+}	Sludge Dairy manure Rice husk	Complex reaction with hydroxyl (-OH) and carboxyl (-COOH) groups, precipitation and complexation Adsorption, ion exchange, and precipitate with PO_4^{3-}, CO_3^{2-} Electrostatic adsorption, ion exchange	Adsorption rate: 38. 2–42.3% Adsorption rate: 57. 7–61.8% Under the same condition: Dairy manure > rice straw	[37], [38], [39]
As^{5+}	Pine needle, maize straw, dairy manure	Electrostatic adsorption.	Under the same condition: Dairy manure > pine needle > maize straw	[27–33]
Cu^{2+}	Biochar	Complexation with functional groups	pH is 6–7	[23–26]
Cd^{2+}	Sugarcane leaves, rice straw, maize straw	Precipitate with CO_3, OH, electrostatic adsorption, precipitation	Under the same condition: rice straw > sugarcane leaves	[34–36]
Hg^{2+}	Irish moss, rice husk, rice straws	Complex reaction with hydroxyl (-OH) and carboxyl (-COOH) groups, ion exchange	Irish moss is better when pH = 5	[40–42]
Cr (VI)	Oak biochar	Complex reaction with hydroxyl (-OH) and carboxyl (-COOH) groups	Cr(VI) is converted into Cr (III) with lower poisonousness and dissolvability	[43–46]

As indicated in Table 1, the remediation system of biochar is diverse for various substantial metal toxins. For a similar HM particle, the adsorption component is distinctive when the biochar is unique. The adsorption impact of biochar on substantial metal particles is affected by numerous elements, such as crude materials of biochar, pyrolysis temperature, pH of soil, and physical and compound properties of HM particles along with the measure of biochar expansion. Studies have shown that under a similar condition, in spite of the fact that plant biochar has the biggest surface region, the adsorption impact of animal fecal biochar on bulk metal particles is superior to stop biochar and plant biochar (Table 1). This stands on the grounds that P-rich in the animal squander biochar can respond to certain HM particles through precipitation or co-precipitation, and make the best commitments during the maintenance interaction. Studies likewise showed that the adsorption impact of animal dung biochar on Pb^{2+} is the awesome three sorts of biochars (Table 1).

Likewise, several researchers contemplated the adsorption impact of various biochars on compound HM (Pb^{2+}, Cu^{2+} and Zn^{2+}, Cd^{2+}) contamination and revealed that the serious adsorption of HM particles can influence the adsorption post-effect of biochar on substantial metals when Pb^{2+}, Cu^{2+}, Zn^{2+}, Cd^{2+} exist together in the dirt [14, 20]. Contrasted with a solitary HM adsorption, the adsorption impact of biochar on complex HMs is altogether decreased. Owing to the restricted adsorption destinations on biochar surface, the degree of serious adsorption rises with the increment of introductory focus [21]. Among the four sorts of substantial metals, Pb^{2+} is most effectively retained. On the one hand, Pb^{2+} has the most grounded joining limit with PO_4^{3-} and CO_3^{2+}. Also, a steady highest electronegativity and lowest hydrated particle span causes Pb^{2+} to respond with oxygenic practical gathering mostly without any problems.

4 Biochar-assisted remediation of HM-contaminated soil

The various wellsprings of biomass feedstocks, and the assorted pyrolysis conditions applied in the arrangement measures, lead to different biochar execution qualities, which may thus influence the cooperation instruments among biochar and HMs. Then again, the best concerns of HMs have been centered around copper, arsenic, cadmium, lead, mercury and chromium.

Table 2 demonstrates the characterization of biochar for the remediation of HM-sullied soil. It is well established that the bountiful surface-effective congregations, mineral substances, basic metal particles, π-electrons, natural matters, and pore constructions of micropores provided by biochar are the compelling restricting locales of HMs. Biochar can ingest or consolidate soil HMs through complexation, decrease, cation trade, electrostatic fascination, and precipitation capacities, or convert HMs

Table 2: Biochar characterization for remediation of HM-sullied soil.

Physicochemical		Surface	Structural/molecular/thermal
Physical	**Chemical**		
pH	Surface area	SEM	TGA (thermal analysis)
Cation exchange capacity	Bulk concentration and pore size	FT-IR and Raman spectroscopy	TEM and XRD (structural)
Electrical conductivity	Particle size circulation	SEM–EDX	NMR and RAMAN (aromaticity)
Boehm titration	Gas chromatography	ZETA potential	X-ray absorption

from inorganic states into natural states, which changes HM versatility and bioavailability and afterward improves soil agronomic advantages [22]. In this manner, the co-operation components among biochar and HMs form basics for the dirt remediation.

4.1 Copper (Cu)

The functional assemblies and negative ζ-potential existing in chicken-compost inferred biochar has been demonstrated with high liking for Cu [23]. The immobilization cycle of Cu is accomplished by expanding soil pH worth and actuating the liming impact to invigorate the complexation of Cu(II) with biochar surface. Moreover, carbonates and hydroxides in the debris of woodchip-determined biochar are the significant adsorbents for Cu(II), which can advance the development of $CuCO_3$ and $Cu(OH)_2$ [24]. Along these lines, the fundamental mark of Cu immobilization is to utilize the alkalinity of biochar to improve soil pH reverence [25]. On the opposite side of the safeguard, the versatility of Cu is exceptionally influenced by the substance of soil-broke dissolved organic carbon (DOC). The *Miscanthus*-determined biochar can expand Cu fixation in soil arrangement. As such, the expansion of Cu focus in soil pore water is the general result of Cu(II) desorption from soil with the type of natural complex [26]. Henceforth, the immobilization/preparation of Cu by biochar ought to be additionally concentrated by the genuine sorts of biochar.

4.2 Arsenic (As)

Phosphorus (P) and arsenic (As) have comparative substance properties and the dirt P-content is a basic factor in controlling the versatility of As [27]. There is a critical positive relationship between the phosphate content and, furthermore, the arsenate content in As-defiled soils. After the dirt is treated with P-containing biochar, the opposition between solvent phosphate and arsenate that happened on the adsorption destination of soil particles advance the desorption of As from soil strong stage and expands the As fixation in pore water, for example, soybean-stover-, pine-needle-, and rice-straw-determined biochar. The job of soil DOC is like phosphorus; however, DOC has an added impact [28]. Scientists have demonstrated that the utilization of rice-straw-determined biochar under the anaerobic conditions expand the plenitude of Fe-lessening microbes (e.g., *Clostridium*, *Bacillus* and *Caloramator* sp.) in paddy soil, advancing the decrement of As(V) adsorbed on the undefined Fe/Al oxides [29]. Increment of DOC content in soils improved the microbial declining impact of As(V), and finally diminished the arrival of As(III) from paddy soil. Under the anoxic conditions, comparative outcomes were acquired in the use of oil palm fiber-determined biochar [30]. Biochar with high aromaticity and alkalinity, furthermore, can amplify the electron transport, and likes humus to advance the microbial decrease of Fe(III). The π-electrons given by the utilitarian gatherings on the outside of poultry droppings-determined biochar could advance the decrease of As(V), which is another significant factor to upgrade the versatility of As [31]. Nonetheless, the Fe-biochar and Mn-biochar can adsorb Fe/Mn oxides of biochar surface, hence, admirably restricting the movement of As into soil arrangement, demonstrating the adequacy of As immobilization by the adjusted biochar [32, 33].

4.3 Cadmium (Cd)

The action of Cd in soils emphatically relies upon soil pH quality [34]. The soluble substances such as CO_3^{2-}, PO_4^{3-} and OH^- contained in biochar generally have solid adsorption and restricting ability to Cd in soils, which makes the free Cd(II) change into $Cd(OH)_2$, $Cd_3(PO_4)_2$ and $CdCO_3$. The high adsorption liking delivered by the cation exchange impacts of soil calcite ($CaCO_3$) with Cd(II) is the primary factor to decrease the bioavailability of compact Cd; obviously, the bountiful useful congregations and enormous explicit surface zone in biochar are additional basics for compact disk immobilization. The corn-straw-inferred biochar diminished 91% of the $CaCl_2$-extractable Cd content in the farmland soil [34]. The application of 40 t ha^{-1} wheat-straw-inferred biochar to paddy soil, independently diminished $CaCl_2$-extractable Cd and DTPA-extractable Cd by 59% and 24%, separately, in the course of recent years [35]. In this way, the post-effects of biochar in expanding soil pH esteem are amazingly successful for the immobilization of Cd. It merits referencing that the fundamental substances in biochar can likewise move the deprotonation of corrosive utilitarian gatherings of biochar, and further upgrade the adsorption limit of Cd [36].

4.4 Lead (Pb)

The immobilization interaction of Pb in soils by biochar is moderately basic. The immobilization interaction of Pb exhibited by soybean-stover-determined biochar is ascribed to the π-cation electron contributor acceptor communication, which happens by the π-electron-rich biochar graphene surface and π-electron-insufficient decidedly charged Pb(II) particle [37]. Moreover, the cation exchange and precipitation responses between fundamental substances (CO_3^{2-}, OH- and other soil-soluble Ca^{2+} and Mg^{2+}) in biochar and Pb(II) can accomplish a critical immobilization impact of Pb like the development of $Pb_3(CO_3)_2(OH)_2$ precipitation [38]. Specifically, the poultry-compost-determined biochar and the sewage-slop-inferred biochar contains plentiful phosphates, intend to form insoluble mixtures, like $Pb_5(PO_4)_3Cl$, $Pb_5(PO_4)_3OH$, and β-$Pb_9(PO_4)_6$, with Pb(II) to lessen the versatility of Pb [37, 38]. The useful assemblies likewise have certain impacts on the immobilization of Pb. The vegetable-squander-determined biochar improved soil pH quality, yet in addition advanced the immobilization of Pb by the solid covalent holding activity of N-containing useful congregations (particularly for $-NH_2$) on the biochar surface, with gradual diminishing of the centralization of NH_4OAc-extractable Pb [39].

4.5 Mercury (Hg)

Hg, as an exceptional metal, has incredible poisonousness in soil; however, biochar is a successful instrument for soil remediation. The carboxyl gathering in a hardwood-determined biochar surface and soil Hg(II) particle build up a coordination response creating a complex of $-COOHg^+$ accelerate, in this way, decreasing the portability of Hg and the thiol functionalities and sulfoxide bunches are ready to respond with Hg(II) particle encouraging the formation of –S(Hg)- and $Hg(OSR_2)^{62+}$ complex [40]. Furthermore, scientists revealed rice-husk-inferred biochar having a higher sulfate focus contrasted to wheat-straw-inferred biochar, which is more successful to advance the mercury–sulfur coordination response and to deliver sulfide precipitation [41]. It is critical to realize that methylation of Hg is an extraordinary natural biogeochemical phenomenon, and once the inorganic Hg in soils is changed over into methylmercury (MeHg), its poisonousness and bioaccumulation will be immensely upgraded. The arrival of rice root exudates decreased the soil pH, in this manner upgrading the methylation of Hg, while the use of basic biochar expanded soil pH and adequately repressed methylation [42]. The sewage-muck inferred biochar with high natural matter substance can escalate the development and action of heterotrophic microorganisms in Hg-debased acidic farmland soil. Notwithstanding, the used pace of MeHg in rice has been altogether diminished, which prompted the restraint of MeHg amassing in rice and viably decreased the bioavailability of natural Hg [42].

4.6 Chromium (Cr)

Chromium (Cr) exists in soils as two valence states, to be specific Cr(VI) and Cr (III). Hence, the immobilization of Cr in soils is a confounded blend interaction of adsorption–decrease precipitation. The association component among biochar and Cr is chiefly shown in the surface adsorption impact of Cr(VI) by the oxygen-containing useful complexes such as C–O, C = O and –COOH in biochar, just as the decreased response of electrons given by biochar [43]. In particular, biochar changes Cr(VI) into Cr(III) with lower poisonousness and dissolvability by adsorption–decrease impacts, and partakes in the development of Cr_2O_3 and additionally $Cr(OH)_3$ precipitation, in order to accomplish the motivation behind Cr immobilization [44]. Chicken-compost-inferred biochar to Cr-spiked soil with low natural matter substance, which essentially expanded the stock of soil's natural carbon and supplements, improves the dirt breath and the microbial action lastly showing a predominant microbial Cr(VI) decreased impact and in turn the Cr biotoxicity [45]. Additionally, the content of Cr(VI) of animal fertilizer-inferred biochar in acidic soil was fundamentally higher than that in basic soil [46].

In brief, numerous HMs can be consistently immobilized in soil by biochar through explicit or ambiguous surface adsorption. It is essential that not just the action of Cu and As are influenced by DOC content in soil, yet in addition to Pb and disk too. This wonder is credited to the way that the high-portion use of biochar in soil brings about the sharp increment of DOC content, prompting the complexation of Pb and disk with DOC, finally expanding the portability and bioavailability of Pb and Cd. The adsorption and cation exchange responses among HMs and biochar debris (e.g., K^+, Na^+, Ca^{2+}, Mg^{2+} and other salt metal particles) are additionally significant variables for the immobilization of HMs.

5 Effect of biochar on bioavailability of HMs in soil

Plants in particular absorb, transport, and acquire HMs from infected soil through roots. Thus, the prime goals of biochar soil remediation are restricting the migration and transformation rates of HMs in soil and lowering their bioavailability, a good way to save HMs from entering into organisms via the meals chain and to remove their poisonous outcomes. The last intention of HM-infected soil remediation is to bloom crop yields on the basis of making sure meals manufacturing safety. On the other hand, the creation of biochar offers a supply of natural matter, N, P, K, Ca, Mg and different vitamins to the soil, which complements soil enzyme and microbial activities. Another way, the plant growth origin domain, soil water maintenance, and saturated hydraulic conductivity may be progressed with the presence of biochar. It increases the plant life biomass and dilutes the content material of HMs in plant tissues to lessen their phytotoxicity.

Scientists reported of expending 24 and 72 t ha^{-1} rice-husk-derived biochar to mercury-infected farmland soils, the Hg content material in rice grains decreased to 10 and 7.2 ng g^{-1} independently, which appreciably inhibited the transportation of Hg from soil to rice grains, and correctly reached the countrywide standard (under 20 ng g^{-1}) [41]. Similarly, implementation of 3% of soybean-straw-derived biochar (hydrothermal carbon at 350 °C) to the arsenic and cadmium co-infected farmland soil decreased the bioaccumulation of As in rice plants through 88% [47]. Besides, the As(III) content material reduced from 3.47 to 0.29 mg kg^{-1}, As(V) reduced from 715 to 150 μg kg^{-1}. The remedy outcomes on Cd had been similar reported. As for cadmium, a discipline trial confirmed that while the utility rates of soybean-straw-derived biochar and rice-straw-derived biochar had been 20 t ha^{-1}, the content material of Cd in rice roots, rice shoots, rice husks, and rice grains reduced through 25. 0–44.1%, 19.9–44.2%, 46.2–70.6% and 25.8–70.9%, respectively. Furthermore, the effectiveness of rice-straw-derived biochar in lowering Cd accumulation in rice grains changed additionally [42]. The immobilization outcomes of soil HMs, issues of HM uptake and accumulation through plant life, enhancement of plant biomass growth, and the dilution impact of HMs in plant tissues are the four pivotal performances of biochar to lessen the bioavailability of HMs.

6 Quality variations of biochar toward soil amendment

Biochar items from various sources shift generally in qualities and capacity limit as a dirt correction [48]. Regular biochar feedstocks stretch out to backwoods trash, crop buildups, food preparing waste, and composts including sewage muck and biosolids (Table 1). These biomass materials are fundamentally unique in natural and debris arrangements, ascribing to the striking quality varieties of the subsequent biochar items. The carbonization (pyrolysis) conditions further impact the quality attributes of biochar. Three boundaries are normally used to control the carbonization conditions: pyrolysis (top) temperature, strong home time, and warming rate, each extending over a wide scope of qualities. The pyrolysis temperature for biochar creation is for the most part in the scope of 300–700 °C. A higher temperature speeds up the carbonization measure, permitting the pyrolytic change of biomass to arrive at a more profound level and be finished in a more limited time [48]. Complete pyrolysis is fundamentally critical to change all feedstock organic carbon (OC) into carbonized, pyrogenic OC (i.e., changed, undefined C design). The strong home time for accomplishing total pyrolysis is resolved by the pyrolysis temperature and the warming rate, going from seconds to days. The warming rate might be pretty much as low as <1 °C s^{-1} furthermore, as high as >200 °C s^{-1}, shifting with the pyrolysis temperature, feed qualities (molecule size, dampness content, thickness), and feed mass stream [49, 50]. Biochar items from deficient pyrolysis likewise contain observable parts of uncarbonized C, with semitransparent character of the precursor resources) [48, 49].

Biochar is permeable, showing a harsh morphological surface with honeycomb-like anatomical or other unpredictable designs under the magnifying instrument. The inalienable micropores (measurement in the scope of 0.8–235 µm and for the most part <22 µm, with normal at 10 µm) incite the material a generally high intrapore volume (e.g., 0.9–1 $cm^3\ g^{-1}$) and low envelope thickness (e.g. 0.2–0.6 g cm^{-3}). A jumbled stacked graphene sheet structure has been proposed for biochar. The aromaticity of biochar for the most part increases while the surface usefulness diminishes as the pyrolysis temperature is raised [48]. This is generally an after-effect of the reformist misfortunes of aliphatic C–H, olefinic C = C, carbonyl, carboxyl, and hydroxyl bunches at a higher pyrolysis temperature [51, 52].

Biochar is the significant result of moderate pyrolysis but then the result of quick pyrolysis (focusing on pyrolysis bio-oil) and gasification (pyrolysis joined with incomplete oxidation – pointing at syngas). Inside every one of the three thermochemical strategies, the carbonization conditions can be changed in accordance with enhanced major item age. Indeed, even with a similar feedstock, biochars from gasification and quick pyrolysis are clearly lower in OC content and higher in debris content than the items from moderate pyrolysis. Generally speaking, biochars from various sources illustrates a pH (in 1:5 strong/water separate) going from 5.4 to 12.4, and mineral debris content from 1.1% to 82.0%. The pH is a far-reaching articulation of the water-dissolvable base cations (Na^+, K^+, Ca^{2+} and Mg^{2+}) and natural acids (e.g., formic corrosive, acidic corrosive, hydroxybutyric corrosive, and benzoic corrosive) in biochar. The base cations start from the mineral debris segments in the feedstock, while the natural acids are produced during biomass pyrolytic change. Biochar items from complete pyrolysis contain negligible natural acids, furthermore exhibiting a pH value ordinarily >7.5 to mirror the presence of essential debris minerals. The pH of biochar, in total, is identified with its mineral debris content; with higher debris content substances exhibiting higher pH. Biochars from higher temperature pyrolysis are typically more prominent in debris content and pH, also, the items got from excrement and rice plant deposits are higher in debris substance and pH than those acquired from wood, furthermore, other plant buildups. The OC substance of biochar went from 12.0% to 93.9%, for the most part diminishing as the debris content expanded. For low-debris feedstocks (e.g., wood and numerous other plant buildups), the inferred biochars showed an expanding pattern in OC content as the pyrolysis temperature was raised, while biochars obtained from debris-rich feedstocks (e.g., excrements, rice plant buildups, and bones) delineated a contrary pattern [49–51]. Biochars additionally contain pretty much the plant supplements N, P, K. The N, P, and K substance of the detailed biochars are 0.5–48.0, 0.1–198.0

and 1.4–91.5 g kg^{-1}, individually, being generally associated with the items' debris substance. A large portion of the feedstock P and K would be recuperated in biochar, yet the immense larger part of feedstock N would be lost to pyrolysis fumes (bio-oil and syngas), particularly at high pyrolysis temperatures (>500 °C). The supplements in biochar are gradually releasable, enhancing the supplement supply of altered soils to help plant development [51]. Sewage sludge and bone-inferred biochars may contain phosphorus at >100 g kg^{-1}, consequently an ideal wellspring of phosphorus to crops. Additionally, biochars created from strength biomass like civil natural solids and tainted soil-developed plant materials may contain poisonous components at the above-edge fixations for use as a dirt alteration [53]. The applications of biochar are represented in Figure 2.

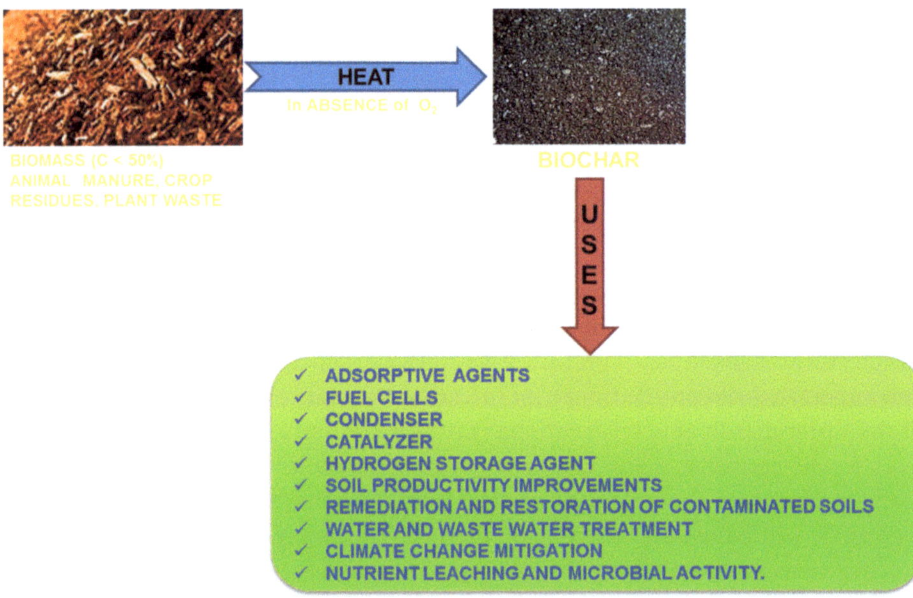

Figure 2: Applications of biochar.

Analysis has uncovered that biochar is for the most part sorptive to polar gases (e.g., H_2S, NH_3, NO_x and unstable natural compounds), water-solvent metal particles (e.g., Cu^{2+}, Pb^{2+}, Zn^{2+} and Cd^{2+}), and different natural toxins such as agrochemicals, antitoxins, and polyaromatic hydrocarbons (PAHs) [51]. A higher pyrolysis temperature in the scope of 300–700 °C favors micropores' improvement while advancing O-containing utilitarian gathering end. Biochars with high debris substance, similar to those got from composts, commonly show low specific surface area (SSA) values, because of debris blockage of the innate micropores. Subsequent to soil application, the SSA and cation exchange capacity (CEC) of biochar may augment step by step after some time because of common persistency [54–56]. Biochars exhibit a CEC estimation of 5–162 cmol kg^{-1}, furthermore, a Brunauer–Emmett–Teller (BET) – SSA estimation of 0.4–448 m^2 g^{-1}. The CEC reveals the limit of biochar to adsorb cationic supplements and metal foreign substances, while the SSA embroils biochar's capacity to hold water, gases, and natural particles. The CEC of biochar is begun essentially from the surface useful assemblies, while SSA is identified with biochar's porosity. As a soil alteration, biochar has a crucial benefit over fertilizer and other crude bio-materials in its high natural obstinacy. The mean home season of biochar in the regular soil climate was assessed at 90–1,600 years [55]. The ecological solidness of biochar starts from its dense sweet-smelling carbon. Biochars with higher aromaticity with noteworthy C buildup are by and large more headstrong [56].

Biochar OC can be ostensibly separated into three divisions: labile, halfway steady and stable [54]. Exploration recommends that >65% of the OC in biochars with H/OC molar proportion <0.7 would stay in common soils 100 years after field application. Biochars arranged at higher pyrolysis temperatures by and large contain a more noteworthy extent of stable OC. The International Biochar Initiative (IBI) suggests the following boundaries for assessing the nature of biochar as a soil revision: pH, lime equality, electrical conductivity (EC), mineral debris content, OC content, SSA, H/OC molar proportion, molecule size conveyance, germination restraint measure, absolute and accessible plant supplements along with the presence of inorganic and natural poisons. The OC content requirements surpassing 10% and the H/OC molar proportion giving up 0.7 for a pyrogenic material qualifies as biochar as per IBI standard defined in 2015. The European Biochar Certificate (EBC) program recommended comparable quality factors for biochar: pH, EC, unstable matter substance, absolute debris content, total carbon (TC) content, SSA, H/OC molar proportion, O/C molar proportion, water content, mass thickness, macronutrients (N, P, K, Ca, and Mg) and substances of HMs and natural impurities. A confirmed biochar item is needed to contain ≥50% TC at ≤0.4 OC, furthermore, ≤0.7 H/OC molar proportions. The substances of Zn, Pb, Cu, Cr, Ni, Cd and Hg are covered at 400, 150, 100, 1.5, 90, 50 and 1.5 mg kg^{-1}, individually, and PAHs, polychlorinated biphenyls, and polychlorinated dibenzo-*p*-dioxin/dibenzofuran (PCDD/F)s at 12, 0.2, and 0.00002 mg kg^{-1}, respectively, defined by EBC in 2012. Considering the instruments and the cycles through which biochar passes encourage soil remediation; biochars chose for soil remediation ought to be weighted more in pH, lime proportionality, SSA, CEC, and supplement (particularly P) content when the TC substance and H/OC molar proportion standards are met [54–56].

7 Potential risks associated with biochar applications

Numerous past investigations have featured the advantages of biochar applications, for example, immobilizing HMs in soils, expanding water holding limit and holding supplements in soils. Be that as it may, hazards related with biochar applications likewise warrant examinations. The harmful substances, namely polycyclic sweet-smelling hydrocarbons (PAHs), HMs, polychlorinated dibenzodioxins contained in biochar might be brought into soils following biochar change, affecting plant development and soil microbial networks [57]. Likewise, the paces of biochar expansion needed for the remediation of HM sullied soils change broadly, going from 0.5% to 30% of all out soil weight [58]. The absence of guidelines concerning utilization of biochar, along with the tirelessness of certain HMs at undeniable levels in soils, may present biological dangers. Challenges still exist in the protected utilization of biochar for soil remediation, in light of its conceivable immediate and roundabout negative consequences for soil environments and people.

7.1 Ecological risk

Various associations have given rules about the greatest allowed groupings of harmful substances in biochar. It was considered in a few surveys that repetitive uses of biochar to soils may not reason antagonistic impacts, in light of the fact that the groupings of poisonous substance in biochars are for the most part beneath the allowed focuses [57, 58]. The expected natural dangers from PAHs and HMs were insignificant as per the danger balance for PAHs along with the potential environmental danger files for HMs when applied with biochar. Notwithstanding, biochars obtained from feedstocks contribute to forming harmful component-rich soils or from sewage slime could contain significant degrees of HMs or other harmful substances [41–44]. This could consequently prompt accumulation of these substances in soils or plants after biochar amendment. Particularly for biochar pyrolyzed at low temperatures, HMs in the first feedstocks may not be volatilized at low temperatures [48–52]. Taking into account the harmful substances delivered from biochar, scientists reported that the expansion of biochar got from pig compost and sewage can deliver HMs (Cu(II), Zn(II), and so forth) to soils [55]. Literature revealed that the convergence of PAHs in biochar was high (5.48 mg kg^{-1}) and prompted moderate or extreme contamination when biochar was applied at high rates [40]. Utilization of matured (biochar that was applied in soils for a period) can upgrade the aggregation of acetochlor in maize, credited to the diminishing in sorption limit of the matured biochar. Therefore, a piece of the adsorbed foreign substances was desorbed and delivered into the dirt arrangement and taken up by plants [52–54]. A more broad assessment of the ongoing and long haul impacts of ecotoxicity of biochar application to soil is essential. Adjacent to the dangers received from harmful substances delivered from biochars, different changes related with biochar revision have likewise been recognized in ongoing investigations, for example, nitrate, PO_4, and sulfate filtering, contribution of broke down natural matter, expanded CH_4 discharge, just as antagonistic consequences for soil life forms, with suggestions on biogeochemistry and biological dangers of

HMs [55–57]. Methane emissions fundamentally expanded after the correction of biochar [13, 42]. Other exploration has revealed that biochar can restrain parasitic development and variety, decline the absolute phospholipid fatty acid contents due to the harmful substances contained in biochar and tend to adsorb microorganisms along with change microorganism-plant flagging, in turn, repressing rice development [42]. Moreover, there are developing confirmations revealing correction of yield residue-derived biochar could upgrade sulfate breaking down natural carbon levels in soils, encouraging microbial creation of methylmercury in soils and subsequently, methylmercury bioaccumulation [42, 49, 51, 52].

7.2 Human health risk

Considering the low thickness of biochar, its application into soils could create dust, especially during the cycle of mechanical aggravations. Scientists showed that dust discharges from soils treated with 2% (w/w) pecan shell biochar were higher than from control soils, and the residue emanations expanded with expanding application rates [40]. In this manner, biochar residue may present dangers to human wellbeing when applied to rural soils. Humans may breathe in biochar dust through the nose and mouth, which could prompt respiratory side effects [30]. Especially, poisonous synthetic substances contained in biochar could instigate human wellbeing chances. It is prescribed to diminish dust openness by applying water to keep up high soil dampness substance during biochar applications or produce biochar by pelleting or prilling [44–47]. The backhanded dangers of biochar to human comfort could be essentially ascribed to openness to the poisonous substances that are moved by biochar. For instance, the PAHs in biochar delivered during pyrolysis prompted their gathering in vegetables, bringing about expanded dangers of human dietary exposure and disease hazards [14, 15]. Furthermore, raised HM levels in crops developed in biochar-altered soils were ascribed to the biochar-contained HMs. The usage of HM-rich biochar fundamentally expanded the Cd(II) substance in plants, even in some marginally contaminated locales [49]. The miniature particles of biochar may likewise, move to more profound soil depths and stance hazards by going about as a mode for the carrier of harmful substances to groundwater when biochar is utilized for soil remediation [22]. Close to the above chances, other unfortunate impacts of biochar have likewise been accounted for. The expansion of biochar has been answered to diminish the germination rate and root length of cabbage and decline the effectiveness of agrochemicals, herbicides and pesticides [38]. This might be due to the biochar's solid adsorption and restricting limit with agrochemicals, diminishing the viability of weeding and insect poison [38]. Exchanges can be made to moderate the potential dangers related with biochar alteration. Selection of a proper pyrolysis temperature and additionally biochar feedstocks can diminish HM focuses in biochar [31, 33, 38, 44]. Pyrolysis can convert labile HMs into less bioavailable structures, namely, co-pyrolysis of slime with other uncontaminated feedstocks or, on the other hand, synthetic compounds can change labile HMs in the muck into additional stable portions [12, 36, 38]. Potassium hydroxide has been utilized to expand the leftover division of Cd(II) in biochars, thus declining its accessibility [51].

8 Conclusions and future prospects

It is an achievable system of green, financial and ecological security to apply biochar to remediate the HM-tainted soil. A few thermochemical transformation innovations have been used to plan biochar, and the presentation qualities of biochar are exceptionally influenced by the kind of feedstock materials, pyrolysis temperature along with the residence time. The immobilization or activation components of HMs incorporate complexation, decrease, cation altercation along with electrostatic enthrallment and precipitation responses. The HM uptake and collection of plants can be smothered by biochar through the variety in soil pH, DOC content and other mineral content, accomplishing the objective of advancing plant development along with reduced HM bioavailability and enhanced soil quality.

Present-day soil natural administration and remediation advances are starting to pay thoughtfulness regarding the drawn-out steadiness of biochar and the natural reactions of toxins. To fulfill the need and application-based outcomes, the future utilization of biochar is anticipated as follows:

- It is normal that the auxiliary natural dangers during biochar creation are focused on the arrangement of HMs, VOCs, PCDD/Fs and PAHs. It is important to screen and pretreat crude biomass materials to eliminate toxic HMs. Furthermore, optimization of the biochar preparation parameters and other interaction boundaries should be restricted to related contaminations. Besides, it is important to set up the administrative frameworks identified with the use of biochar in reasonable designing.
- In spite of the fact that the exploration on soil HM immobilization by biochar has created to research areas and momentary fields, a definitive objective is to extend it to huge scope designing applications. Completion of long-haul situating tests in various kinds of HM-polluted soil should be directed, in order to additionally check the drawn-out strength of biochar and in turn consequences for soil climate (collection, surface potential and thickness greatness), and in conclusion to guarantee the quality and wellbeing of agrarian land.
- Biochar cannot totally remediate the substantial metal sullied soil. Hence, to improve the remediation/improvement impacts of multisubstantial metal sullied soil, the multifunctional biochar materials (e.g., biochar immunized with microorganisms and biochar altered by synthetics and minerals) ought to be slowly put on the phase of designing application.

- The benefits and flaws between the monetary expense (creation) and advantage values (utilization) of biochar should be deliberately estimated. To improve monetary accessibility, simpler production techniques and less expensive origins of crude biomass materials should be found, which could give a platform to mend creation proficiency and diminishing financial weights.

References

[1] Xiao R, Wang S, Li R, Wang JJ, Zhang Z, Soil heavy metal contamination and health risks associated with artisanal gold mining in Tongguan, Shaanxi, China, Ecotoxicol. Environ. Saf., 2017, 141, 17–24.

[2] Gimeno-García E, Andreu V, Boluda R, Heavy metals incidence in the application of inorganic fertilizers and pesticides to rice farming soils, Environ. Pollut., 1996, 92(1), 19–25.

[3] Mahar A, Wang P, Ali A, et al., Challenges and opportunities in the phytoremediation of heavy metals contaminated soils: a review, Ecotoxicol. Environ. Saf., 2016, 126, 111–121.

[4] Li T, Liu Y, Lin S, Liu Y, Xie Y, Soil pollution management in china: a brief introduction, Sustainability, 2019, 11, 556.

[5] Zheng S, Wang Q, Yuan Y, Sun W, Human health risk assessment of heavy metals in soil and food crops in the Pearl River Delta urban agglomeration of China, Food Chem., 2020, 316, 126213.

[6] Dhaliwal SS, Singh J, Taneja PK, Mandal A, Remediation techniques for removal of heavy metals from the soil contaminated through different sources: a review, Environ. Sci. Pollut. Res., 2020, 27, 1319–1333.

[7] Karna RR, Luxton T, Bronstein KE, Hoponick Redmon J, Scheckel KG, State of the science review: potential for beneficial use of waste by-products for *in situ* remediation of metal-contaminated soil and sediment, Crit. Rev. Environ. Sci. Technol., 2017, 47(2), 65–129.

[8] Bolan N, Kunhikrishnan A, Thangarajan R, et al., Remediation of heavy metal(loid)s contaminated soils – To mobilize or to immobilize?, J. Hazard. Mater., 2014, 266, 141–166.

[9] Bashir S, Hussain Q, Akmal M, et al., Sugarcane bagasse-derived biochar reduces the cadmium and chromium bioavailability to mash bean and enhances the microbial activity in contaminated soil, J. Soils Sediment, 2017, 18, 874–886.

[10] Sohi SP, Krull E, Lopez-Capel E, Bol R, A review of biochar and its use and function in soil, Adv. Agron., 2010, 105, 47–52.

[11] Chun Y, Sheng G, Chiou CT, Xing B, Compositions and sorptive properties of crop residue-derived chars, Environ. Sci. Technol., 2004, 38(17), 4649–4655.

[12] Chen B, Zhou D, Zhu L, Transitional adsorption and partition of nonpolar and polar aromatic contaminants by biochars of pine needles with different pyrolytic temperatures, Environ. Sci. Technol., 2008, 42(14), 5137–5143.

[13] Jindo K, Mizumoto H, Sawada Y, Sanchez-Monedero MA, Sonoki T, Physical and chemical characterization of biochars derived from different agricultural residues, Biogeosciences, 2014, 11, 6613–6621.

[14] Campos P, Rosa JMDL, Assessing the effects of biochar on the immobilization of trace elements and plant development in a naturally contaminated soil, Sustainability, 2020, 12, 6025.

[15] Ahmad M, Lee SS, Lim JE, et al., Speciation and phytoavailability of lead and antimony in a small arms range soil amended with mussel shell, cow bone and biochar: EXAFS spectroscopy and chemical extractions, Chemosphere, 2014, 95, 433–441.

[16] Jin HP, Choppala GK, Bolan NS, Chung JW, Chuasavathi T, Biochar reduces the bioavailability and phytotoxicity of heavy metals, Plant Soil, 2011, 348, 439–451.

[17] Lehmann J, Gaunt J, Rondon M, Bio-char sequestration in terrestrial ecosystems-a review, Mitig. Adapt. Strat. Glob. Chang., 2006, 11, 403–410.

[18] Li H, Dong X, Da Silva EB, De Oliveira LM, Chen Y, Ma LQ, Mechanisms of metal sorption by biochars: biochar characteristics and modifications, Chemosphere, 2017, 178, 466–478.

[19] Rees F, Simonnot MO, Morel JL, Short-term effects of biochar on soil heavy metal mobility are controlled by intra-particle diffusion and soil pH increase, Eur. J. Soil Sci., 2014, 65(1), 149–161.

[20] Houben D, Evrard L, Sonnet PS, Beneficial effects of biochar application to contaminated soils on the bioavailability of Cd, Pb and Zn and the biomass production of rapeseed (*Brassica napus* L.), Biomass Bioenerg., 2013, 57, 196–204.

[21] Park JH, Ok YS, Kim SH, et al., Competitive adsorption of heavy metals onto sesame straw biochar in aqueous solutions, Chemosphere, 2016, 142, 77–83.

[22] Wang M, Zhu Y, Cheng L, et al., Review on utilization of biochar for metal-contaminated soil and sediment remediation, J. Environ. Sci., 2018, 63, 156–173.

[23] Meier S, Curaqueo G, Khan N, et al., Chicken-manure-derived biochar reduced bioavailability of copper in a contaminated soil, J. Soils Sediment., 2017, 17, 741–750.

[24] Rechberger MV, Kloss S, Wang S-L, et al., Enhanced Cu and Cd sorption after soil aging of woodchip-derived biochar: what were the driving factors?, Chemosphere, 2019, 216, 463–471.

[25] Gonzaga MIS, Mackowiak C, De Almeida AQ, et al., Assessing biochar applications and repeated *Brassica juncea* L. production cycles to remediate Cu contaminated soil, Chemosphere, 2018, 201, 278–285.

[26] Wagner A, Kaupenjohann M, Biochar addition enhanced growth of *Dactylis glomerata* L. and immobilized Zn and Cd but mobilized Cu and Pb on a former sewage field soil, Eur. J. Soil Sci., 2015, 66, 505–515.

[27] Ahmad M, Ok YS, Kim BY, et al., Impact of soybean stover-and pine needle-derived biochars on Pb and As mobility, microbial community, and carbon stability in a contaminated agricultural soil, J. Environ. Manag., 2016, 166, 131–139.

[28] Kim HB, Kim SH, Jeon EK, et al., Effect of dissolved organic carbon from sludge, Rice straw and spent coffee ground biochar on the mobility of arsenic in soil, Sci. Total Environ., 2018, 636, 1241–1248.

[29] Wang N, Xue XM, Juhasz AL, Chang ZZ, Li HB, Biochar increases arsenic release from an anaerobic paddy soil due to enhanced microbial reduction of iron and arsenic, Environ. Pollut., 2017, 220, 514–522.

[30] Qiao JT, Li XM, Li FB, Roles of different active metal-reducing bacteria in arsenic release from arsenic-contaminated paddy soil amended with biochar, J. Hazard. Mater., 2018, 344, 958–967.

[31] Choppala G, Bolan N, Kunhikrishnan A, Bush R, Differential effect of biochar upon reduction-induced mobility and bioavailability of arsenate and chromate, Chemosphere, 2016, 144, 374–381.

[32] Yin D, Wang X, Peng B, Tan C, Ma LQ, Effect of biochar and Fe-biochar on Cd and As mobility and transfer in soil-rice system, Chemosphere, 2017, 186, 928–937.

[33] Yu Z, Qiu W, Wang F, Lei M, Wang D, Song Z, Effects of manganese oxide-modified biochar composites on arsenic speciation and accumulation in an Indica rice (*Oryza sativa* L.) cultivar, Chemosphere, 2016, 168, 341–349.

[34] Gao X, Peng Y, Zhou Y, Adeel M, Chen Q, Effects of magnesium ferrite biochar on the cadmium passivation in acidic soil and bioavailability for Packoi (*Brassica chinensis* L.), J. Environ. Manag., 2019, 251, 109610.

[35] Rong Q, Zhong K, Li F, et al., Combined effect of ferrous ion and biochar on cadmium and arsenic accumulation in rice, Appl. Sci., 2020, 10(1), 300.

[36] Qian TT, Wu P, Qin QY, Huang YN, Wang YJ, Zhou DM, Screening of wheat straw biochars for the remediation of soils polluted with Zn (II) and Cd (II), J. Hazard. Mater., 2019, 362, 311–317.

[37] Zhao B, Niebuhr AJ, Ly Y, Douangdalangsy K, Effects of soybean stover-derived biochar on microbial community and structure in loess soil, Nat. Environ. Pollut. Technol., 2020, 19(2), 703–710.

[38] Huang D, Deng R, Wan J, et al., Remediation of lead-contaminated sediment by biochar-supported nano-chlorapatite: accompanied with the change of available phosphorus and organic matters, J. Hazard. Mater., 2018, 348, 109–116.

[39] Igalavithana AD, Kwon EE, Vithanage M, et al., Soil lead immobilization by biochars in short-term laboratory incubation studies, Environ. Int., 2019, 127, 190–198.

[40] Wang AO, Ptacek CJ, Blowes DW, et al., Application of hardwood biochar as a reactive capping mat to stabilize mercury derived from contaminated floodplain soil and riverbank sediments, Sci. Total Environ., 2019, 652, 549–561.

[41] Xing Y, Wang J, Xia J, et al., A pilot study on using biochars as sustainable amendments to inhibit rice uptake of Hg from a historically polluted soil in a Karst region of China, Ecotoxicol. Environ. Saf., 2019, 170, 18–24.

[42] Zhang J, Wu S, Xu Z, et al., The role of sewage sludge biochar in methylmercury formation and accumulation in rice, Chemosphere, 2018, 218, 527–533.

[43] Lyu H, Zhao H, Tang J, et al., Immobilization of hexavalent chromium in contaminated soils using biochar supported nanoscale iron sulfide composite, Chemosphere, 2018, 194, 360–369.

[44] Xia S, Song Z, Jeyakumar P, Bolan N, Wang H, Characteristics and applications of biochar for remediating Cr(VI)-contaminated soils and wastewater, Environ. Geochem. Health., 2020, 1543–1567.

[45] Choppala GK, Bolan NS, Megharaj M, Chen Z, Naidu R, The influence of biochar and black carbon on reduction and bioavailability of chromate in soils, J. Environ. Qual., 2012, 41(4), 1175–1184.

[46] Mandal S, Sarkar B, Bolan N, Ok YS, Naidu R, Enhancement of chromate reduction in soils by surface modified biochar, J. Environ. Manage., 2017, 186, 277–284.

[47] Li G, Khan S, Ibrahim M, et al., Biochars induced modification of dissolved organic matter (DOM) in soil and its impact on mobility and bioaccumulation of arsenic and cadmium, J. Hazard. Mater., 2018, 348, 100–108.

[48] Guo M, He Z, Uchimiya SM, Introduction to biochar as an agricultural and environmental amendment, In: Guo M, He Z, Uchimiya SM (ed), Agricultural and Environmental Applications of Biochar: Advances and Barriers, Soil Science Society of America (SSSA) Special Publications, Madison, USA, 2015, Vol. 63, 1–14.

[49] Yang Y, Sheng G, Enhanced pesticide sorption by soils containing particulate matter from crop residue burns, Environ. Sci. Technol., 2003, 37(16), 3635–3639.

[50] Guo M, The 3R principles for applying biochar to improve soil health, Soil Syst., 2020, 4(1), 9.

[51] Tan X, Liu Y, Zeng G, et al., Application of biochar for the removal of pollutants from aqueous solutions, Chemosphere, 2015, 125, 70–85.

[52] Fu P, Hu S, Xiang J, Sun L, Su S, An S, Study on the gas evolution and char structural change during pyrolysis of cotton stalk, J. Anal. Appl. Pyrol., 2012, 97, 130–336.
[53] Buss W, Graham MC, Shepherd JG, Mašek O, Suitability of marginal biomass-derived biochars for soil amendment, Sci. Total Environ., 2016, 547, 314–322.
[54] Keiluweit M, Nico PS, Johnson MG, Kleber M, Dynamic molecular structure of plant biomass-derived black carbon (biochar), Environ. Sci. Technol., 2010, 44(4), 1247–1253.
[55] Zhao Y, Feng D, Zhang Y, Huang Y, Sun S, Effect of pyrolysis temperature on char structure and chemical speciation of alkali and alkaline earth metallic species in biochar, Fuel Process. Technol., 2016, 141, 54–60.
[56] Steiner C, Considerations in biochar characterization, In: Guo M, He Z, Uchimiya SM (ed), Agricultural and Environmental Applications of Biochar: Advances and Barriers, Soil Science Society of America (SSSA) Special Publications, Madison, USA, 2016, 87–100.
[57] Kloss S, Zehetner F, Dellantonio A, et al., Characterization of slow pyrolysis biochars: effects of feedstocks and pyrolysis temperature on biochar properties, J. Environ. Qual., 2012, 41(4), 990–1000.
[58] Mehmood S, Rizwan M, Bashir S, et al., Comparative effects of biochar, slag and ferrous-Mn ore on lead and cadmium immobilization in soil, Bull. Environ. Contam. Toxicol., 2018, 100(2), 286–292.

Srinithya Ravinuthala, R. Nithya, Saprativ P. Das

Application of biochar for wastewater treatment

Abstract: Agricultural food production has surged exponentially over the past few decades to meet the ever-increasing needs of the growing world population, thereby leading to an upsurge of agricultural waste production too. A cost-effective, environmental-friendly method of sustainable agricultural waste management is through the generation of a biomass-pyrolyzed carbonaceous product, "biochar" from these wastes. Biochar has a versatile range of uses – soil remediation and conditioning, carbon sequestration, wastewater treatment and remediation, energy and gas storage. More recently, focus has shifted to the usage of designer or engineered biochars, as they have improved physiochemical and biological properties as compared to pristine biochar and are cost-effective than activated carbon. Biochar designing permits accomplishing biochar properties ideal for explicit applications or potentially under explicit conditions. This chapter focuses on the treatment of various types of wastewater including municipal, industrial, agricultural, and stormwater expending agricultural waste–derived biochar, with emphasis on engineered biochars.

Keywords: biochar, designer biochar, engineered biochar, wastewater treatment, agricultural waste management, pyrolysis

1 Introduction

Agriculture production worldwide caters to about 7 billion people, but due to different cultivation methods, several environmental hazards like the emission of greenhouse gases, eutrophication of aquatic bodies, decrease of soil fertility, biodiversity loss are caused. The field of agriculture has been given great importance due to the rise of poverty and non-safety of food. It is important to raise the cultivation of crops especially in the developing countries. The rise in population denotes that the environmental hazards will also probably increase as there is a need for more food production. It is

Srinithya Ravinuthala, Post-Graduate student, Department of Studies in Biotechnology, University of Mysore, Manasagangothri, Mysuru 570006, Karnataka
R. Nithya, Post-Graduate student, PG & Research Department of Zoology, Justice Basheer Ahmed Sayeed College for Women, Teynampet, Chennai 600018, Tamil Nadu
Saprativ P. Das, Research Engineer, Department of Chemical Engineering, Indian Institute of Technology Bombay, Mumbai 400076, Maharashtra, e-mail: saprativ@iitb.ac.in; saprativdas@gmail.com

https://doi.org/10.1515/9783110734003-009

important to develop different sustainable agricultural methods to produce crops efficiently [1]. Food production is the principal necessity of a human. The nutritional value, assortment, and welfare are taken into account while developing these alternate methods [2]. The current prudence is completely reliable on different nonrenewable energy sources such as coal and gas. Agricultural development is based on different principles like considering the economic situation of the country as well as biophysical constrains [3]. To lessen the amount of people suffering from lack of food by half from 1995 to 2015, the United Nations developed the eight millennium development goals. For meeting the high requirements of food production, different strategies have been adopted; two such are land sharing and land sparing. Land sharing includes amalgamation of biodiversity preservation and crop development on the same field. Land is facilitated in such a way that one part of it is utilized for preservation and another part of it is utilized for crop cultivation, which is known as land sparing. Land sparing helps in forest deforestation as well as land deterioration as durable and lasting methods are aggressively used, whereas land sharing targets certain schemes like organic farming [4].

The untreated wastes from agricultural industries have caused pollution of water, air, and land. The remains of agricultural commodities like fruits, vegetables, poultry, and dairy products after processing are known as agricultural wastes or agro-wastes. Various studies show that around 998 million tons of agricultural waste are generated every year. As mentioned before, rise in agricultural production increases the waste produced usually due to unreasonable administration of exhaustive agricultural techniques. The waste produced is dependent on the farming technique that is used. Tropical climate is a likeable condition for best yield, but also paves way for growth of different insects. Due to this condition, there is a significant requirement of pesticides to get rid of the pests. After the application of pesticides, a part of it is withheld in the soil; a part of it contaminates ponds, lakes, and other water bodies, which cause environmental hazards. Recent studies have been focused on the impacts of agricultural waste. Farming wastes should be utilized such that they are either used for other purposes or should be stored in such way that they do not degrade and get spoiled. Some uses like methane gas production are sourced from agricultural wastes. Methane production from agricultural waste is a two-step process of microbial fermentation. Other techniques include hydro-gasification and hydrolyses. One of the oldest methods of degrading agricultural waste is combustion, but this method highly contributes to degradation of air quality, emission of gases which contribute to global warming. In recent times, agricultural waste management in relation to environment has become a matter of discussion for the government officials while considering new policies. It is a requirement that waste is manufactured into something useful, rather than considering it as an unpleasant material and leaving it to mix with land, water, and air which ultimately causes pollution.

Agricultural waste management consists of totally six steps – production, collection, storage, treatment, transfer, and utilization [5]. These agricultural remains consist of large amounts of nitrogen, phosphorus, and potassium which can be used to improve the soil prolificacy, thus producing nutrition-rich crops providing food security. This alternate method is a farmer-friendly approach as it reduces the cost of fertilizers and also decreases the amount of the waste stocked up in the environment. This overall reduces the impact of agricultural remains on the environment. These alternate methods require considerable recognition among farmers and common people. It reduces the panic and predetermined idea of inconvenience related to reduction of land value and ecological hazards. The effect on environment regarding the agricultural remains is not only the quantity of which is accumulated but also the way it is cleared. Burning of this waste leads to emission of nitrous oxide and carbon monoxide, which causes health hazards to both humans and animals. Converting plant remains to organic fertilizers is one of the best alternate technologies that has been adopted widely. Composting is a technique which also reduces large amount of accumulation of plant residues, and at the same time enhances soil fertility, as well as keeps pathogens and pests under check. Agricultural remains are also consumed as animal feed; the broiler litter is a popular method to feed cattle particularly ruminants.

A more recent, multifaceted application-based agro-waste management approach is by the production of biochar from these wastes. Biochar is a highly carbonaceous product with high energy content and large microscopic surface. Often, agricultural wastes (steams, husks, stalks, etc.) cause disposal problems and hence they could be considered to be used as renewable energy resources. These biomass wastes are converted to biochar using pyrolysis technology. Pyrolysis is technique where biomass is heated to temperatures above 300 °C in a medium where there is no oxygen present (anaerobic condition). This process results in formation of biochar and other gaseous products [6], which are a great source of renewable energy. Biochar contains hydroxyl, carbonyl, carboxyl, and other functional groups along with a large specific surface area (SSA). The sustainability of environment and agriculture can be improved using this new type of absorbent. The absorption capacity of activated biochar is higher than pristine biochar, hence in the recent years, different engineering methods have been developed to maximize its adsorption capacity. Engineered biochars' application varies from wastewater treatment to energy storage. Biochar is also used to make different components of microbial fuel cells (MFCs), which act as a substitute for commercial carbon electrodes [7]. Hence, there is ample scope of future research for the designer biochars in the field of MFCs as well.

This chapter discusses the designing (synthesis) and use of engineered biochar derived from agricultural wastes particularly for the removal pollutants from wastewater, along with the removal mechanisms.

2 Biochar – characteristics and synthesis

2.1 Characteristics

The characteristics of biochar are affected by a number of factors – source of the biochar, pyrolysis temperature and speed (slow or rapid), pretreatments or post-treatments, and so on. Biochar properties can be analyzed by two main aspects: mechanical properties and chemical properties.

2.1.1 Mechanical properties

Biochar's mechanical qualities, especially structure, pore size distribution, and surface area, have a large impact on how it works in water and soil. These physical qualities can impact how biochar adsorbs contaminants in aquatic systems and affects soil systems both directly as well as indirectly. During pyrolyzation, important processing parameters and factors such as the maximum temperature time (HTT) and reaction residence time determine the physical properties of the biochar being formed. Biochar might efficiently contribute its functions in soil, particularly in the case of soil remediation and amendment by decreasing bulk density, increasing pH, changes in water retention capacity, penetration resistance, preventing soil losses, improving retention ability, and aggregation stability of nutrients, as well as crop production. In wastewater pollutant removal, the kinetics and adsorption capacities of removal are greatly influenced. By rising reaction temperatures to the point where any additional temperature increase causes deformation, biochar with an ideal structure may be created. Conjugated aromatic C sheets grow as the reaction temperature is decreased, reducing the formation of structured regular spacing between planes, as well as molecular organization, and increasing interplanar distances [8].

Porosity. Pore size distribution is one of the most important properties of biochar. Biochar is constituted three types of pores based on the pore size: macropores, mesopores, and micropores. Among these, micropores are the main contributors for biochar porosity and density; mesopores play an important role in a variety of liquid–solid adsorption processes. When the pyrolysis temperature rises beyond a particular threshold (e.g., 400 °C), the biochar's SSA expands rapidly. This results in the thermal condensation of organic matter, forming micropores. The heating rate is another component that determines the micropore volume. The biomass melts at a fast pace, at higher heating rate, generating macropores; a moderate heating rate enhances the volume of micropores in the biochar formed. Adsorbed substances are transported into mesopores and micropores via macropores, which act as feeder pores. The volume of macropores is critical when evaluating biochar's potential in sustainable agriculture. Macropores have a smaller surface area than micropores but a larger pore capacity.

Particle size distribution. Biochar particle size is determined by the rigidity of biomass against shrinking and attrition during the pyrolysis process. The particle size of the biochar which is formed is often lower than those of the unpyrolyzed biomass. Smaller biomass particles are needed to improve mass and heat transmissions during pyrolysis processes. In contrast, slow pyrolysis provides larger biochar particles. The generation of biochar with larger particle sizes may be facilitated by increasing the pyrolysis pressure. Melting and subsequent particles' fusion are more evident in pressured pyrolysis, causing swelling and the creation of particle clusters.

Density. Biochar density can be considered of two types: bulk density and solid density. A reduction in solid density goes hand in hand with an increase in bulk density. When compared to macropores and mesopores, the presence of micropores in biochar enhances its density. It is important to keep in mind that pyrolysis temperature and biochar bulk density do not have a significant relationship. Generally speaking, the higher the bulk density of the feedstock utilized, the higher the bulk density of the biochar generated.

Stability. The mechanical strength of biochar is determined by its solid density, aromaticity, and crystallinity. Monolithic carbonized wood biochar has been found to have lower stiffness (37%) and higher strength than virgin wood (28%). According to research, the lowered moduli and hardness of woody biochar increase with increasing temperature, eventually reaching a plateau between 700 and 2,000 °C. Any additional increases in pyrolysis temperature would have a negative impact on the biochar's characteristics. In environmental applications, biochar's capacity to tolerate tear and wear loads is critical. On this premise, biomass which is harder with low ash and high lignin concentration is preferred for producing biochar with superior mechanical qualities. For example, biochar made from eucalyptus has a higher mechanical strength than biochar made from acacia. The temperature of pyrolysis is discovered to be a more important driver of biochar mechanical characteristics than residence time.

2.1.2 Chemical properties

Any kind of pyrolysis technique may chemically transform biomass into biochar with a wide range of characteristics. Biochar qualities are mostly influenced by the feedstock and pyrolysis temperature. The production of biochar drops exponentially as the pyrolysis temperature rises, but the alkalinity (pH) of biochar rises linearly. The thermal breakdown of hydroxyl linkages and other weak links within the biochar structure at high temperatures is responsible for the pH increase. Due to the elimination of acidic functional groups, the cation exchange capacity of biochar has an inverse relationship with pyrolysis temperature [9].

Surface chemistry. The surface chemistry of biochar varies depending on the pyrolysis and feedstock conditions. On the graphene sheet surfaces, heteroatoms containing aromatic rings (H, S, P, O, and N) create numerous functional groups. Electron acceptors or donors are possible functional groups. These groups' concentrations are affected by parameters such as reaction temperature, biomass content, and heating rate [9].

When lignocellulosic biomass is pyrolyzed at temperatures over 400 °C, the H-to-C ratio reduces from around 1.5 to 0.5. This decrease in the H-to-C ratio might be due to changes in elemental concentrations (N, O, H, and C) during thermal decomposition of biomass. At high reaction temperatures and long residence times, the O-to-C and H-to-C ratios in biochar often decrease. The lower the H-to-C ratio, the more graphitic-like the structure of the compounds. Chemical treatment can also be used to change the chemical composition of biochar. Biochar including alkyl C and O-alkyl C with dominant aromatic (aryl) C is produced by biomass pyrolysis at >350 °C. Higher temperatures (>500 °C) and longer residence times enriched aryl C structures at the price of eliminating plant-specific functional group C structures, resulting in increased mass loss and extremely low H-to-C ratios [8].

Molecular composition. Minerals such as P, Ca, Na, Mg, and K are abundant in biochar. The type and content of solid phases, entrapped oils, molecular structure, electrochemical properties, and functional groups on the biochar surface all influence microchemical parameters. The hemolytic breakdown of covalent bonds during pyrolysis of dry biomass releases a large amount of free radicals from inorganic contaminants and structural O within biomass. Low ambient O_2 levels may create free radicals during the early stages of pyrolysis. The production of carboxyl and carbonyl groups, and their subsequent cleavage into CO_2 and CO, continues the process. Biochar residue is created at the end of the pyrolysis process by the recombination of free radical fragments with each other and with the substrate in various ways. The fraction of entrained free radicals in biochar achieves a maximum for a given residence period as the reaction temperature rises (500–600 °C), resulting in pyrophoric biochar. The rate of breakdown and volatilization of anhydro-sugars contained in the tar determines the yield of biochar generated at 300–600 °C. Beyond this temperature, gas will be the dominant product, reducing the production of biocrude oil, tar, and biochar [8].

2.2 Synthesis

The process of thermal degradation of substances where there is no presence or notably reduced presence of oxygen for absolute burning is known as pyrolysis. Gasification is distinguished from pyrolysis as the biomass is treated with steam, air, and so on. In the process of gasification, the biomass is transformed as syngas through cautiously controlling the quantity of oxygen that is exists. The pyrolysis process makes up a wide variety of thermal degradation, so it is strenuous to exactly specify their process. The previous studies show that the pyrolysis process correlates with the process of carbonization in which the char (charcoal) is found to be the chief substance. In recent times, pyrolysis is defined as the process where the products are used in liquid form (bio-oils). The manufacturing of products with liquid is mostly preferred during short pyrolysis times (fast pyrolysis). Conventional (slow) pyrolysis as well as fast (flash) pyrolysis both rely on the basis of utilization of directing determinants (temperature, heating speed, and vapor residence time) [10]. The pyrolysis process differentiates based on the feed's pyrolysis temperature, heating rate, and residence time. The process of conventional pyrolysis is sluggish. More of pyrolysis process has taken place at times which are not quite fast or slow but somewhere in between these two boundaries. The main thing that has to be noted is that regardless or not, vapors and aerosol substances are quickly separated improve liquid development (fast pyrolysis and flash vacuum pyrolysis) or persist in exposure to the solid, which go through the secondary reactions the next response that is to construct add on carbonaceous substance. The framework or the limitation necessary for the pyrolysis process is mentioned in the following sections.

2.2.1 Conventional/slow pyrolysis

It is one of the oldest processes that has been applied to manufacture charcoal. The studies related to understand manufacturing and the characteristics of charcoal have been revised lately. For centuries, slow pyrolysis has been used to make charcoal. The biomass is gradually heated with a temperature of 500 °C where there is no presence of air. The duration of residence of the vapor that are present will differentiate from 5 to 30 min. Vapors that are present in conventional pyrolysis will not get out quickly as in fast pyrolysis.

2.2.2 Fast pyrolysis

Dry feedstock is a prerequisite in fast pyrolysis (<10 wt% moisture), which helps in fast heat transfer and quick temperature rise by heating small quantity of biomass substances (1–2 mm) around 400–500 °C. The duration of vapor residence is minimum 1 s and reaches up to 5 s [11]. Slow pyrolysis distinguishes from fast pyrolysis with a minor difference in their overall process.

2.2.3 Gasification

The process by which the electricity produced through the undeviating effect of heat is known as gasification. The process involves slight burning of solid. The biomass is nursed with the mixture of products, and cautiously done by taking care of temperature, matter size, residence duration, pressure, and gas composition where the catalyst is also recognized. The odor of char is tended to be found in slow pyrolysis compared to fast pyrolysis or gasification. The substances dimensions are generally found to be same in fast and slow pyrolysis chars (7–8 rings per compound). On the other hand, gasification char is generously condensed (17 rings per compound).

2.2.4 Hydrochar

The process by which hydrochars are made differs from how biochars are made. Pyrolysis (slow or fast) or gasification at various temperatures, pressures, and carbonization times produce biochar. Hydrochars are created when biomass is carbonized hydrothermally in water at high temperatures and pressures. The hydrochar process produces a char water–slurry that could be separated easily. Its chemical and physical characteristics are vastly different from that of the original biomass. Agricultural residue hydrochars have been studied and applied for water and soil remediation [12]. The chemical characteristics of biochars from gasification/pyrolysis and hydrochars from HTC were studied by Wiedner et al., who discovered that hydrochars were less stable (alkyl moiety-dominated) than biochars (aromatics-dominated). Hydrochars are not included in the "European Biochar Certificate" (EBC) norm because of their different chemical characteristics [12].

2.2.5 Retification and torrefaction

Retification of wood is a method that produces a wood product which is reddish-brown or chocolate and resistant to biological attack when heated to 230–250 °C. Similarly, at low heating rates between 250 and 280 °C, torrefaction of wood creates a brown or black product with limited strength. Torrefaction raises the energy density of biomass, decreases its weight, and improves the hydrophobicity of wood. Torrefied wood yields range from 67 to 84%, with 77–90% energy yields [13]. The reduced transportation cost of torrefied wood makes it more economical for energy production. Torrefaction and retification do not produce adsorbent chars because only partial biomass decomposition occurs to prevent rot and water loss.

3 Engineered biochar

Engineered biochar is biochar modified to suit particular adsorption properties, enhancing the performance to suit the intended application. This can be done by altering the surface properties (such as the functional groups or the surface area) or porosity (pore size and structure). Biochar can be engineered suitably to meet the desired purpose by a versatile range of physical, chemical or biological methods. Figure 1 depicts the major functional groups added and structural representation of the variations induced by treatments to synthesize engineered biochar [14, 15]. Broadly, engineered biochar undergoes two kinds of treatments – pretreatments and posttreatments.

Pretreatment technologies. Pretreatment with various physical, chemical, and biological methods are some preliminary processes used in producing biochar from a wide range of raw materials. The drying, crushing, sieving, and washing of biomass feedstock are all primary steps of physical pretreatment technology. The lignocellulosic/plant feedstock is normally dried to a consistent weight at 105 °C or other temperatures, then crushed into smaller particles with a hammer mill before being chopped into pieces [16]. Since plant raw materials have different moisture content, a separate drying phase may be necessary for some plant feedstock. The physical pretreatment procedure for biomass feedstock is depending on the biomass composition. Dewatered sludge, for example, is dried overnight in an oven, crushed, sieved, and stored in sealed containers before use; newspapers and cardboard are commonly shredded and blended into a pulp as the feedstock; and paper mill sludge is acid-washed and rinsed with deionized distilled water to obtain mineral-free sludge [17]. Before pyrolysis, algae are normally washed with fresh water and then dried, granulated, or flaked. Chemical pretreatment technology alters the composition and characteristics of the biomass feedstock through chemical processes. Loading chemical precursors or functional agents into the biomass feedstock with the assistance of chemicals or any other functional material is the most often utilized chemical pretreatment approach. This technique involves immersing the biomass feedstock into a chemical solution or a colloidal suspension, followed by drying [18]. After pretreating the biomass feedstock with metal ion solutions such as $FeCl_3$, $AlCl_3$, and $MgCl_2$, biochar-based nanocomposites are made with metal oxyhydroxide (e.g., Fe_3O_4, AlOOH, and MgO) nanoparticles stabilized on the carbon surface, within the pores of the designed biochar. Engineered nanoparticles and natural colloids such as carbon nanotubes, graphene, and clay can also be used to pre-treat biomass. The use of chemicals such as acid, alkali, and oxidants to pretreat the biomass results in engineered biochar with increased surface area, unique pore structure, and enriched surface functional group (SFG), among other things. Biological pretreatment technology is a new idea that uses biological processes to enhance biomass feedstock for the manufacture of tailored biochar [19]. Biologically, activated biochar with enhanced properties can be produced with bacterial treatment of the biomass feedstock, particularly by anaerobic digestion and biofuel processes [20, 21]. Several biomass sources, including sugar beet tailings, bagasse, sludge, and animal manure, underwent anaerobic digestion before being transformed to biochar by slow pyrolysis in another study by Wei et al. Biochar with a greater SSA and improved adsorption efficiency was a result of the anaerobic

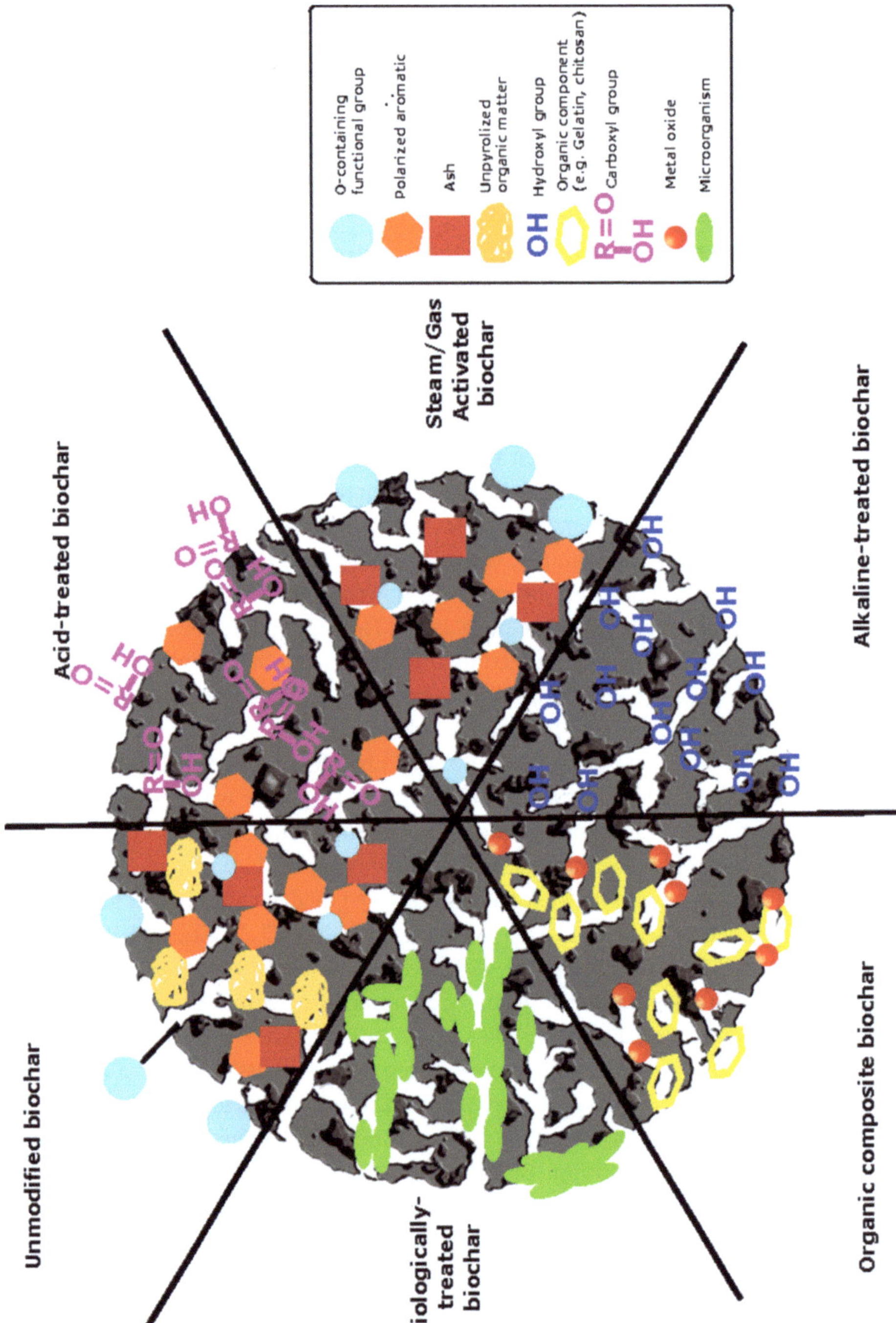

Figure 1: Functional groups and structural engineered biochar.

digestion pretreatment. Another biological pretreatment approach utilized for biochar formation is biomass loaded with high concentrations of minerals, including heavy metals, through bioaccumulation. Wang et al. transformed a heavy metal hyper-accumulating plant to biochar, claiming that this method not only provides a safe way to dispose of hyper accumulators, but also creates value-added biochar nanocomposites. Biochar may be made from biological pretreatment residual materials, which is cost-effective in terms of waste disposal and an environmentally beneficial source of bioenergy [20, 21].

Posttreatment technologies. The addition of functional agents such as SFG and composited nanoparticles to biochar after physical or chemical alteration improves its SSA, pore volume, surface chemistry, negative zeta potential, oxygen-containing functional groups, and adsorption capacity. Magnetization, ball milling, and corrosive chemical treatment are a few of the procedures used [19, 22]. Biochar posttreatment is a relatively young field of study. Such modified biochars have the potential to be cost-effective and environmentally friendly carbon products with a wide range of applications.

These techniques are further discussed in detail in the following sections.

3.1 Physical techniques

Physical techniques are generally used for varying the pore size, increasing the abundance of mesopores and micropores, as well as induction of oxygenic functional groups. They have an advantage of being more cost-effective and also contain no impurities, along with being relatively safer to produce as compared to their chemical-based counterparts.

3.1.1 Ball milling

The use of ball-milling technology to create novel engineered biochar is an efficient engineering method. This method uses the kinetic energy of moving balls to break the chemical bonds, producing nanoscale particles [23]. Ball milling enhances the physiochemical, adsorptive, and other properties of biochar including SSA, pore volume, negative zeta potential, oxygen-containing functional groups, and adsorptive capacity and can be used in various environmental applications. Biochar processed through ball milling can be modified as per the required particle size and surface area properties. Compared to pristine biochar, they show higher adsorption properties. Ball milling exposes the graphitic structure and oxygen-containing functional groups of the biochar by increasing its exterior and interior surface areas. Nano-sized biochar produced by ball milling has shown comparable performance to activated carbo and carbon nanotubes. However, it should be noted that smaller biochar particles are more prone to erosion, which may result in groundwater contamination when used for soil remediation. Biochar should hence be modified to be more stable. Functional groups could also be modified simultaneously along with the surface properties by adding a suitable chemical. This is known as chemical ball milling [15]. By ball milling pristine biochar with ammonium hydroxide, nitrogen-doped biochar can be synthesized. Biochar adsorption ability of acidic carbon dioxide and anionic reactive red is improved by these N groups.

3.1.2 Gas or vapor activation

Gas or vapor activation enhances biochar adsorption properties surface area and aids in porosity formation as volatile compounds such as carboxyl and phenol groups are degraded. As a result, biochar produced by this method is generally less oxidized and less polar. Hence, gas activation is generally preferred for only preliminary treatment. This process is generally used to produce activated carbon, by using either carbon dioxide or steam between temperature 700 and 1,100 °C and develops the porosity of biochar. Steam-activated biochars show almost double the amount of surface area when compared to biochar, allowing to be used in removal of pollutants such as antibiotics, heavy metals, and greenhouse gases [24].

3.1.3 Microwave activation

Microwaves range from wavelengths of 0.01 to 1 m and frequencies of 0.03–300 GHz. Microwaves are employed for pyrolysis at low reaction temperatures between 200 and 300 °C. In this method, there is easier control of temperature, uniform distribution of internal thermal energy, no requirement for drying or shredding of the biomass, and a faster hassle-free process with biochar yields more than 60%. When compared to pristine biochar, the biochar produced through microwave pyrolysis has a higher surface area and more functional groups. Microwave pyrolysis has an advantage that biochar can be obtained at relatively low temperatures of 200 °C with comparatively lower residence times versus conventional pyrolysis methods to obtain biochar of similar properties [8, 24].

3.1.4 Magnetic modification

Biochar used for wastewater treatment has a limitation of having a small particle size and low density, due to which separation becomes difficult after treating the water. Secondary pollution may be caused if the biochar with the adsorbed pollutants is not separated as quickly as possible. Biochar combined with magnetic compounds such as Fe_3O_4, zero-valent iron (FeO), gamma-Fe_2O_3, and $CoFe_2O_4$ could easily be separated using apply a magnetic field. The magnetic nanoparticles require a coupling agent to keep them embedded in the surface of biochar. Amino-terminated 3-triethoxysilylpropylamine is a common coupling reagent used for surface-linking biochar with $Fe3O_4$. Iron oxides also improve the metal binding and cation exchange capacities of the modified biochar. To compensate for the blockage of pores due to surface-linking of the magnetic nanoparticles, biochar is consequently modified by addition of active functional groups such as amino, carboxyl, sulfhydryl, and hydroxyl, which have the capacity to bind to the desired pollutants to be adsorbed. Carboxyl group is known to have a high affinity to metal ions due to the electrostatic interaction of the positive metal ion and negative carboxyl ion as well as the ion exchange capacities [15].

Magnetization of biochar can be achieved by two methods – one method is the conversion of biochar into magnetic material by loading magnetic iron oxide particles (Fe_3O_4, γ-Fe_2O_3, or $CoFe_2O_4$) into the biochar. Because of its high pore volume, SSA, and loaded reductive iron, magnetic zerovalent iron biochar generated from peanut husk, at a pyrolysis temperature of 800 °C has a better removal rate for Cr^{6+}. Another method is by direct chemical coprecipitation of Fe^{3+}/Fe^{2+} onto the biochar surface [18]. The maximum adsorption capacity for metribuzin has been found in magnetic switchgrass biochar, which is made by precipitating iron oxide in an aqueous Fe^{3+}/Fe^{2+} solution. (205 mg g^{-1}, pH = 2) [25].

3.2 Chemical modification

Chemical modification of biochar can be done by either activating biochar in the presence of a chemical with required functional groups or activity and an inert gas, or can be impregnated with a wide range of materials ranging from nanoparticles, organic compounds, metal oxides, clay, and so on, where biochar acts as the scaffold for deposition of the materials. Ultimately in chemical modification, SFGs of biochar are further enhanced or completely new functional groups are introduced.

3.2.1 Chemical activation

Oxidizing modification by acid or alkali treatment. Chemical corrosive treatments with corrosive chemicals alter the biochar surface chemistry. The chemical modified biochar has higher enhanced adsorption capacity, SSA, functional groups, and microporosity. An appropriate oxidizing agent is used to enhance the adsorption capacities by inducing the affinity of negatively charged carbonyl functional groups toward positively charged metal ions (Cd, Cu, Pb, Zn,) and molecules (sulfamethazine and oxytetracycline), particularly after treatment with acids such as

H_2SO_4, H_3PO_4, HCl, and HNO_3. In the case of alkali treatment with reagents such as KOH and NaOH, there are more hydroxyl groups (aromatic functional groups), increased surface electrostatic attraction, and p–p interaction, which are useful for As, Cd, chloramphenicol, and tannic acid. A resulting larger surface area in is due to the demineralization of biochar during treatment. In general, alkaline-treated biochar has a larger surface compared to acid-treated biochar. An appropriate oxidizing reagent can help in producing recalcitrant carboxyl-rich biochar, which could be potentially used for long-term remediation of water and soil [24]. Tang et al. observed an increase in BET surface area, oxygen-containing functional groups, and porosity in municipal sewage sludge biochar by alkali (NaOH)-acid (HNO_3) combined modification, resulting in enhanced tetracycline adsorption [26] H_2O_2 serves as a strong oxidant for modifying biochar. H_2O_2-modified manure biochar has increased oxygen and carboxyl group content exhibiting the capacity to eliminate heavy metals efficiently [27].

Hydrogen peroxide treatment. Hydrogen peroxide being a strong oxidant has lately gained attention as a cost-effective and environmental-friendly chemical modification. Treatment of biochar with hydrogen peroxide greatly enhances the capacity to remove contaminants like heavy metals and ammonium from water without introducing further chemical elements during soil or water remediation [24].

3.2.2 Coating or impregnation of biochar

By coating or impregnating biochar with functional nanoparticles, metal oxides or clay, qualities such as increased surface area, porosity, functional groups, and ion exchange capacities are achieved. Biochar nano or hydroxide composites are synthesized by techniques such as support with functional nanoparticles, pretreatment of the biomass by the use of metal salts or after pyrolysis, and mixing the metal oxide nanoparticles.

Biochar composites coated with nanoparticles combine both the properties of biochar and nanoparticles. The adsorption capacities and thermal stability of biochar as the matrix and large SSA with high adsorption sites of nanoparticles together can result in better removal of pollutants. The limitations of nanoparticles such as poor solubility and low ease of separation are also overcome through these biochar nanocomposites.

Pre-pyrolysis treatment of biomass with chemical reagents, metal oxides, or clay enriches the biomass, followed by pyrolysis, which leads to transformation of metal ions into nanoparticles. The final resulting product would be biochar impregnated with nanometals, a biochar-based nanocomposite.

Metal salts such as $AlCl_3$, $CaCl_2$, $CoCl_2$, $FeCl_3$, $MnCl_2$, $MgCl_2$, $ZnCl_2$, $Co(NO_3)_2$, Zn $(NO_3)_2$, and $KMnO_4$ are used for formation of biochar nanocomposites such as Al_2O_3, CaO, AlOOH, CoO, CO_3O, MgO, FeO_x, MnO_x, and ZnO. Methods such as conventional wet impregnation, direct hydrolysis, evaporative method, and heat treatment are used to produce biochar metal oxide nanocomposites using metal salts, after biomass pyrolysis [15, 24].

3.3 Biological modification

Because of their microscopic dimensions, microorganisms could be used to penetrate biochar pores to colonize and form biofilms, which help them adhere to the surface. Organic compounds adsorbed on to the biochar are further degraded and sequestered to less harmful molecules (at times, value-added products as well such as therapeutic lead compounds) by the various metabolic pathways of the bacteria in the colonies. Heavy metals are adsorbed by a number of species on to their cell membranes. Through this post-modification technique, biochar is biochemically modified due to the colonization of bacteria and allowing adsorption by the biochar and degradation by the bacteria [15]. Biochar obtained from anaerobic digestate residues has also shown better quality in terms of surface area and surface and charge, anionic and cationic exchange capacities, higher pH (alkaline), and hydrophobicity. Biochar from such residues has competent pH value and redox potential values as the feedstock would have previously undergone anaerobic digestion, altering the properties. Due to the effective ion exchange capacities of both anions and cations, such biologically activated biochar has potential to be utilized for remediation of heavy metals and dyes. Coupling this process with biogas production increases the economic value and feasibility of biogas production [28].

4 Engineered biochar for wastewater treatment

SSA and abundant SFG of biochar allow them to be an effective absorbent for the removal of contaminants. Biochars have great potential to remediate the wastewater generated from the domestic, industrial, and agricultural activities, thereby improving the environment quality [19]. The rapid development of agricultural industry has led to the discharge of more pesticides into the farmlands causing agricultural contamination [29]. Researchers have discovered various modified forms of biochar to treat the wastewater. For example, soybean and corn straw biochars showed high removal of atrazine [30], while steam-activated biochar was used for the removal of sulfamethazine [31] and zero valent iron magnetic paper mill sludge biochar (ZVI-MBC) can effectively remove pentachlorophenol (PCP) from the effluent [32]. The feedstock used, functional materials, and target contaminants usually determine the adsorption capacity of the biochar. Toxic heavy metals (As, Cr, Cu, and Pb) in agricultural wastewater are another common concern, and adsorption processes include electrostatic

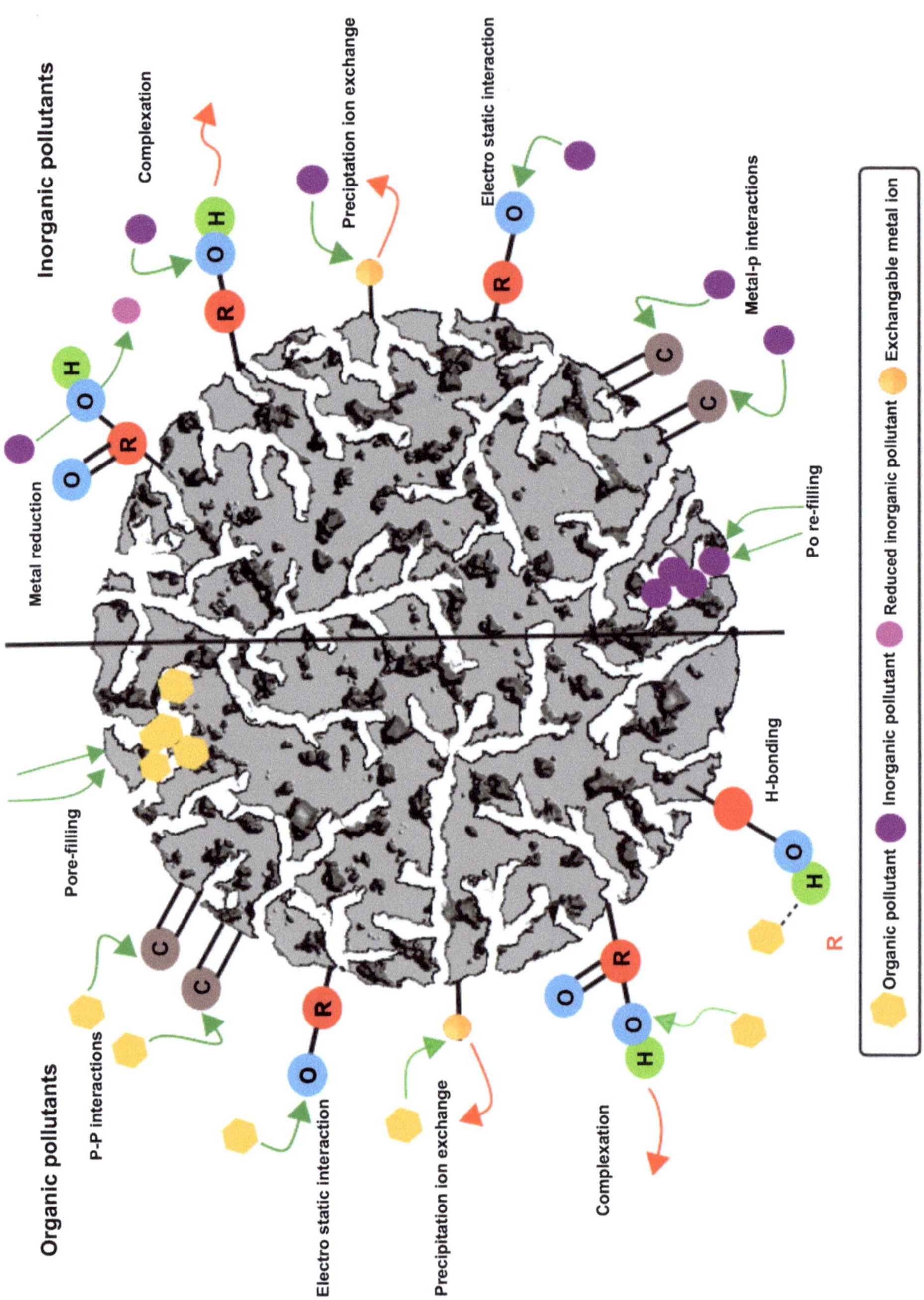

Figure 2: Adsorption mechanisms of organic and inorganic pollutants.

interactions, ion exchange, surface complexation, intermolecular contacts, and p–p interactions [29]. The inner-sphere complexes, hydrophobic effect, p–p interaction, precipitation and ion exchange affect the adsorption capacity of the biochar.

Biochar has a great potential for wastewater treatment due to its good adsorption capacities and economic feasibility. However, pristine biochar has only a limited capacity to adsorb pollutants, and is difficult to separate due to the small size as well as its low density. Hence in the recent years, interest has shifted toward engineered biochar, which has enhanced surface properties, higher surface area and ease of separability [28, 33]. Figure 2 shows the various adsorption mechanisms of organic and inorganic pollutants with the functional groups upon biochar.

4.1 Inorganic pollutant removal

4.1.1 Nitrogen removal

Ammonium is released from various sources such as municipal wastewaters, leeching from agricultural areas due to overuse of ammonium fertilizers, and industrial wastewater discard. These aqueous ammonium ions are a serious environmental hazard, mainly affecting water bodies. Ammonium is known to accelerate eutrophication of water bodies, deplete dissolved oxygen, and also be fatal to aquatic organisms at high concentrations. Excessive ammonium leeching into the soil also affects the quality of groundwater quality too. At present, crystallization, ion exchange, precipitation, biological removal, adsorption, and other such techniques are being used to remove ammonium from wastewater. Adsorption using activated carbon is at present the most widely used method for wastewater treatment.

Khalil et al. have used rice straw-based biochar treated with NaOH and obtained up to 5.16 mg g^{-1} adsorption capacities of ammonium in the form of ammonium chloride solution [34] Nguyen et al. corncob-based biochar used treated with both 0.3 M NaOH and 6.0 M HNO_3 in a fixed-bed column. They had achieved up to 12.83 mg g^{-1} removal efficiency at a bed height of 8 cm, with a flow rate of 3 mL min^{-1} [35]. In both cases, interaction with acidic functional groups such as COO^- was instrumental in ammonium adsorption through mechanisms like ionic exchange. In another study by Siyuan et al., biochar obtained from wheat straw was treated with 1 mol L^{-1} HCl and 1 mol L^{-1} $FeCl_3$. Their study concluded that C–OH group converted into O–C = O played a major role in ammonium adsorption, and was further enhanced by Fe^{3+}/Fe^{2+} redox couple acting as electron shuttles, therefore enhancing adsorption capacities by 14% as compared to pristine biochar [36]. In short it could be concluded that modified biochar with abundant acidic SFGs works efficiently for the removal of ammonium from wastewater.

4.1.2 Phosphorus removal

Phosphorus is a crucial mineral important for the efficient functioning of soil microbial activity, plant growth, and the environmental balance. Although phosphorus is one of the most abundant elements, on Earth, the extensive use of phosphorus for fertilizers, and everyday items such as toothpastes and washing powders have led to predictions that's phosphorus resources may be depleted in the coming 70 to 175 years. The excessive use of phosphorus as fertilizers has led to obvious environmental issues such as eutrophication and endangerment of aquatic species. Therefore, it is necessary to recycle and recover as much of this phosphorus as possible from the various waste water resources such as sludge water, municipal wastewater. Some of the present techniques used for phosphorus recovery include struvite ($MgNH_4PO_4.6H_2O$) precipitation using magnesium salts, crystallization using crystallization cores such as diatomaceous earth, sand, and slag in fluidized bed reactors, ion exchange with porous P-selective media, adsorbents such as bauxite extraction wastes, Fe oxides, fly ash, and blast furnace slag, using biological phosphorus removal microbes like cyanobacteria, algae, and polyphosphate-accumulating bacteria. However, all these methods require a higher amount of phosphorus and may reduce the bioavailability of phosphorus to the environment.

Fenglin et al. treated corn straw biomass with 1 mol L^{-1} ferrous sulfate, followed by pyrolysis at 400 °C, resulting in granular iron scattered upon the biochar surface. A removal efficiency of over 99% with 7.5 mg L^{-1} initial concentration was reported, indicating a great role of the functional groups. Through XPS, FTIR, and XRD analyses, it was concluded that phosphorus was removed through a chain of reactions with a main reaction product $Fe_5(PO_4)_4(OH)_3$. The reactions are as below [37]:

i. $H_2PO^- \rightarrow H^+ + HPO_2^-$
ii. $Fe_3O_4 + 8\,H^+ \rightarrow Fe^{2+} + 2Fe^{3+} + 4H_2O$
iii. $H_2O \leftrightarrow H^+ + OH^-$
iv. $5Fe^{3+} + 4PO^{3-} + 3OH^- \rightarrow Fe_5(PO_4)_4(OH)_3$

A novel chitosan composite wheat straw-based biochar that could work in a wide range of temperatures and pH was synthesized in a study by Huang et al. The biochar was initially soaked in 2% dissolved chitosan acetic acid solution, to obtain the composite. This was followed by cross-linking with glyceraldehyde and formaldehyde to the biochar so the chitosan does not separate when in aqueous solutions. Then it was treated with 0.5 M HCl solution and 0.5 M NaOH to release the protected amino groups due to cross-linking. Glycidyl-trimethylammonium chloride (GTC) was added to form quaternary chitosan, which can adsorb phosphate at both acidic and basic conditions and lanthanum nitrate hexahydrate solutions ($La(NO_3)_3 \cdot 6H_2O$) to enhance the phosphate adsorption capacities. Adsorption took place by three mechanisms – Lewis acid-base interaction, ligand exchange, and electrostatic interaction [38].

4.1.3 Heavy metal removal

Heavy metals released from various industries such as mining, battery manufacturing, electroplating, metal processing and finishing, glass and ceramics, and textile pose a serious threat to the ecosystem. They are especially of concern due to their toxic effects even at low concentrations, and their further accumulation in the higher tropic levels of the food chain through biomagnification. Chemical precipitation, adsorption, ion exchange, electrochemical treatment, membrane technologies, and reverse osmosis are some techniques used for heavy metal removal [39]. Modified biochar for heavy metal removal by adsorption may be synthesized by one-step activation (carbonization and modification steps done simultaneously) or two-step activation (direct pyrolysis of biomass followed by modification methods). Commonly used modification methods include surface oxidation or reduction impregnation with mineral elements or nanometals [40].

In a study by Mosa et al., [41] cotton stalks were used for biochar production and were treated by different reagent – sulfuric acid (SAB), sodium hydroxide (SHB), oxalic acid (OAB), to compare the efficiency of heavy metal ions (Pb^{2+}, Ni^{2+}, Cd^{2+}, and Co^{2+}). Acid-treated biochar showed less removal efficiencies, possibly due to calcite and sulfur deposition formation upon the micropores. Further, on the oxalic acid treated biochar, there was high silicate formation, causing drop in removal efficiency. Silicate ions are tetravalent cations, and cannot participate in ion exchange as easy as divalent cations. On the other hand, NaOH-treated biochar showed high removal efficiencies up to more than 99%, except in the case of cobalt. FTIR analysis showed it may be due to the increased availability of amino and hydroxyl groups.

Bashir et al. [42] used KOH treated rice straw biochar for cadmium removal. Through FTIR and XRD studies, they concluded that the COO^-, OH, and CH_2 SFGs on modified biochar may increase the overall negativity on the biochar surface, hence improving the adsorption capacities of modified biochar by more than three times, as compared to pristine biochar. Their conclusions were similar to those of [41] with regard to the heavy metal adsorption mechanisms of alkaline-treated biochar.

A gelatin-loaded magnetic biochar was used to remove arsenic from aqueous solutions was reported by Zhou et al. The biochar was derived was chestnut shell, the gelatin was added as it increased the number of oxygen-containing functional groups, and therefore the adsorption capacities. The biochar was further treated with 0.02 M $FeCl_3$ and 0.01 M $FeCl_2$ solutions for ease of separation after treatment of the water. They achieved approximately 2.5 times arsenic removal in the modified biochar than pristine biochar because of the enhanced presence of protonated, positive ions and oxygen-containing functional groups [43].

In two different studies by Yanmei et al. [44] and Jiaqin et al. [45], chitosan biochar composites followed by different chemical treatments were used for copper removal. Both reported that the N-containing functional groups (especially amine) of chitosan played a major role in chemisorption of copper.

Table 1: Removal of inorganic pollutants using agricultural waste-based engineered biochar.

Biochar source	Pollutant removed	Pyrolysis temp. (°C)	Activation treatment	Implications	Mechanism of removal	Ref.
Rice straw	Ammonium	500–600	^Soaked in NaOH	Adsorption capacity of 5.16 mg g^{-1}. (Removal efficiency of 43%)	Surface functional groups such as COO^- interact with and adsorb ammonium ions	[34]
Corncob	Ammonium	400	^Soaked in a mixture of 6.0 M HNO_3 and 0.3 M NaOH	Adsorption capacity of 12.83 mg g^{-1} at bed height 8 cm	Ion exchange and electron gravity between functional groups attached on biochar surface	[35]
Wheat straw	Ammonium	450	^Soaked in 1 mol L^{-1} HCl, mixed, oven-dried, followed by 1 mol L^{-1} $FeCl_3$	Adsorption capacity 14% higher than pristine biochar	Fe^{3+}/Fe^{2+} redox couple acted as electron shuttles; C–OH group converted into O-C=O, further converted to N–(C=O)–O after ammonium chemisorption	[36]
Corn straw	Phosphorus	400	*Soaked in 1 mol L^{-1} Ferrous sulfate soln. for 1 h	Removal efficiency of over 99% with 7.5 mg L^{-1} initial conc.	Chemical reaction with of phosphorus with iron oxide functional groups. Main reaction product $Fe_5(PO_4)_4(OH)_3$.	[37]
Wheat straw	Phosphorus	550	^Soaked in 2% acetic acid with dissolved chitosan solution, cross-linked with glyceraldehyde and formaldehyde, followed by addition of glycidyltrimethylammonium chloride (GTC), lanthanum nitrate hexahydrate solutions	Maximum adsorption capacity reaching 109 mg g^{-1}	Three mechanisms for adsorption identified – electrostatic interaction, ligand exchange and Lewis acid-base interaction	[38]

Cotton stalks	Pb^{2+}	550	^Shaking with 0.1 M NaOH (1:100 W/V) at 150 rpm agitation rate for 4 h	More than 99% removal efficiency up to 80 mg L^{-1} initial conc.	Adsorption by hydroxyl and amino groups	[41]
Rice straw	Cadmium	500	^Mixed with 2 M KOH solution and stirred for 1 h	Adsorption capacity of 41.9 mg g^{-1} (more than three times as compared to pristine biochar)	Adsorption through precipitation by complexation and ion exchange	[42]
Chestnut shell	As^{+5}	450	^0.1% dissolved gelatin solution mixed with 0.02 M FeCl3 and 0.01 M FeCl2 solutions	Maximum adsorption capacity 45.8 mg g^{-1} (approximately 2.5 times more than pristine biochar)	Oxygen–containing functional groups greatly enhanced by presence of gelatin played a major role in adsorption	[43]
Bamboo	Cu^{2+}	600	^Soaked in 2% acetic acid with dissolved chitosan solution followed by 1.2% NaOH solution	Removed 58% with initial 30 ppm conc.	Adsorption by chitosan amine functional groups on biochar surface	[44]
Rice straw	Cu^{2+}	450	^Soaked in 2% acetic acid with dissolved chitosan solution, 1.2% NaOH solution. Treated with pyromellitic dianhydride dissolved in dimethyl formamide and mixed.	Adsorption capacity of 70.28 mg L^{-1} (2.5 times higher than pristine biochar)	Chemisorption by carbonyl groups and N-containing functional groups	[45]

*Pre-pyrolysis treatment

^Post-pyrolysis treatment

Table 1 summarizes some of the notable studies that have used agricultural wastes to produce modified biochar for removal of inorganic pollutants.

4.2 Organic pollutant removal

The extensive use of pesticides, pharmaceutical, and other personal care products has led to releasing organic compounds into the environment, especially water, causing a number of environmental and health hazards. Organic compounds are of particular environmental concern as they are more complex than inorganic pollutants, hence more difficult to remove from the environment and lower biodegradability, along with being toxic and potentially carcinogenic. Modified biochar could be used to adsorb these pollutants, or can be used as activators to degrade the pollutants by a more recent technique, activation, and catalysis of advanced oxidation processes (AOPs). Both these processes are described below in detail.

4.2.1 Adsorption of organic pollutants

Organic pollutants such as persistent organic pollutants, dyes, pesticides, antibiotics, and oils can be adsorbed on to the engineered biochar surface by mechanisms such as cation exchange, complexes adsorption, electrostatic attraction, hydrophobic interactions, partition uncarbonized fraction, pore-filling, spectrometer exchange, and p–p electron donor–acceptor interaction [15].

The mechanism of adsorption depends upon the chemical nature of biochar, (the adsorbent surface) and contaminants (adsorbates), as well as the interaction between them. The SFGs and electrical properties, and organic structure of biochar all affect the absorption mechanisms of adsorbents. The biochar surface generally consists of negatively charged functional groups which interact with the positively charged organic compounds through electrostatic attraction. The distance between the two atoms and the atomic size affects the magnitude of the electrostatic interactions. Oxygen-containing functional groups including carboxylic acid, hydroxyl, ketonic, and nitro groups interact with aromatic molecules by p–p electron donor–acceptor interactions. It has been observed that the amount of organic pollutants adsorbed by biochar is directly proportional to the oxygen-containing functional groups upon the biochar surface. Biochar which has low surface oxidation is hydrophobic and is utilized to remove hydrophobic organic substances like oils and lipids through hydrophobic contact [46].

Removal of recalcitrant organic pollutants. Bisphenol A is an endocrine disruptor used as a plasticizer. It causes adverse effects on fauna including human beings even at low concentrations. Zhao et al. studied the efficiency of H_3PO_4-treated biochar from corn straws for the removal of bisphenol A, along with the effects co-adsorption of Cr(VI) and naphthalene, common co-pollutants found along with bisphenol A. They observed that Cr(VI) improved bisphenol A removal effectiveness, perhaps due to increased H-bond interactions between bisphenol A and biochar mediated by bridge bonds of oxygenic groups from chromate [47].

Removal of dyes. Organic dyes used in textile industries released into water bodies are resistant to biodegradation, toxic, and potentially carcinogenic to the aquatic life residing in the water bodies. Methylene blue, a synthetic dye commonly used in the field of biological diagnostics and also as a dye, was removed from aqueous solutions using KOH-treated and H_3PO_4-treated corn stalk–derived biochar in a study by Liu et al. The KOH-treated biochar showed a higher ability to remove methylene blue [48]. In another study by Ren et al., rhodamine A, a dye used in many industries such as textile, food, printing, paper, and leather, was removed by rice straw–based magnetized NaOH-treated biochar [49]. In both studies, there was a significant increase in the surface area as well as pore volume due to alkali treatment, resulting in effective dye removal.

Removal of antibiotics. Pharmaceutical residues are one of the most recalcitrant contaminants which endanger the ecosystems at even low concentrations. The most common class of pharmaceuticals used and excessively misused is antibiotics. Due to excessive misuse of antibiotics, the partially metabolized or unabsorbed antibiotics and their metabolites are being released into the environment causing pollution of water and soil. Antibiotic leeching and contamination into the environment induce the formation of antibiotic-resistant bacteria, as well as cause adverse health issues upon human beings through consumption of accumulated metabolites in plants and water. In a study by Jing et al., methanol-modified rice husk-derived biochar from was used to adsorb tetracycline. The methanol treatment enhanced the number of oxygen-containing groups, resulting in an adsorption capacity of 18.53 mg g^{-1} [50]. Munir et al. used date palm waste–derived biochar to remove chlortetracycline. The biochar was a zero valent iron nanoparticle-based biochar composite, showing high surface area due to the presence of the nanoparticles. It was concluded that multiple mechanisms such as physical and chemical adsorption along with porous diffusion were responsible for chlortetracycline removal [51].

In another study by Afzal et al., ciprofloxacin was removed by using biochar hydrogel beads. The biochar, synthesized from pomelo (grapefruit) peels, was treated with chitosan dissolved in 2% acetic acid solution and was added drop-wise into 1.2% NaOH solution using a syringe to form hydrogel beads. Chitosan played multiple roles in the hydrogel beads – increased the ease of separability after removal and enhanced adsorption due to the additional functional groups supplied by chitosan [52].

A summary of different engineered biochar from agricultural wastes used for organic pollutant removal is shown in Table 2.

Table 2: Adsorption mechanism-based removal of organic pollutants using agricultural waste-based engineered biochar.

Biochar source	Pollutant removed	Pyrolysis temp. (°C)	Activation treatment	Adsorption capacity ($mg\ g^{-1}$)	Mechanism of removal	Ref.
Corn straws	Bisphenol A	300	*Treated with 85% H_3PO_4	476.19	π–π and H-bond interactions	[47]
Corn stalk	Methylene blue	500	^Treated with KOH followed by thermal activation at 700 °C	406.43	Physical interaction, electrostatic interaction, hydrogen bonding and π–π interaction	[48]
Rice straws	Rhodamine B	650	*Treated with 0.2 mol L^{-1} NaOH solution followed by 1.0 mol L^{-1} $FeCl_3$ solution and thermal activation at 50 °C	53.66	H-bonds, electrostatic attraction and π–π stacking	[49]
Rice husk	Tetracycline	450–500	^Treated with 3 mol L^{-1} NaOH solution at a rate of 1:10 (w/v), followed by esterification and reaction between carbonyl groups of biochar and methanol	18.53	Oxygen-containing groups like ester and hydroxyl involved in n–p EDA interactions	[50]
Date palm waste	Chlortetracycline	600	^Soaked in 1 M $FeSO_4$, stirred, 2M $NaBH_4$ solution followed by adding chitosan in 2% CH_3COOH. N_2 purged throughout process	89.05	Chemisorption, H-bonding, inter-particle diffusion, π–π EAD interactions and redox reactions	[51]

Table 2 (continued)

Biochar source	Pollutant removed	Pyrolysis temp. (°C)	Activation treatment	Adsorption capacity (mg g^{-1})	Mechanism of removal	Ref.
Pomelo peels	Ciprofloxacin	450	*Soaked in 2% acetic acid with dissolved chitosan solution, added biochar, and added drop-wise into 1.2% NaOH solution to form hydrogel beads	34.90	Electrostatic, hydrogen bond, hydrophobic and π–π interactions	[52]

*Pre-pyrolysis treatment
^Post-pyrolysis treatment

4.2.2 Degradation of organic pollutants by advanced oxidizing processes

AOPs are considered as one of the most effective techniques to degrade organic pollutants as they result in only non-hazardous, non-toxic, simple molecules such as carbon dioxide and water. Highly reactive free radial species such as •OH, $•SO_4^-$ and $•O_2^-$ are used to degrade organic pollutants by oxidation in AOPs. These highly reactive species are generated through processes such as electrocatalysis, photocatalysis, or superoxide-activation (hydrogen peroxide, ozone, and persulfate (PS)). In Fenton-like reactions, •OH is produced from H_2O_2 in acidic conditions and activated by some iron/copper metal–based compounds. PS which include peroxydisulfate (PDS) and peroxymonosulfate (PMS) have been found to be more efficient as they have higher redox potential, electron affinity as well as lifetime and lower dissociation energy of O–O bond as compared to H_2O_2. Another advantage of PS is that they could be activated in wide ranges of pH, upon versatile metal ions (Fe, Co, Mn), metal oxides (Mn-oxides, Co-oxides), or carbon-based organic substances. Metal-based PS activation, however, possesses a demerit that it may cause secondary pollution and leeching. To overcome this limitation, carbon-based PS activators like carbon nanotubes have been used. Carbon-based PS activators possess advantages such as better surface area and surface properties, porosity, cost-efficacy, sorption ability, high resistance to extreme pH, electrical conductivity, and, most importantly, not posing a risk of secondary pollution [53].

He et al. used pristine sugarcane waste biochar to activate to degrade bisphenol A. The sugarcane wastes were pyrolyzed at 900 °C to obtain an ordered, graphite-like porous structured biochar, with a higher surface area as compared to biochar pyrolyzed at 700 °C and 800 °C. The biochar also had higher amounts of sp2 carbon content that helped in high efficiency for activation of PMS. The sp2 carbon content is responsible for the degree of graphitization in biochar and is directly proportional to the ability of electron transfer and PMS activation, hence the oxidized species production and pollutant degradation. Bisphenol A degradation efficiency through PMS activation by the pristine reached 100% in 60 min [54].

In another study by Rong et al., banana peel–derived magnetic biochar was used to activate PMS for degrading bisphenol A. The modified biochar was prepared by pre-pyrolysis treatment of banana peels with $FeSO_4$ followed by hydrothermal treated at 180 °C for 6 h, centrifugated, and pyrolyzed at a temperature of 600 °C overnight. Complete bisphenol A removal was achieved in 20 min. Both the biochar matrix and functional groups contributed in oxidative species production. In particular, the biochar matrix contributed to PS activation indirectly through electron transfer and water oxidation. The active sites with functional groups directly interacted with PS by weakening the O–O bond. Persistent free radicals upon the biochar surface have long half-lives due to their resonance-stabilized radicals and are generated when biochar is thermally treated. These radicals are known to be intrinsic redox centers, and contribute to sulfate radical generation via the electron-transfer process. γ-Fe2O3 nanoparticles promoted SO_4 generation. The reactions contributing to the radical production are shown below [55]:

$$(\text{Biochar}_{\text{surface}})\text{–OOH} + S_2O_8{}^{2-} \rightarrow SO_4{}^{-} + (\text{Biochar}_{\text{surface}})\text{–OO}\cdot + HSO_4{}^{-}$$

$$(\text{Biochar}_{\text{surface}})\text{–OH} + S_2O_8{}^{2-} \rightarrow SO_4{}^{-} + (\text{Biochar}_{\text{surface}})\text{–O}\cdot + HSO_4{}^{-}$$

$$SO_4{}^{-} + H_2O \rightarrow H_SO_4{}^{-} + OH\cdot$$

$$SO_4{}^{-} + OH^{-} \rightarrow SO_2{}^{4-} + OH\cdot$$

17β-estradiol, a major natural estrogen known for its seriously high endocrine-disrupting activity in the class of endocrine-disrupting chemicals, was degraded by PS activated by a metal-free, carbon dioxide-activated cellulose-based biochar. The biochar, derived from peanut shells, was pyrolyzed at 400 °C, and treated with 0.15 M K_2FeO_4. A complete 100% degradation efficiency was reached within 90 min, by PS activation. Here, K_2FeO_4 exhibited various roles – activating and magnetizing agent, along with being a graphitization and redox catalyst [56]. Mechanisms of radical generation were similar to those reported in [55].

A metal-free, carbon dioxide-activated cellulose-based biochar was used to activate PS for degrading phenol was reported by Sun et al [57]. Biochar obtained by pyrolysis at 500 °C was carbon dioxide–activated by placing in a tube furnace with carbon dioxide being pumped in at a rate of 500 mL min^{-1} and thermally activated at 950 °C for 1 h. Up to 99.5% removal efficiency of phenol was reached in over 60 min. Unlike the typical radical-based mechanism normally seen in PS activation [55, 56], the carbon dioxide–activated biochar exhibited a non-radical mechanism that involves electron transfer reaction pathway and singlet state oxygen generation.

5 Regeneration of engineered biochar

Regeneration of biochar is the reverse process of adsorption, consisting of two principles: adsorbate desorption and decomposition. The reusability of any biochar sorbent is a very crucial factor taken into consideration especially at industrial level, as it greatly affects cost-efficiency. Different methods are used for regenerating different modified biochar based on their composition.

Thermal regeneration. At high temperatures, the adsorbate decomposes and eventually escapes the biochar pores as it is smaller. This method is considered to be one of the most cost-effective as well as effective methods for desorption. Also, it has been found that the pyrolysis temperature is directly proportional to the efficiency of regeneration.

Microwave irradiation regeneration. Microwave heating uniformly heats biochar both at the surface and inside and also preserves the original porous structure of the biochar porous carbon matrix, reducing the decomposition, as compared to other conventional heating methods. There is also better efficiency and control over the heating process. The electromagnetic field applied produces heat, which decomposes the adsorbates, while the microwaves induce polarization, making them escape from the pores.

Solvent regeneration. Based on the principle of adsorption equilibrium, solvent regeneration uses change in conditions like temperature and pH for desorbing the adsorbate from biochar. In acid-base regeneration method, inorganic acids (sulfuric acid or hydrochloric acid) or alkalis (sodium hydroxide) are used for desorption. When organic solvents (acetone, methanol, and benzene) are used, it is referred to as organic solvent regeneration. This method is especially economical when inorganic solvents are used.

Supercritical fluid regeneration. Supercritical fluids are materials that possess properties between liquids and gases in the state above critical temperature and pressure. By adjustment of the operating pressure, supercritical fluids are used as extractants to separate adsorbents. Although still at experimental stage, this technique has several advantages such as recovering adsorbate without any change in the physical or chemical properties, low biochar loss and operating temperatures, short operating cycles, and no solvent pollution. Limitations of this technique include high costs for equipment and high-pressure resistance. Commonly used supercritical fluids are carbon dioxide and water [46].

6 Conclusion

Biochar shows great potential for waste management as a sustainable option. Although studies continue to improve the performance of biochar, it is to be noted that the exact environmental fate of biochar is still not completely defined. Only when full-fledged standardization of synthesis methods, application of green chemistry methods, and long-term cost-benefit analysis are taken into serious consideration, biochar could be scaled up to full potential.

References

[1] Clark M, Tilman D, Comparative analysis of environmental impacts of agricultural production systems, agricultural input efficiency, and food choice, Environ. Res. Lett., 2017, 12(6), doi: 10.1088/1748-9326/aa6cd5.

[2] Notarnicola B, Sala S, Anton A, McLaren SJ, Saouter E, Sonesson U, The role of life cycle assessment in supporting sustainable agri-food systems: A review of the challenges, J. Clean. Prod., 2017, 140, 399–409, doi: 10.1016/j.jclepro.2016.06.071.

[3] Sarkar N, Ghosh SK, Bannerjee S, Aikat K, Bioethanol production from agricultural wastes: An overview, Renew. Energy, 2012, 37(1), 19–27, doi: 10.1016/j.renene.2011.06.045.

[4] Phalan B, Onial M, Balmford A, Green RE, Reconciling food production and biodiversity conservation: Land sharing and land sparing compared, Science (80-.)., 2011, 333(6047), 1289–1291, doi: 10.1126/science.1208742.

[5] Obi FO, Ugwuishiwu BO, Nwakaire JN, Agri. Wastes, 2016, 35(4), 957–964.

[6] Suman S, Gautam S, Biochar derived from agricultural waste biomass act as a clean and alternative energy source of fossil fuel inputs, Energy Syst. Environ., 2018, doi: 10.5772/intechopen.73833.

[7] Chakraborty I, Sathe SM, Dubey BK, Ghangrekar MM, Waste-derived biochar: Applications and future perspective in microbial fuel cells, Bioresour. Technol., 2020, 312, no. May, 123587, doi: 10.1016/j.biortech.2020.123587.

[8] Kazemi Shariat H, Panahi MD, Ok YS, Nizami AS, Khoshnevisan B, Mussatto SI, Aghbashlo M, Tabatabaei M, Lam SS, A comprehensive review of engineered biochar: Production, characteristics, and environmental applications, J. Clean. Prod., 2020, 270, 122462, doi: 10.1016/j.jclepro.2020.122462.

[9] Akhil D, Lakshmi D, Kartik A, Vo DVN, Arun J, Gopinath KP, Production, Characterization, Activation and Environmental Applications of Engineered Biochar: A Review, no. 0123456789, Springer International Publishing, 2021.

[10] Czernik S, Bridgwater AV, Overview of applications of biomass fast pyrolysis oil, Energy Fuels, 2004, 18(2), 590–598, doi: 10.1021/ef034067u.

[11] Lima IM, Boateng AA, Klasson KT, Physicochemical and adsorptive properties of fast-pyrolysis bio-chars and their steam activated counterparts, J. Chem. Technol. Biotechnol., 2010, 85(11), 1515–1521, doi: 10.1002/jctb.2461.

[12] Wiedner K, Rumpel C, Steiner C, Pozzi A, Maas R, Glaser B, Chemical evaluation of chars produced by thermochemical conversion (gasification, pyrolysis and hydrothermal carbonization) of agro-industrial biomass on a commercial scale, Biomass Bioenerg, 2013, 59, 264–278, doi: 10.1016/j.biombioe.2013.08.026.

[13] Antal MJ, Grønli M, The art, science, and technology of charcoal production, Ind. Eng. Chem. Res., 2003, 42(8), 1619–1640, doi: 10.1021/ie0207919.
[14] Sizmur T, Fresno T, Akgül G, Frost H, Moreno-Jiménez E, Biochar modification to enhance sorption of inorganics from water, Bioresour. Technol., 2017, 246, 34–47, doi: 10.1016/j.biortech.2017.07.082.
[15] Kazemi H, Panahi S, Dehhaghi M, Sik Y, Nizami A, Khoshnevisan B, Mussatto SI, Aghbashlo M, Tabatabaei M, Shiung S, A comprehensive review of engineered biochar: Production, characteristics, and environmental applications, J. Clean. Prod., 2020, 270, 122462, doi: 10.1016/j.jclepro.2020.122462.
[16] Wang B, Lehmann J, Hanley K, Hestrin R, Enders A, Ammonium retention by oxidized biochars produced at different pyrolysis temperatures and residence times, RSC Adv., 2016, 6(48), 41907–41913, doi: 10.1039/c6ra06419a.
[17] Cho DW, Kwon G, Yoon K, Tsang YF, Ok YS, Kwon EE, Song H, Simultaneous production of syngas and magnetic biochar via pyrolysis of paper mill sludge using CO2 as reaction medium, Energy Convers. Manag., 2017, 145, 1–9, doi: 10.1016/j.enconman.2017.04.095.
[18] Tan XF, Liu YG, Gu YL, Xu Y, Zeng GM, Hu XJ, Liu SB, Wang X, Liu SM, Li J, Biochar-based nano-composites for the decontamination of wastewater: A review, Bioresour. Technol., 2016, 212, 318–333, doi: 10.1016/j.biortech.2016.04.093.
[19] Wang B, Gao B, Fang J, Recent advances in engineered biochar productions and applications, Crit. Rev. Environ. Sci. Technol., 2017, 47(22), 2158–2207, doi: 10.1080/10643389.2017.1418580.
[20] Inyang M, Gao B, Pullammanappallil P, Ding W, Zimmerman AR, Biochar from anaerobically digested sugarcane bagasse, Bioresour. Technol., 2010, 101(22), 8868–8872, doi: 10.1016/j.biortech.2010.06.088.
[21] Yao Y, Gao B, Wu F, Zhang C, Yang L, Engineered biochar from biofuel residue: Characterization and its silver removal potential, ACS Appl. Mater. Interfaces, 2015, 7(19), 10634–10640, doi: 10.1021/acsami.5b03131.
[22] Mohamed BA, Kim CS, Ellis N, Bi X, Microwave-assisted catalytic pyrolysis of switchgrass for improving bio-oil and biochar properties, Bioresour. Technol., 2016, 201, 121–132, doi: 10.1016/j.biortech.2015.10.096.
[23] Lyu H, Gao B, He F, Ding C, Tang J, Crittenden JC, Ball-milled carbon nanomaterials for energy and environmental applications, ACS Sustain. Chem. Eng., 2017, 5(11), 9568–9585, doi: 10.1021/acssuschemeng.7b02170.
[24] Wang B, Gao B, Fang J, Wang B, Recent advances in engineered biochar productions and applications, Crit. Rev. Environ. Sci. Technol., 2022, 47(22), 2158–2207, doi: 10.1080/10643389.2017.1418580.
[25] Essandoh M, Wolgemuth D, Pittman CU, Mohan D, Mlsna T, Adsorption of metribuzin from aqueous solution using magnetic and nonmagnetic sustainable low-cost biochar adsorbents, Environ. Sci. Pollut. Res., 2017, 24(5), 4577–4590, doi: 10.1007/s11356-016-8188-6.
[26] Tang L, Yu J, Pang Y, Zeng G, Deng Y, Wang J, Ren X, Ye S, Peng B, Feng H, Sustainable efficient adsorbent: Alkali-acid modified magnetic biochar derived from sewage sludge for aqueous organic contaminant removal, Chem. Eng. J., 2018, 336, no. September 2017, 160–169, doi: 10.1016/j.cej.2017.11.048.
[27] Wang Y, Liu R, H2O2 treatment enhanced the heavy metals removal by manure biochar in aqueous solutions, Sci. Total Environ., 2018, 628–629, 1139–1148, doi: 10.1016/j.scitotenv.2018.02.137.
[28] Wang B, Gao B, Fang J, Technology recent advances in engineered biochar productions and applications, Crit. Rev. Environ. Sci. Technol., 2018, 0(0), 1–50, doi: 10.1080/10643389.2017.1418580.

[29] Wei D, Li B, Huang H, Luo L, Zhang J, Yang Y, Guo J, Tang L, Zeng G, Zhou Y, Biochar-based functional materials in the purification of agricultural wastewater: Fabrication, application and future research needs, Chemosphere, 2018, 197, 165–180, doi: 10.1016/j.chemosphere.2017.12.193.

[30] Zhao X, Ouyang W, Hao F, Lin C, Wang F, Han S, Geng X, Properties comparison of biochars from corn straw with different pretreatment and sorption behaviour of atrazine, Bioresour. Technol., 2013, 147, 338–344, doi: 10.1016/j.biortech.2013.08.042.

[31] Rajapaksha AU, Vithanage M, Ahmad M, Seo DC, Cho JS, Lee SE, Lee SS, Ok YS, Enhanced sulfamethazine removal by steam-activated invasive plant-derived biochar, J. Hazard. Mater., 2015, 290, 43–50, doi: 10.1016/j.jhazmat.2015.02.046.

[32] Devi P, Saroha AK, Synthesis of the magnetic biochar composites for use as an adsorbent for the removal of pentachlorophenol from the effluent, Bioresour. Technol., 2014, 169, 525–531, doi: 10.1016/j.biortech.2014.07.062.

[33] Mohan D, Sarswat A, Ok YS, Pittman CU, Organic and inorganic contaminants removal from water with biochar, a renewable, low cost and sustainable adsorbent – A critical review, Bioresour. Technol., 2014, 160, no. October, 191–202, doi: 10.1016/j.biortech.2014.01.120.

[34] Khalil A, Sergeevich N, Borisova V, Removal of ammonium from fish farms by biochar obtained from rice straw: Isotherm and kinetic studies for ammonium adsorption, Adsorpt. Sci. Technol., 2018, 36(5–6), 1294–1309, doi: 10.1177/0263617418768944.

[35] Nguyen LH, Vu TM, Le TT, Trinh VT, Tran TP, Ammonium removal from aqueous solutions by fixed-bed column using corncob-based modified biochar, Environ. Technol., 2017, 0(0), 1–10, doi: 10.1080/09593330.2017.1404134.

[36] Wang S, Ai S, Nzediegwu C, Kwak J, Islam S, Li Y, Chang SX, Bioresource technology carboxyl and hydroxyl groups enhance ammonium adsorption capacity of iron (III) chloride and hydrochloric acid modi fi ed biochars, Bioresour. Technol., 2020, 309, no. April, 123390, doi: 10.1016/j.biortech.2020.123390.

[37] Liu F, Zuo J, Chi T, Wang P, Yang B, Removing phosphorus from aqueous solutions by using iron- modi fi ed corn straw biochar, 2015, 9(6), 1066–1075, doi: 10.1007/s11783-015-0769-y.

[38] Huang Y, Lee X, Grattieri M, Yuan M, Cai R, Macazo FC, Minteer SD, Modi fi ed biochar for phosphate adsorption in environmentally relevant conditions, Chem. Eng. J., 2020, 380, no. July 2019, 122375, doi: 10.1016/j.cej.2019.122375.

[39] Godwin PM, Pan Y, Xiao H, Afzal MT, Progress in preparation and application of modified biochar for improving heavy metal ion removal from wastewater, J. Bioresour. Bioprod., 2019, 4(1), 31–42, doi: 10.21967/jbb.v4i1.180.

[40] Wang L, Wang Y, Ma F, Tankpa V, Bai S, Guo X, Wang X, Science of the total environment mechanisms and reutilization of modi fi ed biochar used for removal of heavy metals from wastewater: A review, Sci. Total Environ., 2019, 668, 1298–1309, doi: 10.1016/j.scitotenv.2019.03.011.

[41] Mosa AA, El-Ghamry A, Al-Zahrani H, Selim EM, El-Khateeb A, Chemically modified biochar derived from cotton stalks: Characterization and assessing its potential for heavy metals removal from wastewater, Env. Biodiv. Soil Secur., 2017, 1, 33–45.

[42] Bashir S, Zhu J, Fu Q, Hu H, Comparing the adsorption mechanism of Cd by rice straw pristine and KOH-modified biochar, Environ. Sci. Pollut. Res., 2018, 25(12), 11875–11883.

[43] Zhou Z, Liu Y, Liu S, Liu H, Zeng G, Tan X, Yang C, Ding Y, Yan Z, Cai X, Sorption performance and mechanisms of arsenic(V) removal by magnetic gelatin-modified biochar, Chem. Eng. J, 2016, 314, doi: 10.1016/j.cej.2016.12.113.

[44] Zhou Y, Gao B, Zimmerman AR, Fang J, Sun Y, Cao X, Sorption of heavy metals on chitosan-modified biochars and its biological effects, Chem. Eng. J., 2013, 231, 512–518.

[45] Deng J, Liu Y, Liu S, Zeng G, Tan X, Huang B, Tang X, Wang S, Hua Q, Yan Z, Competitive adsorption of Pb(II), Cd(II) and Cu(II) onto chitosan-pyromellitic dianhydride modified biochar, J. Colloid Interface Sci., no. Ii, 2017, doi: 10.1016/j.jcis.2017.07.069.
[46] Dai Y, Zhang N, Xing C, Cui Q, Sun Q, Chemosphere The adsorption, regeneration and engineering applications of biochar for removal organic pollutants: A review, Chemosphere, 2019, 223, 12–27, doi: 10.1016/j.chemosphere.2019.01.161.
[47] Zhao N, Zhao C, Lv Y, Zhang W, Du Y, Hao Z, Zhang J, Adsorption and coadsorption mechanisms of Cr(VI) and organic contaminants on H3PO4 treated biochar, ECSN, no. Vi, 2017, doi: 10.1016/j.chemosphere.2017.08.016.
[48] Liu L, Li Y, Fan S, Preparation of KOH and H 3 PO 4 Modified Biochar and Its Application in Methylene Blue Removal from Aqueous Solution, Processes, 2019, 7(891).
[49] Ren Z, Chen F, Wang B, Song Z, Zhou Z, Ren D, Magnetic biochar from alkali-activated rice straw for removal of rhodamine B from aqueous solution, 2020, 25(4), 536–544.
[50] Jing X, Wang Y, Liu W, Wang Y, Jiang H, Enhanced adsorption performance of tetracycline in aqueous solutions by methanol-modified biochar, Chem. Eng. J., 2014, 248, 168–174, doi: 10.1016/j.cej.2014.03.006.
[51] Ahmad M, Usman ARA, Rafique MI, Al-wabel MI, Engineered biochar composites with zeolite, silica, and nano-zerovalent iron for the efficient scavenging of chlortetracycline from aqueous solutions, 2019, 15136–15152.
[52] Zaheer M, Sun X, Liu J, Song C, Wang S, Javed A, Science of the total environment enhancement of cipro fl oxacin sorption on chitosan / biochar hydrogel beads, Sci. Total Environ., 2018, 639, 560–569, doi: 10.1016/j.scitotenv.2018.05.129.
[53] Mian MM, Liu G, Activation of peroxymonosulfate by chemically modified sludge biochar for the removal of organic pollutants: Understanding the role of active sites and mechanism, Chem. Eng. J., 2019, 123681, doi: 10.1016/j.cej.2019.123681.
[54] He L, Liu Z, Hu J, Qin C, Yao L, Zhang Y, Piao Y, Sugarcane biochar as novel catalyst for highly efficient oxidative removal of organic compounds in water, Chem. Eng. J., 2021, 405, no. September 2020, 126895, doi: 10.1016/j.cej.2020.126895.
[55] Rong X, Xie M, Kong L, Natarajan V, Ma L, Zhan J, The magnetic biochar derived from banana peels as a persulfate activator for organic contaminants degradation, Chem. Eng. J., 2019, 372, no. April, 294–303, doi: 10.1016/j.cej.2019.04.135.
[56] Zhang P, Tan X, Liu S, Liu Y, Zeng G, Ye S, Catalytic degradation of estrogen by persulfate activated with iron-doped graphitic biochar: Process variables effects and matrix effects, Chem. Eng. J., 2019, 378, no. May, 122141, doi: 10.1016/j.cej.2019.122141.
[57] Sun C, Chen T, Huang Q, Zhan M, Li X, Yan J, Activation of persulfate by CO 2 -activated biochar for improved phenolic pollutant degradation: Performance and mechanism, Chem. Eng. J., 2020, 380, no. April 2019, 122519, doi: 10.1016/j.cej.2019.122519.

Mohit Aggarwal, Neelancherry Remya

Removal of the pharmaceutical compounds from wastewater by biochar

Abstract: Water pollution from pharmaceutical compounds is a critical global challenge. The pharmaceutical compounds and their subsequent metabolites are termed emerging contaminants. The detection of these compounds in surface water and groundwater as well as conventional wastewater treatment plant effluent necessitates a sustainable solution to remove these pollutants to protect water resources. Multiple biological, chemical, and physical methods have been employed to clean the wastewater from the pharmaceuticals. However, they are either low efficiency or expensive. Recently, adsorption has gained prominence in removing pharmaceutical contaminants due to its high efficiency and low capital requirement. Several investigations are conducted to produce low-cost but highly efficient adsorbents for pharmaceutical removal. Biochar, a solid carbonized by-product obtained from biomass pyrolysis, is considered a cheap replacement to commercially available materials to clean the wastewaters. Biochar with or without modification is used for the removal of pharmaceutical contaminants and their metabolites. The biochar modifications such as alkali/acid modification, magnetic biochar, metal impregnated biochar, microwave-activated biochar, the improvement in biochar properties with these modifications, and enhanced adsorptive removal of pharmaceutical compounds by modified biochar will be discussed in this chapter.

1 Introduction

Water pollution from emerging contaminants like pharmaceuticals remains a global challenge urging a sustainable solution to remove these pollutants to protect water resources. Pharmaceutical compounds have been detected in wastewater, treatment plant effluent, surface, and groundwater in concentrations ranging from ng L^{-1} to μg L^{-1} [1–4]. Pharmaceuticals include anti-inflammatory drugs like diclofenac, ibuprofen, and ketoprofen. Antibiotics such as sulfamethoxazole and trimethoprim, antiepileptic drugs like carbamazepine, beta-blockers, propranolol, nervous stimulants like caffeine, estrogens such as 17α-ethinylestradio and estriol, and lipid regulators like clofibric acid and gemfibrozil [5].

Mohit Aggarwal, Neelancherry Remya, Indian Institute of Technology Bhubaneswar, Odisha 752050, India, e-mail: remya@iitbbs.ac.in

https://doi.org/10.1515/9783110734003-010

Potential adverse effects of pharmaceuticals in the environment depend on factors such as concentration, exposure time, bioaccumulation potential, biodegradability, and lipophilicity [6, 7]. These compounds pose a potential threat to humans, other living organisms, and the environment in many ways, which include hindering of prostaglandin synthesis and histological alterations in gills by diclofenac, cardio abnormalities caused by ibuprofen, reduced growth and decreased ability to catch prey in animals owing to fluoxetine exposure, inhibition of brain aromatase caused by fadrozole, lower plasma androgen concentration, acute toxicity resulted from clofibrate, reduced growth, altered heart rate and drop in glucose production triggered by propranolol, and chronic toxicity effect by sulfamethoxazole [8].

Many pharmaceuticals are not removed efficiently using conventional wastewater treatment methods. Thus pharmaceuticals in significant concentration are detected in effluents from traditional wastewater treatment plants [9]. Naproxen (2-(6-methoxy-2-naphthyl) propionic acid), a non-steroidal anti-inflammatory drug with analgesic, antipyretic, and anti-inflammatory effects, has been detected up to a concentration of 2.3 mg L^{-1} in effluent from conventional wastewater plants [10]. Carbamazepine, an antiepileptic drug at a concentration of ~ 2.1 mg L^{-1}, was detected in the treated effluents [11].

Advanced technologies such as photocatalysis [12], nanofiltration [3], ultrasonication [3], membrane filtration [3], adsorption [13], ultraviolet (UV)/chlorine and UV/ H_2O_2 [14]. Fenton and photo-Fenton reactions [15], electrochemical oxidation [16], and ozonation [17, 18] are employed for efficient removal of pharmaceuticals from wastewater. However, these techniques possess disadvantages like high operational costs, toxic by-product formation, and use of chemicals [3]. Meanwhile, adsorption techniques have gained increasing attention for removing such contaminants due to their high efficiency, simple operation method, and nongeneration reaction by-products [19]. However, a common disadvantage of adsorption is the high cost of commercially available sorbent material. The unmet need for cost-effective and highly efficient pharmaceutical adsorbents has encouraged low-cost adsorbent production from readily available waste materials [20].

Recently, the application of biochar as an adsorbent has attracted significant attention. Biochar is a highly porous carbonaceous material derived by the pyrolysis of waste biomass under limited oxygen-limited conditions. Biomass, a by-product of the agricultural process, is abundant in nature and contains celluloses, hemicelluloses, lignin, and other carbohydrates. Several studies have reported the effective removal of pharmaceutical contaminants and their metabolites using biochar [21]. Besides, various modifications improved biochar's adsorption property and posttreatment separability [22–24].

2 Production of biochar

A wide range of energy crops such as switchgrass, forest residues such as grain crops [20], organic waste such as animal manure, yard waste, kitchen waste, sewage sludge, and agricultural waste [25] is used for the production of biochar. The properties of biomass feedstock, particularly the proximate characteristics, that is, ash content (AC), fixed carbon (FC), volatile matter (VM), and moisture content (MC), ultimate characteristics (carbon, hydrogen, nitrogen, sulfur, and oxygen content), component composition (cellulose, hemicellulose, and lignin), calorific value, particle size, and density are significant in thermal conversion processes of biomass to biochar [20]. A higher biochar yield was observed during the biomass conversion with more lignin and less cellulose content [3]. However, the VM, MC, and particle size affected the obtained biochar characteristics [26].

In addition to the biomass composition, biochar's yield and property depend on the conversion technology and operating conditions used for biochar production [27]. Biochar formed at higher temperatures (lower oxygen to carbon ratio) is expected to be Π-donors, while BCs formed at lower temperatures are expected to be Π-acceptors [26]. The proposed temperature-dependent structure transition of biochar include (i) transition biochar having crystalline nature of the biomass preserved, (ii) amorphous biochar, a random mixture of thermally changed molecules, and emerging aromatic polycondensates, (iii) composite biochar having poorly structured graphene stacks fixed in their amorphous phase, and (iv) turbastic biochar dominated by chaotic graphite crystallites [28]. Various technologies used to convert biomass include torrefaction, hydrothermal carbonization (HTC), slow pyrolysis, fast pyrolysis, gasification, and flash carbonization [20].

The combustion process, pyrolysis, and gasification in traditional charcoal kilns or modern retorts or furnaces [20]. Pyrolysis is classified by the pyrolysis material's residence time and pyrolytic temperature into slow, fast, and flash pyrolysis. The heating method adopted includes heating by burning fuels, electrical heating, or microwave heating [3]. Slow pyrolysis is carried out at a low temperature and slow heating rate. On the other hand, fast pyrolysis employs biomass heating to a temperature of ~ 650 °C with a rapid heating rate (100–1,000 °C s^{-1}). Slow pyrolysis is conducted with a relatively high residence time compared to fast pyrolysis. High residence time causes secondary reactions. These mainly include the interaction of tar on the surface of biochar which causes the charring of the tar [7].

In gasification, biomass is oxidized in a chamber at a temperature of 800 °C at elevated or atmospheric pressure. Gas is the main product that is produced in gasification. Very little or no biochar is formed during the process [6]. In HTC, the biomass is heated for several hours at 200–250 °C temperature under high atmospheric pressure in liquid to yield solid fraction as the main by-product along with some liquids and gases [7]. In flash carbonization, the biomass, a flash fire starts at high pressure (at about 1–3 MPa) below the packed biomass bed. The fire goes from the bottom to the top through the biomass bed. Around 0.7–1.4 g of air per gram of biomass is added into the container from the top to the bottom of the reactor. The reactor temperature is kept between 330 and 650 °C for around 30 min [29]. Both solid and gaseous products are formed in this process.

3 Modification of biochar

Biochar is used to remove various pollutants from wastewater by adsorption with high efficiency and lower cost than other expensive methods. To modulate biochar properties for environmental purposes, different chemical and physical modification methods were adopted to modify biochar [30]. Table 1 gives some examples of biomass used in biochar production and the modification method employed to improve the property of biochar to improve the removal efficiency [31]. The chemical modification includes oxidizing agent modification, alkaline modification, acid modification, metal salts modification, and physical modification comprising steam and gas purging [30].

3.1 Acid modification

The primary purpose of the acid modification is to add acid functional groups on the biochar surface, remove metal impurities, and increase the surface area (Figure 1). The acids used are sulfuric acid, phosphoric acid, hydrochloric acid, citric acid, nitric acid, and oxalic acid. Acid modification using nitric acid (2 M) adds phenolic and carboxylic groups on biochar derived from bamboo (Zhang et al. [33]). Acid modification with 1 M of hydrochloric acid decreased AC from 30 to 12%. It adds hydrophobic adsorption sites for the adsorption of pollutants. The modification increased the surface area of reed-derived biochar from 59 to 88 $m^2\ g^{-1}$ after treatment due to the removal of surface impurities [30]. But, the surface area of biochar derived from rice straw, when treated with 2% sulfuric acid, reduced from 72 to 57 $m^2\ g^{-1}$ [31]. The type of biomass, acid, and its concentration, can also change the surface area of the biochar.

3.2 Alkali modification

Similar to acid modification, alkali modification also increases the surface area and introduces the functional groups that are oxygen-rich [25]. The common alkaline agents used for biochar modification include potassium hydroxide and sodium hydroxide. Potassium hydroxide increased the surface area of municipal solid waste-derived biochar, obtained through pyrolysis from 14.3 to 49.2 $m^2\ g^{-1}$. It also added several oxygen-containing functional groups on the biochar surface [31]. NaOH

Table 1: Modification of biochar and properties of modified biochar.

Biomass used	Biochar preparation method	Operating conditions			Modification	Biochar properties		Reference
		Temperature (°C)	Time (h)	Heating rate (°C/min)		Before	After	
Woodchips[1]	Pyrolysis	600	2	10	–	AC (%): 4.8 BSA (m^2 g^{-1}): 369.1 PV (cm^3 g^{-1}): 0.1		[28]
Woodchips[2]	Pyrolysis	600	2	10	–	AC (%): 6.5 SA (m^2 g^{-1}): 207.8 PV (cm^3 g^{-1}): 0.1		
Woodchips[3]	Pyrolysis	600	2	10	–	AC (%): 3.0 BSA (m^2 g^{-1}): 7.7 PV (cm^3 g^{-1}): 0.1		
Pharmaceutical sludge	Pyrolysis	450	1	20	Ultrasonication	BSA (m^2 g^{-1}): 215.0 adsorption capacity (mg g^{-1}): 94.7	BSA (m^2 g^{-1}): 319.8 adsorption capacity (mg g^{-1}): 157.4	[26]

(continued)

Table 1 (continued)

Biomass used	Biochar preparation method	Operating conditions			Modification	Biochar properties		Reference
		Temperature (°C)	Time (h)	Heating rate (°C/min)		Before	After	
Novel lignocellulosic waste biomass (functionalized biochar (WpOH))	Pyrolysis	180–500	35–60	10	Alkali modification under microwave irradiation followed by acid neutralization	C (%): 52.4 O (%): 37.0 H (%): 7.6 N (%): 2.6	C (%): 88.4 O (%): 10.1 H (%): 0.9 N (%): 0.6	[7]
Douglas biochar (a waste by-product of timber industry)	Gasification	900–1,000	–	–	Magnetisation with $FeCl_3$ – $FeSO4.7H_2O$ and alkali (NaOH) modification	C (%): 76.3 H (%): 1.8 N (%): 1.8 O (%): 3.2 AC (%): 6.8 Fe (%): 1.2 BSA ($m^2\ g^{-1}$): 468.2 LSA ($m^2\ g^{-1}$): 747.6 PV ($cm^3\ g^{-1}$): 0.2 PD (nm): 1.4	C (%): 58.5 H (%): 1.2 N (%): 1.5 O (%): 5.8 AC (%): 32.9 Fe (%): 21.1 BSA ($m^2\ g^{-1}$): 322.0 LSA ($m^2\ g^{-1}$): 516.3 PV ($cm^3\ g^{-1}$): 0.1 PD (nm): 1.3	[29]

Pure glucose (innovative spherical biochar)	HTC and pyrolysis	600–900	3	–	HTC followed by pyrolysis	C (%): 77.0 O (%): 22.0 N (%): 1.1 BSA ($m^2\ g^{-1}$): 1,033 PV ($m^3\ g^{-1}$): 1.1 PD (nm): 4.2	C (%): 80.1 O (%): 19.9 N (%): 0 BSA (m^2/g): 1,292 PV ($m^3\ g^{-1}$): 0.7 PD (nm): 2.2	[34]
Forest and agricultural residues, industrial by-products and municipal solid waste	Hydrothermal synthesis	50–160	8–24	–	Co-precipitation of salts layered double hydroxide onto biochar	C (%): 44.97 H (%): 6.01 O (%): 35.34 N (%): 3.52 BSA ($m^2\ g^{-1}$):539.0 PV ($m^3\ g^{-1}$): 0.4 PD (nm): 29.9	Improvement in specific surface area, surface functional groups, structure heterogeneity. C (%): 20.49 H (%): – O (%): 45.86 N (%): – BSA ($m^2\ g^{-1}$):404.76 PV ($m^3\ g^{-1}$): 0.8 PD (nm): 81.0	[35]
Engineered tea-waste	Pyrolysis	700	2	7	Steam activation	C (%): 72.0 O (%): 08.30 N (%): 01.08	Enhancement of surface properties C (%): 63.11 O (%): 10.85 N (%): 01.46	[6]

(continued)

Table 1 (continued)

Biomass used	Biochar preparation method	Operating conditions			Modification	Biochar properties		Reference
		Temperature (°C)	Time (h)	Heating rate (°C/min)		Before	After	
CG	Slow pyrolysis	700	2	–	–	BSA ($m^2\ g^{-1}$): 16.3 PV ($cm^3\ g^{-1}$): $1.1E^{-02}$, PD (nm): 3.6, pH: 10.9	–	[3]
GB						BSA ($m^2\ g^{-1}$): 5.9 PV ($cm^3\ g^{-1}$): $6.59E^{-03}$, pH: 9.9	–	

[1]Obtained from Wakefield Agricultural Carbon LLC.
[2]Obtained from Cool Planet LLC.
[3]Obtained from Green Texan Organic Farms LLC; GB: Guayule bagasse; CG: Cotton gin waste; AC: Adsorption capacity; PV: Pore Volume; BET Surface Area; PD: Pore Diameter.

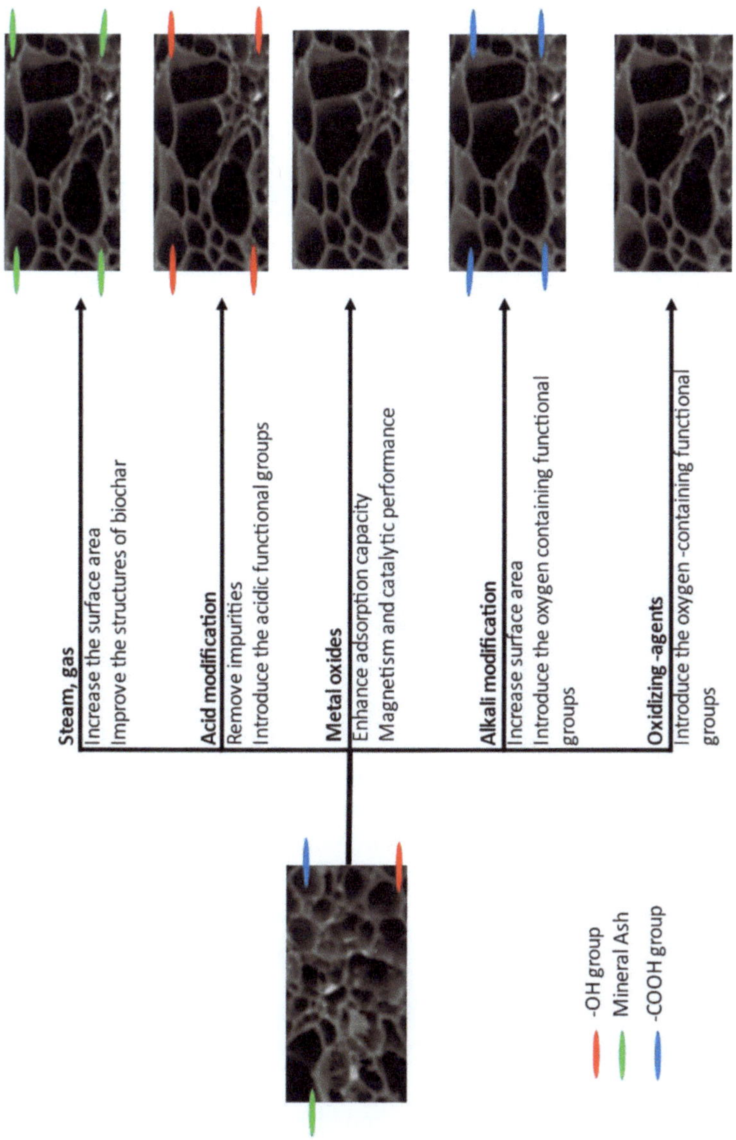

Figure 1: Different modifications of biochar.

modified biochar (MBC) derived from pomelo peel was prepared for Mn^{2+} adsorption. The results showed an increase in surface area from 0.259 to 2.407 $m^2\ g^{-1}$ [32].

3.3 Oxidizing agents modification

The modification using oxidizing agents such as hydrogen peroxide and potassium permanganate caused an increase in oxygen-containing functional groups on the surface of biochar [31]. For example, hydrogen peroxide modification of the pine-wood-derived biochar decreased the pH of biochar from 7.16 to 5.66 and increased oxygen-containing functional groups on biochar surface [33]. An improvement in the oxygen-containing functional groups, especially carboxyl groups, was observed in the peanut hull-derived biochar prepared by hydrothermal carbonization upon hydrogen peroxide modification [25]. On the other hand, an increase in surface area from 101 to 205 $m^2\ g^{-1}$ was observed in hickory wood-derived biochar [31].

3.4 Metal salts or metal oxides modification

Metal salts or metal oxides modification can be accomplished via two ways: (i) mixing of metal salts or metal oxides with biomass followed by pyrolysis of the mixture and (ii) pyrolysis of biochar to produce biochar followed by incorporating metal salts or metal oxides by soaking and chemical reactions under certain conditions [31, 33]. The incorporation of metal salts or metal oxides on the biochar surface significantly altered the surface charge of biochar and improved its catalytic property (Figure 1). Besides, the modifications with iron salts or iron oxide improved the magnetic property of biochar, which enabled the easy magnetic separation of biochar from aqueous systems after the treatment process [29].

3.5 Steam modification

Steam modification is a physical modification method in which steam purging is done to the biochar obtained through different ways [30]. In the steam modification, the oxygen present in the water molecule is passed onto the free active sites of the biochar surface. As a result, hydrogen is produced. This gas reacts with carbon on the biochar surface and forms surface hydrogen complexes [3]. Steam causes the devolatilization and the formation of crystalline carbon in the biochar. It also releases the trapped partially combusted products formed in the initial pyrolysis process (Figure 1). Hence, steam modification improves the biochar surface area and its structure [31].

4 Application of biochar in pharmaceuticals wastewater treatment

Removing different pharmaceutical compounds requires different modifications in the biochar due to the complexity of the compounds. The mechanisms of removal and the efficiency of modified biochar for removing different pharmaceutical compounds from wastewater is given in Table 2. One study used biochar made from wood chips sourced from three different industries to study the removal of ibuprofen and sulfamethoxazole and observed that the removal efficiency varied from 0.4 to 75% for ibuprofen and 0 to 65% for sulfamethoxazole. The adsorption capacity of the biochar depends on their pore volume, AC, charge, specific surface area, hydrophobicity, π-energy, and the nature of pharmaceuticals [28]. Another study compared the adsorption capacities of biochar obtained from the pyrolysis of guayule and cotton at different temperatures to remove sulfapyridine, docusate, and erythromycin. Biochar obtained from cotton showed a very promising removal rate of 98% for docusate, 70% for sulfapyridine, and 74% for erythromycin. On the other hand, biochar obtained from guayule showed a moderate removal rate for all three compounds at around 50% [3].

Table 2: Removal of pharmaceutical compounds using biochar.

Biochar source	Pharmaceutical compounds	Removal mechanism	Removal efficiency (%)	Adsorption capacity ($mg\ g^{-1}$)	Reference
Wood chips[1]	Ibuprofen/ sulfamethoxazole	Adsorption	75/65	–	[28]
Wood chips[2]			0.4/10		
Wood chips[3]			5/0		
Pharmaceutical sludge	Tetracycline	Adsorption	71	157.7	[26]
Functionalized biochar (WpOH)	Naproxen	Adsorption, hydrogen bonding	73.14	38.8	[7]
Magnetic Fe_3O_4/Douglas fir biochar	Caffeine	Adsorption, electrostatic, hydrogen bonding	95	75.1	[29]
Magnetic Fe_3O_4/Douglas fir biochar	Ibuprofen	Adsorption, electrostatic, hydrogen bonding	90	39.9	
Magnetic Fe_3O_4/Douglas fir biochar	Acetylsalicylic acid	Adsorption, electrostatic, hydrogen bonding	70	149.9	

Table 2 (continued)

Biochar source	Pharmaceutical compounds	Removal mechanism	Removal efficiency (%)	Adsorption capacity (mg g^{-1})	Reference
Spherical biochar	Paracetamol	van der Waals forces, electrostatic attraction, hydrogen bonding formation, n–π interaction	99.8	268	[34]
Biochar/LDH	Diclofenac sodium	Electrostatic attraction, interlayer anions exchange, surface adsorption, π–π attraction	80	2114.4	[36]
	Diclofenac sodium	Interlayer anions exchange, ligand exchange, surface adsorption		1118.2	
	Carbon tetrachloride	Interlayer anions exchange, ligand exchange, surface adsorption		90.2	
	Sulfamethoxazole	π–π interactions, hydrophobic interactions, hydrogen bonding		2.2	
	Paracetamol	Surface adsorption, π–π and hydrophobic interactions		17.7	
	Caffeine	Surface adsorption, π–π and hydrophobic interactions		26.2	
Engineered tea-waste biochar	Caffeine	π–π, dipole–dipole, hydrogen bonding, and electrostatic attraction interactions	95.67	15.4	[6]

Table 2 (continued)

Biochar source	Pharmaceutical compounds	Removal mechanism	Removal efficiency (%)	Adsorption capacity (mg g^{-1})	Reference
CG	Sulfapyridine	Interlayer anions exchange, π–π and hydrophobic interactions, and surface adsorption	70	0.7	[3]
	Docusate	Ligand exchange, π–π and hydrophobic interactions, and surface adsorption	98	2.1	
	Erythromycin	Interlayer anions exchange, surface adsorption, π–π and hydrogen bonding	74	1.9	
GB	Sulfapyridine	Interlayer anions exchange, π–π and hydrophobic interactions, surface adsorption, and hydrogen bonding	50	0.1	
	Docusate	π–π and hydrophobic interactions, surface adsorption, exchange, and hydrogen bonding	51	1.0	
CGGB	Sulfapyridine, docusate, and erythromycin	π–π and hydrophobic interactions, surface adsorption, and hydrogen bonding	53	1.4	

[1]Obtained from Wakefield Agricultural Carbon LLC.
[2]Obtained from Cool Planet LLC.
[3]Obtained from Green Texan; Organic Farms LLC; GB: Guayule bagasse; CG: Cotton gin waste; LDH: Layered double hydroxide.

5 Conclusion

The composition and properties of every pharmaceutical compound are different. Hence, removing these compounds from the wastewater is exceedingly complex and expensive. Adsorption using biochar is a very cost-effective and easy method to treat pharmaceutical wastewater. To increase its efficiency, some modifications are done to biochar to remove specific pharmaceutical wastewater components. These modifications increase the surface area and add some function group, and enhance biochar's surface properties, increasing the removal efficiencies of biochar many times.

In some cases, efficiency increases up to four times after modifications. Biochar is modified mainly by oxidizing agent modification, acid modification, alkali modification, carbonaceous materials modification, metal salts modification, and physical modification comprising gas and steam purging. By using these modifications, many pharmaceutical compounds are removed with ease and excellent efficiency. Due to this process being cheap compared to other methods, it becomes economical.

References

[1] Fatta-Kassinos D, Meric S, Nikolaou A, Pharmaceutical residues in environmental waters and wastewater: Current state of knowledge and future research, Anal. Bioanal. Chem., 10-Jan-2011, 399(1), 251–275, Springer.
[2] Patel M, Kumar R, Kishor K, Mlsna T, Pittman CU, Mohan D, Pharmaceuticals of emerging concern in aquatic systems: Chemistry, occurrence, effects, and removal methods, Chem. Rev., 27-Mar-2019, 119(6), 3510–3673, American Chemical Society.
[3] Ndoun MC, Elliott HA, Preisendanz HE, Williams CF, Knopf A, Watson JE, Adsorption of pharmaceuticals from aqueous solutions using biochar derived from cotton gin waste and guayule bagasse, Biochar, Mar. 2021, 3(1), 89–104.
[4] Van De Steene JC, Stove CP, Lambert WE, A field study on 8 pharmaceuticals and 1 pesticide in Belgium: Removal rates in waste water treatment plants and occurrence in surface water, Sci. Total Environ., Jul. 2010, 408(16), 3448–3453.
[5] Martín J, Camacho-Muñoz D, Santos JL, Aparicio I, Alonso E, Occurrence of pharmaceutical compounds in wastewater and sludge from wastewater treatment plants: Removal and ecotoxicological impact of wastewater discharges and sludge disposal, J. Hazard. Mater., Nov. 2012, 239–240, 40–47.
[6] Keerthanan S, Bhatnagar A, Mahatantila K, Jayasinghe C, Ok YS, Vithanage M, Engineered tea-waste biochar for the removal of caffeine, a model compound in pharmaceuticals and personal care products (PPCPs), from aqueous media, Environ. Technol. Innov., Aug. 2020, 19, 100847.
[7] Paunovic O, Pap S, Maletic S, Taggart MA, Boskovic N, Turk Sekulic M, Ionisable emerging pharmaceutical adsorption onto microwave functionalised biochar derived from novel lignocellulosic waste biomass, J. Colloid Interface Sci., Jul. 2019, 547, 350–360
[8] Khan AH, et al., Impact, disease outbreak and the eco-hazards associated with pharmaceutical residues: A Critical review, Int. J. Environ. Sci.Technol., 10-Feb-2021, 1–12, Springer Science and Business Media Deutschland GmbH.

[9] Omil F, Suárez S, Carballa M, Reif R, Lema JM, "Criteria for Designing Sewage Treatment Plants for Enhanced Removal of Organic Micropollutants," Springer, Dordrecht, 2010, pp. 283–306.
[10] Sebok Á, Vasanits-Zsigrai A, Palkó G, Záray G, Molnár-Perl I, Identification and quantification of ibuprofen, naproxen, ketoprofen and diclofenac present in waste-waters, as their trimethylsilyl derivatives, by gas chromatography mass spectrometry, Talanta, 2008, 76(3), 642–650.
[11] Nielsen L, Zhang P, Bandosz TJ, Adsorption of carbamazepine on sludge/fish waste derived adsorbents: Effect of surface chemistry and texture, Chem. Eng. J., 2015, 267, 170–181.
[12] Mirzaei A, Chen Z, Haghighat F, Yerushalmi L, Removal of pharmaceuticals and endocrine disrupting compounds from water by zinc oxide-based photocatalytic degradation: A review, Sustain. Cities Soc., 2016, 27, 407–418.
[13] Solanki A, Boyer TH, Physical-chemical interactions between pharmaceuticals and biochar in synthetic and real urine, Chemosphere, Mar. 2019, 218, 818–826.
[14] Dao YH, Tran HN, Tran-Lam TT, Pham TQ, Le GT, Degradation of paracetamol by an UV/chlorine advanced oxidation process: Influencing factors, factorial design, and intermediates identification, Int. J. Environ. Res. Public Health, Dec. 2018, 15(12).
[15] Audino F, Conte LO, Schenone AV, Pérez-Moya M, Graells M, Alfano OM, A kinetic study for the Fenton and photo-Fenton paracetamol degradation in an annular photoreactor, Environ. Sci. Pollut. Res., Feb. 2019, 26(5), 4312–4323.
[16] He Y, *et al.*, Hydrophobic networked PbO2 electrode for electrochemical oxidation of paracetamol drug and degradation mechanism kinetics, Chemosphere, Feb. 2018, 193, 89–99.
[17] Villota N, Lombraña JI, Cruz-Alcalde A, Marcé M, Esplugas S, Kinetic study of colored species formation during paracetamol removal from water in a semicontinuous ozonation contactor, Sci. Total Environ., Feb. 2019, 649, 1434–1442.
[18] Ziylan-Yavaş A, Ince NH, Catalytic ozonation of paracetamol using commercial and Pt-supported nanocomposites of Al2O3: The impact of ultrasound, Ultrason. Sonochem., Jan. 2018, 40(no. Pt B), 175–182.
[19] Delgado N, Capparelli A, Navarro A, Marino D, Pharmaceutical emerging pollutants removal from water using powdered activated carbon: Study of kinetics and adsorption equilibrium, J. Environ. Manage., Apr. 2019, 236, 301–308.
[20] Nartey OD, Zhao B, Biochar Preparation, Characterization, and Adsorptive Capacity and Its Effect on Bioavailability of Contaminants: An Overview, vol. 2014. Hindawi Publishing Corporation, 2014.
[21] Peiris C, Gunatilake SR, Mlsna TE, Mohan D, Vithanage M, Biochar based removal of antibiotic sulfonamides and tetracyclines in aquatic environments: A critical review, Bioresour. Technol., 01-Dec-2017, 246, 150–159, Elsevier Ltd.
[22] Bushra B, Remya N, Biochar from Pyrolysis of Rice Husk Biomass – Characteristics, Modification and Environmental Application, 2020.
[23] Jing XR, Wang YY, Liu WJ, Wang YK, Jiang H, Enhanced adsorption performance of tetracycline in aqueous solutions by methanol-modified biochar, Chem. Eng. J., 2014, 248, 168–174, Jul.
[24] Karunanayake AG, *et al.*, Rapid removal of salicylic acid, 4-nitroaniline, benzoic acid and phthalic acid from wastewater using magnetized fast pyrolysis biochar from waste Douglas fir, Chem. Eng. J., 319, 75–88, 2017.
[25] Liu C, *et al.*, Preparation of acid- and alkali-modified biochar for removal of methylene blue pigment, ACS Omega, 5(48), 30906–30922, Dec 2020.
[26] Liu H, Xu G, Li G, The characteristics of pharmaceutical sludge-derived biochar and its application for the adsorption of tetracycline, Sci. Total Environ., 2020, 747, 141492, Dec.

[27] Sahoo D, Remya N, Influence of operating parameters on the microwave pyrolysis of rice husk: Biochar yield, energy yield, and property of biochar, Biomass Convers. Biorefinery, 2020.
[28] Lin L, Jiang W, Xu P, Comparative study on pharmaceuticals adsorption in reclaimed water desalination concentrate using biochar: Impact of salts and organic matter, Sci. Total Environ., 2017, 601–602, 857–864, Dec.
[29] Liyanage AS, Canaday S, Pittman CU, Mlsna T, Rapid remediation of pharmaceuticals from wastewater using magnetic Fe3O4/Douglas fir biochar adsorbents, Chemosphere, 2020, 258, 127336, Nov.
[30] Huang WH, Lee DJ, Huang C, Modification on biochars for applications: A research update, Bioresour. Technol., Elsevier Ltd, 319, 01-Jan-2021, 124100.
[31] Wang J, Wang S, Preparation, modification and environmental application of biochar: A review, J. Clean. Prod., 01-Aug-2019, 227, 1002–1022, Elsevier Ltd.
[32] An Q, Miao Y, Zhao B, Li Z, Zhu S, An alkali modified biochar for enhancing Mn2+ adsorption: Performance and chemical mechanism, Mater. Chem. Phys., Jul. 2020, 248, 122895.
[33] Zhang M, *et al*, Evaluating biochar and its modifications for the removal of ammonium, nitrate, and phosphate in water, Water Res., 01-Nov-2020, 186, 116303, Elsevier Ltd.
[34] Tran HN, *et al*, Innovative spherical biochar for pharmaceutical removal from water: Insight into adsorption mechanism, J. Hazard. Mater., Jul. 2020, 394, 122255.
[35] Zubair M, Ihsanullah I, Abdul Aziz H, Azmier Ahmad M, Al-Harthi MA, Sustainable wastewater treatment by biochar/layered double hydroxide composites: Progress, challenges, and outlook, Bioresour. Technol., 01-Jan-2021, 319, 124128, Elsevier Ltd.
[36] Lins PVS, *et al.*, Adsorption of a non-steroidal anti-inflammatory drug onto MgAl/LDH-activated carbon composite – Experimental investigation and statistical physics modeling, Colloids Surf. A Physicochem. Eng. Asp., Feb. 2020, 586, 124217.

Sweety Kaur, Ajay Kumar, Gurmeen Rakhra, Richa Arora

Application of biochar for wastewater treatment

Abstract: Biochar is a high-carbon content, eco-friendly, cost-effective, and highly efficacious adsorbent reagent and has attained uplifting interest in wastewater treatment sector due to its distinctive properties like carbon sequestration and ability of cationic exchange. It can be synthesized by utilizing different feedstocks with methods including pyrolysis, gasification, and hydrothermal carbonization that can further be modified by treatments with acids, alkalis, oxidizing agents, metallic salts, carboniferous substances, and purging using steam and gas, and their selection relies upon the economical implementing fields. Many research studies have been carried out in biochar application of adsorb livestock wastewater pollutants due to its cheap preparation process, largely defined surface area, evenly firm structure, greater pore size, and abundance of functional groups. Moreover, biochar can be recycled because of its greater treatment efficacy and recovery potency. This review investigates the structure, preparation strategy using different feedstocks, and adsorption mechanism of biochar for various pollutants from wastewater livestock.

Keywords: biochar, carbon-rich, pollutants, adsorption, modification, wastewater treatment

1 Introduction

At present, the world population rate is 7.7 billion which is expanding rapidly and is estimated to reach around 10 billion by 2050 [1]. Expansion of population will result in increasing demand of essentials like water, food, and energy resources that are already in constricted proportion because of contending demands of such limited sources worldwide. In the near future, there will be need for uplifting production of food, treating water resources and substitutive sustainable energy resources with least usage, thereby declining the effects on environment [2]. There is also need for radical shift in the management and treatment of disposed sewage wastewater resources. Earth's water bodies are getting depleted due to constant liberation of huge amount of organic and inorganic pollutants including toxic dyes, heavy metals, frothing detergents, pharmaceutical waste, and industrial and municipal effluent directly into the water reservoirs [3]. As major proportion of such pollutants are extremely stable in the atmosphere or else gets converted into recalcitrant form, abandoned release of these toxic pollutants is a matter of concern as they can exhibit adverse impact on environment's ecology [4]. There are many traditional methods

https://doi.org/10.1515/9783110734003-011

that are employed in different parts of world for removing toxic pollutants from wastewater involving simultaneous coagulation and flocculation, adsorption of pollutants using activated carbon, membrane separation, reverse osmosis, precipitation, ion-exchange chromatography, electrolytic treatment, and solvent extraction for eliminating organic and inorganic pollutants [5]. Nevertheless, these conventional techniques exhibit a wide range of pitfalls extending from ineffectiveness to eliminating pollutants at comparatively low concentration and to absolute conversion of pollutants into biodegradable or low toxic derivatives, energy and chemical consuming, complexity of technique, greater maintenance and operation capitals, and so on [6]. As an adsorbent, biochar exhibit as a cost-effective wastewater treatment substitute to activated carbon used in removing pollutants by adsorption. Biochar usage in treating wastewater approaches the recent concerns with wastewater management strategies [7]. Biochar is a cost-effective, low-carbonated biomaterial, regarded as an efficient alternative to activated carbon that eradicates various organic pollutants like agrochemicals, pharmaceuticals, polycyclic aromatic hydrocarbons, polychlorinated biphenyls, volatile compounds, and aromatic dyes along with some inorganic pollutants like heavy metals and ammonia from solid, liquid, or gaseous phases [8]. Biochar is a derivative of thermochemical conversion including pyrolysis, gasification, torrefaction and hydrothermal carbonation of carbon-rich feedstocks like agricultural and forest residues, algal feedstock, fertilizers, biofuel plants, and digestates at very high temperature of 300–900 °C, under partial anaerobic conditions [9]. Biochar exhibits various diversified properties including greater adsorption ability, greater defined surface area, and less pore size and ability of exchanging ions [10]. The inconsistency and prevalence of any particular reaction are regulated by defined physicochemical features of biochar that are related to two major factors, that is, type of biomass and pyrolytic conditions employed during its synthesis. These two parameters immensely alter the physicochemical features like surface area dimensions, pH, and elemental configuration, hence, complete biochar surface property also. Such alterations in biochar properties exhibit important role on its convenience and effectivity during remediation of specific contaminants [11]. Potential implementation of biochar in soil helps in remediating contaminants from soil and enhances the property of soil. Biochar enhances physical properties like soil's water holding ability, O_2 and moisture content, chemical properties like carbon sequestration and immobility of contaminants, and biological properties like diversity and abundance of microorganisms [12]. Eventually, biochar properties promote carbon sequestration in soil, decline in greenhouse gases emissions, and hence providing overall upgradation in soil texture [13]. Biochar also exhibit an novelistic ability to bind with polar compounds via charged functional groups that facilitates to restrain heavily toxic metals and agro-based chemicals on its upper surface and prevents their movement into the agricultural crops [14]. Therefore, biochar is considered as a promising and efficient technology for eliminating toxic heavy metals and lowering the load of organic pollutants that can be an idealistic adsorbent for treatment and management of wastewater due to its greater stability, cost-effective, micro-porosity, ecofriendly, and easy to prepare and handling ability [15]. This review investigates the structure, preparation strategy using different feedstocks, and adsorption mechanism of biochar for various pollutants from wastewater livestock.

2 Structural arrangement and elemental constitution of biochar

The characteristic properties and potential application of any constituent or material can be accomplished by analyzing the structural arrangement and elemental composition of that particular material. Different forms of feedstocks derived from various sources are collected and converted into high-carbon microporous substance, by the process of pyrolysis and carbonation that exhibit fine porous structural arrangement and greater aromatization degree [16]. The attributes of crude feedstocks and the carbonation method employed for the process of synthesis resembles nearly the characteristic features of biochar produced [17]. The framework of biochar and its derived products mainly relies on the arrangement, inclusions, and content of cellulose and lignin from various substrates. The carbonization temperature required for the synthesis of biochar is also a crucial parameter impacting the structural arrangement of biochar. With increase in the temperature of carbonization, structure of aromatic carbon and pore-size of the biochar increase; that is, carbonization temperature is directly proportional to the structure and pore-size of biochar [18]. But, when the carbonization temperature goes beyond 700 °C, few microporous structures on top layer of biochar material may be damaged and moreover, when temperature reaches above 800 °C, the carbonaceous backbone structure of structure becomes volatile in nature [19].

The basic constituents of biochar majorly comprise elements like C, H, O, N, P, S, K, Ca, Mg, Na, and Si, among which the carbon exhibits a higher content of about more than 60%, followed consecutively by hydrogen and oxygen [20]. The carbon present in biochar is basically in aromatic form that gets precipitated into an uneven stack of stabilized aromatic rings. Different carbon-derived compounds involved are fatty acids, esters, phenolic compounds, and alcoholic compounds, along with compounds relevant to fulvic and humic acid, which are mainly found in crude biochar, biochar derived from low-temperature pyrolysis and cattle dung biochar [21]. Comparatively, nitrogen exhibits low proportion in biochar that is mainly found on their surface within heterocyclic structural linkage of C-N [16]. The content of P in biochar is even much lower than that of nitrogen and is inversely associated with the temperature of carbonization. The accessibility of P alters greatly due to high value of pH and P-containing Ca and Mg synthesized during carbonization process [22]. The contents of K, Ca, Mg, and Na varies in different types of biochar, comprising the highest proportion in cattle dung–derived biochar, followed by biochar derived from agro- and forest residues. Therefore, the elemental constitution and functional activity of biochar are relevant to crude feedstocks, environmental conditions of carbonization process, temperature, and pH [23].

3 Production of biochar using different feedstocks

The production of biochar is carried out by pyrolysis of biomasses from different resources, which is considered to be an ancient Egypt originated process of thermal treatment of feedstocks [24]. Different waste materials derived from agricultural field, food industry, and forest residues are the major feedstocks for biochar including pellet and chips of food, residues and bagasse from agro-crops, distillate grains, pressed oil and pulp cakes, and rice and wheat husk. Biochar production can also be derived from feedstocks derived from lignocellulose materials like wastewater sludge, poultry manure, and dairy wastes [25]. The yielded end products of pyrolysis include hydrocarbon-rich oil and artificial gas together with blended hydrocarbon gases and biochar material with varying amount of their fraction depending upon the range of temperature, pressure, residence time, and so on [26, 30]. Fast pyrolysis or thermal processing is considered to be the most effective process for production of biofuels and apparently gasification is employed for the production of syngas, thereby also generating a large amount of heat and energy. Due to high rate of heating and less residence duration, fast pyrolysis favors production of higher amount of oils. Slow pyrolysis process is also considered to be effective one for the production of biochar with yield rate of 35% from dry weight of biomass. In comparison to fast pyrolysis, slow pyrolysis method generates greater amount of biochar due to slow rate of heating and larger residence time [27]. Production of biochar comprises of three major steps including pre-pyrolytic phase, main pyrolytic phase and generation of carbon-rich soil end-products. The first pre-pyrolytic step exhibits atmospheric temperature of 200 °C that causes evaporation of moisture and low volatile compounds resulting in breakdown of linkages and production of hydrogen peroxide, carboxylic and ketone groups. The second step of main pyrolysis exhibits temperature of 200–500 °C that undergoes devolatilization and decomposition lignocellulosic components, that is, cellulose and hemicellulose, at a very rapid rate. The final step encompasses ambient temperature of >500 °C and includes breakdown of lignin and other organic substances with strongest chemical bond [28]. Pyrolytic temperature has direct correlation with modifications in structural and physicochemical properties of biochar and the temperature exhibit great impact on physicochemical properties like pH, functional groups, and surface area. High temperature of pyrolysis results in enhanced surface area dimensions, carbonaceous proportions, pH, volatile compounds, and decline in cation exchange capacity and amount of functional groups of biochar [29].

Table 1: Different phases of pyrolytic temperature for biochar production [30].

Pyrolytic temperature (°C)	Range of pressure	Proportion of formed end-products (%)	
		Biochar	Other by-products (oil and gas)
400–600 °C (fast pyrolysis)	Vacuum pressure to atmospheric pressure	75%	25%
350–800 °C (slow pyrolysis)	Atmospheric pressure	30%	70%
700–1,500 °C (gasification)	Atmospheric to high temperature	5%	95%
775–1,025 °C (flash pyrolysis)	High pressure	10–15%	85–90%
<350 °C (hydrothermal carbonization)	Atmospheric pressure	50–80%	–
200–320 °C (torrefication)	Atmospheric pressure	80%	20%

4 Physicochemical features of biochar

4.1 pH

Biochar is mainly alkaline in nature in reference to inorganic minerals like carbonates, phosphates, and ashes formed and also susceptible to crude substances and pyrolysis-carbonization temperature for biochar synthesis. For instance, the pH of biochar derived from leguminous plants residues will be comparatively greater than that of non-leguminous derived biochar and within similar circumstances cattle manure–derived biochar will exhibit higher pH value than that of biochar derived from herbaceous plant and forest residues [20]. With varying different pyrolysis and carbonization temperature, whenever the temperature increases, the pH of the biochar also increases because of the volatilization of organic acids and degradation of anionic functional groups like carboxyl group and phenolic hydroxyl group [31].

4.2 Stability

Biochar exhibits a high degree of carboxylic acid esterification and structural aromatization, carbonaceous matter, less solubility, enhanced stabilization, greater boiling point, and high resistance to physical, chemical, and biological disintegration [32]. These properties of biochar allow it to remain under the soil for millions of years within spontaneous ecological conditions [30].

4.3 Volatile matter

Pyrolysis temperature affects the structural arrangement of biochar because of the liberation of volatile compounds and generation of transitional blends followed by their volatilization and as the temperature increases the amount of volatile matter decreases [33]. It was found that when there is an increase in the pyrolysis temperature, the amount of volatile matter decreases by 60.8% to 14.9% for the biochar derived from the branches of apple tree as the increasing temperature causes disintegration of volatiles into low molecular weight liquids and gases other than that of biochar. Moreover, increasing temperature may lead to evaporation of –OH groups and breakdown of lignocellulosic components, that is, cellulose and lignin [34]. Based on different types of compounds present, the high-temperature biochar differs from low-temperature biochar in having pyranones, ethers, and quinines, whereas low-temperature biochar comprises hydroxyl, carbonyl, carboxyl, and hemiacetal compounds. It was also reported that with increasing temperature, the structure of cellulose from feedstocks also undergoes chemical modifications; that is, during the initial phase of process of pyrolysis with temperature of 250 °C, the proportion of biochar is dominant over the proportion of oligomers. At 290 °C, the biochar comprises mainly phenolic groups and furans and above 290 °C, it contains alkylated furans, benzenoid, and concentrated aromatic compounds [35]. The content of volatile matter in biochar also impacts the consistence of the biomaterial along with its nitrogen accessibility, plant development, and its adsorption ability [36].

4.4 Specific surface area

The particular surface area of biochar mainly lies within the diversified array of 1.5–500 $m^2\ g^{-1}$ that expands as the pyrolysis temperature increases under a definitive limit [37]. At moderately low temperatures via thermal degradation of feedstocks, the volatile compounds, viscous tars, and other derivatives occupy the biochar's interior pore structure, hence decreasing the surface area. With increase in temperature, these compounds disrupt into volatile gases and release out, hence minimizing the size of pores but increasing the count of pores within the biochar leading to microporous structure and greater specific surface area. Nevertheless, the specific surfaces will ultimately reach plateau with elevated temperature, and when the temperature passes its critical point, the specific area declines, because of the breakdown of microporous structure of biochar and with expansion of micropores into macropores [38].

4.5 Surface functional group and cation exchange capacity (CEC)

Biochar consists of a large amount of surface functional groups like carboxyl, hydroxyl, and carbonyl groups. Many functional groups are oxygen-rich or alkaline that offers biochar with great adsorption capacity, hydrophobicity, and ion exchange ability [39]. The amount of functional groups present on the surface of biochar is dependent on the temperature of carbonization. With increase in the carbonization temperature, carboxyl, hydroxyl, and carbonyl linkage proportion decreases in biochar, whereas the number of alkyl groups increases, and overall the amount and density of functional groups declines with increasing carbonization temperature. The cation exchange capacity (CEC) of biochar depends on the type of feedstock and pyrolysis temperature. Carbonization process increased the cation exchange capacity of biochar because of incomplete degradation of cellulose by consuming the oxygenated functional groups including carboxyl, hydroxyl and carbonyl groups. Under a defined range, biochar cation exchange capacity declines with uplifting temperature, followed by the breakdown of O_2-containing groups and the decrease of negative charge on the surface of biochar and oxygen to carbon ratio [40]. Whereas, on the other side, inclination in temperature leads to increase in amount of alkali metals like Mg, Ca, and K within the biochar, resulting in enhancement in cation exchange capacity [41].

4.6 Carbon and ash content

The carbon and ash content of biochar inclines with increasing temperature of pyrolysis. High amount of carbon depicts that biochar comprises a definitive amount of natural plant remnants like cellulose. It was observed that uplifted pyrolysis temperature leads to increase in ash content by 5.7–18.7% that is due to constant accumulation of inorganic components and organic matter remnants [42].

5 Mechanism of biochar adsorption for wastewater pollutants

Adsorption is a process in which the adsorbate is linked with the surface of adsorbent till the equivalence is accomplished. The major phases involved in the process of adsorption are: (1) physical adsorption wherein the adsorbate accumulates on the surface of adsorbent, followed by the second phase, that is, (2) precipitation and complexation where the adsorbate collects on the adsorbent's layer surface, and (3) filling up of pores in which the adsorbate is concentrated into pores of adsorbent [43]. The adsorption process is carried out in three stages mitigating three primary zones. In the initial stage, no adsorption occurs which is called as clean zone. The second stage is called the massive transfer zone where the adsorption starts and is under progress. The final stage is where the equilibrium is attained which is also called as exhausted or saturated zone [44]. During the process, when the saturated zonal volume increases the clean zone area decreases. The massive transfer zone may get affected if the adsorbate's concentration is increased and if the concentration remains in equilibrium; then it doesn't have impact on mass transfer zone and it continues till the revolutionary point, that is, a point where the adsorbent is exhausted. There are different mechanisms involved in the removal of toxic metallic components including sorption, ion-exchange, electrostatic interaction, precipitation, and complexation. Surface sorption is a physical mechanism in which generation of chemical bonds takes place through diffusion of metal ions in the sorbent pores, and this pore volume and sorbent surface area rely on the carbonization temperature [45]. Electrostatic interaction of metals on biochar includes interaction between charged biochar material and metal ions in order to restrict the mobility of toxic metals [46]. Ion exchange capacity is the exchange of protons and ionized cations with salts on the surface of biochar. The adsorption capacity for removal of heavy metals of this mechanism is dependent on the size of contaminants and surface functional groups of biochar [47]. Precipitation is one the major processes which can be employed for eliminating inorganic pollutants onto biochar that involves production of precipitates within the solution or upon the surface layer of adsorbing material specifically for biochar developed from breakdown of lignocellulosic components [48]. The processes involved in the adsorption of organic pollutants include pore filling, hydrophobic interactions, partitioning, and electrostatic interaction. In partitioning, the adsorbate substance diffuses within the pores of non-carbonized biochar and of carbonized crystalline and graphene proportions of biochar. Pore filling mechanism involves mesopores of 2–50 nm and micropores of <2 nm where the organic contaminants settles. This mechanism absolutely relies on the nature, biochar type, and polarity of organic contaminants. Electrostatic interaction is one of the most crucial mechanisms that include adsorption of ionizable organic compounds to biochar's positive charged surface via electrostatic interaction [49].

6 Future perspectives of biochar

Biochar is a sustainable source with ability to overcome wastewater decontamination issue [Figure 1]. It can be consumed as an adsorbent with high adsorbing ability for a broad range of pollutants, but knowledge of various processes involved in adsorption

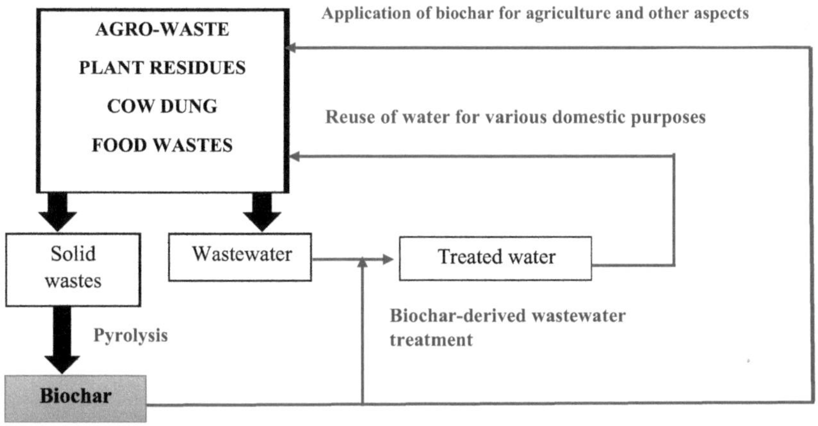

Figure 1: A vicious flow of production of biochar and its implementation in wastewater treatment and other purposes.

is still inadequate. As the application of biochar for wastewater treatment relies on properties of biochar like functional groups and surface area, on the abundance and type of biomass, and on the parameters of thermal conversion implied for biochar production, further studies and research are required to be done for modulating the properties of biochar for advanced applications. Modification of biochar is also a crucial field that is to be explored in order to increase the capacity of biochar for elimination of particular pollutants. Moreover, it is also required to study the different modification methods and mechanisms responsible for adsorption with their corresponding reagents. Biochar exhibit various applications in order to resolve ecological issues and can also be indulged in the pretreatment processes defined for the removal of toxic compounds followed by the subsequent biological treatment.

References

[1] Pokharel A, Acharya B, Farooque A, Biochar-assisted wastewater treatment and waste valorization, Appl. Biochar Environ. Safety, 2020, 221.

[2] Foereid B, Biochar in nutrient recycling – the effect and its use in wastewater treatment, Open J. Soil Sci., 2015, 39–44.

[3] Zulfiqar M, Samsudin MF, Sufian S, Modelling and optimization of photocatalytic degradation of phenol via TiO2 nanoparticles: an insight into response surface methodology and artificial neural network, J. Photochem. Photobiol. A. Chem., 2019, 112039.

[4] Bogusz A, Oleszczuk P, Dobrowolski R, Application of laboratory prepared and commercially available biochars to adsorption of cadmium, copper and zinc ions from water, Bioresourc. Technol., 2015, 540–549.

[5] Ejraei A, Aroon MA, Saravani AZ, Wastewater treatment using a hybrid system combining adsorption, photocatalytic degradation and membrane filtration processes, J. Water Process. Eng., 2019, 45–53.

[6] He J, Li Y, Cai X, Chen K, Zheng H, Wang C, Zhang K, Lin D, Kong L, Liu J, Study on the removal of organic micropollutants from aqueous and ethanol solutions by HAP membranes with tunable hydrophilicity and hydrophobicity, Chemosphere, 2017, 380–389.

[7] Perez-Mercado LF, Lalander C, Berger C, Dalahmeh SS, Potential of biochar filters for onsite wastewater treatment: effects of biochar type, physical properties and operating conditions, Water, 2018, 1835.

[8] Jung C, Oh J, Yoon Y, Removal of acetaminophen and naproxen by combined coagulation and adsorption using biochar: influence of combined sewer overflow components, Environ. Sci. Pollut. Res., 2015, 10058–10069.

[9] Kambo HS, Dutta A, A comparative review of biochar and hydrochar in terms of production, physico-chemical properties and applications, Renew. Sust. Energ. Rev., 2015, 359–378.

[10] Ahmad M, Rajapaksha AU, Lim JE, Zhang M, Bolan N, Mohan D, Vithanage M, Lee SS, Ok YS, Biochar as a sorbent for contaminant management in soil and water: a review, Chemosphere, 2014, 19–33.

[11] Ronsse F, Van Hecke S, Dickinson D, Prins W, Production and characterization of slow pyrolysis biochar: influence of feedstock type and pyrolysis conditions, Gcb Bioenergy, 2013, 104–115.

[12] Gul S, Whalen JK, Thomas BW, Sachdeva V, Deng H, Physico-chemical properties and microbial responses in biochar-amended soils: mechanisms and future directions, Agri. Ecosys. Environ., 2015, 46–59.

[13] Windeatt JH, Ross AB, Williams PT, Forster PM, Nahil MA, Singh S, Characteristics of biochars from crop residues: potential for carbon sequestration and soil amendment, J. Environ. Manag., 2014, 189–197.

[14] Bolan N, Kunhikrishnan A, Thangarajan R, Kumpiene J, Park J, Makino T, Kirkham MB, Scheckel K, Remediation of heavy metal (loid) s contaminated soils–to mobilize or to immobilize?, J. Hazard. Mater., 2014, 141–166.

[15] Fang W, Wang Q, Han D, Liu P, Huang B, Yan D, Ouyang C, Li Y, Cao A, The effects and mode of action of biochar on the degradation of methyl isothiocyanate in soil, Sci. Total. Environ., 2016, 339–345.

[16] Lehmann J, Gaunt J, Rondon M, Bio-char sequestration in terrestrial ecosystems–a review, Mitig. Adapt. Strateg. Glob. Change, 2006, 403–427.

[17] Zhao B, O'Connor D, Zhang J, Peng T, Shen Z, Tsang DC, Hou D, Effect of pyrolysis temperature, heating rate, and residence time on rapeseed stem derived biochar, J. Clean. Prod., 2018, 977–987.

[18] Chen W, Meng J, Han X, Lan Y, Zhang W, Past, present, and future of biochar, Biochar, 1, 75–87.

[19] Huang H, Wang YX, Tang JC, Zhu WY, Properties of maize stalk biochar produced under different pyrolysis temperatures and its sorption capability to naphthalene, 2014, 1884–1890.

[20] Yuan JH, Xu RK, Zhang H, The forms of alkalis in the biochar produced from crop residues at different temperatures, Bioresour. Technol., 2011, 3488–3497.

[21] Bruun EW, Ambus P, Egsgaard H, Hauggaard-Nielsen H, Effects of slow and fast pyrolysis biochar on soil C and N turnover dynamics, Soil Biol. Biochem., 2012, 73–79.

[22] Cao X, Harris W, Properties of dairy-manure-derived biochar pertinent to its potential use in remediation, Bioresour. Technol., 2010, 5222–5228.

[23] Silber A, Levkovitch I, Graber ER, pH-dependent mineral release and surface properties of cornstraw biochar: agronomic implications, Environ. Sci. Technol., 2010, 9318–9323.
[24] Mohan D, Pittman JCU, Steele PH, Pyrolysis of wood/biomass for bio-oil: a critical review, Energy & fuels, 2006, 848–889.
[25] Shakya A, Agarwal T, Poultry litter biochar: an approach towards poultry litter management–a review, Int. J. Curr. Microbiol. Appl. Sci., 2017, 2657–2668.
[26] Cheah S, Jablonski WS, Olstad JL, Carpenter DL, Barthelemy KD, Robichaud DJ, Andrews JC, Black SK, Oddo MD, Westover TL, Effects of thermal pretreatment and catalyst on biomass gasification efficiency and syngas composition, Green Chem., 2016, 6291–6304.
[27] Tomczyk A, Sokołowska Z, Boguta P, Biochar physicochemical properties: pyrolysis temperature and feedstock kind effects, Rev. Environ. Sci. Bio. Technol., 2020, 191–215.
[28] Cárdenas-Aguiar E, Gascó G, Paz-Ferreiro J, Méndez A, The effect of biochar and compost from urban organic waste on plant biomass and properties of an artificially copper polluted soil, Int. Biodeter. Biodeg., 2017, 223–232.
[29] Denyes MJ, Parisien MA, Rutter A, Zeeb BA, Physical, chemical and biological characterization of six biochars produced for the remediation of contaminated sites, JoVE, 2014, e52183.
[30] Sun X, Han X, Ping F, Zhang L, Zhang K, Chen M, Wu W, Effect of rice-straw biochar on nitrous oxide emissions from paddy soils under elevated CO2 and temperature, Sci. Total Environ., 2018, 1009–1016.
[31] Chintala R, Mollinedo J, Schumacher TE, Malo DD, Julson JL, Effect of biochar on chemical properties of acidic soil, Arch. Agronomy Soil Sci., 2014, 93–404.
[32] Leng L, Huang H, Li H, Li J, Zhou W, Biochar stability assessment methods: a review, Sci. Total Environ., 2019, 210–222.
[33] Tag AT, Duman G, Ucar S, Yanik J, Effects of feedstock type and pyrolysis temperature on potential applications of biochar, J. Anal. Appl. Pyrol., 2016, 200–206.
[34] Shi-Xiang Z, Ta N, Xu-Dong W, Effect of Temperature on the Structural and Physicochemical Properties of Biochar with Apple Tree Branches as Feedstock Material, Energies, 2017, 1293.
[35] Antal MJ, Grønli M, The art, science, and technology of charcoal production, Indus. Eng. Chem. Res, 2003, 1619–1640.
[36] Zimmerman AR, Abiotic and microbial oxidation of laboratory-produced black carbon (biochar), Environ. Sci. Technol., 2010, 1295–1301.
[37] Li J, Cao L, Yuan Y, Wang R, Wen Y, Man J, Comparative study for microcystin-LR sorption onto biochars produced from various plant-and animal-wastes at different pyrolysis temperatures: influencing mechanisms of biochar properties, Bioresour. Technol., 2018, 794–803.
[38] Brown RA, Kercher AK, Nguyen TH, Nagle DC, Ball WP, Production and characterization of synthetic wood chars for use as surrogates for natural sorbents, Org. Geochem., 2006, 321–333.
[39] Antón-Herrero R, García-Delgado C, Alonso-Izquierdo M, García-Rodríguez G, Cuevas J, Eymar E, Comparative adsorption of tetracyclines on biochars and stevensite: looking for the most effective adsorbent, App. Clay Sci., 2018, 162–172.
[40] Suliman W, Harsh JB, Fortuna AM, Garcia-Pérez M, Abu-Lail NI, Quantitative effects of biochar oxidation and pyrolysis temperature on the transport of pathogenic and nonpathogenic Escherichia coli in biochar-amended sand columns, Environ. Sci. Technol., 2017, 5071–5081.
[41] Kalinke C, Oliveira PR, Oliveira GA, Mangrich AS, Marcolino-Junior LH, Bergamini MF, Activated biochar: preparation, characterization and electroanalytical application in an alternative strategy of nickel determination, Anal. Chim. Acta, 2017, 103–111.
[42] Rafiq MK, Bachmann RT, Rafiq MT, Shang Z, Joseph S, Long R, Influence of pyrolysis temperature on physico-chemical properties of corn stover (Zea mays L.) biochar and feasibility for carbon capture and energy balance, PloS One, 2016, e0156894.

[43] Fagbohungbe MO, Herbert BM, Hurst L, Ibeto CN, Li H, Usmani SQ, Semple KT, The challenges of anaerobic digestion and the role of biochar in optimizing anaerobic digestion, Waste Manage., 2017, 236–249.

[44] De Ridder M, De Jong S, Polchar J, Lingemann S, Risks and opportunities in the global phosphate rock market: Robust strategies in times of uncertainty, 2012.

[45] Kumar A, Kumar A, Sharma G, Naushad M, Stadler FJ, Ghfar AA, Dhiman P, Saini RV, Sustainable nano-hybrids of magnetic biochar supported g-C3N4/FeVO4 for solar powered degradation of noxious pollutants-Synergism of adsorption, photocatalysis & photo-ozonation, J. Clean. Prod., 2017, 431–451.

[46] Mukherjee A, Zimmerman AR, Harris W, Surface chemistry variations among a series of laboratory-produced biochars, Geoderma, 2011, 247–255.

[47] Rizwan M, Ali S, Qayyum MF, Ibrahim M, Zia-ur-Rehman M, Abbas T, Ok YS, Mechanisms of biochar-mediated alleviation of toxicity of trace elements in plants: a critical review, Environ. Sci. Pol. Res., 2016, 2230–2248.

[48] Cao X, Ma L, Gao B, Harris W, Application of bio-char for removal of toxic metals, Environ. Sci. Technol., 2009, 3285–3291.

[49] Zheng H, Wang Z, Zhao J, Herbert S, Xing B, Sorption of antibiotic sulfamethoxazole varies with biochars produced at different temperatures, Environ. Pol., 2013, 60–67.

Tejveer Singh, Marzuqa Quraishi, Aindree, Mayur Thakre, Piyush Kumar Gupta, Aberam Natarajan, Mohit Sahni, Kanupriya, Meenal Gupta, Ankit Kumar, Soumya Pandit

An overview on the application of biomass-derived biochar in the treatment of wastewater

Abstract: Water is an essential molecule that sustains life on the planet and is crucial for our survival. Water bodies are incessantly being contaminated with effluents from industrial plants and factories, causing increased concern about the water's quality. Biomass is becoming one of the popular energy resources being an eco-friendly and efficient alternative to traditional methods. When the biomass is subjected to high temperatures under controlled oxygen levels, the decomposition process results in the inception of a solid product, rich in carbon, having ideal adsorption properties, called biochar. It has myriads of environmental applications, but this chapter focuses on lignocellulosic biomass-derived biochar (LBDB) in decontamination of wastewater (WW), from multiple pollutants such as heavy metals, dyes, pharmaceutical compounds, pesticides, and so on. It describes the different feedstock materials required, its physicochemical characterization, the various conditions for the production process (pyrolysis), and the challenges associated with biochar application, requiring detailed research and feasible approaches to effectuate it, thereby helping the readers understand the importance of biochar in WW treatment, explicitly via LBDB, being a promising solution.

Keywords: biochar, pyrolysis, lignocellulosic biomass, heavy metal adsorption, wastewater treatment

1 Introduction to biochar

Biochar is nothing but charcoal processed by a biological process called pyrolysis of biomass in the absence of oxygen (O_2). The generation journey of biochar and charcoal are the same to a great extent where both products are borrowed from the carbonaceous feedstock through a pyrolysis process, but the major difference resides within their initial products and their end application. For instance, the biochar that contains high porosity, high nutrient content, and water storage capability is applied for a purpose other than combustion like, as a growth promoter, soil conditioner, in

Note: The authors Tejveer Singh and Marzuqa Quraishi have contributed equally.

https://doi.org/10.1515/9783110734003-012

water purification, green habitats, and a lot more. On the other hand, charcoal can be used as energy fuel for heat generation [1].

Lignocellulose biomass (LB) materials have large-scale commercial applications in the production of valuable products due to their large amount, high availability, and low price. Biochar is the product of LB which is obtained through thermal decomposition under limited oxygen conditions, making it a sustainable potential adsorbent toward the removal of wastewater pollutants. It is one of the low-cost natural sorbents based on sustainable development. Biochar is produced by pyrolysis (thermal decomposition) brought about by high temperatures in the absence of oxygen, which can be used for energy-related purposes associated with environmental conservations and agriculture. The precursor used for biochar production is wood, manure, or leaves which are heated under high temperatures in a closed container with low or no oxygen [2, 3].

As of now, biochar adsorption innovation is step by step applied in the field of wastewater (WW) treatment. It has been shown that biochar has a great adsorption impact on common contaminations for animal WW, for example – natural poisons, substantial metals, nitrogen, and phosphorus. After polarization, biochar with great attractiveness is isolated from liquid, which is more appropriate for animal WW, contrasted with business-initiated carbon items [4, 5].

The increasing damage to the environment has made researchers cogitate on livable bioproducts that are made from biomass, and most importantly are renewable like methyl esters (biodiesel), biosolvents, biogas, and biosorbents. The problems arise from the management strategies of the sewage sludge and vague disposal, which are released after WW treatment, polluting the water by the persistent release of profuse contaminants such as aromatic hydrocarbons, heavy metals, DDT (pesticides), Aroclor (polychlorinated biphenyl), and pharmaceutical products [6]. The most valuable materials for biochar production are lignocellulosic as stated by the European Biochar Certificate, wherein LB comprises two carbohydrate polymers, cellulose and hemicelluloses, and an aromatic polymer, lignin. This lignocellulosic biomass is known to be the most economical, as well as a highly renewable natural resource in the whole world. They are an alternative to fossil fuel that makes them essential for the survival of humankind. Since these are inclusive of wood, agricultural, or arboriculture waste together, they make a mixture of natural polymers that are based on hemicellulose, lignin, and cellulose with a lot more molecules with them. Therefore, lignocellulosic feedstock refers to the overabundance of renewable bioresource on the earth that are useful for several applications in human welfare.

In lignocellulosic biomass, the content of these integrates fluctuates depending upon the type of biomass; habitually, cellulose is 40–50%, hemicelluloses – 20–40%, and lignin is 10–40% for biomass of plant material origin, together with a dash of ash and extractives. The newspaper biomass consists of 40–50% cellulose, hemicellulose 25–40%, and lignin 18–30%, whereas swine waste biomass has 6% of cellulose, 28% of hemicellulose, and 0% of lignin. The biochar from the newspaper, paper towel, and demolition wood has properties that are much the same as unprocessed wood biochar. The biochar is profitably stuck to highly competent and cost-effective products for the adsorption of a manifold of pollutants. This property of adsorption toward the removal of pollutants from aqueous solutions is accredited to the presence of high surface area, poriferous structure, and upgraded surface chemical properties of biochar [7, 8]. This chapter intends to highlight biochar, its applications, core issues, benefits of generating and using biochar, the biochar in the market, as well as its future aspects.

2 Mechanism of pyrolysis

Biomass that undergoes breakdown by heat under anaerobic conditions and produces liquid, gaseous, and solid products is termed pyrolysis. Transformation of biomass composed of holocellulose and lignin is used for the production of fuels and chemical. The first step is the degradation of hemicellulose at the temperature of 470–530 K, then the cellulose is broken down at around the temperature range of 510–620 K, finally, followed by the lignin to be pyrolyzed at temperatures of 550–770 K [9]. To obtain more yields of liquid products via pyrolysis of the biomass, the conditions required are low temperature, high heating rate, and short gas residence time process. The lower the temperature and heating rate process, the higher the char manufacture. The production method for pyrolytic oil, also known as bio-oil, is by Pyrolysis of lignocelluloses biomass materials. Mostly, carbon is released as a residue in extreme pyrolysis cases like carbonization. The widely used industrial processes like gasification or combustion employ pyrolysis as the first step. This converts the biomass into biochar and further wastes into biodegradables substances. There are various types of pyrolysis like methane and hydrolysis [10].

In pyrolysis, the materials are heated more than their decomposition temperature leading to the breakdown of the chemical bonds in its molecule. By combining smaller fragment, a larger molecular mass, and the amorphous covalent solid residue is formed. Furthermore, other processes may take place besides pyrolysis like combustion, hydrolysis, or other chemical processes due to the presence of some amount of oxygen, water, or other substances [11]. In the firewood burning, the manufacture of charcoal in the traditional method, and the steam cracking of crude oil is by intentional addition of these items. To avoid chemical side reactions (such as combustion or hydrolysis), the starting material is heated in a vacuum, which lowers the boiling point of the by-products, thus enhancing their recuperation.

The pyrolysis mechanism of lignin is complex compared to cellulose and hemicellulose. A precise decomposition of glycoside bond breakdown near 350 °C is indicated, whereas, there is no precise decomposition in lignin. At 200–400 °C the primary decomposition takes place, whereas, at temperatures higher than 400 °C, the secondary reaction is classified in the overall reaction. 4-substituted guaiacols from G-lignins and 4-substituted syringes from S-lignins are the aromatic compounds formed in the primary pyrolysis stage. It is due to the aromatic methoxy groups that they are stable in the primary stage of pyrolysis, but at a temperature range of 400–450 °C, they become highly reactive [12]. Sidechains are mainly formed by the unsaturated alkyl groups along with a smaller number of saturated alkyls groups.

Secondary pyrolysis reactions occur when the temperature increases from 400 to 450 °C. This causes the transformation of the guaiacols to catechols and *o*-cresols along with phenols. The formation of polycyclic aromatic hydrocarbons (PAH) increases when the temperature rises above 700 °C, along with phenols and o-cresols are present since they are stable at high temperatures. The pyrolysis of lignin forms major primary products like monomeric phenolic compounds with phenol, 2,6-dimethoxy phenol, 2-methoxy-4-vinyl phenol, and 4-vinyl phenol [13]. This whole process is summarized in Figure 1 as a schematic overview of the pyrolysis cycle of biomass.

Since the lignin structure is complex, the mechanism of the pyrolysis is difficult to acknowledge. Many studies have accomplished to identify the initial reactions on lignin model compounds. The free radical reaction is the main process that takes place in the thermal degradation of lignin, which has been proved by the biomass and the surface-immobilized model compounds of the products from the lignin pyrolysis since the ether can dissociate easily.

Char formation and development of low molecular mass products are mainly by the breakdown of functional groups in the slow thermal decomposition of lignin [14]. A connection between the type and number of functional groups using lignin samples prepared by different isolation and abundance of volatile products has been observed. The decomposition of terminal CH_2OH groups decomposes by the release of water, and formaldehyde is observed by the relationship between the aliphatic OH-group content and the formaldehyde as well as the water evolution. The formation of monomeric or dimeric lignin products represents 40–50% of the total pyrolysis yield and β-4-O bonds are generally cleaved, with partially or fully removed aliphatic side groups [15].

3 Biochar in feedstock

Biomass waste material can also be the source for the production of biochar. There should be no competition created among the other land uses, for example, the production of food or pull out of the land in the immaculate state in the production of biochar from the biomass waste material. Biomass waste material like crop residues as well as yard, food and forestry wastes, and animal manure can be used for the production of biochar. The production of biochar from the biomass debris gives plenty benefits such as the removal of the pollutant like CO_2 and methane that are formed by the burning of the large amounts of agricultural, municipal, and forestry biomass. As a by-product of producing energy from biomass, biochar has been obtained. Sewage sludge and industrial or landfill debris should not be present in the feedstock, since these contain unacceptable levels of toxins such as heavy metals.

To study all the aspects of feedstock supply in a local community/business, since the feedstock availability changes from year to year, it is vital in setting up a biochar operation. In a developing country, areas had key side benefits of biochar production where feedstocks are often used as heating and/or cooking fuel. Biochar that consists of animal litter contains high potassium content than the one obtained from wood; the latter has high carbon content. The pyrolysis condition affects the nutrient properties of the biochar. Thus, the biochar is tested batch by batch to identify its specific properties.

Feedstock like switchgrass is less used compared to feedstock residue since the former requires cost input for the harvesting of the product whereas the residue can easily be obtained from municipal biomass waste, crop residues, and biogases.

The potential nutrient value is lost from using feedstock for biochar production rather than as a direct fertilizer on the field. Certain areas consider chicken litter as debris and require a disposal cost. This drawback can be solved by using chicken litter in other areas where it can be utilized as a direct fertilizer.

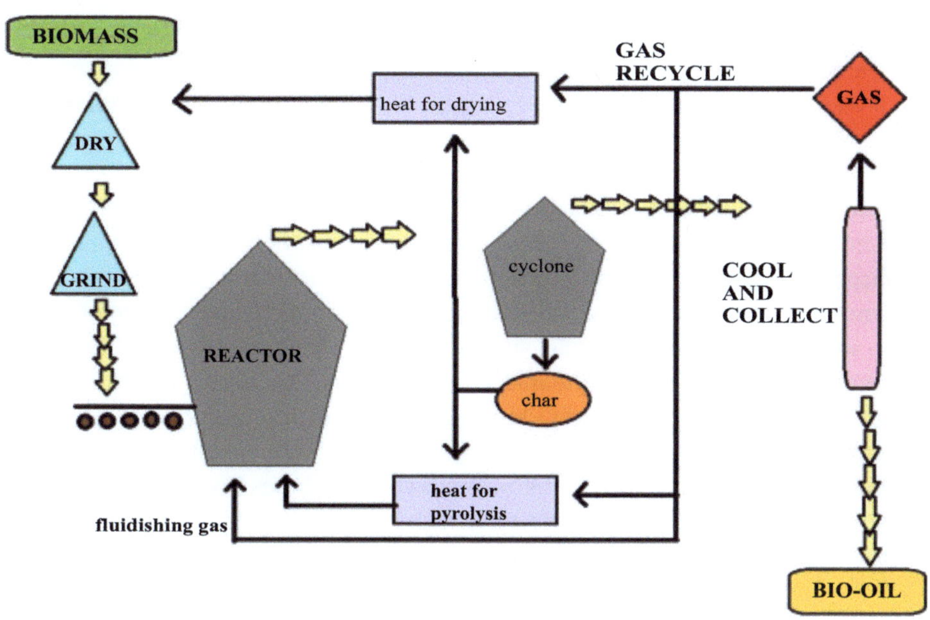

Figure 1: Pyrolysis cycle of biomass.

To reduce the transportation cost for feedstock transport, the biomass density is reduced by chipping and is converted into the form of pellets before being transported. This method has been used for the feedstock where the biomass is far from where it will be used since this leads to a very high-cost investment.

Careful storage and the use of a drier are needed that requires energy and manpower based on the feedstock choice. Thus the process of drying is conducted passively; previously pyrolysis batches of the feedstock were the source of energy for the drying process.

4 Characterization of biochar

Biochar derived from the individual chemical trait feedstock has continually proven to differ notably each spatially and temporarily. Production of biochar is mainly through the changes that might occur in the concentration of elements such as carbon, oxygen, hydrogen sculpture and nitrogen, and their related ratios. The particle sample might be carbonized, and the unstable matter is expelled because the constant carbon is the stable flammable residue. Thus, it is used as an estimate of the wide variety of carbonaceous materials a good way to be yielded from a stable sample. Aromaticity and maturation are determined by hydrogen/carbon and oxygen/carbon ratio. When the canola oil is transesterified with a catalyst based upon biochar, it shows the result of element analysis with reduced hydrogen/carbon and oxygen/carbon ratio along with an increase in pyrolysis temperature. To easily measure the level of pyrolysis and oxidative adjustment of biochar in soil and solution system, we can know through the elemental ratio of oxygen/carbon oxygen/hydrogen carbon/hydrogen.

Areas get exaggerated with a rise of carbon depletion, regardless of the thermolysis temperature. The exhaust of carbon has the foremost important outcome on the surface area expansion. The physical properties like surface area increase with continuance and shift temperature. Magnetic biochar comprises enough amount of iron oxide with the presence of a smaller surface area and abundant temporary pores, aiding in adsorbing the organic pollutants and phosphates. The biochar has surfaces (internal and external surface areas) less than others with a larger pore radius. To determine the acidic and basic functional groups on biochar, Boehm titration is used in which the biochar is titrated in the presence of strong bases (Hydrocarbon, carbon, hydroxide group, and ethoxides) or strong acids (sulfuric acid, hydrochloric acid, and nitric acid). The tartan undergoes titration with strong acid or base once more to evaluate the fraction that reacted. This is carried out to estimate the relative amount of certain functional groups (base equilibration) or basic functionalities (acidic equilibration), the contrast in the number bases and acid is required [2].

To measure functional groups that break down at a very high pH, while immobilization of pollutants takes place at moderately low pH, is done via ethoxide equilibration that is commonly excluded. Although there seems to be major insufficiency when a remarkable number of bio-oils or mineral surfaces are present, the Boehm titration works better for aqua-phobic biochar. Non-hydrothermal condition is observed in the preparation of biochar through carbonization of biomass feedstock, major regions of changes that can be observed are dehydration, pyrolysis, and carbonization.

Scanning electron microscopy (SEM) is the microscopic technique used to detect macro-porosity and physical morphology of solid substances. The intrinsic properties of the feedstock, its water-holding and chemisorption ability of waste materials in soil and solution systems, is a factor for macro-porous structure of biochar produced from cellulose plant material. The identifiable honey-comb-like structure is formed in the biochar that manifests at various pyrolytic temperatures because of the tubular structures which emanated from plant cells. Biochar has a high BET (Brunauer, Emmett, and Teller) surface area ergo the built pores. The capacity to store water in the soil and adsorb pollutant may decrease due to the surface of low temperatures of biochar that can be aqua-phobic. For the regulation and release of fertilizer nutrients, the biochar is preferably produced at low pyrolytic temperature. Furthermore, the importance of pyrolytic temperature is shown by the high temperature that leads the material analogous to AC in environmental remediation. In addition, the characteristic of nature quality and potential use of biochar is affected by specific properties of biomass feedstock. Its size is affected by the ratio of exposure to the total surface area of biochar. Biochar devised by low pyrolytic temperature is stronger than higher temperature products, but it has certain disadvantages like brittle pores are abraded into fine parts once mixed into the soil [2].

5 Applications of biochar

Pyrolysis of LB and various organic materials produce biochar which is a carbonaceous substance [16]. Biochar has many applications, but only a few have caught the attention of scientists worldwide, which are:

1. Mostly it is seen as a useful substance for energy production related activities which can be economical and environment friendly.
2. Biochar can be used for wastewater treatment.
3. Further, it can also be used as additional material for compost or fermentation of methane [17].
4. For detoxification purpose biochar has been using a border scale because of its excellent adsorption qualities. It shows a high affinity with non-polar materials, surface functional groups and high surface porosity [18]

5. Recent researchers show that biochar could be used to prepare microbial fuel cell components [19, 20]. The development of biochar-based electrodes, cathode catalyst, and MFC membrane has emerged as a new field with great potential for the future [21].

The first pollutants to be known in the industry are dyes or color; these are hazardous to the biological system of the water. Weed activated by nitric acid biochar helps with the methylene blue adsorption from aqueous solution under pH of 7.5 and temperature 50 °C. Biochar produced with the use of eucalyptus raw dust and activated with citric acid shows the largest adsorption capacity due to the presence of more carboxyl acid. Sunflower seed pyrolyzed biochar used for chemical activation using the chemical zinc chloride also determines the absorption of methylene blue. Furthermore, biochar contains advantages like high carbon without sulfur and ashes offer biomass as a fine pioneer. Reactive 19 dyes are being removed by switch gas biochar modified with cationic surfactant tetra-decyl-trimethyl ammonium bromide [22]. Methylene blue adsorption onto biochar is described better by the adsorption kinetic model and pseudo-second-order. The absorption of dye onto biochar represents the chemisorption activity. Langmuir model describes better methylene blue adsorption compared to Freundlich. The thermodynamic parameter of the adsorption data from experiments suggested endothermic and voluntary nature. The excessive heating rate produced biochar with low surface area, pure volume, and decreased yard because of fast depolymerization at the surface of biochar. Biochar activated with the usage of sulfuric acid represented better capability for methylene blue elimination.

Biochar had a notable functional group and higher porous shape preferred for adsorptions. Biochar includes hydro product in the course of pyrolysis at the biomass. The oxygen concentration is progressively reduced with the thermal treatment due to the putrefaction of the oxygen surface onto biochar. The powder X-ray diffraction sample of the biochar from thermal treatment showed a completely extensive and occasional density top round 25° constitute the amorphous carbon with a weakly crystalline structure. Biochar is highly effective toward dye removal at 3.0 pH. The electrostatic interaction among the biochar derived from the residues of crop and methyl violence increases as the pH of the solution gets increased. Biochar has a functional group phenolic-OH, at separation results to the uplift net negative alternate at the surface of biochar. WW pollutants like pesticides or herbicides are also removed via adsorption elimination with the use of biochar.

The influence of the coexistence of organic matter and salt at the elimination of sulfamethoxazole has a look at the usage of biochar. Biochar uptake capability toward the elimination of pharmaceutical effected with the aid of using the chemical and bodily feature function along with natural extent surface vicinity ash content material hydrophobicity charge and pharmaceutical speciation. Electrostatic interaction specifically makes contributions to the absorption of sulfamethazine [23].

The biochar had been determined toward the absorption of imidacloprid and atrazine from the aqueous solution. The biochar component parameters like diameter, pH, polarity, and weak acid fraction impact the adsorption of defoliant. The biochar adsorption affinity in the direction of simazine and atrazine elimination increased with the reduction of solid or ratio of solution. The temperature of pyrolysis had a direct impact on the biochar floor area. The biochar received from special pyrolysis temperature presence specific mechanism for the absorption (Figure 2). The better concentration cation including calcium (Ca) and potassium (K) use to promote the absorption of 2,4-dichlorophenol.

The inorganic element inside the biochar provides the rate and the residence time has a direct impact on the manufacturing of biochar. The biochar amendment via extraordinary activation techniques suggested with the improved physical properties as well as chemical properties inclusive of the specific floor area, distribution of the functional group. The greater studies at the optimization method for the most biochar manufacturing yield, cost factor evaluation for biochar manufacturing, diverse element study, lively investigation, modelling, and biochar re-production [24].

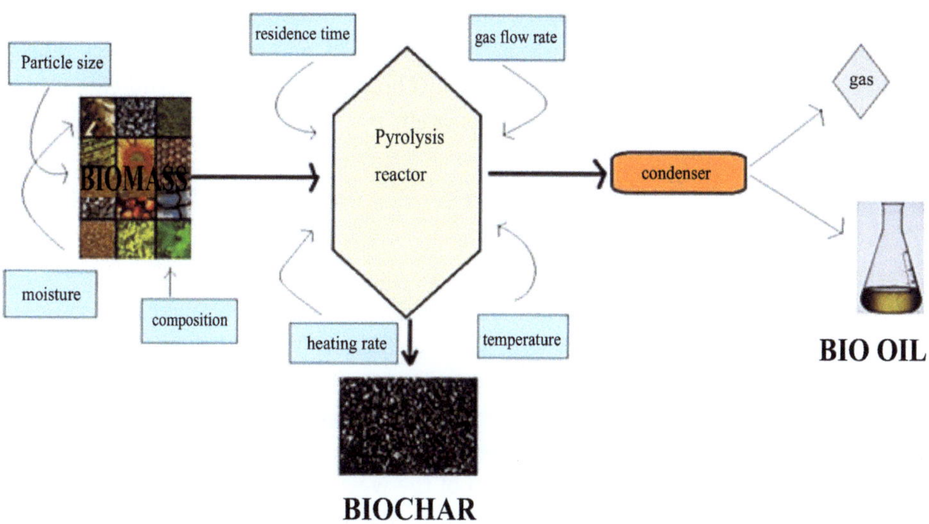

Figure 2: Production and application of biochar.

6 Biochar for power generation

The production of biochar has always involved partial burning; the segment of a plant with carbon captured in it is immediately released. But other portions with more than 50% of the carbon remain in them with its new form as biochar. Not only this, but during this process plant materials tend to release some energy too, and this released energy can be captured and used by us. To capture and use this pure energy is simple, heating. It is so possible to produce electricity from this, but the most difficult is to make a liquid fuel that can be transported, but not impossible [25].

There is a thermochemical conversion process known as pyrolysis that tends to convert biomass into biochar as well as a liquid fuel. The obtained biochar is generally rich in carbon and also has the potential of being pollution-free. The gas produced through pyrolysis could be used as a fuel in the industries which have combustion processes, but that can be possible only if we check the amount of methane and carbon dioxide it emits (see Figure 3). For example, methanol, which is called methyl alcohol, can be obtained from the conversion of syngas; cellulosic ethanol is obtained from the fermentation of grasses, wood, and crop residues. Such as that ethanol is produced by corn grain fermentation. Some oils are drawn out from wood and then combined with alcohol with some catalyst to the process known to be alcoholysis, which gives rise to diesel fuel.

In Japan, people have discovered a way to solve the problem of eliminating wood waste material. They developed equipment from which they can generate electricity from biochar (wood waste material). The biochar electricity generation system needs 150 kilograms of waste wood every hour, and each hour it generates a lot of electricity that can power up about 100 houses for one hour and can subsequently produce 15 kilograms of biochar as well.

7 Biochar in adsorption

The rapid blooming of urbanization and industrialization in evolved countries causes a high level of environmental risk which affects human health through the WW produced from industries in large quantities, thus massively impacting the country's economy. The alternative for the activated carbon in the removal of waste material is biochar. The biochar that had undergone different pyrolysis at different temperatures had been characterized for the adsorption of the waste material. The specific character that had been generated into the biochar by these pyrolysis temperatures significantly influences the adsorption behavior [26].

There are two types of adsorption: physical adsorption and chemisorption. The lower the temperature, the adsorption had been carried out by the electrostatic interaction. Whereas the higher the temperature, the adsorption is conducted by the hydrogen bonding and π–π interactions. Biochar is considered the cheapest and effective adsorption method due to its microporous structure. The physiochemical properties of the biochar rely upon the raw material that has been used and the condition of the preparation as well as the pyrolysis temperature [27].

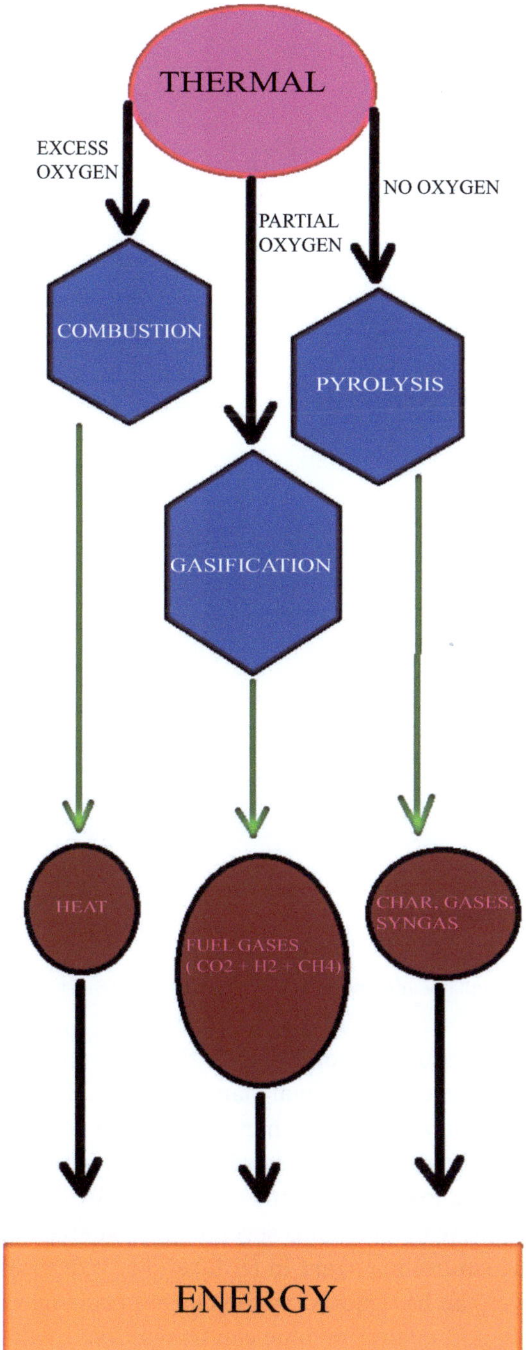

Figure 3: Pathways for conversion of biomass to energy.

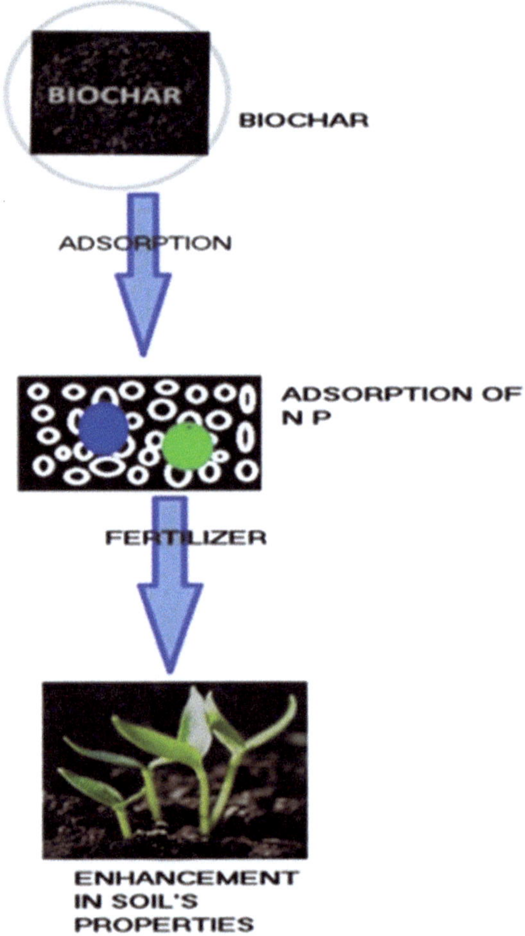

Figure 4: Improvement of soil's properties through the adsorption of macro elements by Biochar.

Biochar is rich in carbon which in the form of solid help in the adsorption of organic and inorganic material. Biochar made up of animal manure is useful for the adsorption of pharmaceutical and toxic metals. This biochar shows better adsorption compared to the activated carbon. Furthermore, the sawdust biochar which is an agricultural waste material adsorb dyes, toxic metals, and salts from WW. It also helps in the adsorption of the industrial WW that contains several toxic materials. It can also be used for the release of the controlling agent in fertilizer. For industrial and domestic effluents comprising dyes can be removed by the biochar made up of the orange peel [28]. Whereas safranin and methylene blue can be removed by the biochar made up of rice husk. Biochar is an eco-friendly adsorbent.

8 Traditional uses of biochar

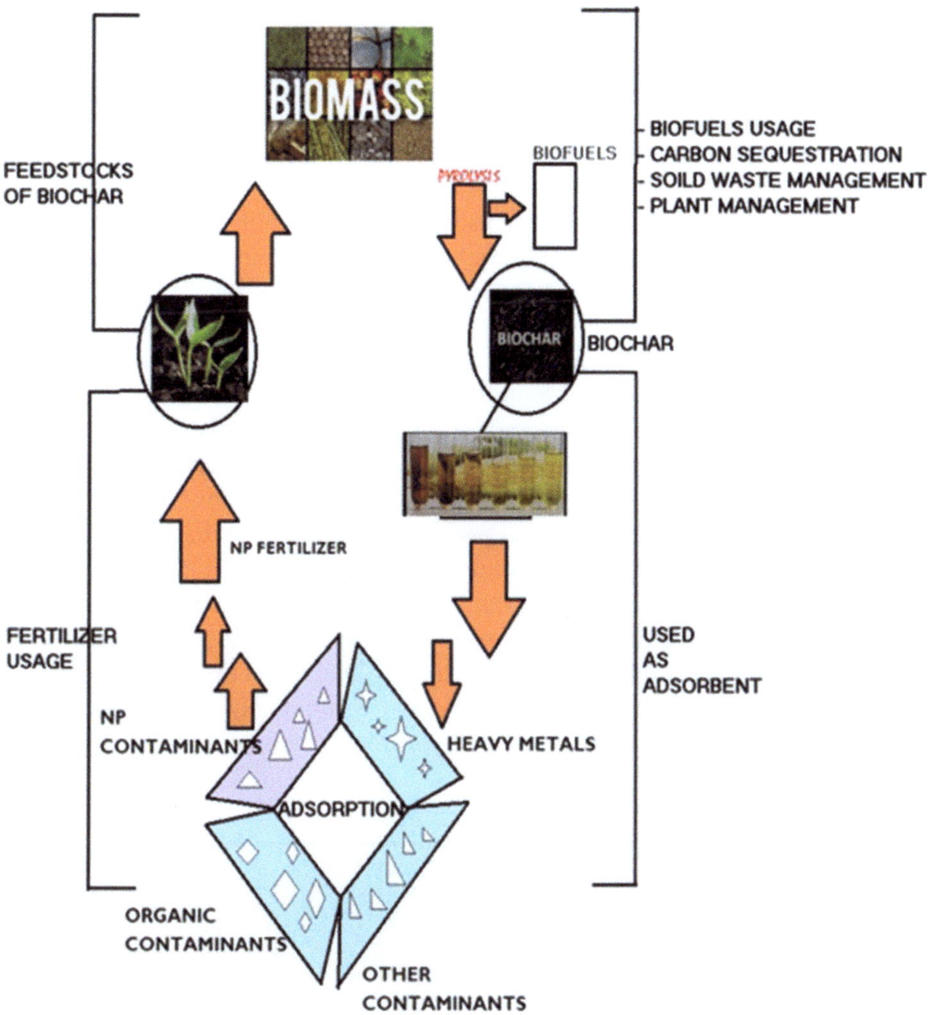

Figure 5: Application of biochar (removal of pollutants from aqueous solution).

Fortunately, biochar manufacturing and its good results on soil have a long-drawn history as well as prodigious ease of dispensation worldwide. The use of biochar for soil nutrient remembrance and development is not a new idea; it has been practiced in the Brazilian Amazon for over 200 decades ago. The ancient civilizations of the Amazon made the rich black fertile soil by transforming the poor-quality red soil, using biochar, which makes it an ancient mechanism for enhancement in soil fertility for the agriculture process. Not only ancient Amazonian civilizations used biochar, but also civilization in West Africa, Peru, Ecuador, and Australia are found using biochar in agriculture at its best [29].

Biocharculture (bio + char + culture) is a vast characterization of charcoal that covers all facets that are related to its application. Biochar culture is a holistic approach that means it provides support that focuses on a person's wellness, the health of the entire body, as well as mental health. So, this has been historically examined, consistently practiced, reasonably priced, sustainable for the environment, globally acceptable plan, and is also practiced as a form of cultural exchange in which some people with other cultural beliefs start assuming their beliefs, rituals and practices without immolating own beliefs and culture. Biochar culture is socially responsible, as they act best in terms of environment, economics, ethics, and human welfare. Biochar is acquired as an outgrowth of several activities such as burning crop residues either accidentally, intentionally, or naturally (like usage of cookstoves with burning biomass as a fuel). Biochar has been also used for sanitation purposes for instance; people use biochar for cleaning toilets, their utensils, for cleaning their teeth. Also, some unique types of bricks used for construction are produced by using biochar. In Japan, people use bamboo biochar for their traditional use. In addition, biochar serves numerous purposes as a commercial product rather than as a traditional product, that includes its application in the fields like water purification, green habitats, livestock such as cows and pigs, soil management, biomass energy, and food before eventually landing on the fields for the user to enhance the soil quality. In the end, biochar is ultimately utilized in the soil, helping to label the issues of global warming through the separation of the carbon atom.

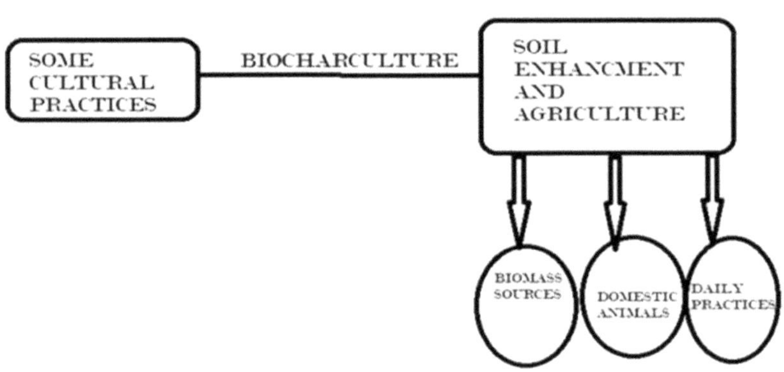

Figure 6: How biocharculture practices affect the soil's properties and subsequently agriculture.

Adapting biochar culture has several benefits like:
1) Using biochar recoups the debased soils.
2) Usage of biochar for water conservation.
3) Using biochar can lessen emissions and increase the reclusion of carbon.
4) Using biochar during farming assure the crop from climate unevenness and drought.
5) Usage of biochar in agriculture reduces the impact of perilous pesticides.
6) The making of biochar by converting biomass is eco-friendly.
7) There you can see an increase in crop yield.
8) It increases the available phosphorus, nitrogen, carbon, and pH for the plants.
9) At last, biochar has its lasting impacts for more than 1,000 years.

In August 2014, Sai Bhaskar N. Reddy wrote a book on biochar culture named *Biocharculture*", in which he wrote how biochar will help in the development as well as the environment. Biocharculture is unification of two practices, particularly traditional practices and local practices [30]. These practices should be emboldened and local communities should get encouraged to use and produce biochar with the raw materials (made from the blend of natural or organic materials) that are locally or easily available and are appraised as sustainable.

9 Core issues

Biochar serves a bunch of benefits to the environment or says to ameliorate the nature that has been spoiled by us.

9.1 Water management

In water management, there are various roots of water getting polluted, like marine dumping, industrial waste being dumped into nearby water bodies, accidental oil leakage that soak up the carbon pollutants from the air, the oceans absorbing plenty of carbon emissions from man-made inventions, leakages in sewer lines leading to these sewer waste into water bodies, and mining activities having harmful effects on groundwater as they use tremendous amounts of water for draining mine.

Biochar can abolish several contaminants/pollutants from the aqueous solutions because of its magnificent adsorption ability. Biochar water treatment has a handful of possible goodness as compared to some other cheap methods like boiling, chlorination, sand filtration, and solar disinfection. Biochar is low budget and sustainably made, moreover, it can be willingly accessed as a biomaterial and makes it suitable for the masses that cannot afford it.

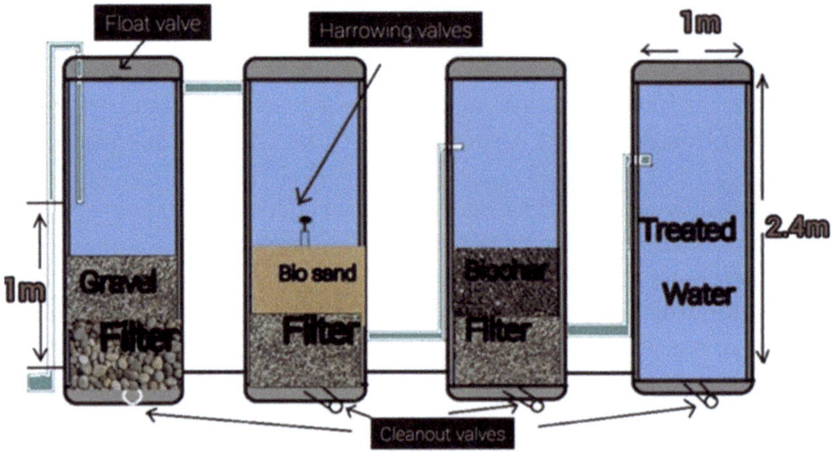

Figure 7: Biochar water filter.

9.2 Climate change

Biochar can remove carbon dioxide from the atmosphere for some increased period, furthermore offering other benefits to the environment. Biochar has potential benefits for climate change. Engulfing biochar in the soil develops fertility, aiding in crop growth as well as improving some other properties of soil. This unique hypothesis of using biochar on a large scale for climate change mitigation has been traced for several years. Biochar's climate change mitigation potential comes from its steady decomposition than the underdone biomass from where it is produced, thus taking down the rate at which carbon that is photosynthetically fixed, is reinstated to the atmosphere. Biochar has an overall impact on climate change mitigation depending on the other secondary mechanism that can also reduce nitrous oxide (N_2O) emissions from soil, by conversion of biomass into biochar.

The stored carbon in biochar is uncooperative, from which one can increase the water-holding and nutrient-holding capacities of soil which constantly results in raised growth of plants. This intensified productivity is a good reaction that enhances the amount of carbon dioxide (CO_2) that is removed from the atmosphere (carbon negative) in the form of carbon sequestration.

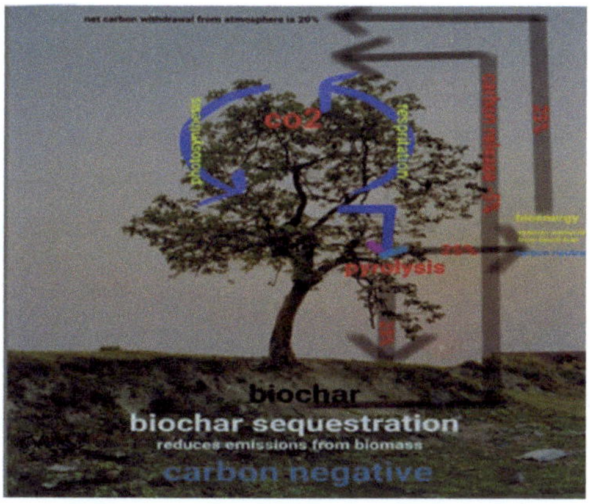

Figure 8: Biochar slows climate change (carbon sequestration).

10 Benefits of biochar

Deliberate putting the charcoal on the ground with cooking fires and kitchen debris increases the fertility and the carbon content. This conversion of carbon into a stable form and cleaner than the other form of charcoal is done by biochar. When plant matters like (leaves, trunks, roots), manures, or other organic material are heated in a no or low level of oxygen environment biochars are produced. Biochar can be in any shape like sticks, pellets, or dust. The unique ability of the biochar is to allure and hold the moisture, nutrients, and agricultural chemical along with nutrients that are difficult to hold like nitrogen and phosphorus due to their physical and chemical nature. The emissions of about 50–80% of CO_2 and N_2O had been decreased by the biochar-enriched soils since they can hold the gasses. A secured habitat for the microorganisms and fungi had been provided by biochar's vast surface area and complex porous structure. The symbiotic relationship can be formed with fibers of a plant root that allows plants and certain fungi for the greater uptake of nutrients. Compared to any form of organic matter that is commonly applied to soil, it is clear that biochar is more persistent. Due to the long-term persistence of biochar, there are also some benefits like water retention, overall fertility of the soil, and nutrient retention compared to some fertilizers that are commonly used. Biochar will not be broken down like some other organic soil amendments and has resistance toward chemical degradation and microbial degradation, mainly when buried due to its fairly inert property [31]. A significant quantity of nutrients is retained from its origin by the biochar made up of manure.

The ability of the biochar to attract and hold the soil nutrient helps to reduce the requirements of the fertilizers which cannot be achieved by common organic matter. Cost of fertilizers (either organic or chemical) is decreased and is eventually preserved in the soil for a long period. Additionally, indirect climate change benefits by reducing the need for fossil fuel chemical fertilizers had been provided by the biochar. By adding the biochar to the soil it helps to increase food production and crop yield as well as to enhance plant growth. Furthermore, the areas with exhaust soil finite resources, which are organic and inadequate water and entry to agrochemical fertilizers are sustainable. It takes about a few years to observe the results as not all soil react in the same way to biochar. Biochar can increase crop yields up to four folds in soil with poor carbon content.

A 25–50% increase in the growth of many vegetables and winter wheat was observed in the mixture of biochar plus chemical fertilizer compared to chemical fertilizer alone [32]. The key point is that not all the biochar are similar since the biochar has key chemical and physical properties according to the type of feedstock used and the condition during the process of pyrolysis, for example, biochar that has been manufactured from manure has a high level of nutrient content compared to the biochar manufactured from cuttings (pieces) of wood, though the latter would have a high percentage of persistence for a longer period. Although the biochar looks similar it works accordingly.

Contaminated soils comprising toxic heavy metals can be repaired by biochar. It can "adsorb" or mop up the metals and other toxic waste due to its incredibly large surface area and spongy nature and its alkalinity. The tendency to de-acidify the soil has become a contribution to decontamination, since some toxic metals for plants, like aluminum, can be dissolved in soils with pH levels below 5.0 (acidic soils). Biochar helps the plant to increase the resistance toward the pathogens since the soil acidity plays a key point, in the soils which have a pH value of 5.5 or less is difficult to grow the plant as the nutrients drain from such soil (leaching). The optimum pH that prevents it is between 5.5 and 7 that will result in the bacteria that can change and release nitrogen from any organic matter (and fertilizers). The structure of the soil is also affected by the pH mainly in clay soil. Clay soil is granular and easy to work within the optimum pH. The soil tends to become claggy and difficult to cultivate if it is either extremely acid or alkaline clay (refer Table 1).

Biochar is a more environmentally friendly method that helps to reduce global warming than regular reengineering. To extract the carbon from the atmosphere becomes fascinating since the amount of carbon dioxide in the atmosphere increases and causes changes in climate. CO_2 can be withdrawn from the atmosphere effectively by conversion of biomass into biochar, and burying it underground at enough scale and with rapid reduction of emission of greenhouse gasses, thus mitigating the atmospheric concentrations of CO_2. The hindrance to the intention is the high cost of biochar. The World Health Organization (WHO) said that even the small quantity could cause some serious health issues related to the nervous system, immune system, and digestive systems, mainly for fetuses and children since it is one of the element or chemical from ten elements or chemicals that causes the most serious public health concern. The country's largest source of mercury pollution is the power plants using coal as a fuel, which emits 90% of the amount of mercury that can be reduced by biochars.

Table 1: Removal of heavy metals using biochar-based adsorbents.

Heavy metal	Types of biochar	pH	Temperature	Condition
Copper	Peat moss	–	500 °C	Highly developed foramina's structure or carbonization condition or higher pH
	Pinewood	–	700 °C	–
	Peanut	3.5 to 5.0	400 °C	Higher phosphate presence Adsorption and surface precipitation
Lead	Peanut	3.0 to 8.5	400–1,000 °C	More number of carboxyl functional group/cation exchange
	Pistachio shell	–	550 °C	Higher carbon content/surface area/porous structure/porosity/distribution of pores
Cadmium	*Triticum aestivum*	–	–	Adsorption strongly influenced by solution pH/ Langmuir isothermal model/biochar functional group/metal ion precipitation/existence of minerals such as apatite calcite
	Peat moss	–	800 °C	Highly developed foraminous structure/ carbonization condition/higher pH
	Corn straw	4.0 to 7.0	–	Surface perplexed with active adsorption sites/ precipitation of inorganic anions
Chromium	Potassium hydroxide	–	700 °C	Large surface area/surface functional group/ polycyclic aromatic hydrocarbons sheets

11 Biochar in market

The key growth driver of the industry might be the increasing use of products in the form of organic food and its capacity to increase the fertility of the soil and the growth of the plant, the waste material such as agriculture debris, forest debris wood debris and animal waste, manure from charcoal after the controlled heating of the material in biochar, to reduce hazardous element used in charcoal-derived biochar, the reduction of moisture level, soil leaching and fertilizer is prevented by this. Asia Pacific

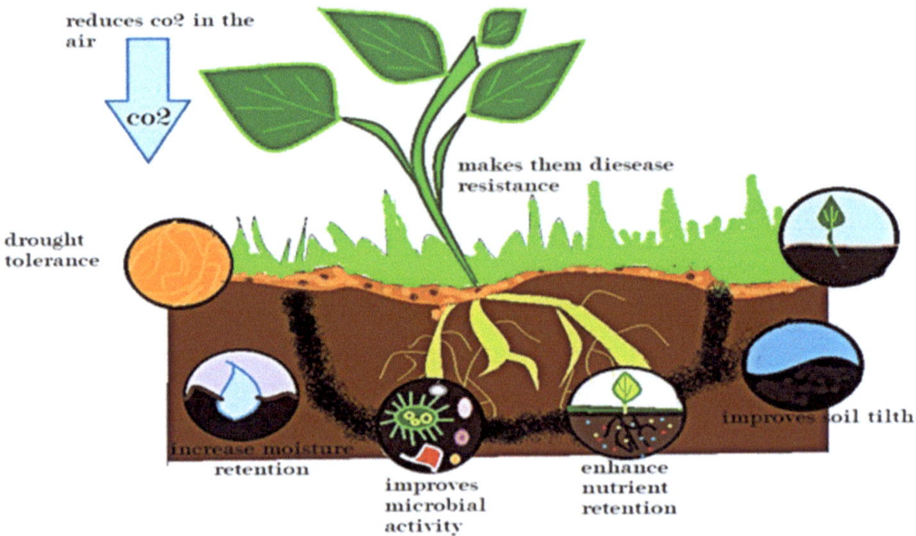

Figure 9: Benefits of biochar [2].

and Middle Eastern countries are expected to grow at a slower rate due to a lack of awareness regarding this product's benefits and its long-term advantages.

High capital investment is needed for the production of the biochar with high quality that causes the companies to exit from the market in the past few years. Biochar is not potential for the agricultural sector but it has high potential in other sectors, for example, as a fabric additive in the textile industry and to shield against electromagnetic radiation in the electronic industry. It does also expect demand from the food sector that may help in the improvement of biochar market growth. The water treatment facility required by the emerging economy becomes the reason for the use of biochar in water treatment complement for the market growth over the forecast period by the biochar production using biogas and crop residue.

According to the technical insight, pyrolysis is efficient in the production process compared to other various manufacturing technology like Gasification shows the rise in usage due to the growing need for electricity generation, but is predicted to lose the market share in the agricultural application since the process is unable to produce stable biochar.

Farmers are still not made much aware of the products and benefits. The demand in the overall market is expected to rise through general farming since it is a major segment. Researcher group and institution-owned to increase the knowledge among farmer. The consumption in livestock farming which includes various farming method has increased due to the ability of the product to provide essential nutrients and maintain the health of livestock. To grow exponentially over the coming years, factors in organic farming have been complemented in this sector. The government and private institutions help with the use of biochar into mixed farming, zero-fill age farming, and biodynamic agriculture.

According to the regional sight, significant growth with increasing demand for organic food and high intake of meat had been witnessed by North America since it has the higher consumption of biochar. This region had higher awareness of biochar compared to others and the knowledge about the products and their benefits is well known by the farming community too. Low feed costs for livestock may expand the consumption in this sector creating huge awareness about the product as an outcome; this is how the United States has been generating the highest revenues in the world.

The key factor for the growth of this market is the rising utilization of livestock feed, the increase in the demand for biochar in the greenhouse remediation, and the rapidly growing agricultural industry.

12 Conclusion

Land degradation causes high-level demand for the sustainable practice of crop production for an ever-growing population. Biochar reduces acidity and makes strides on the availability of nutrient as a reparative for the contaminated agricultural soil, increasing soil fertility. Increasing the crop addition of biochar to soil is the best practice to prevail over any biotic soil stress. The activity between soil plan water brings a better photosynthetic action and increase nitrogen and water use efficiency, generating a beneficial impact in favor of the use of biochar. Biochar has great possibilities for improving the properties of soil, microbial abundance, biological nitrogen fixation, and plant growth. The long-term carbon sink restoration acts as a soil addendum for carbon sequestration. It also acts as a control tool that can reach multiple targets, including alleviation of global climate change, high-value utilization of waste resources, environmental remediation, and food safety.

Due to the tough interaction between soil, its inherent contaminants, and biochar, not all biochar gives ideal or good results. The content and composition of a contaminant in fresh biochar must be considered from the biochar derived different feedstock and pyrolysis condition. Evolving standard methods for extricating and measuring this contaminant are a crucial and vital feature in this.

Declaration of competing interest

The authors declare that they have no known competing financial interests or personal relationships that could have appeared to influence the work reported in this chapter.

References

[1] Rangabhashiyam S, Suganya E, Selvaraju N, Varghese LA, Significance of exploiting non-living biomaterials for the biosorption of wastewater pollutants, World J. Microbiol. Biotechnol., 2014, 30, 1669–1689, doi: https://doi.org/10.1007/s11274-014-1599-y.

[2] Nartey OD, Zhao B, Biochar preparation, characterization, and adsorptive capacity and its effect on bioavailability of contaminants: An overview, Adv. Mater. Sci. Eng., 2014, 2014, doi: https://doi.org/10.1155/2014/715398.

[3] Luo X, Zhang Z, Zhou P, Liu Y, Ma G, Lei Z, Synergic adsorption of acid blue 80 and heavy metal ions (Cu2+/Ni2+) onto activated carbon and its mechanisms, J. Ind. Eng. Chem., 2015, 27, 164–174, doi: https://doi.org/10.1016/j.jiec.2014.12.031.

[4] Yu J, Zhao Y, Li Y, Utilization of corn cob biochar in a direct carbon fuel cell, J. Power Sour., 2014, 270, 312–317, doi: https://doi.org/10.1016/j.jpowsour.2014.07.125.

[5] Rangabhashiyam S, Balasubramanian P, The potential of lignocellulosic biomass precursors for biochar production: Performance, mechanism and wastewater application – A review, Ind. Crops Prod., 2019, 128, 405–423, doi: https://doi.org/10.1016/j.indcrop.2018.11.041.

[6] Gupta RK, Dubey M, Kharel P, Gu Z, Fan QH, Biochar activated by oxygen plasma for supercapacitors, J. Power Sour., 2015, 274, 1300–1305, doi: https://doi.org/10.1016/j.jpowsour.2014.10.169.

[7] Gallezot P, Conversion of biomass to selected chemical products, Chem. Soc. Rev., 2012, 41, 1538–1558, doi: https://doi.org/10.1039/c1cs15147a.

[8] Chemerys V, Baltrėnaitė E, influence of intrinsic properties of lignocellulosic feedstock on adsorptive properties of biochar, J. Environ. Eng., 2018, 144, 04018075, doi: https://doi.org/10.1061/(asce)ee.1943-7870.0001420.

[9] Leng LJ, Yuan XZ, Huang HJ, Wang H, Bin Wu Z, Fu LH, Peng X, Chen XH, Zeng GM, Characterization and application of bio-chars from liquefaction of microalgae, lignocellulosic biomass and sewage sludge, Fuel Process. Technol., 2015, 129, 8–14, doi: https://doi.org/10.1016/j.fuproc.2014.08.016.

[10] Kim P, Johnson A, Edmunds CW, Radosevich M, Vogt F, Rials TG, Labbé N, Surface functionality and carbon structures in lignocellulosic-derived biochars produced by fast pyrolysis, Energy Fuels, 2011, 25, 4693–4703, doi: https://doi.org/10.1021/ef200915s.

[11] Meyer S, Glaser B, Quicker P, Technical, economical, and climate-related aspects of biochar production technologies: A literature review, Environ. Sci. Technol., 2011, 45, 9473–9483, doi: https://doi.org/10.1021/es201792c.

[12] Islam MA, Auta M, Kabir G, Hameed BH, A thermogravimetric analysis of the combustion kinetics of karanja (Pongamia pinnata) fruit hulls char, Bioresour. Technol., 2016, 200, 335–341, doi: https://doi.org/10.1016/j.biortech.2015.09.057.

[13] Dias D, Lapa N, Bernardo M, Godinho D, Fonseca I, Miranda M, Pinto F, Lemos F, Properties of chars from the gasification and pyrolysis of rice waste streams towards their valorisation as adsorbent materials, Waste Manag., 2017, 65, 186–194, doi: https://doi.org/10.1016/j.wasman.2017.04.011.

[14] Liu Z, Zhang FS, Wu J, Characterization and application of chars produced from pinewood pyrolysis and hydrothermal treatment, Fuel, 2010, 89, 510–514, doi: https://doi.org/10.1016/j.fuel.2009.08.042.

[15] Emami Taba L, Irfan MF, Wan Daud WAM, Chakrabarti MH, The effect of temperature on various parameters in coal, biomass and CO-gasification: A review, Renew. Sustain. Energy Rev., 2012, 16, 5584–5596, doi: https://doi.org/10.1016/j.rser.2012.06.015.

[16] Nanda S, Dalai AK, Berruti F, Kozinski JA, Biochar as an exceptional bioresource for energy, agronomy, carbon sequestration, activated carbon and specialty materials, Waste Biomass Valorization, 2016, 7, 201–235, doi: https://doi.org/10.1007/s12649-015-9459-z.

[17] Saletnik B, Zagula G, Bajcar M, Tarapatskyy M, Bobula G, Puchalski C, Biochar as a multifunctional component of the environment-a review, Appl. Sci, 2019, 9, 1139, doi: https://doi.org/10.3390/app9061139.

[18] Hou J, Zhang X, Liu S, Zhang S, Zhang Q, Critical A, Review on bioethanol and biochar production from lignocellulosic biomass and their combined application in generation of high-value byproducts, Energy Technol., 2020, 8, 2000025, doi: https://doi.org/10.1002/ente.202000025.

[19] Wang J, Wang S, Preparation, modification and environmental application of biochar: A review, J. Clean. Prod., 2019, 227, 1002–1022, doi: https://doi.org/10.1016/j.jclepro.2019.04.282.

[20] Shaikh R, Rizvi A, Quraishi M, Pandit S, Mathuriya AS, Gupta PK, Singh J, Prasad R, Bioelectricity production using plant-microbial fuel cell: present state of art, South African J. Bot., 2020, 000, 2020, doi: https://doi.org/10.1016/j.sajb.2020.09.025.

[21] Chakraborty I, Sathe SM, Dubey BK, Ghangrekar MM, Waste-derived biochar: applications and future perspective in microbial fuel cells, Bioresour. Technol., 2020, 312, 123587, doi: https://doi.org/10.1016/j.biortech.2020.123587.

[22] Angin D, Altintig E, Köse TE, Influence of process parameters on the surface and chemical properties of activated carbon obtained from biochar by chemical activation, Bioresour. Technol., 2013, 148, 542–549, doi: https://doi.org/10.1016/j.biortech.2013.08.164.

[23] Kalderis D, Kayan B, Akay S, Kulaksiz E, Gözmen B, Adsorption of 2,4-dichlorophenol on paper sludge/wheat husk biochar: process optimization and comparison with biochars prepared from wood chips, sewage sludge and HOG fuel/demolition waste, J. Environ. Chem. Eng., 2017, 5, 2222–2231, doi: https://doi.org/10.1016/j.jece.2017.04.039.

[24] Sewu DD, Boakye P, Woo SH, Highly efficient adsorption of cationic dye by biochar produced with Korean cabbage waste, Bioresour. Technol., 2017, 224, 206–213, doi: https://doi.org/10.1016/j.biortech.2016.11.009.

[25] Feng D, Zhao Y, Zhang Y, Zhang Z, Zhang L, Gao J, Sun S, Synergetic effects of biochar structure and AAEM species on reactivity of H2O-activated biochar from cyclone air gasification, Int. J. Hydrogen Energy., 2017, 42, 16045–16053, doi: https://doi.org/10.1016/j.ijhydene.2017.05.153.

[26] Dorez G, Ferry L, Sonnier R, Taguet A, Lopez-Cuesta JM, Effect of cellulose, hemicellulose and lignin contents on pyrolysis and combustion of natural fibers, J. Anal. Appl. Pyrol., 2014, 107, 323–331, doi: https://doi.org/10.1016/j.jaap.2014.03.017.

[27] Daneshvar E, Vazirzadeh A, Niazi A, Sillanpää M, Bhatnagar A, A comparative study of methylene blue biosorption using different modified brown, red and green macroalgae – Effect of pretreatment, Chem. Eng. J., 2017, 307, 435–446, doi: https://doi.org/10.1016/j.cej.2016.08.093.

[28] Eom IY, Kim JY, Kim TS, Lee SM, Choi D, Choi IG, Choi JW, Effect of essential inorganic metals on primary thermal degradation of lignocellulosic biomass, Bioresour. Technol., 2012, 104, 687–694, doi: https://doi.org/10.1016/j.biortech.2011.10.035.

[29] Torabian S, Qin R, Noulas C, Lu Y, Wang G, Biochar: an organic amendment to crops and an environmental solution, AIMS Agric. Food, 2021, 6, 401–405, doi: https://doi.org/10.3934/AGRFOOD.2021024.

[30] Fassinou WF, Van De Steene L, Toure S, Volle G, Girard P, Pyrolysis of Pinus pinaster in a two-stage gasifier: influence of processing parameters and thermal cracking of tar, Fuel Process. Technol., 2009, 90, 75–90, doi: https://doi.org/10.1016/j.fuproc.2008.07.016.

[31] De Mendonça FG, Da Cunha IT, Soares RR, Tristão JC, Lago RM, Tuning the surface properties of biochar by thermal treatment, Bioresour. Technol., 2017, 246, 28–33, doi: https://doi.org/10.1016/j.biortech.2017.07.099.
[32] Fisher T, Hajaligol M, Waymack B, Kellogg D, Pyrolysis behavior and kinetics of biomass derived materials, J. Anal. Appl. Pyrol., 2002, 62, 331–349, doi: https://doi.org/10.1016/S0165-2370(01)00129-2.

G. Subbulakshmi, R. Thiruneelakandan, G. Padmapriya

Biochar remediation techniques: efficient and eco-friendly tool for sustainable environment

Abstract: In recent years, disposal of wastes from various sources like domestic, agriculture, and industries has caused serious environmental hazards and economic problems. Contaminated soil remediation using biochar has been intensively studied in ex situ and in situ techniques with acquired valuable results and versatile effects. Biochar has been exploiting as a good alteration to bring to a halt of heavy metals in contaminated soils. Biochar-based remediation technology has developed rapidly. This method offers an approach that may be less environmentally disruptive and more cost-effective, though the efficacy of biochar in metal contamination process depends on biochar properties and metal species. This chapter brings to light the margin of biochar application as an efficient improvement and propose further research directions to make sure the safe and sound use of biochar as adaptation remediation system of polluted soil.

Keywords: waste, environmental pollution, soil contamination, remediation, biochar

1 Introduction

Biochar, resulting from thermo chemical conversion of biomass, has been observed as an economically low-cost, sustainable method and widely applied in agriculture and ecology fields. The present tendency about waste management is a crucial point on recycling and the revival of waste as new materials. A number of processes have been explored for effective management of solid waste and composting is one potential method in managing the organic nutrient residues originating from agricultural residues, as it is biodegradable. A newer branch of technology called "biochar technology" is an emerging method to solve various environmental problems from waste management to land (soil) improvement. The growing speed of agricultural wastes due to the large scale of urbanization and an outcome of economic development has become a trouble that produces the massive quantities of waste and causes a serious environmental crisis which is not easy for management. Management of agricultural waste has grown to be one of the major problems. The major components of agricultural solid wastes are biodegradable organics. After harvesting, large amount of agricultural waste are generated and left in the field for natural degradation which takes several months. Fresh organic waste materials cannot be

https://doi.org/10.1515/9783110734003-013

applied to soil until they have been suitably iostabilized, because function of immature organic materials to soil may influence plant growth due to nitrogen malnourishment and production of noxious metabolites These declined are also disposed off simply by burning in the field which results in loss of nutrients as well as causes atmospheric pollution due to the emission of toxic gases [1]. Uncontrolled and improper handling can often lead to many situations where agricultural waste can become an environmental issue. In order to mitigate this problem, agricultural waste was predigested and utilized for the production of biochar. The need to recycle organic agricultural wastes is not just for environmental issues but also for economical and sustainable advantages. Reprocess of waste to biochar provides a good option for small and marginal farmers to produce organic manure locally for use in their farms. It is an environment friendly method to recycle agricultural wastes and to produce enriched product. Converting agriculture waste (Figure 1) into biochar through pyrolysis was an appropriate management for agriculture waste and became a trendy focus in recent time. Currently, studies have demonstrated that biochar, used as a soil amendment, may improve soil fertility and crop yields [2].

Figure 1: Crop residue.

Moreover, the biochar produced from these agriculture wastes shows signs of tremendous performance in the elimination of contaminants from wastewater and soil. This chapter may help to recognize future guidelines in biochar research and applications. Furthermore, modernized reactors and technologies applied for competent manufacture of biochar for environmental remediation have been widely premeditated recently. The conclusion may help the new researchers to confiscate the explore boundary in the biochar field.

2 Nutritive value of biochar

Biochar in future will be one of the best, technological sound techniques for treatment of solid waste. Traditional agriculture is currently characterized by excessive

Figure 2: Biochar.

inputs of chemical fertilizers, pesticides, and herbicides, due to insufficient application of organic fertilizers. The excess use of chemical fertilizers and pesticides has resulted in numerous negative effects on the environment, including water, soil degradation of soil quality, and losses of agricultural biodiversity. These soils often show high toxicity and low percentage base saturation [3]. The application of the biochar (Figure 2) enriches the soil microorganism, plant growth (size of leaf, height, width, and weight), and nutrient content of the yield. The high concentrations of biochar may delay plant growth due to the concentration of soluble salts. As a result, biochar should be applied at required quantity to produced higher yield. The nutrients content in biochar vary depending on the waste materials that are being used for the preparation. Studies have shown that biochar produced from crop residues such as corn cob and rice husk often have low nitrogen [4] and phosphorus [4] contents. If the waste materials are heterogeneous one, there will be wide range of nutrients available in the compost. If the waste materials are homogenous one, there will be only certain nutrients are available.

Nutrients in biochar (Figure 3) are in readily available form and are released within a month of application. When added to acidic soils, biochar like other organic fertilizers (green manure, animal manure, and compost) influences soil characteristics such as soil pH and bulk density [5].

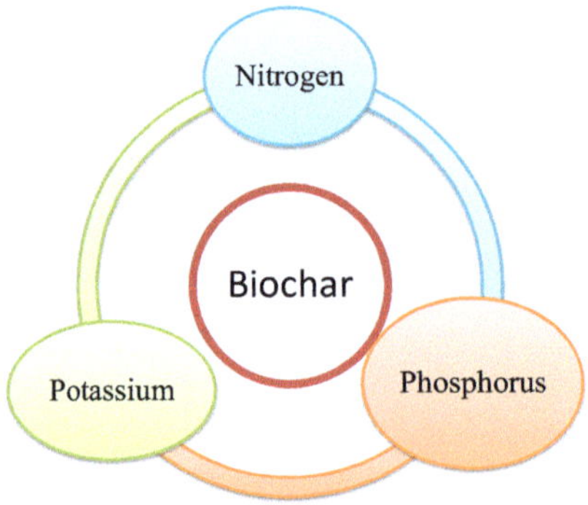

Figure 3: Nutrients cycle in biochar.

3 Environmental benefits of biochar

When organic matter decays, it releases carbon dioxide into the atmosphere. Production of biochar can prevent this addition of carbon dioxide by converting organic waste into a highly stable form of carbon that does not seep out reverse into the atmosphere when obscured in the soil. Biochar may facilitate the mineralization of organic pollutants by enhancing soil microbial activities. The reviews of studies over biochar additions to soil indicate that biochar can increase crop yield, suppress greenhouse gas (GHG) emissions from soil [6], reduce fertilizer requirements, and increase soil organic carbon stocks [7].When biochar is added to soil can sequester carbon in the soil for many years, reducing the amount of carbon dioxide present in the air and improving the liveliness of the soil over the years. Biochar's high adsorption competence also allows it to eliminate unwanted organic and inorganic contaminants from a soil and water.

4 Nutrient accessibility in soils

Soil particles are generally negatively charged and so bind positively charged nutrient ions (cations). Biochar alteration can increase cation exchange capacity (CEC), which strengthens the nutrient retention capacity [8]. Biochar enriched with carboxyl and phenolic carbon groups has a higher CEC (Figure 4), thus presenting a higher capacity to absorb nutrients. This is a useful indicator of soil fertility because it shows the soil's ability to supply three important plant nutrients. The larger the CEC value, the more cations the soils can absorb and make available to plants. Clay plus biochar be a magnet for nutrient ions and hold them on its surface. This prevents leaching and makes the nutrients available for plant roots. The clay–biochar combination means soil can hold more water during dry periods.

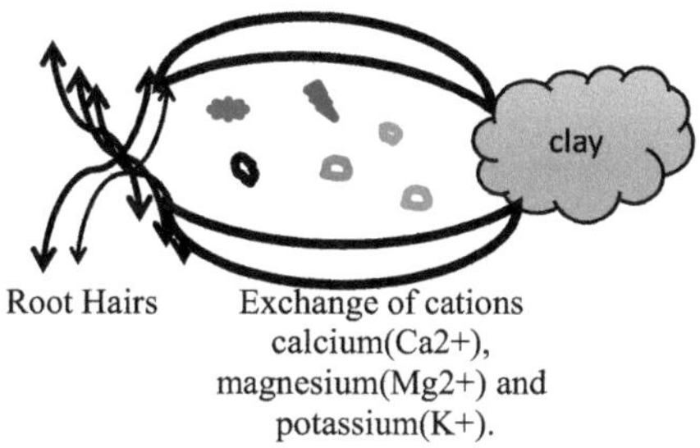

Figure 4: Cation exchange capacity – biochar.

Figure 4 is a schematic diagram showing exchanging of cations between the root hairs and the clay particle, and the movement of these cations from soil solution to roots (rhizosphere) for uptake, and capable of chelating or binding to minerals and transport them into cells. The most important feature of biochar lies in their ability to bind insoluble metal ions, oxides, and hydroxides, and to release them slowly and continually to plants when required.

4.1 Increase in the production of crop

Boon to farmers: Organic farming method will help improve the quality of land and reduce air and water pollution. Biochar: New approach to modern agriculture practices for sustainable food productivity. It is a new technology for increasing the rice production, reducing the cost of cultivation, and increasing the agricultural income. The addition of organic residues to agricultural soils has been used as a practical alternative to reduce environmental impacts and improve soil quality and health.

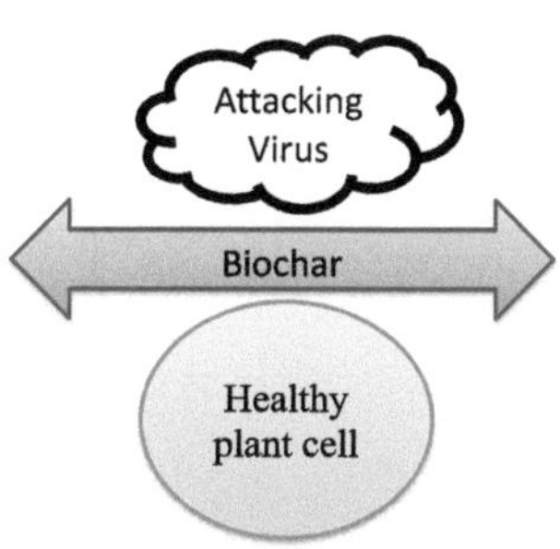

Figure 5: Biochar blocking a virus particle.

There have been several reports that biochar can induce excellent plant growth and enhance crop production compared to chemical fertilizers. Plant's response to biochar showed much better results (Figure 5) than any other commercial potting or rooting media. Biochar can also influence a number of physical, biological, and chemical processes of soil which contain their bearings on plant's growth. Biochar use in rice systems has been advocated as a potential strategy to reduce GHG emissions from soils [9], enhance soil carbon (C) stocks and nitrogen (N) retention [10], and improve soil function and crop productivity.

Figure 6: Growth of panicle using biochar.

Figure 7: Growth of panicle using chemical fertilizer.

These results revealed that an increase in plant growth(Figure 6 & Figure 7) could probably be due to improvement in the physicochemical properties of soil; increase in enzymatic activity; increase in microbial population, diversity, and activity; easy availability of macro- and micronutrients; and also an increase in plant growth hormones by application of biochar.

5 Biochar – organic farming mitigates climate change?

Reduces GHG, especially nitrous oxide, as no chemical nitrogen fertilizer is used and nutrient losses are minimized. Stores carbon in soil and plant biomass by the structure of organic matter, hopeful and frightening the approval of primary ecosystems. Biochar is one of the outcomes of scientific experiments, which has an important role in achieving sustainable agriculture and the environment [11].

6 Conclusion

Biochar is the best method in which agricultural waste can be recycled. Recycling is truly an eco-friendly technology through which one can convert all organic waste into a product which is rich in nutrient content and can replace chemical fertilizer. From the research, one can safely conclude that the biochar carried out was safe for agriculture use and could be very beneficial for increasing the productivity of our crops by using natural fertilizers rather than chemical fertilizers. It could be concluded from this study that application of biochar had an optimistic result on the soil physical, chemical, and biological properties, escalating plant cover and declining the soil losses.

References

[1] Draper K, Tomlinson T, The potential for biochar to deliver greater sustainability for coffee cultivation and processing: a white paper. Ithaka Institute for Carbon Intelligence and International Biochar Initiative, 2015, 4.

[2] Agegnehu G, Am B, Bird Mi NPA, Benefits of biochar, compost and biochar-compost for soil quality, maize yield and greenhouse gas emissions in a tropical agricultural soil, Sci. Environ., 2016, 543, 295–306.

[3] Sade H, Meriga B, Surapu V, Gadi J, Sunita MSL, Suravajhala P, Kishor PK, Toxicity and tolerance of aluminum in plants: tailoring plants to suit to acidsoils, Biometals, 2016, 29(2), 187–210.

[4] Deenik JL, Cooney MJ, The potential benefits and limitations of corn cob and sewage sludge biochars in an infertile oxisol, Sustainability, 2016, 8(2), 131.

[5] Djousse BMK, Allaire SE, Munson BA, Quantifying the influence of Eucalyptus bark and corncob biochars on the physical properties of an oxisol under maize cultivation in Dschang Cameroon. Biochar 5th North American Symposium, 2016.

[6] Cayuela ML, Van Zwieten L, Singh BP, Jeffery S, Roig A, Sánchez-Monedero MA, Biochar's role in mitigating soil nitrous oxide emissions: A review and meta-analysis, Agric. Ecosyst. Environ., 2014, 191, 5–16.

[7] Kameyama K, Shinogi Y, Miyamoto T, Agarie K, Estimation of net carbon sequestration potential with farmland application of bagasse charcoal: Life cycle inventory analysis through a pilot sugarcane bagasse carbonisation plant, Aust. J. Soil Res., 2010, 48, 586–592.

[8] Gao S, Hoffman-Krull K, Bidwell AL, DeLuca TH, Locally produced woodbiochar increases nutrient retention and availability in agricultural soils of the San Juan Islands, USA, Agric. Ecosyst. Environ., 2016, 233, 43–54.

[9] Qin X, Li Y, Wang H, Liu C, Li J, Wan Y, Gao Q, Fan F, Liao Y, Long-term effect of biochar application on yield-scaled greenhouse gas emissions in a rice paddy cropping system: A four-year case study in south China, Sci. Total. Environ., 2016, 569–570, 1390–1401.

[10] Sun H, Feng Y, Xue L, Mandal S, Wang H, Shi W, Yang L, Responses of ammonia volatilizationfrom rice paddy soil to application of wood vinegar alone or combined with biochar, Chemosphere, 2020, 242, 125247.

[11] Lu L, Yu W, Wang Y, Zhang K, Zhu X, Zhang Y, Chen B, Application of biochar-based materials in environmental remediation: From multi-level structures to specific devices, Biochar, 2020, 2, 1–31.

Dipita Ghosh, Subodh Kumar Maiti

Mechanism of removal of contaminants by modified biochar

1 Introduction

Mining activities cause the complete destruction of soil and in turn the ecosystem [1]. In the process of removing the desired mineral material, original soil is lost, or buried by wastes. The extraction activities usually result in large amounts of tailings or overburden materials, which lack the basic soil properties and contain potentially toxic elements. These tailings and overburden materials cannot provide favorable medium for plant growth because of the pH, low levels of organic matter and nutrients, degraded pedological structure, low water availability, and high concentrations of harmful heavy metals [2]. Biochar is a black charcoal like mass prepared by the pyrolysis of biomass and improves the physico-chemical biological, and nutritional properties of a derelict and degraded soil [3, 4]. Biochar has been reported to have a pH range of 8–11, cation exchange capacity of 27–485 $cmol_{(+)}$ kg^{-1}, surface area of 140–336 $m^2\ g^{-1}$, high porosity ranging around 0–1.32 $cm^3\ g^{-1}$, ash content of 0.3–77%, a high water holding capacity of 75–257%, and specific surface area (10–400 $m^2\ g^{-1}$) [5]. Due to these properties, biochar has the basic ability of adsorption of potentially toxic elements and organic pollutants.

Adsorption is a function of other physio-chemical properties such as pore size, surface area, functional groups, and cation exchange capacity. All these properties are dependent on feedstock material and biochar production condition [6]. A biochar produced at high temperature generally has more aromatic carbon, large surface area, and more porosity; however, the yield is often compromised at higher production condition [7, 8]. Therefore, a novel technique is needed to improve biochar yield and adsorption capacity. Modification of biochar is known to improve the physio-chemical and biological properties of biochar and improve its pollutant remediation potential (Figure 1). Methods of biochar modification include physical modifications such as steam activation and gas purging [9–12], chemical modification such as acid/base modification, surfactant modification, surface coating with nano-tubes and graphene [13–15], and biological modification such as by microbial action and utilization of digestion residue [16, 17]. While modifying the biochar, certain characteristics such as stability should not be affected as it might alter the basic recalcitrant nature of the biochar [18]. Also very little research has been done on the production of designer biochar in large scale for mine soil remediation.

The current chapter aims to broaden the focus on the scope of biochar modification that fulfils the need of contamination remediation, improves soil physiochemical

https://doi.org/10.1515/9783110734003-014

properties and the mechanism involved, explore and establish links between biochar modification, economic viability and large scale applicability, and whenever possible identify knowledge gaps and research needs.

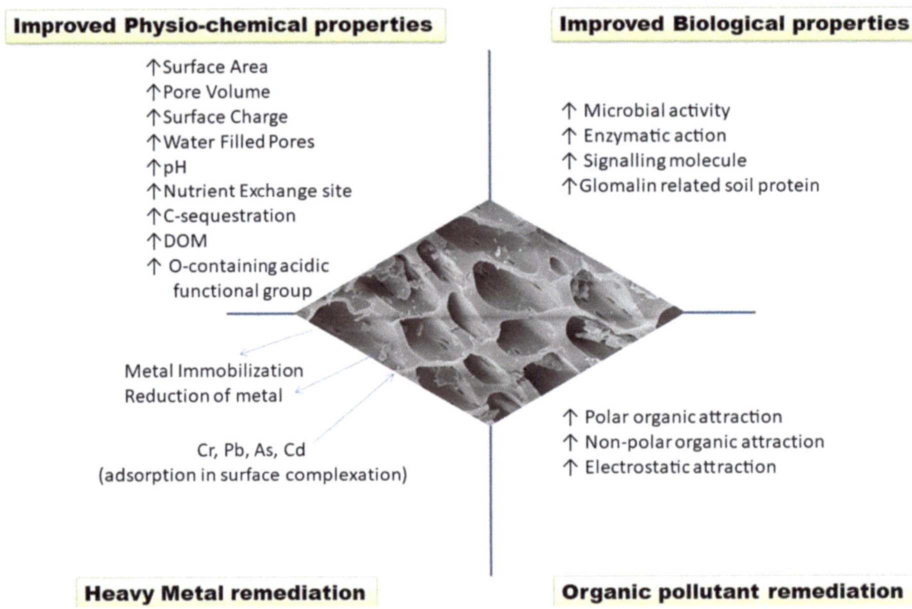

Figure 1: Impact of modification/engineering on biochar properties.

2 Biochar modification

The available literature suggests that biochar can be engineered or modified for the production of designer biochar by chemical, physical, and biological means (Figure 2).

2.1 Physical modification

The physical modification of biochar is a two-step sequential process based on carbonization in an oxygen limited environment [6]. In sequential process of carbonization, the unstructured areas of biomass are firstly pyrolyzed at high temperature which opens the fine pores. These pores when activated by physical modification methods possess expanded pores with large surface area. It has been reported that physical/mechanical means of modification are comparatively more cost feasible than chemical methods. However, chemical method is more effective means of

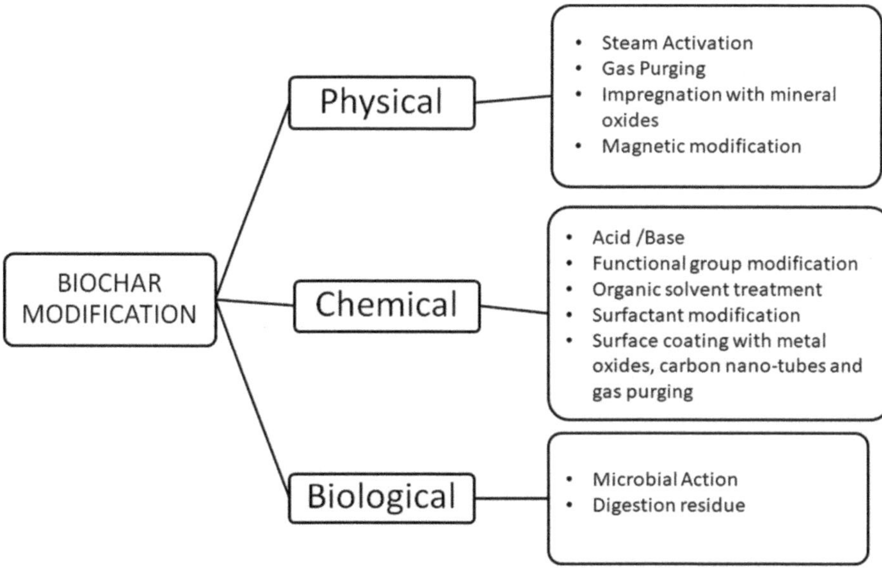

Figure 2: Classification of biochar modification methods.

modification [19]. The most widely used agents of physical modification for oxidizing the biochar surface are steam and carbon dioxide, which are purged into the sample for activation. Studies based on physical modification of biochar have been summarized in Table 1.

2.1.1 Steam activation

Steam activation is used to upgrade the quality of conventional biochar which has high surface area and improved carbonaceous structures. A number of studies have reported the beneficial use of steam activation on biochar properties (Table 1). For steam activation, a pyrolyzed biochar is treated with steam which creates an environment similar to the process of gasification; this promotes devolatilization and the formation of crystalline carbon in biochar surface [19].

During pyrolysis, some of volatile organic carbons remain trapped due to the incomplete combustion of the feedstock. Steam influx leads to the release of these volatile matters; this alters the surface properties of the biochar. The reaction mechanism for steam activation of biochar includes an exchange of oxygen from the H_2O molecule to the carbon surface site which leads to the formation of C(O) and H_2 gas. The H_2 generated then reacts with the biochar C to form surface complexes [C(H)] and hydrogen gas (equations (1)–(3)). The partial combustion of carbon leads to the formation of carbon monoxide, this can be further converted to CO_2 and H_2 in the presence of steam:

$$C + H_2O \rightarrow C(O) + H_2 \quad (1)$$

$$2C + H_2 \rightarrow C(H) + H_2 \quad (2)$$

$$CO + H_2O \rightarrow CO_2 + H_2 \quad (3)$$

Kwak et al. [9] studied the effect of steam activation in sawdust of white spruce, canola, wheat straw, and manure pellet biochar. The study reported that modification by steam increased the surface area and increased the adsorption capacity of Pb(II) up to the maximum value of 195 mg g^{-1}. Rajapaksha et al. [18] studied the effect of steam-activated *Sicyos angulatus* biochar on the adsorption of pharmaceuticals namely, sulfamethazine present in soils and water. The steam-activated biochar showed the highest sorption capacity of 37.7 mg g^{-1} for sulfamethazine. Steam activation increased the sorption capacity by 55% compared to pristine biochar produced at the same temperature. However, in another study conducted by Shim et al., [20] it was observed that steam modified *Miscanthus* biochar induced acute toxicity to the test plant (*Daphnia magna*), increased aromaticity by steam activation might be responsible for such anomaly. Thus, steam activation of *Miscanthus* biochar produced at 500 °C may not be appropriate for Cu sorption and toxicity reduction.

2.1.2 Gas purging

Biochar activated in the presence of CO_2 has been reported to be beneficial for the remediation of contaminants from water and soil (Table 1). The formation of pores occurs by Boudouard reaction, where CO_2 gets dissociatively chemisorbed to form surface oxides and carbon monoxide (equation (4)) [6]. The surface oxide, C (O), is then desorbed from the biochar surface, which further leads to pore formation (equation (5)). CO produced can further react with the surface carbon to form complex on the biochar surface (equation (6)).

$$C + CO_2 \rightarrow C(O) + CO \quad (4)$$

$$C + C(O) \rightarrow CO \quad (5)$$

$$C + CO \rightarrow C(O) \quad (6)$$

Shao et al., [11] used CO_2-activated corn cob biochar to study its effect on flue gas adsorption. The study reported that after CO_2 activation, the specific surface area of biochar increased to 755.35 $m^2\ g^{-1}$ from its initial value of 56.91 $m^2\ g^{-1}$. This was found to be effective in removal of SO_2 from the flue gas. Sun et al., [12] exposed biochar with CO_2 (500 mL min^{-1}) in a furnace for activation. CO_2-activated biochar was found to be an economically feasible and efficient catalyst for persulfate activation and the degradation of phenol and chlorophenols. Franciski et al., [21] used barley malt bagasse CO_2-activated biochar to initiate methylene blue remediation. The study reported that CO_2-activated biochar possessed a typical mesoporous surface with an area around 80 $m^2\ g^{-1}$. As consequence, maximum adsorption capacity of CO_2–biochar was 161 mg g^{-1} for methylene blue.

2.2 Chemical modification

2.2.1 Acid/base

Acid modification of biochar is known to improve its surface properties significantly (Table 1). Acid treatment acts as catalyst and accelerates the dehydration and elimination reaction by decreasing the pyrolysis temperature [22]. However, a major drawback of acid modification includes the corrosion of micropore wall which sometimes affect the adsorption of contaminants [6]. Phosphoric acid is the most popular acid modification technique due to its cost effectiveness and the ability to improve the phosphorous content in the biochar [22–24]. Biswas and Siddiqi [22] reported that pine cone biochar on treatment with H_2SO_4 and H_3PO_4 improved the specific surface area to 472 from initial value of 145 $m^2\ g^{-1}$. Treatment of biochar with acid increases the surface COO – functional group which is responsible for adsorption of Pb from the substrate. Maximum Pb^{2+} adsorption capacities for H_2SO_4 modified biochar was 283 mg g^{-1} and 321 mg g^{-1} for H_3PO_4 treatment. Similarly, Peng et al., [23] conducted a sorption study on the remediation of Cu (II) and Cd (II) by H_3PO_4 modified biochar prepared from pine sawdust. Functional groups such as -COOH),-OH, P = O and P = OOH on modified biochars were larger than those of pristine, these forms complexes with Cu (II) and Cd (II) and immobilizes them. In another adsorption experiment, Zeng et al., [24] reported that H_3PO_4 treated Eucalyptus biochar immobilized 99.76% of Cr (VI), which was higher than pristine biochar removal rate of 25.24%. Sahin et al., [25] reported that poultry manure biochar treated with HNO_3 and/or $HNO_3 + H_3PO_4$ increased the water-soluble P, K, Ca, Mg, Fe, Zn, Cu, and Mg and increased plant-available nutrients and plant growth.

Alkali treatment uses hydroxides and oxides to activate the reaction sites present on the biochar surface. Activation in the presence of hydroxides especially in the case of KOH involves a complex reaction. KOH reacts with the carbon atoms present on biochar surface to form K_2CO_3 which leads to the formation of irregularity on the biochar surface and help in the formation of intercalation compounds [6]. Liu et al., [26] reported that carbon content of the KOH-treated walnut shell and wood powder biochar was the highest and has a large specific surface area of 983.75 $m^2\ g^{-1}$. Similar studies by An et al., [27] on NaOH treated *Pomelo* peel biochar reported that modification increased the pore volume of biochar which promoted the absorption of Mn^{2+}. Alkaline treatment introduces –COO– and CO_3^{2+} on biochar surface which then reacts with Mn^{2+} to form $-COOMn^+$ and $MnCO_3$. The main mechanism for the above reaction includes cation exchange and surface precipitation. Similarly, Shin et al., [28] reported that NaOH modified spent coffee grounds biochar had improved physicochemical characteristics such as specific surface area and pore volume, which helps improve the adsorption of micropollutants. More than 45% removal efficiency of micropollutants was observed by modified biochar compared to pristine biochar of ≤18.5%.

2.2.2 Surfactant modification

Surfactants are chemical agents that are known to reduce the surface tension between a solid and in a liquid medium. Surfactants are also used as detergents, dispersants, and foaming agents [6]. The dispersant nature of surfactant is used for biochar modification for environmental remediation. Surfactants can be cationic and anionic in nature; reagents are used depending on the nature of the contaminant which needs to be remediated [18]. Studies based on the surfactant modification of biochar have been reported in Table 1.

Li et al. [29] studied the use of cationic surfactant namely Cetyltrimethyl Ammonium Bromide for the modification of *Auricularia auricula* dreg biochar. According to the study, cationic surfactant modification improved hydrophilic and hydrophobic groups, such as –OH, PO_4^{2-}, CO_3^{2-}, and –COO which helps improve the adsorption capacity of biochar for Cr(VI). It was observed that surfactant modification increased the adsorption rate and quantity for Cr(VI) by 8.0%. In another study conducted by Kuang et al. [30], sodium dodecyl sulfonate, sodium lauryl sulfate, and hexadecyl trimethyl ammonium bromide were used to modify the activated carbon for methylene blue remediation. The study reported that the anionic surfactants (sodium lauryl sulfate) improved the adsorption of the cationic dye; however the adsorption was reduced by cationic surfactant (hexadecyl trimethyl ammonium bromide). The reactions between methylene blue dye with sodium lauryl sulfate are shown in the following equations:

$$RSO_3Na + H^+ \rightarrow RSO_3H + Na^+ \tag{7}$$

$$RSO_3H + M^+ \rightarrow RSO_3M + Na^+ \tag{8}$$

$$ROH + M^+ \rightarrow ROM + H^+ \tag{9}$$

2.2.3 Surface coating with metal oxides, carbon nanotubes

Carbon nanotubes (CNTs) are known to be efficient for the removal of contaminates due to its high surface area and nanoporous structure. However, the cost constraint associated with nano-tubes makes them unpopular and inconvenient for large scale application. Therefore, biochar can act as a carrier of CNTs to develop novel, efficient, and recyclable sorbents for contaminant remediation. Zhu et al. [31] reported that nickel and nitrogen hybrid CNT modified sugarcane bagasse biochar showed a high removal capacity of 824.4 mg g^{-1} for Cr(VI) and a stronger ability to reduce to Cr(III). Similarly, Inyang et al. [14] reported that hybrid multiwalled CNT-coated biochar developed by varying concentrations of carboxyl-functionalized CNT solutions (0.01% and 1% w/w) can be excellent material for a sustainable wastewater treatment alternative for carbon nanotubes. In another study, Inyang et al. [15] reported that CNT-modified biochars had the highest methylene blue sorption capacity of 6.2 mg g^{-1}.

Graphene has a special 2-d structure and a number of unique properties, such as electroconductivity, mechanical strength, thermal conductivity, and surface area [18]. Similar to CNT, the cost and re-use limitation makes it less popular in water and wastewater treatment. Composite of graphene covering onto the biochar surface can overcome these disadvantages and provide a possible solution of the current challenges. Pine wood waste biochar coated with graphene can remove Cu^{2+} at 10.6 mg g^{-1}, while the Langmuir adsorption capacities were reported to be 18.4 mg g^{-1} [13]. In another study, graphene oxide–coated sawdust biochar had the ability to remove up to 500 mg L^{-1} of Cd(II) solution efficiently [32]. Tang et al. [34] reported that wheat straw biochar coated with grapheme improved the surface complexation between mercury and C–O, C = C, –OH, and O = C–O functional groups, which was responsible for mercury removal from the substrate.

2.3 Biological modification

Biological modification techniques include methods of engineering in which the biochar is treated either by microbes through anaerobic digestion or bacterial conversion or trans-abdominal treatment (Table 2).

2.3.1 Microbial action

Microbial action for biochar modifications includes the anaerobic digestion of biomass with the help of anaerobic bacteria for the conversion of organic matter to digestate and biogas. *Escherichia coli* and other bacterial strains have been often used to produce biofuel and bio-product from organic biomass [17, 35]. Biochar prepared by pyrolysis of anaerobically digested residue has been proposed as a beneficial product that could be used for environmental remediation [36, 37]. The study reported that the modified biochar showed higher pH, increased surface area, cation exchange capacity, and hydrophobicity as well as a more negative surface charge compared to unmodified biochar. The microbial action alters the redox potential and pH values of the feedstock biomass which imparts these properties to the biochar [36, 37].

According to Yao et al. [39], anaerobic digestion of bagasse biochar pyrolyzed at 600 °C was an effective adsorbent to remove sulfamethoxazole and sulfapyridine. The study reported that pi–pi interaction between the contaminants and modified biochar was the mechanism involved for contamination remediation. Organic dyes such as methylene blue have also been reported to be adsorbed (9.50 mg g^{-1}) by anaerobic digested food waste and Chinese silver grass biochar [40].

Anaerobically modified biochar has also been reported to remediate heavy metals such as Ag^+, Pb^{2+}, Cu^{2+}, Ni^{2+}, and Cd^{2+} [17, 37, 41]. Bagasse stillage treatment by *Escherichia coli* (*E. coli*) KO11 pre-pyrolysis was found to be effective in the removal of Ag^+ from aqueous solutions with a Langmuir maximum capacity of 90.06 mg g^{-1}. The main mechanism involved in Ag^+ sorption includes reduction and surface adsorption [17]. Similarly, dairy waste and sugar beet when anaerobic digested was effective in removing heavy metals such as Pb^{2+}, Cu^{2+}, Ni^{2+}, and Cd^{2+} [37]. Inyang et al. [41] reported that sugarcane bagasse on anaerobic digestion had lead absorption capacity of 653.9 mmol kg^{-1} compared to non-modified biochar 395.3 mmol kg^{-1}.

2.3.2 Digestion residue

Digestion residue refers to the technique in which the feedstock undergoes transabdominal transformation. It is a typical abdominal digestion process in livestock which has the potential to make full use of protein, amino acid, sugar, and other nutrients in fodder. This modification technique ensures the reuse of fodder effectively. Tao et al. [16] hypothesized that through transabdominal treatment, the corn straw feedstock physicochemical properties are enhanced by microbial action. It was observed that this modification significantly increased surface area (4.24–56.58 $m^2\,g^{-1}$), oxygen-containing functional group, and mineral components such as calcium carbonate and potassium chloride. The maximum Cd sorption capacity (Q_m) obtained from Langmuir model for was thrice compared to unmodified biochar. The main mechanism involved in this Cd(II) sorption by modified biochar includes precipitation with minerals, complexation and ion exchange with oxygen-containing functional groups.

Table 1: Physical and chemical methods of biochar modification, methods employed, and applicability.

Biochar feedstock	Modification method	Application	References
Physical modification			
Steam activation			
Sawdust of white spruce, canola, wheat straw, and manure pellet	Steam activation by deionized water, after 1 h of steam activation was replaced with N_2 gas	Modification by steam increased the adsorption capacity of Pb(II) It improved the surface area and caused the maximum Q_{max} Pb of 195 mg g^{-1} for canola straw biochar pyrolyzed at 700 ℃	[9]
Sicyos angulatus	The pyrolyzed samples were steam-activated at a flow rate of 5 mL min^{-1} for 45 min, at 300–700 °C.	– The steam-activated biochar showed the highest sorption capacity of 37.7 mg g^{-1} for sulfamethazine. – A 55% increase in sorption capacity compared to unmodified biochar produced at the same temperature.	[10]
Miscanthus	Biochar was placed in a preheated steam (500–700 °C) then heated to activation temperature of 800 ℃.	– Steam-modified biochar induced acute toxicity to the test plant (*Daphnia magna*), due to increased aromaticity upon steam activation. – Activation of biochar produced from *Miscanthus* at a temperature of 500 ℃ may not be useful for Cu sorption and toxicity reduction.	[20]

(continued)

Table 1 (continued)

Biochar feedstock	Modification method	Application	References
Gas purging			
Corncobs	Biochar was exposed to CO_2 (500 mL min^{-1}) and the system was maintained at 600 °C for 60 min	– CO_2 activation increased the specific surface area of biochar from 56.91 to 755.35 m^2 g^{-1}. – Increased surface area removes SO_2 from the flue gas.	[11]
N/A	Biochar was exposed to CO_2 (500 mL min^{-1}) in a furnace to start the activation for 1 h.	CO_2-activated biochar was found to be a low-cost and efficient catalyst for persulfate activation and the degradation of phenol and chlorophenols.	[12]
Barley malt bagasse	CO_2-activated biochar were prepared through N_2 pyrolysis followed by CO_2 activation	– CO_2-activated biochar was a typical mesoporous material with surface area around 80 m^2 g^{-1}. – The maximum adsorption capacity of methylene blue by CO_2–biochar was 161 mg g^{-1}.	[21]
Chemical modification			
Acid modification			
Pine cone	H_2SO_4 and H_3PO_4	– The specific surface area increased from 145 to 472 m^2 g^{-1} by acid modification. – Acid treatment causes an increase of the COO^- functional group on the adsorbent surface which are responsible for adsorption. – Maximum Pb^{2+} adsorption capacities were obtained by H_2SO_4 and H_3PO_4, treated biochars as 283 and 321 mg g^{-1}, respectively	[22]

Pine sawdust	H_3PO_4	– Modified biochars sorbed Cu(II) and Cd(II) much more strongly. – Functional groups such as -COOH), -OH, P = O, and P = OOH on modified biochars were larger than those of pristine. These functional groups forms complexes with Cu(II) and Cd(II) and immobilizes them.	[23]
Eucalyptus	H_3PO_4	– The acid-modified biochar removed 99.76% of Cr(VI), which was higher than pristine biochar removal rate of 25.24%. – The surface area of modified biochar increased to 1265.56 $m^2\ g^{-1}$ compared to pristine biochar at 253.25 $m^2\ g^{-1}$.	[24]
Poultry manure	$HNO_3 + H_3PO_4$	– Acids enriched the biochar with phosphate, nitro groups and nitrate. – Biochar modified with HNO_3 and/or $HNO_3 + H_3PO_4$ after pyrolysis increased the water-soluble essential nutrients and increased plant-available nutrients and plant growth	[25]
Base modification			
Pomelo peels	NaOH solution	– Modification improved pore structure of biochar, which was beneficial for the absorption ability for Mn^{2+} – Modification process introduced –COO– and CO_3^{2+} on biochar, which would could for complexes by cation exchange and surface precipitation.	[27]
Walnut shell and wood powder	KOH solution	Carbon content of the KOH-treated biochar was the highest and has a large specific surface area of 983.75 $m^2\ g^{-1}$	[26]
Spent coffee grounds	NaOH solution	– Modified biochar had improved physicochemical characteristics such as specific surface area and pore volume, which helps improve the adsorption of micropollutants. – >44.5% removal efficiency of micropollutants was observed in modified biochar compared to pristine biochar of ≤18.5%.	[28]

(continued)

Table 1 (continued)

Biochar feedstock	Modification method	Application	References
Surfactant modification			
Auricularia auricula dreg	Cetyltrimethyl ammonium bromide	– Surfactant modification improved the adsorption capacity of biochar for Cr(VI) by providing hydrophilic and hydrophobic groups. – Modification improved the adsorption rate and quantity for Cr (VI) by 8.0%.	[29]
Activated carbon	Sodium lauryl sulfate, sodium dodecyl sulfonate and hexadecyl trimethyl ammonium bromide	– The adsorption of methylene blue by anionic surfactants (sodium lauryl sulfate) was significantly improved. – Adsorption performance of methylene blue biochar modified by cationic surfactant (hexadecyl trimethyl ammonium bromide) was reduced.	[30]
Coating with carbon nanotubes			
Sugarcane bagasse	Nickel and nitrogen hybrid carbon nanotubes	Modified biochar had a high removal capacity of 824.4 mg g^{-1} for Cr(VI) and a stronger ability to reduce to Cr(III).	[31]
Hickory chips and sugarcane bagasse biomass	Carbon nanotubes	Effective wastewater treatment alternative.	[14]
Hickory chips and sugarcane bagasse biomass	Carbon nanotubes	Carbon nanotube–modified biochars had the highest methylene blue sorption capacity of 6.2 mg g^{-1}	[15]

Coating with graphene			
Pine wood waste	Graphene	Cu^{2+} removal capacities were found to be 10.6 mg g^{-1}. Langmuir adsorption capacity was 18.4 mg g^{-1}	[13]
Sawdust	Graphene oxide	The modification had the ability to remove up to 500 mg L^{-1} of Cd (II) solution efficiently.	[32]
Wheat straw	Graphene	The surface complexation between mercury and oxygen containing functional groups was responsible for mercury removal.	[34]
Impregnation with mineral oxides			
Sludge	Transition metals and rare earth elements (Fe, Ce, La, Al, Ti)	– Improved catalytic function shortened the reaction time by biochar modification. – Degradation efficiencies of tetracycline by the biochars were 51.7% (blank-biochar), 90.7% (Fe-biochar), 69.0% (Ce-biochar), 59.9% (La-biochar), 58.0% (Al-biochar), 58.0% (Ti-biochar).	[42]

Table 2: Biological methods of biochar modification, method, and applicability.

Biochar feedstock	Modification method	Application	References
Microbial action			
Bagasse	Anaerobic digestion	Digested bagasse biochar prepared was effective adsorbent to remove sulfamethoxazole and sulfapyridine.	[39]
Bagasse stillage	Escherichia coli (*E. coli*) KO11 treatment pre-pyrolysis	– Modified biochar can efficiently remove Ag^+ from aqueous solutions with a adsorption capacity of 90.06 mg g^{-1}. – The high sorption of Ag^+ onto the biochar by reduction and surface adsorption mechanisms.	[17]
Dairy waste and sugar beet	Anaerobic digestion	Modified biochar was effective in removing heavy metals such as Pb^{2+}, Cu^{2+}, Ni^{2+}, and Cd^{2+}	[37]
Food waste and Chinese silver grass	Anaerobic digestion	Increased adsorption capacity of methylene blue by 9.50 mg g-1.	[40]
Sugarcane bagasse	Anaerobic digestion	The lead absorption capacity of modified biochar was 653.9 mmol kg^{-1} compared to non-modified biochar 395.3 mmol kg^{-1}.	[41]
Raw sugar beet tailings	Anaerobic digestion	Anaerobic digestion enhances the phosphate adsorption ability of biochar.	[36]
Sugarcane bagasse	Anaerobic digestion	Modified biochar had higher surface area, pH, cation exchange capacity, hydrophobicity, and more negative surface charge.	[43]
Digestion residue			
Corn straw silage	Trans-abdominal transformation in cow	– This modification increased the surface area by 4.24–56.58 m^2 g^{-1} – Increased oxygen-containing functional group and mineral components such as calcium carbonate and potassium chloride.	[16]

3 Novel engineering methods for modification of biochar

The engineering methods for the modification of biochar are the alteration at molecular and macromolecular levels for improving the physicochemical characteristics of the biochar. Biochar properties such as surface area, porosity, cation exchange capacity, and surface functional groups are enhanced by these engineering methods and a highly efficient biochar is obtained [6]. A few recently developed novel engineering techniques includes ultrasonication, ball milling, magnetic particle impregnation, composite fabrication, and microwave and plasma treatment (Table 3).

The energy of ultrasound helps unclog the pores, breaks the carbon structure of biochar, and increases microporous area. Amin and Chetpattananondh [44] reported the use of ultrasonic engineering for *Chlorella* sp. biochar for heavy metal remediation. The study reported that the maximum adsorption capacities of Cr(VI), Zn(II), and Ni(II) for untreated and ultrasonic modified biochar were 15.94, 17.62, and 24.76 mg g^{-1} and 18.86, 21.31, and 27.45 mg g^{-1}, respectively. About 1.1–1.3 times greater removal efficiencies of heavy metals were observed compared to untreated biochar.

Ball milling method combines two materials by means of some mechanical force. This engineering method increases the activity of materials by breaking through the choke point of the oxide layer and reducing the size of particles. This process enhances electrostatic interaction, cation–π interactions and surface complexation. This leads to the increment in the adsorption capacity compared to unmilled biochar. Tang et al. [45] used pyrite and biochar composite by simply ball milling the two components and applying in aqueous solution for the removal of Cr(VI). The dominant mechanism by which ball milling remediates Cr(VI) incudes adsorption, reduction, and surface complexation. The study reported that reduction/precipitation was responsible for 92.25% of Cr(VI) remediation while the rest 8.75% was removed by adsorption/surface complexation.

Magnetic biochar has been successfully used as a catalyst in the catalytic degradation of organic pollutants in contaminated soil and water. He et al. [46] reported the use of corn straw biochar treated with $FeCl_3$ solution and agitated and sonicated for 2 h. The treated biomass was then slow pyrolysis at 600 °C for 1 h. The experimental result showed that the engineered biochar exhibited excellent As(V) adsorption efficiency of 6.80 mg g^{-1} compared to 0.017 mg g^{-1} for unmodified biochar. Magnetizing the biochar improved the surface area, increased functional groups and thermal stability.

The fabrication of composite biochar helps prevent the aggregation of nanoparticles, improves the retention time of magnetic properties, and enhances metal precipitation. These properties play vital role in soil remediation, especially in the removal of heavy metals. Rahman et al. [47] used sewage treatment biomass for the preparation of Zr–Fe biochar composites. Zr–Fe composite biochar removed As(V) from solution, with a maximum adoption capacity of Zr-biochar and Zr–Fe biochar composites of 33.1 and 62.5 mg g^{-1} (pH 6), respectively, while only 15.2 mg g^{-1} for pristine biochar. The biochar composites also showed excellent reusability and stability over four cycles.

Microwave being an electromagnetic wave acts as a source for thermal energy for the activation of biochar. This enhances the surface properties, has more yield, abundant functional groups and high surface area due to abundance of pores. According to Foong et al. [48] in microwave treatment for modified biochar, heat is transferred more efficiently as the heat is generated within the core of biomass by microwave interaction. The reaction mechanisms of microwave heating are ionic conduction, dipole polarization, and interfacial polarization.

Plasma treatment is a technique which is used to modify the surface properties of a material. When a gas is exposed to a high energy source, it is often gets ionized to form molecular or ionic fragments. This engineering method activates the surface and improves the adhesion properties of the biochar. The application of plasma has gained popularity in the field of environmental remediation especially in the removal of mercury. Zhang et al. [49] studied the impact of Cl_2 plasma modification on tobacco stem biochar for the remediation of mercury. In this modification method, C–Cl and carboxyl groups are responsible for adsorption of mercury in the form of mercurous chloride and oxide, while the unmodified biochar physically adsorbs atomic mercury. C–Cl is the main functional group involved in mercury removal. It was reported that modified tobacco biochar had Hg removal efficacy of 99.1% at the beginning of the experiment and 79.1% after 100 min of reaction time.

Table 3: Engineering methods used for biochar modification and their effects on the biochar properties.

Engineering method	Effect on biochar	References
Ultrasonication	Increases the micropores present in the biochar and increases surface area.	[44]
Ball milling	Increased surface functional groups, improved zeta potential compared to pristine biochar	[45]
Magnetic particle impregnation	Improved surface area and ferromagnetic properties of biochar	[46]
Composite fabrication	Pore volume and percentage of elements such as O, S, Mg, and Ca increase. Carbon content decreases.	[47]
Microwave treatment	Increase in surface area, pore volume, and oxygen containing functional groups on the surface	[48]
Plasma treatment	Active surface functional groups containing oxygen surface increases.	[49]

4 Conclusions

The concept of waste valorization for the production of biochar form invasive weeds and industrial biomass wastes has gained popularity recently. Biochar is an efficient tool for various applications such as in environmental remediation, restoration, contamination remediation, and treatment. However, a simple pyrolysis method often leads to a lower yield and inferior biochar properties. Modification of biochar using various technology results in alteration of surface properties such as surface area, surface charge, functional groups, and pore distribution and volume. The mechanism involved has been discussed widely and most of the studies are mostly limited to the laboratory scale. There is a limited knowledge regarding the large scale production of engineered biochar and their field application. Therefore, extensive field-based study especially in the mine-degraded land is essential to promote large-scale application of modified biochar to remediate degraded land and promote vegetation growth.

References

[1] Maiti SK, Ecorestoration of Coal Mine Degraded Lands, Springer, New Delhi, 2013.

[2] Ussiri DAN, Jacinthe PA, Lal R, Methods for determination of coal carbon in reclaimed minesoils: a review, Geoderma, 2014, 214–215(2014), 155–167, doi: http://dx.doi.org/10.1016/j.geoderma.2013.09.015.

[3] Ghosh D, Maiti SK, Can biochar reclaim coal mine spoil?, J. Environ. Manage., 2020, 272, 111097, doi: https://doi.org/10.1016/j.jenvman.2020.111097.

[4] Mohapatra S, Kumar M, Karim AA, Dhal NK, Biochars Evaluation for Chromium Pollution Abatement in Chromite Mine Wastewater and Overburden of Sukinda, Odisha, India, 2020.

[5] Omondi MO, Xia X, Nahayo A, Liu X, Korai PK, Pan G, Quantification of biochar effects on soil hydrological properties using meta-analysis of literature data, Geoderma, 2016, 274, 28–34, doi: http://dx.doi.org/10.1016/j.geoderma.2016.03.029.

[6] Akhil D, Lakshmi D, Kartik A, Viet D, Jayaseelan NV, Production, characterization, activation and environmental applications of engineered biochar: a review, In: Environmental Chemistry Letters, Springer International Publishing, 2021, doi: https://doi.org/10.1007/s10311-020-01167-7.

[7] Ghosh D, Masto RE, Maiti SK Ameliorative effect of *Lantan*a camara biochar on coal mine spoil and growth of maize (*Zea mays*). Soil Use Manag. 2020 March, 1–14.

[8] Xie T, Reddy KR, Wang C, Yargicoglu E, Spokas K, Characteristics and applications of biochar for environmental remediation: a review, Crit. Rev. Environ. Sci. Technol., 2015, 45(9), 939–969.

[9] Kwak J, Islam S, Wang S, Ashagre S, Naeth MA, El-din MG, et al., Biochar properties and lead (II) adsorption capacity depend on feedstock type, pyrolysis temperature, and steam activation, Chemosphere, 2019, 231, 393–404, doi: https://doi.org/10.1016/j.chemosphere.2019.05.128.

[10] Rajapaksha UA, Vithanage M, Ahmad M, Seo D, Cho J, Lee S, et al., Enhanced sulfamethazine removal by steam-activated invasive plant-derived biochar, J. Hazard. Mater., 2015, 290, 43–50, doi: http://dx.doi.org/10.1016/j.jhazmat.2015.02.046.
[11] Shao J, Zhang J, Zhang X, Feng Y, Zhang H, Zhang S, Enhance SO_2 adsorption performance of biochar modified by CO_2 activation and amine impregnation, Fuel, 2018, 224, 138–146, doi: https://doi.org/10.1016/j.fuel.2018.03.064.
[12] Sun C, Chen T, Huang Q, Zhan M, Li X, Yan J, Activation of persulfate by CO_2 -activated biochar for improved phenolic pollutant degradation: performance and mechanism, Chem. Eng. J., 2020, 380, 122519, doi: https://doi.org/10.1016/j.cej.2019.122519.
[13] Samaraweera H, Pittman CU, Venkatesh R, Thirumalai KG, Barbary E, Perez F, et al., Characterization of graphene / pine wood biochar hybrids: potential to remove aqueous Cu^{2+}, Environ. Res., 2021, 192, 110283, doi: https://doi.org/10.1016/j.envres.2020.110283.
[14] Inyang M, Gao B, Zimmerman A, Zhang M, Chen H, Synthesis, characterization, and dye sorption ability of carbon nanotube – biochar nanocomposites, Chem. Eng. J., 2014, 236, 39–46, doi: http://dx.doi.org/10.1016/j.cej.2013.09.074.
[15] Inyang M, Gao B, Zimmerman A, Sorption and Cosorption of Lead and Sulfapyridine on Carbon Nanotube-Modified Biochars, 2015, 1868–1876.
[16] Tao Q, Chen Y, Zhao J, Li B, Li Y, Tao S, et al., Enhanced Cd removal from aqueous solution by biologically modified biochar derived from digestion residue of corn straw silage, Sci. Total Environ., 2019, 674, 213–222, doi: https://doi.org/10.1016/j.scitotenv.2019.03.438.
[17] Yao Y, Gao B, Wu F, Zhang C, Yang L, Engineered biochar from biofuel residue: characterization and its silver removal potential, ACS Appl. Mater. Interfaces, 2015, 7(19), 10634–10640.
[18] Rajapaksha AU, Chen SS, Tsang DCW, Zhang M, Vithanage M, et al., Engineered / designer biochar for contaminant removal / immobilization from soil and water: potential and implication of biochar modification, Chemosphere, 2016, 148, 276–291, doi: http://dx.doi.org/10.1016/j.chemosphere.2016.01.043.
[19] Rajapaksha AU, Ok YS, El-Naggar A, Kim H, Song F, Kang S, et al., Dissolved organic matter characterization of biochars produced from different feedstock materials, J. Environ. Manage., 2019, 233, 393–399, doi: https://doi.org/10.1016/j.jenvman.2018.12.069.
[20] Shim T, Yoo J, Ryu C, Park Y, Jung J, Effect of steam activation of biochar produced from a giant Miscanthus on copper sorption and toxicity, Bioresour. Technol., 2015, 197, 85–90, doi: http://dx.doi.org/10.1016/j.biortech.2015.08.055.
[21] Franciski MA, Peres EC, Godinho M, Perondi D, Foletto EL, Collazzo GC, et al., Development of CO_2 activated biochar from solid wastes of a beer industry and its application for methylene blue adsorption, Waste Manag., 2018, 78, 630–638, doi: https://doi.org/10.1016/j.wasman.2018.06.040.
[22] Biswas S, Siddiqi H, Preparation and Characterization of Raw and Inorganic Acid-Activated Pine Cone Biochar and Its Application in the Removal of Aqueous-Phase Pb^{2+} Metal Ions by Adsorption, Water Air Soil Pollut., 2020, 231(1), 1–17.
[23] Peng H, Gao P, Chu G, Pan B, Peng J, Xing B, Enhanced adsorption of Cu (II) and Cd (II) by phosphoric acid-modified, Environ. Pollut., 2017, 229, 846–853, doi: http://dx.doi.org/10.1016/j.envpol.2017.07.004.
[24] Zeng H, Zeng H, Zhang H, Shahab A, Ef fi cient adsorption of Cr (VI) from aqueous environments by phosphoric acid activated eucalyptus biochar, J. Clean. Prod., 2021, 286, 124964, doi: https://doi.org/10.1016/j.jclepro.2020.124964.
[25] Sahin O, Taskin M, Kaya E, Atakol O, Emir E, Inal A, et al., Effect of acid modification of biochar on nutrient availability and maize growth in a calcareous soil, Soil Use Manag., 2011, 2017, 1–10.

[26] Liu C, Wang W, Wu R, Liu Y, Lin X, Kan H, et al., Preparation of Acid- and Alkali-Modi fi ed Biochar for Removal of Methylene Blue Pigment, ACS omega, 2020, 5(48), 30906–30922.

[27] An Q, Miao Y, Zhao B, Li Z, Zhu S, An alkali modified biochar for enhancing Mn^{2+} adsorption: performance and chemical mechanism, Mater. Chem. Phys., 2020, 248, 122895, doi: https://doi.org/10.1016/j.matchemphys.2020.122895.

[28] Shin J, Lee Y, Lee S, Kim S, Ochir D, Park Y, et al., Single and competitive adsorptions of micropollutants using pristine and alkali-modified biochars from spent coffee grounds, J. Hazard Mater., 2020, 400, 123102, doi: https://doi.org/10.1016/j.jhazmat.2020.123102.

[29] Li Y, Wei Y, Huang S, Liu X, Jin Z, Zhang M, et al., Biosorption of Cr (VI) onto *Auricularia auricula* dreg biochar modi fi ed by cationic surfactant: characteristics and mechanism, J. Mol. Liq., 2018, 269, 824–832, doi: https://doi.org/10.1016/j.molliq.2018.08.060.

[30] Kuang Y, Zhang X, Zhou S, Adsorption of methylene blue in water onto activated carbon by surfactant modification, Water, 2020, 12(2), 587.

[31] Zhu D, Shao J, Li Z, Yang H, Zhang S, Chen H, Nano nickel embedded in N-doped CNTs-supported porous biochar for adsorption-reduction of hexavalent chromium, J. Hazard. Mater., 2021, 416, 125693, doi: https://doi.org/10.1016/j.jhazmat.2021.125693.

[32] Chakraborty V, Das P Graphene oxide – coated pyrolysed biochar from waste sawdust and its application for treatment of cadmium-containing solution: batch, fixed-bed column, regeneration, and mathematical modelling. 2021.

[33] Tang J, Lv H, Gong Y, Huang Y, Preparation and characterization of a novel graphene/biochar composite for aqueous phenanthrene and mercury removal, Bioresour. Technol., 2015, 196, 355–363, doi: http://dx.doi.org/10.1016/j.biortech.2015.07.047.

[34] Wang B, Gao B, Fang J, Technology Recent advances in engineered biochar productions and applications, Crit. Rev. Environ. Sci. Technol., 2018, 0(0), 1–50, doi: https://doi.org/10.1080/10643389.2017.1418580.

[35] Yao Y, Gao B, Inyang M, Zimmerman AR, Cao X, Pullammanappallil P, et al., Biochar derived from anaerobically digested sugar beet tailings: characterization and phosphate removal potential, Bioresour. Technol., 2011, 102(10), 6273–6278, doi: http://dx.doi.org/10.1016/j.biortech.2011.03.006.

[36] Inyang M, Gao B, Yao Y, Xue Y, Zimmerman AR, Removal of heavy metals from aqueous solution by biochars derived from anaerobically digested biomass, Bioresour. Technol., 2012, 110, 50–56, doi: http://dx.doi.org/10.1016/j.biortech.2012.01.072.

[37] Yao Y, Zhang Y, Gao B, Chen R, Wu F, Removal of sulfamethoxazole (SMX) and sulfapyridine (SPY) from aqueous solutions by biochars derived from anaerobically digested bagasse, Environ. Sci. Pollut. Res., 2018, 25(26), 25659–25667.

[38] Sun L, Wan S, Luo W, Biochars prepared from anaerobic digestion residue, palm bark, and eucalyptus for adsorption of cationic methylene blue dye: characterization, equilibrium, and kinetic studies, Bioresour. Technol., 2013, 140, 406–413, doi: http://dx.doi.org/10.1016/j.biortech.2013.04.116.

[39] Inyang M, Gao B, Ding W, Pullammanappallil P, Andrew R, Inyang M, et al., Enhanced lead sorption by biochar derived from anaerobically digested sugarcane bagasse enhanced lead sorption by biochar derived from anaerobically digested sugarcane bagasse, Sep. Sci. Technol., 2011, 37–41.

[40] Bao D, Li Z, Tang R, Wan C, Zhang C, Tan X, Metal-modified sludge-based biochar enhance catalytic capacity: characteristics and mechanism, J. Environ. Manage., 2005, 2021(284), 112113, doi: https://doi.org/10.1016/j.jenvman.2021.112113.

[41] Inyang M, Gao B, Pullammanappallil P, Ding W, Zimmerman AR, Biochar from anaerobically digested sugarcane bagasse, Bioresour. Technol., 2010, 101(22), 8868–8872, doi: http://dx.doi.org/10.1016/j.biortech.2010.06.088.

[42] Amin M, Chetpattananondh P, Biochar from extracted marine *Chlorella* sp. residue for high efficiency adsorption with ultrasonication to remove Cr (VI), Zn (II) and Ni (II), Bioresour. Technol., 2019, 289, 121578, doi: https://doi.org/10.1016/j.biortech.2019.121578.
[43] Tang J, Zhao B, Lyu H, Li D, Development of a novel pyrite / biochar composite (BM-FeS 2 @ BC) by ball milling for aqueous Cr (VI) removal and its mechanisms, J. Hazard. Mater., 2021, 413, 125415, doi: https://doi.org/10.1016/j.jhazmat.2021.125415.
[44] He R, Peng Z, Lyu H, Huang H, Nan Q, Tang J, Synthesis and characterization of an iron-impregnated biochar for aqueous arsenic removal, Sci. Total Environ., 2018, 612, 1177–1186, doi: https://doi.org/10.1016/j.scitotenv.2017.09.016.
[45] Rahman A, Lamb D, Mahmudur M, Bahar M, Sanderson P, Abbasi S, et al., Removal of arsenate from contaminated waters by novel zirconium and zirconium-iron modified biochar, J. Hazard. Mater., 2021, 409, 124488, doi: https://doi.org/10.1016/j.jhazmat.2020.124488.
[46] Foong SY, Keey R, Yang Y, Wang Y, Nai P, Yek Y, et al., Valorization of biomass waste to engineered activated biochar by microwave pyrolysis: progress, challenges, and future directions, Chem. Eng. J., 2020, 389, 124401, doi: https://doi.org/10.1016/j.cej.2020.124401.
[47] Zhang H, Wang T, Sui Z, Zhang Y, Norris P, Plasma induced addition of active functional groups to biochar for elemental mercury removal, Plasma Chem. Plasma Process, 2019, 39(6), 1449–1468, doi: https://doi.org/10.1007/s11090-019-10019-4.

Swapnila Roy

Application of biochar in wastewater treatment

Abstract: Biochar is a low-cost, carbonaceous substance, having high capability to decontaminate wastewater. The application of biochar is very interesting due to a simple preparation method with various physicochemical properties. The removal efficiency of chemical contaminants from wastewater depends on nature of surface, particle radius, and functional groups; physical properties of biochar depend on the preparation method. Usually the pyrolysis method at high temperature yielding biochar having higher surface area is very effective in removal of organic pollutants, but the pyrolysis method at low temperature yielding biochar having lower surface area is very effective in removal of inorganic pollutants. Biochar has been exclusively used for the decontamination of heavy metals and microorganisms. It is also used as a catalyst in the dye degradation process. This review chapter highlights the overview of applications of biochar in wastewater treatment.

Keywords: biochar, thermal analysis, modification, adsorption, wastewater treatment

1 Introduction

The water resources within the world square measured being deteriorated because of the continual discharge of an outsized variety of chemical contaminants like textile dyes, toxic metals [1], prescribed drugs, fertilizers, and chemical industries and municipal waste. A large part of these contaminants square measure extremely toxic in nature and chemically convertible in other form. The heavy release of those toxic substances [2] may be a serious issue which can affect negatively on living system. Many typical technologies have been utilized for the remediation of wastewater [3] together such as adsorption, membrane filtration, reverse diffusion, chemical precipitation, ion-exchange, solvent extraction, and floatation for the removal of inorganic pollutants. However, these technologies suffer from a spread of disadvantages stretching from inefficiency to get rid of pollutants at low concentration to fully converting pollutants into various products which require high energy and hazardous chemicals, complicated methods, expensive maintenance and operation, and so on. An associate degree and treatability process may meet each socio-economical need to be surveyed and effective application. The easily available cheap materials embedded in

Swapnila Roy, School of Science &Technology, Glocal University, UP-247121, India, e-mail: swapnilaroy@gmail.com

https://doi.org/10.1515/9783110734003-015

treatability method may reduce the cost and efficiency of method [4]. Biochar is considered to be an environment-friendly substance which is generated from different organic wastes such as agricultural and biological residues [5] which have extensive applications in environment. Organic wastes can be regenerated into carbon by various methods together with shift whereas chars from gasification, torrefaction, and so on. Owning to its increased properties like wealthy carbon content, increased area, high cation/anion exchange capability, and stable structure, biochar and its activated derivatives were reported as potential decontaminant for remediation like heavy metals and pathogens and different organic contaminants [6]. Research studies revealed that from continuous batch experiments, biochar can be exclusively applied for treatment of wastewater but there is lack of application of statistical optimization method. The capacity of biochar for remediation mainly depends on removal properties of chemical pollutants which relate to their various physical and chemical properties such as composition, surface area, pore size, surface functional groups, and ion exchange capacity; all these properties differ with the procedure and method of operation [7]. Since biochar exhibited exclusive application for the treatment of wastewater, the analysis of worse impact should also be clarified. This review chapter highlights procedures, chemical modifications, applications, and uses of biochar from various thermal conversion processes in order to focus on remediation of chemical and biological contaminants [8]. This chapter represents the recent advancement in the application of biochar for treatment of wastewater as filtration media, catalytic support, and its action during anaerobic digestion of wastewater.

2 Conventional thermal methods

2.1 Pyrolysis

Pyrolysis is the significant heating process applied for transforming organic wastes into carbon-rich substances utilized for wastewater treatment. There are many factors such as carbonization temperature and time, heating rate, and nitrogen flow rate, which control the process of pyrolysis [9, 10], and we can predict the characteristics and application of carbon-rich compounds: solid part as biochar, condensable liquid as bio-oil, and various noncondensable gases such as CO, CO_2, CH_4, and H_2.

Below 500 °C biochar has predominant properties such as small pore size, low surface area, and presence of large number of oxygen-containing functional groups, which is applicable for the remediation of inorganic pollutants. Increasing pyrolysis temperature drives the rise of pH of biochar as a result of higher amount of ash content. As a consequence of thermal process in medium temperatures with higher heating time in anaerobic condition, high biochar is yielded (30%) followed by the cracking reactions which reduce condensable liquid products.

2.2 Hydrothermal carbonization (HTC)

HTC is the important thermochemical conversion [11] process used for the production of high energy density materials. During this thermochemical conversion, there is a decrease in the pH which is highlighted because of various organic acids such as acetic, levulinic, formic, and lactic acids, which accelerate the catalytic reaction of organic compounds without using acid. This carbonization process follows various mechanisms in different phases such as solid phase, aqueous phase, and a gaseous phase (CO_2). The physicochemical properties of pyrolyzed biochar are dependent on reaction temperature, heating time, pressure, and ratio of water/biomass. Different hydrochar materials such as pyrolyzed switchgrass, swine solids, cassava slag, banana peels, and phloroglucinol have been investigated as significant adsorbents for the remediation of different chemical contaminants.

2.3 Gasification

Gasification is a high heating chemical process conducted by the transformation [12] of biomass or different organic substances into a mixture of gases which contains H_2, CO, CO_2, and negligible amount of hydrocarbons such as methane, a solid char, and a liquid phase tar. In the gasification process, tar and char are unexpected products. The highly investigated study is the preparation of biochar which has been highlighted exclusively, but on the other hand research studies did not reveal that much about the properties of biomass gasification. Nevertheless, a sound knowledge of the char properties and its correlation with the gasification technology are significant from the perspective of wastewater treatment.

2.4 Torrefaction

Torrefaction is a conventional thermal pretreatment [13] of biomass which is performed to improve the physicochemical and thermochemical properties of biomass; it enables the energy densification and the homogenization of biomass. The torrefaction is performed generally at slow heating rates under atmospheric pressure at temperatures ranging from 200 °C to 300 °C and without or with limited oxygen supplier. Torrefaction is dominantly used for biofuel production, while pyrolysis is the dominant process to prepare biochar for water and wastewater treatment. Compared to biochar, torrefaction-char may contain higher amount of oxygen-containing functional groups, attributed to the much lower temperatures used during the thermal process. At the lower reaction temperature, simple preparation method, and high percentage of yield, the char resulted from torrefaction is considered to be as potential oxygen-rich-char adsorbent. More attention should be paid to its application of decontamination.

Table 1: Thermal conversion process and products [14–17].

Method	Temperature	Heating rate	By-products (%)		
			Biochar	Bio-oil	Syngas
Fast pyrolysis	400–1,000 °C	Very fast	12	75	13
Slow pyrolysis	350–980 °C	Slow (<10 °C/min)	27–37	19–29	25–41
hydrothermal carbonization (HTC)	180–300 °C	Slow	50–80	5–20	2–5
Gasification	700–1,500 °C	Moderate–very fast	10	5	85
Torrefication	200–300 °C	Slow (<10 °C/min)	80	0	20

3 Chemical modification of biochar

Due to the presence of functional groups, small surface area, and pore volume shown by biochar, the results of dry pyrolysis exhibited that during carbonization these materials are plugged with black tar material; after chemical or physical treatment, few properties are improved for its different environmental applications. As a consequence of physical process, biochar is heated to at high temperature under conditional atmosphere. There are remarkable changes in physical characteristics such as changes in surface area, pores volume, and surface chemical properties containing surface functional groups, hydrophobicity, and polarity. The most exclusive oxidizing agent used for biochar modification is water in vapor phase. The following chemical reactions are involved during steam activation:

$$C + H_2O = CO + H_2 \tag{1}$$

$$CO + H_2O = CO_2 + H_2 \tag{2}$$

The reaction between steam and carbon results in the decomposition of tar, inducing the development of new micropores. This tendency is enlightened with the rising of temperature. At high temperatures, there is remarkable decrease in mass yield which is not economically feasible. Biochar is incorporated with different chemical reagents such as acids, bases, and oxidants (hydrogen peroxide and potassium permanganate) during chemical activation. The acid modification is subjected to mainly introduce acid functional groups on the surface of biochar. Research studies revealed that [18] reed-derived biochar is modified by hydrochloric acid, which results in enrichment of hydrophobic adsorption sites for the adsorption of pentachlorophenol. So the acid modification results in the modification of the surface area and the pore structure of biochar which depends on the type and concentration of the acid [19]. As a consequence of alkaline modification, the surface area and the oxygen-containing functional groups on the surface are enhanced. Some research studies revealed that [20] activated biochar can be synthesized from a mixture of olive solid wastes and olive mill wastewater using KOH as an activating agent, where high surface area and micropore volume showed high adsorption capacity. Few research studies revealed that [21] municipal solid wastes – derived biochar – modified using KOH exhibited an enhancement in the surface area, as a consequence there is an increase in As(V) removal efficiency.

4 Biochar as an efficient adsorbent

4.1 Organic pollutants

Different biochar synthesized products have been efficiently applied as adsorbents for organic contaminants in water and wastewater. This part is focused mainly in aqueous solutions of phenols and dyes. Many biochars at higher temperatures were extensively used for methylene blue and Congo red [22] decontamination. The observed results highlighted that methylene blue, denoted as cationic dye, was mostly adsorbed than Congo red, denoted as anionic dye. Usually, higher adsorption capacity was observed for biochars synthesized at higher temperatures due to the significantly increased surface area. The comparative study of effectiveness of straw-based biochar with commercial activated carbon for dye removal was estimated [23]. Research studies revealed that carbon-based materials exhibited high adsorption efficiency toward two common dyes such as reactive brilliant blue and rhodamine B at pH values of 3.0 and 6.5 due to enhanced π–π interactions by the surface functional groups.

4.2 Heavy metals

Research studies revealed that biochar is the potential adsorbent for the removal of inorganic toxic metals from wastewater. It has been exclusively reported that the adsorption capacity of heavy metals by biochar mostly depends on physical properties and decontaminants. Biochar is considered as an effective adsorbent for the adsorption of heavy metals because of the presence of phenolic, hydroxyl, and carboxyl groups (functional groups), and its porosity and large surface area. Another research study showed that [24] biochar prepared from rice-straw was investigated toward Cd^{2+} adsorption.

5 Adsorption mechanism

The heterogeneity of biochar surface is driven by different adsorption mechanisms. The adsorption mechanism depends on the physicochemical characteristics of the adsorbent surface [25]. According to the research study [26], the main part of adsorption could be considered as physical adsorption, where the adsorbate is dispersed on the surface of the adsorbent, the precipitated part generates cover on the adsorbent surface, and the porosity is characterized by the reaction of the adsorbate into the adsorbent. At high temperatures, synthesized biochar contains functional groups which are less polar and more aromatic in nature so that those are less suitable for the removal of polar organic contaminants. In many cases, hydrogen bonding is catalyzed as a result of the electrostatic force of repulsion between anionic organic compounds and biochar. The penetration of nonpolar contaminant into hydrophobic sites is facilitated in the presence of hydrogen bonding.

6 Parameters affecting the adsorption of pollutants

6.1 Biochar characteristics

The adsorption efficiency of an adsorbent is driven by its micropore volume. Usually, based on the size, pores are divided into micropores, mesopores, and macropores, and it mainly depends on the biochar production temperature. Micropores are observed as the most abundant within the biochar structure, and due to its surface area, it has high adsorptive capacity. Hydrophobicity of biochars are considered as the main contributor are insoluble adsorbates while hydrophilicity of biochar are less effective due to the presence of oxygen-containing functional groups on biochar surface allows the penetration of water via hydrogen bonding. Finally, there is water-adsorbate competition on the surface sites.

6.2 Solution pH

The solution pH is a significant part which drives the adsorption process by affecting surface charge of the adsorbent and the degree of ionization of the adsorbate [27]. The competition between various cations in solution decreases in the higher pH greater than the point of zero charge. In that case, the deprotonation of phenolic and carboxylic groups on its surface results in the anionic presence on the surface.

6.3 Action of adsorbent

The dose of adsorbent also has a remarkable effect on the adsorbent–adsorbate equilibrium of a system. The removal efficiency of organic and inorganic pollutants increases by increasing the dose of adsorbent as well as more adsorption sites.

6.4 Temperature

The adsorption efficiency of biochar has been observed to be affected by the temperature of the system. Many research studies revealed that the adsorption capacity increased with the increase in temperature. It is inferred an endothermic nature of the adsorption process. Some studies showed that [28] the adsorption of due onto potassium hydroxide (KOH)-activated biochar increases with temperature rising because of the endothermic adsorption process. As a consequence of increasing temperature, there is the increase in the mobility nature of dye as well as increase in the porosity of the adsorbent.

7 Environmental application of biochar for wastewater treatment

7.1 Role as support media in anaerobic digestion

Anaerobic digestion is an important treatment process which is reported as an eco-friendly method for waste management due to its capacity for biological remediation.

Conducting the anaerobic digestion method in stable conditions with efficient decontamination is a significant part for researchers as some ingredients of the metabolic products can probably prevent microorganisms for their activity. According to the different proposals so far accepted for the barriers of anaerobic reactors such as the acclimation of bacterial cells, the adoption of thermophilic operating conditions and the reduction of the concentration of inhibitors by dilution. The application of packing materials in this digestion has been observed as an effective approach for decreasing the bioavailability of inhibitors toward large bacterial mass within the digester. Biochar is considered as a bio-stable and an abundant biomass, which is an important substitute to adsorbents. Its contribution in this digestion process is enhancing day by day.

7.2 Role as filtration support media

Biochar can be applied in different water treatment processes. For example, it can be embedded from a biochar column to bio-sand filter systems for wastewater treatment. The application of biochar filtration systems has immense potential for the remediation of particulate matter, heavy metals, and also reduction of pathogen loads. Research studies showed that total suspended solid and organic matter decontamination, biochar filter could be performed by various steps [29] which will be started with sedimentation of suspended solids in water layer and the filtration of larger particles on the filter surface. In that case, smaller particles are easily removed by adsorption in filter zones, whereas organic matter is hydrolyzed under anaerobic conditions. The removal efficiency of the biochar filter increases with increasing attachment sites on surface. This will help in the establishment of biofilm and the bacterial deposition within the filter column.

7.3 Catalytic role in heterogeneous oxidation

Biochar-based catalysts have been mostly applied in various systems for the removal of pollutants in wastewater in catalytic ozonation processes, such as Fenton-like reactions and photocatalytic systems. Some research studies revealed that [30] the catalytic efficiency of biochar synthesized from biomass have large pores containing hydroxyl and phenolic surface functional groups. It is shown that higher catalytic efficiency of biochar at a pH 10.0 which has reported a 58% catalytic efficiency and a 71% dye solution mineralization after 1 h reaction time.

According to the literature survey, it is revealed that the remarkable capability of activity of carbon materials toward different oxidants create reactive oxygen species for the decontamination of organic pollutants.

8 Future perspectives of biochar

Biochar has immense potential to reveal various environmental issues regarding waste water treatment. Owing to high adsorbing capacity for contaminants, biochar is used as efficient adsorbent for the different adsorption mechanisms [31]. The most effective application of biochar for wastewater treatment depends on the presence of functional groups and surface adsorption, on the availability of raw materials, and on the conditions of thermal conversion process adopted for the biochar synthesis. The chemical modification of biochar modification is the highlighted area which has to be exploited to increase the biochar capacity for decontamination. The correlation between thermal transformation, reaction of biochar for wastewater pollutants can enhance its application for very complex compounds for example biological remediation for wastewater treatment. Overall, for biological treatment, biochar is considered to be embedded in the treatment processes adopted for the removal of pollutants.

Different research studies revealed that different modification process and the various mechanisms following modification agents are identified. The relationship between the conditions of thermal conversion and biochar behavior toward wastewater treatment can facilitate formation of complex compounds which obstruct different reactions followed for wastewater treatment. The stability of biochar is one of the significant factors which are considered in the application of biochar. The stability of biochar mainly depends on its aromatic character and degree of aromatic condensation [32]. In the application of wastewater treatment, the major contributor is carbon derived from biochar, which can increase the amount of carbon in the solution for treatment. Moreover, sludge-derived biochar contains heavy metals which can be also concern for scientists. From research studies, it is [33] proved that biochar-derived organic compounds dissolve into the aqueous solution because of biochar instability. Due to the variable stability of biochar, it is applied as support for catalyst which is generally decreased after few cycles because of the variation in the structure of carbon.

Table 2: Comparative study of different biochars used for wastewater treatment.

Feedstock	Preparation condition	Removal efficiency (%)
Waste wood pellets	Gasification, 520 °C	17–75 (heavy metals) [34]
Forestry wood waste	Pyrolysis, 700 °C	60–93 (total phosphorus) [35]
Rice husk	Gasification	92–99 (*E. coli*) [36]
Miscanthus	Pyrolysis, 850 °C	23 (*E. coli*) [37]

9 Conclusion

This chapter highlighted a summary of different biochar preparation methods and adsorption techniques toward remediation of wastewater treatment. The removal efficiency of biochar in aqueous solutions is correlated to their physical and chemical characteristics dependent on starting materials, the thermal process, and the conditions of synthesis. The chemical modification of biochar through activation methods were observed for the adjustment of various surface functional groups, which can increase adsorption area of biochar. Biochar has been exclusively applied for the applications of wastewater treatment because of its versatile nature which contains the adsorption of toxic and carcinogenic substances from the aqueous solutions. Biochar can play a catalytic role and media support for immobilized microorganisms and adsorbent of unwanted compounds in anaerobic digestion.

References

[1] Bogusz A, Oleszczuk P, Dobrowolski R, Application of laboratory prepared and commercially available biochars to adsorption of cadmium, copper and zinc ions from water, Bioresour. Technol., 2015, 196, 540–549.

[2] Sadat M, Hashemi H, Eslami F, Karimzadeh R, Organic contaminants removal from industrial wastewater by CTAB treated synthetic zeolite Y, J. Environ. Manag., 2019, 233, 785–792.

[3] Fu F, Wang Q, Removal of heavy metal ions from wastewaters: a review, J. Environ. Manag., 2011, 92, 407–418.

[4] Vunain E, Masoamphambe EF, Mpeketula PMG, Monjerezi M, Etale A, Evaluation of coagulating efficiency and water borne pathogens reduction capacity of Moringa oleifera seed powder for treatment of domestic wastewater from Zomba, Malawi, J. Environ. Chem. Eng., 2019, 7, 103–118.

[5] Rizwan M, Ali S, Qayyum MF, Ibrahim M, Ziaurrehman M, Abbas T, Ok YS, Mechanisms of biochar-mediated alleviation of toxicity of trace elements in plants: a critical review, Environ. Sci. Pollut. Res., 2016, 23, 2230–2248.

[6] He J, Li Y, Cai X, Chen K, Zheng H, Wang C, Zhang K, Lin D, Kong L, Liu J, Study on the removal of organic micropollutants from aqueous and ethanol solutions by HAP membranes with tunable hydrophilicity and hydrophobicity, Chemosphere, 2017, 174, 380–389.

[7] Yang W, Wang Z, Song S, Han J, Chen H, Wang X, Sun R, Cheng J, Adsorption of copper(II) and lead(II) from seawater using hydrothermal biochar derived from Enteromorpha., Mar. Pollut. Bull., 2019, 149, 110586.

[8] Park J-H, Wang JJ, Meng Y, Wei Z, DeLaune RD, Seo D-C, Adsorption/desorption behavior of cationic and anionic dyes by biochars prepared at normal and high pyrolysis temperatures, Colloids Surf. A Physicochem. Eng. Asp., 2019, 572, 274–282.

[9] Arni SA, Comparison of slow and fast pyrolysis for converting biomass into fuel, Renew. Energy., 2018, 124, 197–201.

[10] Duku MH, Gu S, BenHagan E, Biochar production potential in Ghana – A review, Renew. Sust. Energ. Rev., 2011, 15, 3539–3551.

[11] Peng P, Lang YH, Wang XM, Adsorption behavior and mechanism of pentachlorophenol on reed biochars: pH effect, pyrolysis temperature, hydrochloric acid treatment and isotherms, Ecol. Eng., 2016, 90, 225–233.

[12] Ahmad M, Lee SS, Dou X, Mohan D, Sung JK, Yang JE, Ok YS, Effects of pyrolysis temperature on soybean stover and peanut shell-derived biochar properties and TCE adsorption in water, Bioresour. Technol., 2012, 118, 536–544.

[13] Zhang CY, Ho SH, Chen WH, Fu YJ, Chang JS, Bi XT, Oxidative torrefaction of biomass nutshells: evaluations of energy efficiency as well as biochar transportation and storage, Appl. Energy., 2019, 235, 428–441.

[14] Arni SA, Comparison of slow and fast pyrolysis for converting biomass into fuel, Renew. Energy., 2018, 124, 197–201.

[15] Duku MH, Gu S, Ben Hagan E, Biochar production potential in Ghana – A review, Renew. Sust. Energ. Rev., 2011, 15, 3539–3551.

[16] Libra JA, Ro KS, Kammann C, Funke A, Berge ND, Neubauer Y, Titirici -M-M, Fühner C, Bens O, Kern J, et al., Hydrothermal carbonization of biomass residuals: a comparative review of the chemistry, processes and applications of wet and dry pyrolysis, Biofuels, 2011, 2, 71–106.

[17] Bridgwater AV, Review of fast pyrolysis of biomass and product upgrading, Biomass Bioenerg., 2012, 38, 68–94.

[18] Kaetzl K, Lübken M, Uzuna G, Gehring T, Nettmann E, Stenchly K, Wichern M, On-farm wastewater treatment using biochar from local agroresidues reduces pathogens from

irrigation water for safer food production in developing countries, Sci. Total Environ., 2019, 682, 601–610.
[19] Zhou Z, Liu Y, Liu S, Liu H, Zeng G, Tan X, Yang C, Ding Y, Yan Z, Cai X, Sorption performance and mechanisms of arsenic(V) removal by magnetic gelatin-modified biochar, Chem. Eng. J., 2017, 314, 223–231.
[20] Enaime G, Ennaciri K, Ounas A, Baçaoui A, Seffen M, Selmi T, Yaacoubi A, Preparation and characterization of activated carbons from olive wastes by physical and chemical activation: application to indigo carmine adsorption, J. Mater. Environ. Sci., 2017, 8, 4125–4137.
[21] Jing XR, Wang YY, Liu WJ, Wang YK, Jiang H, Enhanced adsorption performance of tetracycline in aqueous solutions by methanol-modified biochar, Chem. Eng. J., 2014, 248, 168–174.
[22] Park J-H, Wang JJ, Meng Y, Wei Z, DeLaune RD, Seo D-C, Adsorption/desorption behavior of cationic and anionic dyes by biochars prepared at normal and high pyrolysis temperatures, Colloids Surf. A Physicochem. Eng. Asp., 2019, 572, 274–282.
[23] Qiu Y, Zheng Z, Zhou Z, Sheng GD, Effectiveness and mechanisms of dye adsorption on a strawbased biochar, Bioresour. Technol., 2009, 100, 5348–5351.
[24] Liu L, Fan S, Removal of cadmium in aqueous solution using wheat straw biochar: effect of minerals and mechanism, Environ. Sci. Pollut. Res., 2018, 25, 8688–8700.
[25] Rosales E, Meijide J, Pazos M, Sanroman MA, Challenges and recent advances in biochar as lowcost biosorbent: from batch assays to continuous-flow systems, Bioresour. Technol., 2017, 246, 176–192.
[26] Pignatello JJ, Interactions of Anthropogenic Organic Chemicals with Natural Organic Matter and Black Carbon in Environmental Particles. Biophysico-Chemical Processes of Anthropogenic Organic Compounds in Environmental Systems, John Wiley Sons, Inc., Hoboken, NJ, USA, 2011, 1–50.
[27] Kılıc M, Mutlu ÇKÖÇ, Pütün AE, Adsorption of heavy metal ions from aqueous solutions by bio-char, a by-product of pyrolysis, Appl. Surf. Sci., 2013, 283, 856–862.
[28] Enaime G, Ennaciri K, Ounas A, Baçaoui A, Seffen M, Selmi T, Yaacoubi A, Preparation and characterization of activated carbons from olive wastes by physical and chemical activation: application to indigo carmine adsorption, J. Mater. Environ. Sci., 2017, 8, 4125–4137.
[29] Kaetzl K, Lübken M, Uzuna G, Gehring T, Nettmann E, Stenchly K, Wichern M, On-farm wastewater treatment using biochar from local agroresidues reduces pathogens from irrigation water for safer food production in developing countries, Sci. Total Environ., 2019, 682, 601–610.
[30] Moussavi G, Khosravi R, Preparation and characterization of a biochar from pistachio hull biomass and its catalytic potential for ozonation of water recalcitrant contaminants, Bioresour. Technol., 2012, 119, 66–71.
[31] Varjani S, Joshi R, Srivastava VK, Ngo HH, Guo W, Treatment of wastewater from petroleum industry: current practices and perspectives, Environ. Sci. Pollut. Res., 2019.
[32] Wiedemeier DB, Abiven S, Hockaday WC, Keiluweit M, Kleber M, Masiello CA, McBeath AV, Nico PS, Pyle LA, Schneider MPW, et al., Aromaticity and degree of aromatic condensation of char, Org. Geochem., 2015, 78, 135–143.
[33] Huang M, Li Z, Luo N, Yang RY, Wen J, Huang B, Zeng G, Application potential of biochar in environment: insight from degradation of biocharderived dom and complexation of dom with heavy metals, Sci. Total Environ., 2019, 646, 220–228.
[34] Reddy KR, Xie T, Dastgheibi S, Evaluation of biochar as a potential filter media for the removal of mixed contaminants from urban storm water runoff, J. Environ. Eng., 2014.
[35] Kaetzl K, Lübken M, Nettmann E, Krimmler S, Wichern M, Slow sand filtration of raw wastewater using biochar as an alternative filtration media, Sci. Rep., 2020, 10, 1229.

[36] Lau AYT, Tsang DCW, Graham N, Ok Y, Yang X, Li X, Surface-modified biochar in a bioretention system for Escherichia coli removal from stormwater, Chemosphere, 2017, 169, 89–98.
[37] Kaetzl K, Lübken M, Uzuna G, Gehring T, Nettmann E, Stenchly K, Wichern M, On-farm wastewater treatment using biochar from local agroresidues reduces pathogens from irrigation water for safer food production in developing countries, Sci. Total Environ., 2019, 682, 601–610.

Dharm Pal, Vandana Gupta

Biochar as an adsorbent for removal of pharmaceutical compounds from wastewater

Abstract: Pharmaceutical wastewater comprises several organic compounds including therapeutic and cosmatologic drugs, diagnostic agents, nutraceuticals, and surfactants. Due to severe adverse effects the removal of such contaminants from wastewaters is essential. The efficient and cost effective way to remove pharmaceutical compound from aqueous media is the use of alternative adsorbent material such as biochar. Biochar can be produced from agrowaste biomass generated from different anthropogenic activities.

Recently, biochar is gaining more demand due to its inherent properties such as high carbon sequestration, soil property enhancer, energy production, and environmental remediation. Biochar is a stable carbon-rich material and act as an efficient adsorbent for sorption of organic contaminants owing to its hydrophobic nature with higher surface area and volume of micropores resulted in high-temperature pyrolysis. Moreover, the oxygen containing group on the surface of biochar exhibit strong binding ability through heterogeneous surface chemistry. However, along with surface functional groups, and the size of the pollutants; removal efficiency also depends on surface area, pore size distribution of the target molecules. Further, feed nature and method of preparation govern the surface morphology and properties of biochar.

This chapter is focused on production and characterization of biochar, mechanism, equilibrium, and kinetics of adsorption with systematic overview of pharmaceutical contaminants removal from wastewater using biochar.

Keywords: adsorptive removal, biochar, pyrolysis, pharmaceutical contaminants, waste water treatment

1 Introduction

Wastewater is usually defined as the waste or contaminants carried with water that is discharged from institutions, laboratories, domestic, agricultural, commercial, and industrial establishment. The presence of contaminants in water can create nuisance to environments. Pathogenic microorganism, organic and toxic compound in untreated wastewater can dwell in human intestinal tract, affect the aquatic life, environment, and vegetation as well. It can also be carcinogenic or mutagenic. New research, developments, urbanization, advancement, and technologies are numerous causes of wastewater generation.

https://doi.org/10.1515/9783110734003-016

Nowadays, innovations in pharmaceutical industries are in boom and necessary also. Recent advances, research, and innovation for prevention, control, and detection of pharmaceuticals are essential to facilitate and help human and living creature. Use of pharmaceuticals compounds has become indispensable component of a healthy life and society. It also deals with the production of large spectrum of chemical compounds and used as medicine such as antibiotics, antipyretics, analgesics, antidepressant, chemotherapeutical agents, contraceptives, diagnostic agents, dermatological, cosmetological drugs, and personal care products. [1]. On the other hand, economy related to pharmaceutical industries depends upon the utilization and acquisition of new drugs and agents. This acquisition increases with exponential rate and related to the change in new habits, heathcare treatments, and consumption of chronic medicines.

Pharmaceutical compounds (PC) became one of the emerging contaminant which is released in to environment due to anthropogenic activities. Due to its environmental and health impacts PCs are also denoted as new emerging contaminant by United State Environmental Protection Agency (USEPA). Moreover, new PCs are continuously introduced due to high and endless demand, therefore, as an emerging contaminant it requires constant monitoring as well as toxicological studies.

2 Pharmaceutical wastewater

PCs are one of the emerging contaminants which include steroids, surfactant, flame retardant, pesticide, industrial additives, phenolic compounds, organic compounds, hormones, gasoline additives, and nanomaterial. Presence of PCs in wastewater was identified in last few decades. Its presence is detected in sewage, ground water, surface water estuaries, atmosphere, river sediments, and drinking water. Transmission of PC in nature may be due to various anthropogenic activities, emission, precipitation, and sedimentation [2]. Disposal of expiry dated medicines, discharge of used medicinal agents after end of treatments, indiscriminate uses also responsible for the existence of PC in surface, ground and wastewater. Groundwater may be contaminated by effluent infiltration, rural, domestic, or industrial sewage line. Sediments due to occurrence of active and insoluble species deposition, discharge and improper dumping on land and water bodies may be reason for river and seawater contaminations. Use of fertilizer and antiparasitic agents also cause contamination of water by agricultural activities.

Pharmaceutical contaminants bearing water or wastewater can be regarded as three classes on the basis of their sources or generator (Figure 1). These three classes are large generators, small generators, and microgenerators [1]. Large generator includes industries like pharmaceutical, personal care product, and disinfectant. Small-scale generators involve laboratories, research institutes, hospitals, and diagnostic centers whereas, microscale generators deal with the agricultural and domestic discharges.

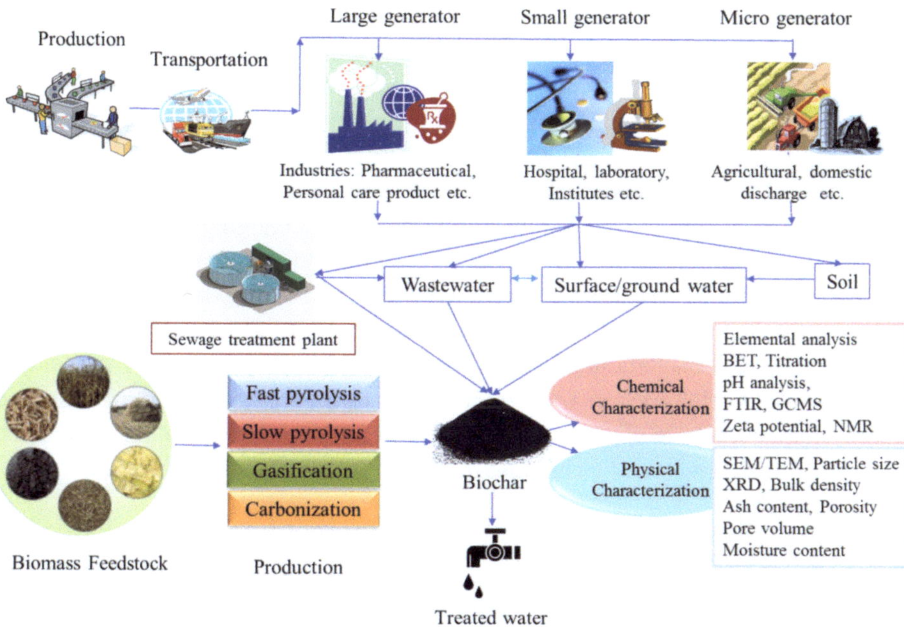

Figure 1: Pathway of PC's generation, distribution in environment, biochar production, and characterization for treatment of PC contaminant water and wastewater.

In the past few years, PCs have been frequently detected in wastewater treatment plants (sediments, treated effluents, and sludge), surface and ground water (medicines, veterinary pharmaceuticals, processing, production, secondary compounds, etc.), and consumable water sources [1]. Numerous PC are categorized and listed in Table 1. All the PCs may have different destination and appearance in the environment due to different physiochemical properties, structural characteristics, biodegradability, lipophilicity, and photosensitivity [3]. Therefore, it has several adverse effects on human, environment, and aquatic life.

2.1 Adverse effects caused by pharmaceutical contaminants

Adverse effects caused by pharmaceutical contaminants involve the aquatic toxicity, endocrine disorders, development of resistance and multi-resistance in pathogenic bacteria and genotoxicity. Some of the PCs are persistent in treated water due to high solubility, polarity, very small size, and easy mobility in environment. In this way, PC can return to human body through food chain or water cycle and potentiate the heath hazards [4]. More than 200 PC in surface, ground water, and sewage are reported in literature, which are still under consideration for study on discharge levels, effects, and fate on human health and environment. Toxicity data of several PC such as psychiatric drugs (fluoxetine and diazepam), lipid regulator (fenofibric, fenofibrate, and gemibrozil), analgesic or anti-inflammatory (ibuprofen and mefenamic acid), and antibiotics (erythromycin, amoxicillin, clarithromycin, sulfamethoxazole, ofloxacin, azithromycin, and tetracycline) depicted to cause acute and chronic affects [5]. Presence of hormones or estrogenic PC in water can cause variation in vitellogenin and estradiol levels in aquatic creature. Few PC in water can also affected the microbial communities by changing the metabolization of different carbon sources, carbon cycle, metabolic diversity of soil community, and other ecological functions. Moreover, PC can cause lung deceases, cancer, allergies, brain disfunction, and improper digestive function in humans. Production of antibiotic resistant bacteria in humans is also of major concern, which in turn affects and resists the body function and response toward the treatment and prevention of various diseases. Due to all these adverse effects of PCs, special attention is needed to establish efficient treatment techniques, disposal limits, and regulations.

2.2 Treatment techniques

Impact of PC on environment, human, aquatic, and wild life compels the development of efficient treatment techniques. Conventional methods of wastewater treatment plants are majorly focused on removal of contaminants, organic matter, and nutrients from wastewater. Thus, complete removal of PC in conventional wastewater treatment plant is quite challenging due to PC's complex structure, extremely low concentration, different hydrophobicity, microorganism influencing nature, and different interaction mechanism [4]. Some of the PCs are persistent while drinking water treatment process due to some physiochemical properties such as high soluble with water, very small size, polarity. Some of the PC treatment techniques are membrane technology, biological treatment methods, phytoremediation, constructed wetland, carbon nanotubes, adsorption process, advance oxidation process, and ozonation. [6].

Structure of carbon nanotube resembles graphite that can remove PC, for example, amoxicillin, ibuprofen, and norfloxacin by adsorption mechanism. Membrane process can be selectively removed the diclofenac, caffeine, bisprenol, naproxen, ketoprofen etc. by virtue of pore size, hydrophobicity, and surface charges. Various membranes such as nanofiltration, RO, microfiltration, and ultrafiltration have different removal efficiency toward the PC. Biological treatment process involves the activated sludge process, bioreactor (anaerobic and aerobic), and membrane bioreactors. Biological treatment of PC exhibits maximum removal efficiency as compared to other process. Almost 100 percentage removal of ketoprofen, diclofenac, triclosan, naproxen, and ibuprofen is reported by this method. Seeding, sludge disposal, and control of microbial growth are the issues which limits its application. Constructed wetland, biofilter, or phytoremediation is another technique for treatment of PC from wastewater which comes under the green technology. This process involves vegetation, soil, microorganism, and substrate for treatment of wastewater [1]. Various PC with different treatment techniques are shown in Table 1.

These techniques are efficient to some extent as compare to convention wastewater treatment process, though they have some drawbacks like concentration polarization, large land requisition, sludge disposal and handling, adsorption loading and disposal, sanitation, and high cost. Among all the techniques adsorption using activated carbon can be justified as comparatively economical, efficient and widely used technique. Main concern associated with adsorption process is the high cost of commercially available adsorbent.

Biochar produced by locally available and waste biomass could be a potential alternative of costly commercial adsorbent. Moreover, biochar can be customized as per the requirement. Biochar exhibits excellent adsorption capacity due to active surface functional groups, high surface area, and extensive pore size distribution. High efficiency, easy and economic preparation methodology, and various physicochemical characteristics of biochar along with solid waste management of biomass make it more demanding methodology for the treatment of wastewater. Function of biochar, preparation methodology, and characteristics are discussed in the subsequent sections of this chapter.

3 Biochar as an adsorbent

Biochar is a solid carbonaceous material that can be produced by thermochemical conversion of biomass in presence of limited oxygen. In past few decade, biochar has been gaining more attention due its multifunction nature such as high carbon sequestration, enhances soil property, energy production, and environmental remediation. On the other hand, it is also considered to be the most economical alternative of commercially available adsorbent.

3.1 Preparation of biochar

3.1.1 Feed stock

Biochar can be produced by various feedstock sources (Figure 1). which are basically categorized as primary produced biomass such as bioenergy crops and waste biomass [7]. Physiochemical properties of biomass have significant effect on properties of biochar for instance high lignin content biomass resulted in more porosity in biochar, presence of oxygen containing groups such as –OH, C = O, C–O contributes in negatively charged surface, complex heteropolysaccharide groups yield in electron donor functional group in biochar [8, 9]. Waste biomass as feedstock for biochar production gains more attention due to the advantage of food security and cost effectiveness. Waste or by-product biomass feedstock includes industrial waste/by-product, agricultural residue, forest residue, domestic residue, and sewage sludge [10–12]. Utilization of residual biomass of agriculture-forest-domestic materials resulted in production of low cost biochar. Several non-conventional materials such as used tyre, newspaper scraps, algae, and bones are also reported as potential feedstock for biochar production [8].

3.1.2 Preparation methodology

Biochar is usually prepared by pyrolysis or carbonization of biomass [13]. Pyrolysis process involves the burning of fuel by electrical heating or microwave heating. Pyrolysis or carbonization process can be carried out in traditional earthen kiln and modern biochar retorts. The process of pyrolysis can be carried out in different temperature, residence time, and heating rate. It can be distinguished as torrefaction (pyrolysis at low temperature), intermediate, slow or fast pyrolysis, gasification, flash or hydrothermal carbonization. Slow pyrolysis process is usually conducted at low to moderate temperature for long residence time in continuous mode, whereas fast pyrolysis takes place at high temperature (<650 °C) with rapid heating rate [14]. Secondary reaction such as charring or reaction of tar can resulted due to prolonged residence time. Hydrothermal carbonization of biomass occurs at low temperature (180–250 °C) under autogenous pressure to lower down the hydrogen and oxygen content through decarboxylation and dehydration [14]. This process produces solid, liquid, and gaseous products. Hydrothermal carbonization process is also refers as wet carbonization or wet pyrolysis process due to the carbonization of liquid biomass. Whereas, flash fire is utilized beneath the packed bed of biomass in flash carbonization process which takes place at the pyrolytic temperature of 330–650 °C, less than 30 min residence time and elevated pressure of about 1–3 MPa [15]. Among all the mentioned techniques, slow pyrolysis, flash carbonization, torrefaction, and hydrothermal carbonization yield more solid products as compared to gasification and fast pyrolysis process [16]. Numerous biochar feedstocks for adsorption of PCs are listed in Table 1.

3.2 Characterization of biochar

Properties of biochar and raw material can be characterized in three modes including physiochemical, surface, and structural or molecular characterization. Various characterization techniques are reported in literature. Pollutant removal ability and other applications can be predicted by the characterization of biochar. Morphological studies, pore size and volume, specific surface area, elemental composition

Table 1: Various pharmaceutical compounds (PCs) with their treatment techniques and different low cost feedstocks for biochar preparation.

Category	Pharmaceutical compounds	Treatment techniques	Adsorption on biochar (feedstock)
Analgesic and anti-inflammatory	Aspirin Acetaminophen Diclofenac Ibuprofen Indomethacin ketoprofen Naproxen Nimesulide Paracetamol phenazone Salicylic acid	UV disinfection, oxidation, microbial digestion, ultrafiltration, ozonation, coagulation, adsorption	Municipal solid waste, wheat straw, saw dust, pitch pine saw dust, wild plum kernel, etc.
Antibiotic	Ampicillin Amoxicillin Carbamazepine Cephalexin Ciprofloxacin Doxycycline Difloxacine Erythromycin Lincomycin Metronidazole Nalidixic acid Ofloxacin Sulfametoxazole Tetracicline Trimethoprine	Ozonation, adsorption, activated sludge process, anaerobic digestion, UV disinfection, advance oxidation, sedimentation, ultrafiltration, nanofiltration	Municipal solid waste, wheat straw, bamboo, sawdust, corn husk, bagasse, reed straw, tea waste, etc.
Antidepressants and psychiatric	Amitriptyline Carbamazepine Diazepam Doxepin Fluoxetine Imipramine Oxazepam Mianserin Nordiazepam Pipamperone Thioridazine	Anaerobic digestion, activated carbon adsorption, membranes separation	Bagasse, reed straw, saw dust, corn husk, bamboo, municipal solid waste, etc.

Table 1 (continued)

Category	Pharmaceutical compounds	Treatment techniques	Adsorption on biochar (feedstock)
Anticonvulsants	Carbapentin Carbamazepine Dilantin Phenobarbital Primidone	Anaerobic digestion, oxidation, membrane filtration, adsorption	Saw dust, agricultural waste, bovine bone, bagasse, etc.
Antineoplastics	Cyclophosphamide Epirubicin Ifosfamide Methotrexate Tamoxiline	UV disinfection, adsorption, oxidation, nanofiltration	Bagasse, saw dust, corn stalks, etc.
Beta-blocker	Acebutolol Atenolol Metoprolol Nadolol Propranolol Pindolol Sotalol	Microbial degradation, anaerobic sludge digestion, ultra-nanofiltration, adsorption	Tea waste, cotton straw, kenaf bar, saw dust, etc.
Cardiac	Amlodipine Captopril Metoprolol	Ozonation, oxidation, adsorption	Saw dust, rice husk, vegetable peel, etc.
Hormones	Androstenedione Estriol Estrone Mestranol Progesterone Testosterone	Activated sludge process, ultrafiltration, adsorption	Bagasse, saw dust, banana peel, etc.
Lipid regulators and metabolite	Bendroflumethiazide Benzafibrate Clorfibric acid Clofibrate Furosemide Gemfibrozil Simvastatin	Oxidation, activated sludge process, anaerobic digestion, activated carbon adsorption	Bamboo, rice husk, cotton straw, saw dust, etc.
Diagnostic agents	Diatrizoate acid Iopromide Iomeprol	Oxidation, UV disinfection, adsorption, ultrafiltration, ozonation	Pomela peel, saw dust, cotton straw, etc.

Table 1 (continued)

Category	Pharmaceutical compounds	Treatment techniques	Adsorption on biochar (feedstock)
Disinfectants	Bromoprene Chloroprene Triclosan Phenylphenol Chlorocresol Chloroxylenol	Activated sludge process, anaerobic digestion, adsorption	Husk and paper sludge, oil palm fiber, etc.
Sunscreen agents	Ethylhexyl methoxycinnamate Oxybenzone Octocrylene	Activated sludge process, oxidation, adsorption, anaerobic digestion,	Bagasse, reed straw, saw dust, etc.
Insecticide, herbicide and fungicide	Carbamate Pyrethroid Phenylpyrazole Organochlorine Benzimendazole Dicarboximide Strobilurin Triazole Anilide Benzothiazinone Chloroacetamide Phenylamide Urea Triazine	Activated carbon adsorption, advance oxidation, phytoremediation, membrane filtration	Bamboo, corn husk, saw dust, fruit's peel, cotton straw, lotus seed, etc.

analysis, acidic and alkaline nature (pH value), functional group analysis, surface charge are some of the important analysis of biochar and its raw material. Scanning electron microscope (SEM), Fourier transform infrared spectroscopy (FTIR), Raman spectra, Energy dispersive X-ray analysis (EDX), Brunauer-Emmett-Teller (BET) analysis are some analytical techniques which provide the information about the surface properties of biochar. Morphological studies such as macro-micro porosity and surface smoothness and roughness can be easily determined by SEM micrographs. It also provides a rough estimation of pore size. Different pyrolytic temperature can results in different intrinsic architecture of biochar such as needle, spherical, honey comb type, and tubular. BET analysis is based on the adsorption of liquid N_2 on pores of biochar with respect to appropriate pressure range. Hence on this basis calculation of effective pore size, area, and volume (micropore, total pore, and mesopore) and pore size distribution can be determined [16].

Acidic and basic functional group can be estimated by Boehm titration and FTIR analysis. Boehm titration involves equilibration of biochar in strong acids/ bases and thereafter titration of extract with strong acid and base to find the reacted fraction. Elemental analysis is an important parameter to find the degree of maturation and aromaticity by determining the H/C and O/C ration. Moreover, rise in pyrolytic temperature can increase the carbon content and reduce the O and H contents in biochar. Elemental composition can be examined by CHNS elemental analyzer and XRD analysis. Combustion of sample in presence of pure O_2 (CHNS analyzer) is used to find the oxygen content of biochar. Whereas, X-ray beam under vacuum condition is applied on sample to find the phase and elemental mapping using XRD analysis. Surface charge and pH of biochar are an important property of biochar which influences the sorption capacity. It can be analyzed by zeta potential analyzer with different pH range solution. pH of biochar strongly dependent on pyrolytic temperature. High temperature resulted in high pH and high ash content and vice versa. Positive and negative surface charge can be estimated on the basis of point of zero charge w.r.t. pH. Surface is considered to be negatively charge when pH of solution is above the more than point of zero charge and can bind cations. Whereas, pH value less than point of zero charge than surface of biochar behaves like positively charged that can efficiently bind the anions [13]. In this way, synthesized biochar can be characterized and applied for the suitable applications.

3.3 Sorption mechanism of biochar

Mechanism of adsorption strongly depends upon some of the properties of biochar such as specific surface area, porosity, surface functional group, surface charge, elemental composition, pH, and pyrolysis temperature. Chemisorption and physical sorption can be the dominant mechanism for separation of PC from water or wastewater. Chemi or physiosorption mechanism generally involves hydrogen bonding, functional group, π- π electron interaction, pore filling, electrostatic interaction, hydrophobic effect, ion exchange, and bridging interaction [17].

3.3.1 Hydrogen bonding and functional group

Hydrogen bonding is one of the important mechanisms for the adsorption of PCs on the biochar. Presence of functional group such as carboxyl, anhydride, and phenolic hydroxyls is responsible for the formation of hydrogen bond with the PCs on the biochar surface. Hydrogen bonding and diffusion of sorbate molecules is accountable for the synergetical adsorption of antibiotic. Moreover, charge assist hydrogen bond is also responsible for the adsorption of sulfonamides antibiotic [18]. Some of the PCs like dibutyl phthalate form hydrogen bond through the water molecules.

3.3.2 π–π electron interaction, ion exchange, and bridging interaction

Biochar is reported as π-electron donor, whereas most of the PC are π-electron acceptor, hence electron donor acceptor mechanism may contribute to enhancement of PCs adsorption. Surface structure of biochar is a dominating characteristic of PC adsorption for instance grapheme like biochar surface may sorb planar aromatic compound by noncovalent π–π electron interaction. Whereas functionalized biochar (enriched with C = C, -OH, -COOH groups) acts as strong π electron donor or acceptor due to the presence of hydroxyl or carboxyl group, respectively. Adsorption by π–π electron interaction is also resulted by some other chemical interactions such as resonance, lone pair electron donors, π electron acceptors, and electron resonance [18]. One of the examples of PC adsorption by π–π electron interaction is adsorption of tetracycline compound on bagasse-based biochar [15].

3.3.3 Pore filling

Pore filling mechanism is related with the surface properties of biochar such as microporous surface area and pore volume. Carbonized organic matter in biochar contributes as adsorbent whereas non-organic matters are assumed as absorption phase. The sorption of PCs on biochar takes place by both adsorption (carbonized organic matter) and absorption (non-carbonized organic matter). Hence, sorptive uptake of biochar can be estimated by the relative non-carbonized and carbonized fractions, their surface properties and bulk properties [17]. Example includes micropore filling adsorption of low-molecular-size sulfamethoxazole on the biochar.

3.3.4 Electrostatic interaction

This type of mechanism is dominated in sorption of ionic and ionizable PCs, which in turn dependent on surface charge of biochar. Degree of ionization and surface charge usually depends upon the solution pH. One of the examples for adsorption by electrostatic interaction is the adsorption of ciprofloxacin on biochar. Ciprofloxacin consists of amino and carboxyl groups [18]. At lower pH, biochar surface becomes negatively charged and facilitate electrostatic attraction toward cations, hence, enhances the sorption rate. Similarly, electrostatic repulsion contributes in enhanced sorption rate due to increase in ionic strength of sorbate solution [18].

3.3.5 Hydrophobic effect

This mechanism incorporates hydrophobic adsorption and partitioning. Generally, biochars have amorphous structure and considerable amounts of soft aliphatic provinces which resulted in portioning of hydrophobic PCs. Adsorption of simazine and polychlorinated biphenyls on biochar are example of adsorption due to hydrophobic portioning. All the listed mechanism for adsorption of PC on biochar is proposed in Figure 2.

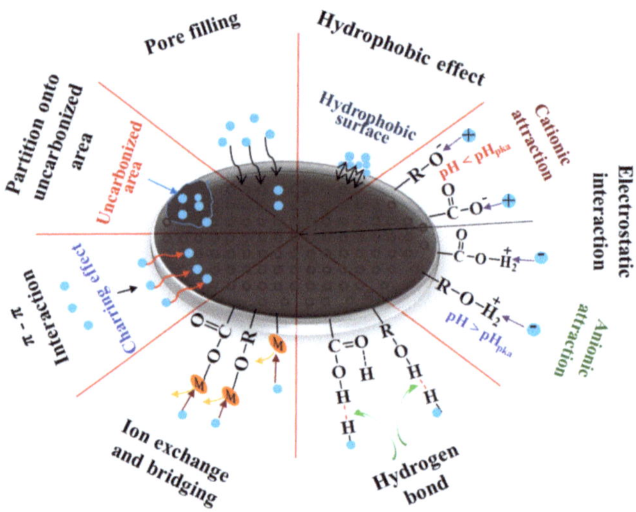

Figure 2: Schematic of PC adsorption mechanism on biochar.

3.4 Adsorption equilibrium isotherms

Adsorption equilibrium can be described by adsorption isotherms. Adsorption isotherms are essential parameter for optimization of adsorbent, concentration, solution pH, adsorbate–adsorbent interaction, contact time, and so on [18]. Adsorption isotherms models are derived on the basis of binding affinity or binding strength on the adsorbent surface. Langmuir isotherms, Freundlich, Langmuir–Freundlich (L-F) isotherms, and dual mode isotherms are commonly reported for the PC removal studies [8]. Mathematical expression for all the mentioned isotherms is represented in Table 2.

Langmuir isotherm model is widely used mechanistic model with some assumptions. These assumptions include fixed number of adsorption site, equal size, and shape of all vacant sites, no lateral interaction or steric hindrance between adjacent site and adsorbed molecules, monolayer adsorption.

Freundlich isotherm model deals with the heterogeneous surface and multilayer sorption. Varied affinity of sorption site and occupancy of stronger binding sites are the main assumptions of this model.

L-F isotherm is a versatile model which is also known as Sip's equation. It has tendency to simulate both Langmuir and Freundlich adsorption behavior. It can predict saturation at high concentration. Thermodynamic inconsistency at zero surface coverage is one of the main drawbacks of this model because it can not exhibit a finite Henry's constant like Langmuir. Dual-mode (L-F) model deals with the sorption of adsorbate on heterogeneous surface with assumption of two different surface of contrasting bonding energy. Among all the models, Langmuir and Freundlich models have been reported as the most effective models for adsorption of fluoroquinolone antibiotics and levofloxacin on biochar.

3.5 Adsorption kinetics and thermodynamics

On the basis of adsorption isotherm or equilibrium isotherms, numerous mathematical models have been developed to describe the adsorption kinetics. These models are pseudo-first-order model, pseudo-second-order model, Elovich model and intraparticle diffusion model (as represented in Table 2). Lagergren pseudo-first-order model perhaps describes the rate of adsorption in liquid phase system. This model can be applicable to a limited time of adsorption thereafter a complex model is required for accurate data representation. Chemisorption process is generally represented by pseudo-second-order model. Second order reaction rate depends on the amount of adsorbate on the adsorbent surface at equilibrium. Elovich model is adopted for the chemisorption on heterogeneous surfaces of adsorbent. It also evaluates the homogeneity of adsorbent surface for whole sorption reaction time by deviating from the linearity which reflects the change in surface reactivity. Intra particle diffusion model depends on external and diffusional mass transport rate as well as adsorption on active surface. Intraparticle diffusion usually occurs in micro- and macroporous adsorbent surface.

Adsorption kinetics of PCs on biochar is mostly studied on pseudo-first- and -second-order models. Most of the cases of removal of PCs follow the pseudo-second-order model whereas energetically heterogeneous adsorption conform the Elovich model. Adsorption kinetics data of amoxicillin on biochar well fitted to the pseudo-first and -second as well as Elovich models where chemisorption of adsorbate on adsorbent is assumed. Similarly, adsorption kinetics of sulfonamide and chloramphenicol antibiotics on functional biochar can be best represented by pseudo-first- and -second-order and intraparticle diffusion model from water and wastewater.

Thermodynamics of adsorption has an essential role in adsorption, as it characterizes the adsorption mechanism and equilibrium. The characterization of adsorption

Table 2: Adsorption isotherms and kinetic models for adsorption of pharmaceutical compounds on biochar.

Adsorption isotherm		Kinetic models	
Langmuir isotherm	$Q = \frac{Q_{max}.KC}{(KC+1)}$	Pseudo-first-order	$q_t = q_e\left(1 - e^{-k_1 t}\right)$
Freundlich isotherm	$Q = K_f C^n$	Pseudo-second-order	$q_t = \frac{k_2.q_e{}^2 t}{(1 + k_2.q_e t)}$
Langmuir–Freundlich (L-F) isotherm	$Q = \frac{Q_{max}.(KC)^n}{(1+(KC)^n}$	Elovich	$q_t = \frac{1}{\beta}\ln(\alpha\beta t + 1)$
Dual (Langmuir–Langmuir (L-L) isotherm) isotherm	$Q = \frac{Q_{max,1}.K_1 C}{(K_1 C+1)} + \frac{Q_{max,2}.K_2 C}{(K_2 C+1)}$	Intraparticle diffusion	$q_t = K_1 t^{0.5} + W$

*Q (adsorption capacity), Q_{max} (total adsorption site per unit mass of adsorbent), C (concentration of adsorbate), n (Freundlich linearity constant), K_f (Freundlich equilibrium constant), K (Langmuir adsorption constant), q_t and q_e (amount of adsorbate sorbed at time t and equilibrium), k_1 and k_2 (first- and second-order sorption rate constants), α and β (Elovich sorption and desorption rate constant).

mechanism and equilibrium is dependent on some thermodynamic parameters such as change of heat or enthalpy (ΔH), entropy (ΔS), and Gibb's free energy (ΔG). All these thermodynamic properties (at absolute temperature T) can be correlated as follows:

$$K_e = q_e / C_e \tag{1}$$

$$\Delta G^0 = \Delta H^0 - T\Delta S^0 \tag{2}$$

$$\Delta G^0 = -RT \ \ln K_e \tag{3}$$

$$\ln K_e = \left(T\Delta S^0 - \Delta H^0\right) / \ RT \tag{4}$$

where K_e, R, q_e, and C_e are equilibrium adsorption constant, ideal gas constant, amount and concentration of adsorbate at equilibrium respectively. Adsorption equilibrium constant can be obtained by various isotherm models. Value of ΔH^0 and ΔS^0 can be estimated by slope and intercept of ln K_e versus $1/T$ curve. Increase in q_e with temperature resulted in positive ΔH^0 which signifies the endothermic reaction where as negative value represents the exothermic nature of reaction. Similarly, positive ΔG^0 and ΔS^0 depicts spontaneous sorption process and affinity of carbon sorbent. Values of ΔG^0 also distinguish the different sorption process for instance physical sorption occurs at –20 to 0 kJ mol^{-1}, whereas –400 to –80 kH mol^{-1} is reported for chemisorption [8].

3.6 Influencing factors

Biochar can be considered as highly efficient, eco-friendly, and low cost adsorbent due to the presence of different carbon and inorganic fractions. Properties of biochar may have great influence of pyrolysis temperature, type of feed stock, residence or pyrolytic time, and thermochemical conversion techniques [18]. These parameters also effect the removal efficiency of biochar toward the targeted PCs. Effect of such parameters on biochar properties and adsorption are discussed in this section.

3.6.1 Type of feedstock

Different biomass exhibits different characteristics in terms of cellulosic compound, lignin, volatile matter, ash content, crystalline, or amorphous phase. Micro or meso porous structured biochar can be produced by biomass of lower ash content, reduced crystalline size. Heterogeneous and homogenous surface structure is also dependent upon the type of feedstock. Heterogeneous structure possesses high adsorption capacity as compared to homogenous structure. Presence of inorganic minerals contributes in surface functional groups on biochar. Therefore the feedstocks have significant influence on properties of biochar.

3.6.2 Pyrolysis temperature

Biochar's basic characteristics are highly dependent on pyrolysis temperature and time. Increase in pyrolysis temperature resulted in decrease in amount of elemental oxygen and hydrogen, H/C ration, and O/C ratios due to the significant condensation reaction thereby indicating deoxygenation and dehydrogenation of biomass. Moreover, increase in hydrophobicity and decrease in polar group is also resulted due to the low O/C ratios. Temperature of pyrolysis also based on the type of feedstock as dissociation and depolymerization of lignocellulosic compounds are temperature dependent. For instance, biomass collected form sewage or poultry manure requires less temperature for the dissociation of biomass due to the absence of lignocellulosic compounds. Whereas plant-based biomass consists of lignocellulosic compounds that require high temperature (160–900 C) for dissociation and depolymerization biomass [19]. High-temperature biochar usually contents more aromatic compounds and condensed structure but less graphitic carbon and carboxyl functional groups.

3.6.3 Solution pH

The solution pH has significant effect on the surface charge of biochar. It also contributes in the protonation and deprotonation phenomena of biochar surface functional groups. Surface charge of biochar controls the mechanism, rate, and capacity of adsorption of PC on biochar. Typically, solution at value of pH 3 exhibited the maximum removal efficiency due to formation of atomic hydrogen. Range of pH form pH 4 to pH 7 also shows high removal efficiency due to the adequate H^+ concentration for the atomic hydrogen formation. Moreover, isoelectric point or pKa values also reveal the rate of adsorption or removal efficiency. As above this value surface of biochar acts as negatively charge and below that point, positive charge surface contributes for the separation or adsorption of PC depending upon the ionic nature of PCs [19].

3.6.4 Physiochemical properties of PCs

Different PC exhibits different physiochemical properties such as hydrophobicity, large molecular weight, and small molecular size. The performance of biochar also get influenced by such properties. For example ibuprofen adsorption is slow process which attributed to the hydrophobicity and involvement of less extensive site due to small molecular structure. Whereas diclofenac exhibits high adsorption rate due to chemisorption on biochar [19].

4 Recent advances and future challenges (Li)

Recent developments of biochar for removal of PCs are mainly focused on the modification of surface morphology, enhancement of numbers of functional groups, and improvements in surface charges. Gases (CO_2) or steam is used to treat the biochar in order to increase the porosity, surface area, micropore volume, and sorption site in biochar. Use of acidic and alkaline treatment is also carried out to improve the surface morphology. Moreover, research work is also going on for the preparation biochar composites by impregnating biochar with specific materials. Embedding specific metal oxides with porous biochar resulted in enhanced surface charge. Development of biochar with increased surface oxygen containing functional groups (carboxylic, hydroxylic, phenolic group, etc.) is associated with the use of acidic/alkaline solution and potential of chemical binding with PCs. Improvement of biochar's properties by incorporation of surface amino groups, magnetization, and biofilm formation is new frontier of modification [20].

There are more than thousands of PCs are available commercially among which less than 2% has been detected and examined for treatments. Hence, more efforts are needed toward the detection, sensitivity, toxicity, and priorities of hazardous PCs. Furthermore, effects of low level and chronic exposure of PCs on sensitive subpopulation should be assessed. Removal of PCs from water and wastewater is paramount due to the failure of conventional water and wastewater treatment plant for PC removal. Although the biochar adsorption has potential toward the efficient removal of PCs, it has some challenges which should be rectified to give new direction to the biochar-based adsorption. One of the major challenges is associated with the production of biochar. Feedstock for biochar required pretreatment (crushing, cleaning, drying, etc.) before pyrolyzation, which intern increases the production cost as compared to the conventional activated carbon. Hence, future research can be focused on the optimization of production process and maximization of applicability. Another concern of future study is the selection of appropriate feedstock for removal of targeted compound, that is quite complex and dependent on adsorption parameters (initial PCs concentration, pH, contact time, ionic strength, etc.). Modification, innovation, and preparation of biochar for adsorption of specific PC are the most challenging tasks, which need highly focused research activity.

5 Conclusion

Presence of PCs in water, wastewater, and environment is explicit. Many anthropogenic activities are responsible for the discharge of PCs in environment. The major concern with PCs is linked to the adverse effects on living beings and environments, which associated with the exposure to large, medium, and small spaces of time. There is a necessity for the legislation that focuses on PC contaminants, its discharge limits and restricts the release of PCs into the environment. These compounds are difficult to detect due to its nano-gram appearance. Measurement of PCs in water and wastewater is relatively new matter. Therefore, utilization of some techniques for its detection and removal is essential. Among various technologies, biochar-based adsorption process presents great potential toward the treatment of PC contaminated water and wastewater. Moreover, different models can be utilized to predict the mechanism, kinetic, equilibrium, and rate of adsorption of PCs on biochar. Biochar based adsorbent is eco-friendly, economical, easy to produce. The conversion of biomass into biochar provides waste biomass management, protecting the environment and attaining the sustainable use of resources. Therefore, diverse nature of biochar makes it more potential alternative for the treatment of wide spectrum of PCs from water and wastewater.

References

[1] Oliveira MD, Frihling BEF, Velasques V, Filho FJCM, Cavalheri PS, Migliolo L, Pharmaceuticals residues and xenobiotics contaminants: Occurrence, analytical techniques and sustainable alternatives for wastewater treatment, Sci Total Environ., 2020, 705(135568), 1–10.

[2] Wilkinson JL, Hooda PS, Barker J, Barton S, Swinden J, Ecotoxic pharmaceuticals, personal care products, and other emerging contaminants: A review of environmental, receptor-mediated, developmental, and epigenetic toxicity with discussion of proposed toxicity to humans, Crit. Rev. Environ. Sci. Technol., 2016, 46(4), 336–381.

[3] Melo SAS, Trovo AG, Beautitz IR, Nogueira RFP, Degradação de fármacos residuais por processos oxidativos avnçados, Quim Nova, 2009, 32(1), 188–197.

[4] Couto CF, Lange LC, Amaral MCS, A critical review on membrane separation processes applied to remove pharmaceutically active compounds from water and wastewater, J. Water Process Eng., 2018, 26, 156–175.

[5] Verlicchi PC, Al-Aukidy M, Zambello E, Occurrence of pharmaceutical compounds in urban wastewater: removal, mass load and environmental risk after a secondary treatment-A review, Sci Total Environ., 2012, 429, 123–155.

[6] Wang J, Wang S, Removal of pharmaceutical and personal care products (PPCPs) from wastewater: A review, J. Environ. Manage., 2016, 182, 620–640.

[7] Angin D, Effect of pyrolysis temperature and heating rate on biochar obtained from pyrolysis of safflower seed press cake, Bioresour. Technol., 2013, 128, 593–597.

[8] Inyang MI, Gao B, Yao Y, Xue Y, Zimmerman A, Mosa A, Pullammanappallil P, Ok YS, Cao X, A Review of Biochar as a Low-Cost Adsorbent for Aqueous Heavy Metal Removal, Crit. Rev. Environ. Sci. Technol., 2015, 1–56.

[9] Aksu Z, Isoglu IA, Removal of copper(II) ions from aqueous solution by biosorption onto agricultural waste sugar beet pulp, Process Biochem., 2005, 40, 3031–3044.

[10] Lu H, Zhang W, Yang Y, Huang X, Wang S, Qiu R, Relative distribution of Pb2+ sorption mechanisms by sludge derived biochar, Water Res., 2012, 46(3), 854–862.

[11] Tsai W, Liu S, Hsieh C, Preparation and fuel properties of biochars from the pyrolysis of exhausted coffee residue, J. Anal Appl. Pyrolysis, 2012, 93, 63–67.

[12] Cao X, Harris W, Properties of dairy-manure-derived biochar pertinent to its potential use in remediation, Bioresour. Technol., 2010, 101(14), 5222–5228.

[13] Braghiroli FL, Bouafif H, Hamza N, Neculita CM, Koubaa A, Production, characterization, and potential of activated biochar as adsorbent for phenolic compounds from leachates in a lumber industry site, Environ. Sci. Pollut. Res., 2018, 25, 26562–26575.

[14] Meyer S, Glaser B, Quicker P, Technical, economical, and climate-related aspects of biochar production technologies: a literature review, Environ. Sci. Technol., 2011, 45(22), 9473–9483.

[15] Oliver DP, Pan YF, Anderson JS, Sorption of pesticides by a mineral sand mining by-product, neutralised used acid (NUA), Sci Total Environ., 2013, 442, 255–262.

[16] Nartey OD, Zhao B, (2014, Biochar Preparation, Characterization, and Adsorptive Capacity and Its Effect on Bioavailability of Contaminants: An Overview, Adv. Mater. Sci. Eng., 2014 (715398), 1–12.

[17] Cheng N, Wang B, Wu P, Lee X, Xing Y, Chen M, Gao B, Adsorption of emerging contaminants from water and wastewater by modified biochar: A review, Environ. Pollut., 2021, 273 (116448), 1–14.

[18] Li L, Zou D, Xiao Z, Zeng X, Zhang L, Jiang L, Wang A, Ge D, Zhang G, Liu F, Biochar as a sorbent for emerging contaminants enables improvements in waste management and sustainable resource use, J. Cleaner Prod., 2019, 210, 1324–1342.
[19] Krasucka P, Pan P, Ok YS, Mohan D, Sarkar B, Oleszczu P, Engineered biochar – A sustainable solution for the removal of antibiotics from water, Chem. Eng. J., 2021, 405(126926), 1–27.
[20] Wang X, Guo Z, Hu Z, Zhang J, Recent advances in biochar application for water and wastewater treatment: a review, Peer J, 2020, 8, e9164. doi: http://doi.org/10.7717/peerj.9164.

Vijaya Geetha B., Shreenidhi K.S., Vihaa Sriee M.G., Rashminiza A.

Removal of the pharmaceutical compounds from wastewater by biochar

Abstract: An escalating figure of pharmaceuticals has become apparent in the surroundings, and the issue has become a human and environmental health concern because they are not fully degraded in wastewater treatments of industries. Owing to biochar's stability, surface area, and surface-enriched functional groups, they have been exploited in various wastewater effluents. Being organic in nature, it has been found to be used as fertilizers and amends the soil properties also. Practically, biochar has been proved to efficiently and effectively remove heavy metal contaminants. Apart from heavy metals, nitrogen and phosphorus contents are also being removed using biochar. Nanobiochar extracted from the pine wood was employed in the elimination of carbamazepine detected in the wastewater treatment plants. The BC derived from guayule bagasse was found to adsorb the antibiotics like erythromycin, sulfapyridine (SPY), and compounds like docusate and convert wastewater irrigation compatible. In this chapter, we explore the possibilities of using biochar in removing of pharmaceutical compounds from wastewater.

Keywords: pharmaceuticals, wastewater, biochar, removal, agricultural waste, biomass

1 Biochar

1.1 Introduction

There has been a growing demand for medical and personal products among the world population. Cosmetics comprise synthetics. Pharmaceuticals are used to treat diseases while personal care products are to increase the standard of life. They survive in anionic, cationic, or neutral conditions. The increasing and continuous use

Vijaya Geetha B., Assistant Professor (SS), Department of Biotechnology, Rajalakshmi Engineering College (Autonomous), Affiliated to Anna University, Thandalam, Chennai 602105, Tamil Nadu, e-mail: vijayageetha.b@rajalakshmi.edu.in
Shreenidhi K.S., Assistant Professor, Department of Biotechnology, Rajalakshmi Engineering College (Autonomous), Affiliated to Anna University, Thandalam, Chennai 602105, Tamil Nadu
Vihaa Sriee M.G., Rashminiza A., Postgraduate, M.Tech student, Department of Biotechnology, Rajalakshmi Engineering College (Autonomous), Affiliated to Anna University, Thandalam, Chennai 602105, Tamil Nadu

https://doi.org/10.1515/9783110734003-017

of these chemicals gives an alarming signal to the conservation of the surrounding and sound heath of humans [1]. The accumulation of antibiotics in the environment will lead to rapid increase in antibiotic resistance genes which results in harmful repercussions to the environment [2].

Pharmaceuticals and personal care products (PPCPs) can enter the surroundings from the wastewater treatment plants, or they may be nonsteroidal anti-inflammatory drugs, antiepileptic carbamazepine, and so on [3]. The biochar (BC) prepared from different resources of plant-based bagasse is used as a well-organized adsorbent for removing pollutants from the surroundings [4].

The differences in parameters such as functional groups, pH, porosity, and surface area make biochar an appropriate adsorbent material for targeting many contaminants. Adsorption process is considered to be the most advanced process of separation of pollutants from water due to its efficiency, minimum biological sludge, cheap adsorption material, and so on [4]. The adsorption of pharmaceutical substances using biochar gained importance because of its cost-effectiveness, simplicity, and competency. To overcome the limitations of separating pharmaceutical wastes from water, magnetic nanoparticles are incorporated into BCs [5]. The integration of magnetic nanoparticles in the carbon precursor can be implemented by pre- or post-coating reaction [6].

The antibiotics get adsorbed to the BC surface by the functional groups and the entrapment of adsorbate takes place as the result of filling the pores of the BC. Thus, the fusion of antibiotics with adsorbents showed enhanced performance in the removal of antibiotic contaminants from the water bodies [7].

The aromatic character of the BC pyrolyzed at 300 °C is less when compared to aromatic character of the BC pyrolyzed at 700 °C. This leads to a decrease in hydrophilic nature which in-turn favors adsorption of organic contaminants from the aquatic environment [8].

Steam-activated BC has high capacity to adsorb the organic pollutants from the aquatic environment. This type of BC pyrolyzed at 700 °C showed the highest adsorption capacity (37.73 mg g^{-1}) for sulfamethazine [8].

1.2 Pharmaceutical pollution and its impacts

There are several antibiotics used in various fields for the treatment of bacterial infections [9]. The human body cannot metabolize most of the antibiotics and hence they are excreted and are released from humans, animals, and other sources like hospitals and pharmaceutical companies into the environment [10]. The accumulation of micropollutants in water bodies and the environment, which are nonbiodegradable and bioaccumulative in character, poses a serious threat to the global community.

The pollutants comprise pesticides, metals, pharmaceuticals, cosmetics, and so on in small amounts to larger amounts. A notable increase in the pollutants is cosmetics.

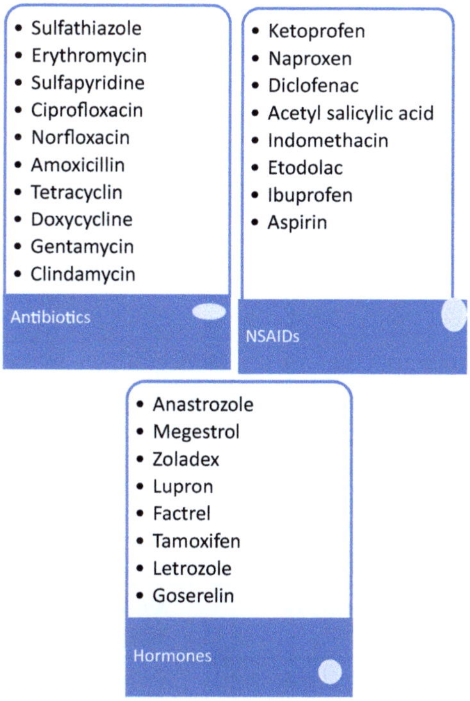

Figure 1: Different types of antibiotics, nonsteroidal anti-inflammatory drugs and hormones.

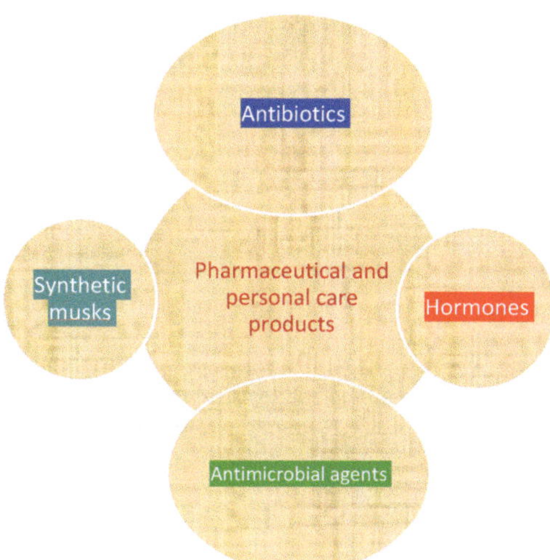

Figure 2: Pharmaceuticals and personal care products.

Pharmaceutical products enter the water bodies through various means and ways, for example wastewater treatment, effluent from pharma industries, animal excreta, and anthropogenic activities carried out by humans in water bodies, and sewage discharge in ponds and lakes. These pharmaceutical products bring destructive effects on humans and animals. It may cause inhibition of embryonic cell growth in humans, debilitated liver and gills in fish, and repress gram-negative bacteria [11].

1.3 Biosorbents

Biosorbent materials are used to remove pollutants that are not biodegradable in nature. Microbes, agricultural remains, tree barks, wood chips, and so on are common biosorbents. They are eco-friendly, cost-effective, and contamination free. These sorbents are used for the removal of pharmaceuticals, metals, dyes, and other organics.

Table 1: Preparation and production of biochars [7].

Antibiotics	Biochar preparation	Biochar feedstock	Adsorption efficiency
Tetracycline (TC)	– Biomass is pretreated with clays suspension: montmorillonite (MMT) and red earth clay (RE) – $Fe_2(SO4)_3$ – $ZnCl_2$ and $FeCl_3.6H_2O$ and their mixture – $FeCl_3.6H_2O$; pyrolysis and posttreatment BC with $FeCl_3.6H_2O$ – Mg/Al solution, then with mixture of NaOH with Na_2CO_3 – $CuCl_2$ $2H_2O$ followed by reduction with $NaBH_4$	– Municipal solid waste – Treatment plant sludge – Sawdust – Corn husk – Bagasse – Solid digest from rice straw	Varies from 4% to 100% based on the pretreatment process
Norfloxacin (NOR)	Bio-mass is pretreated with montmorillonite suspension – First pretreatment of biomass with $FeCl_3.6H_2O$, and second impregnation with $KMnO_4$	– Wheat straw – Potato stems and leaves	Varies from 6% to 75% based on the pretreatment process

Table 1 (continued)

Antibiotics	Biochar preparation	Biochar feedstock	Adsorption efficiency
Ciprofloxacin (CPX)	– Pretreatment of biomass with bentonite suspension – First pretreatment biomass with $FeCl_3.6H_2O$ and second impregnation with $KMnO_4$	– Municipal solid waste – Potato stems and leaves	Varies from 7% to 100% based on the pretreatment process
Sulfamethazine (SMT)	– Pretreatment of biomass with graphene oxide suspension – Posttreatment of biochar impregnation with $FeCl_3.6H_2O$ followed by reduction with $NaBH_4$	– Bamboo sawdust – Cornstalk	Varies from 3% to 30% based on the pretreatment process
Sulfapyridine (SPY)	– Pretreatment of biomass (HC/SB) with carbon nanotubes (CNTs) suspension and surfactant (SDBS) + CNT suspension	– Hickory chips (HC) – Sugarcane bagasse (SB)	Varies from 10% to 80% based on the pretreatment process
Levofloxacin (LEV)	– Pretreatment of biomass by impregnation with $FeCl_3.6H_2O$ – Pyrolysis and posttreatment BC with $FeCl_3.6H_2O$	– Corn husk	Varies from 50% to 100% based on the pretreatment process
Enrofloxacin (ENR)	– First pretreatment of biomass with $FeCl_3.6H_2O$ and second impregnation with $KMnO_4$	– Potato stems and leaves	Varies from 60% based on the pretreatment process
Sulfamethoxazole (SMX)	– Posttreatment biochar impregnation with the suspension of CuO, Fe_2O_3, ZnO followed by hydrothermal treatment – Treatment with HCl; post-treatment introduction of TiO_2 by sol–gel method followed by calcination	– Bamboo – Reed straw	Varies from 20% to 100% based on the pretreatment process

Because of their complex chemical nature, it is difficult to find a suitable sorbent to carry out efficient adsorption. In order to find a cheaper and effective sorbent, different modified adsorbents are evaluated. The application of the sorbents depends upon the contamination level, chemical nature of the PPCPs and processing costs [12].

Alternative nonconventional, inexpensive sorbents like clay minerals, solid agricultural wastes, synthetic materials, and by-products of industries were used [12]. The composition of the cellular surface and kinetic equilibrium are factors responsible for the removal of contaminants.

1.4 Synthetic sorbents

Synthetic sorbents can improve the porosity of the adsorbent. Though there are several advanced techniques used for the removal of cosmetics in the marine environment, synthetic sorbents get rid of the drawbacks of using other low-cost materials. Polymeric sorbents show high resistance to fouling and last up to regeneration cycles [13].

1.5 Pharmaceuticals and personal care products

PPCPs improve personal health care of individuals and also helps to uplift the growth or health of livestock. Traces of cosmetics have been detected in water bodies and surroundings throughout the world, producing adverse effects on coral reefs, fish, amphibians, and so on. They may enter into the environment directly or indirectly from households, hospitals, wastewater treatments plants which lead to soil pollution. Inappropriate discarding of expired drugs, dumping of biological wastes in the water bodies may also add to pollution. Considerable aggregates of antibiotics and other pharmaceuticals were introduced in water bodies [1].

Medicaments used in veterinaries, or additives in animal food too, are discharged into the land or soil or open water bodies and may affect terrestrial organisms directly. PPCPs get dissolved easily in the surroundings and they do not evaporate at normal temperatures and enter the soil and water bodies. The most common pharmaceuticals detected were nonprescription, prescription medications, natural and synthetic hormones, and an assortment of used or unused antibiotics. Besides the identified aggregates of human medications, diffusion from pharmaceuticals used in agriculture was added to the global pollution. PPCPs affect aquatic ecosystems and those habitats that were exposed to this polluted water [1].

Medications contain weak acids or bases that are moderately hydrophobic, whereas BCs have intermediate degrees of hydrophobic in nature. Cosmetics are bioactive complex molecules and they can sustain in all the three forms namely anionic, neutral, cationic, or zwitterionic molecules depending upon the environmental conditions [1]. Adsorbent efficiency of the biochar always depends on the size and molecular weight of the cosmetic molecules.

1.6 Technologies in removing pollutions from the environment

There are conventional treatment processes for the removal of pharma compounds. Recently developed technologies for treating pharma wastewater include membrane separation, irradiation, nanoparticle treatment, bioaugmentation, and hybrid technology. Many number of techniques such as membrane separation, electrochemical separation, biological processes, and ozonation were also used in the removal of pollution. Electrochemical advanced electrochemical processes like photoelectric catalysis and electro-Fenton reaction can emend the contaminated water by the degeneration of medications.

However, the abovesaid processes can generate more toxic and biorecalcitrant by-products than the parent PPCPs can generate [14]. By ozonation technique, complete removal of paracetamol with 30% mineralization can be initiated, thereby liberating intermediate products such as oxalic, glyoxylic, ketoaldonic, and formic acids. Biological treatments like fungi can remove antidepressants (citalopram), antibiotics (sulfamethoxazole (SMX)), anti-inflammatory drugs (ibuprofen, naproxen), and antiepileptics (carbamazepine), and so on [15].

One technique used for the removal of pollution was adsorption through BC. BC-based adsorption was simple, eco-friendly, and economical to follow as a new technique to remove cosmetic pollution.

Competitive sorption occurs between solid–water interface, which can influence the affinity of the adsorbent and adsorption feature for each pollutant [8, 11].

1.7 Factors affecting adsorption

The very important factors affecting adsorption were pH, temperature, cationic and anionic strengths of the polluted area, and the factors responsible for nonadsorption. The PPCP adsorption is different under different environmental conditions [8, 11].

The chemical nature of cosmetics was influenced by the surrounding pH. With increasing pH, in the surrounding, the dominant form of cosmetics will be anionic. At high pH values, electrostatic repulsion occurs between the biochar surface and the cosmetic pollution, resulting in less adsorption of pollutants [16].

In general, the cationic and anionic form of compounds in wastewater may influence the removal process of pharmaceutical products. Adsorption process can be overemphasized by the interference of organic or inorganic compounds. Adsorption process was influenced by the target cosmetic substance and nature of the adsorption material [16].

1.8 Biomass

Biomass composed of cellulose, lignin, and hemicellulose. Thermal decomposition of biomass in anaerobic conditions will yield BC, which is a carbonaceous material. Biomass is converted into BC by intermolecular and intramolecular reactions with higher thermal stability, by following procedures, namely, pyrolysis, hydrothermal carbonization, and gasification [17].

Lignocellulosic biomass refers to plant residues which are not classified as feed or food. It may be agricultural residues, forest residues, and herbaceous biomass. Agricultural residues include straw, sugarcane bagasse, and husk. Forest residues include roots, wood chips, and sawdust. Herbaceous biomass includes switchgrass and elephant grass [4].

1.8.1 Steps involved in biomass conversion

- Pyrolysis
- Hydrothermal carbonization
- Gasification
- Torrefaction

1.8.2 Pyrolysis

Pyrolysis is a thermochemical method of decomposing biomass, at more than 400 °C in the absence of oxygen [6]. During pyrolysis gaseous products such as CH_4, CO_2, H_2, moisture and CO are released [17]. The conversion of BC depends on the temperature, pressure, and residential time of the material used. Low temperature and longer residence time are ideal conditions for a larger volume of BC conversion. High furnace temperature, low pressure, and constant heating rate give a yield of BC with increased carbon content and high surface area. Biochar can be stimulated physically, chemically, and biologically [4].

1.8.2.1 Pyrolysis temperatures

- Cellulose is thermally stable and has semicrystalline chains. It decomposes at a temperature of 305–375 °C.
- Hemicellulose is a branched polymer and has short side chains. It decomposes at a temperature of 200–350 °C.
- Lignin has phenolic structures that decompose at 250–500 °C [4].

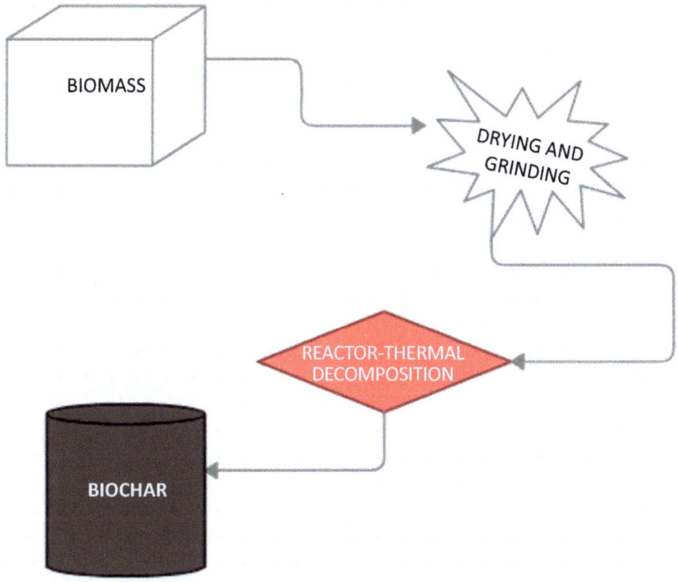

Figure 3: Mechanism of biochar production.

1.8.3 Hydrothermal carbonization

This is a process in which the biomass is dissolved in a sealed container which contains water and subjected to heat at high temperatures, namely, 300 °C for nearly 16 h. Water content is removed and the processing procedure resulted in the formation of BC. It has more chemical functional groups. Further the nature of the BC is determined by the factors, namely temperature, residence time, and pressure [4].

1.8.4 Gasification

During the process of gasification, syngas and char are obtained. Gasification comprises two steps. Drying is employed to remove the excess moisture content. Oxidation and combustion reactions followed which produces CO and CO_2 and water [4].

1.8.5 Torrefaction

The torrefaction consists of three types. In steam torrefaction method, the biomass is treated with steam for 10 min at 260 °C. In wet torrefaction or hydrothermal carbonization, biomass with water is processed for 5 min to 4 h at a temperature range of 180–260 °C. Oxidative torrefaction method is carried out by treating the biomass with oxidizing agents being used to produce heat energy. After product formation, room temperature is restored [4].

1.9 Properties of biochar

The characteristics of BC are determined by the rate of heating, feedstock, temperature, and residential time. BC surface is assumed to be negatively charged and the rate of adsorption increases with functional groups. There are variations that occur in the porosity and size of the surface area. The BC surface is considered to be heterogeneous in nature due to the carbonized and non-carbonized compounds [4].

BC has immobilization properties. BC possessing of small particle size possess large surface area and show better removal result, and the removal time required was found to be less. The methods of adsorption include hydrophobic, electrostatic interaction or attraction, hydrogen bonding, and pore filling [4].

2 Applications

BC is used in the adsorption of pollutants from the environment. The biocatalytic activity has more functional groups at the surface. Hence, it requires a larger surface area to act on the molecules.

2.1 Four steps of adsorption

The adsorption of antibiotics with the use of BC involves four main steps:
- Transportation of adsorbate molecules from the solution to the BC material
- Diffusion of the adsorbate takes place through the liquid film surrounding the BC molecule
- Diffusion of the adsorbate in the pores, that is, transport of adsorbate molecules from the surface of BC particles along the solid pore surface to active sites and adsorption as a result of adsorbate-BC interactions [7].

2.2 Mechanism of adsorption

The mechanism of adsorption takes place is directly related to the interaction between the BC and the pollutants. The adsorption of pollutants occurs by the following methods, namely, physisorption–electrostatic attraction, repulsion, pore diffusion, H-bonding and hydrophobic, and chemisorption–electrophilic interaction [4].

2.3 Adsorbent for removal of hazardous pollutants

BCs derived from wood and straw materials can be used in the removal of imidazoline from effluents. BCs obtained from rice straw, cattle and poultry manure, cow dung, and corn stover were used in the removal of atrazine [18].

The ketonic, carboxylic acid, and nitro groups that were present at the surface of BC act as electron acceptors and interact with aromatic molecules of the pollutants enhances adsorption. Adsorption of pollution takes place through electrostatic attraction and interaction, hydrophobic interaction, pore-filling, hydrogen bonding, and complex adsorption. The rate of adsorption increases with an increase in functional groups, specifically with oxygen [4].

2.4 Removal of tenacious organic pollutants

BC obtained from bamboo and corn stalk biomass is used for the adsorption of polycyclic aromatic hydrocarbons [19]. The functional group of the catalyst, that is O–H, plays an important role for the adsorption of NOR, and C=O and –OH are responsible for ammonium adsorption.

2.5 Clay minerals

Clay mineral-based adsorbents were used in removing PPCPs due to their large surface area and high cationic exchange property. They adsorb both organic and inorganic pollutants from the surface. Mostly, a variety of pharmaceuticals vary in their acidity in different solutions with different pH values. Ibuprofen, diclofenac, ketoprofen, carbamazepine, BPA, and triclosan were some of the pharmaceuticals which can be adsorbed in natural clay minerals. The changes in acidity and pH determine their mechanism of adsorption toward natural clay minerals [19].

3 Modification of biochars

Fe, Fe_2O_3, and Fe_3O_4 are the most commonly introduced particles into BCs by way of chemical co-precipitation or pyrolysis activation [20]. Modifications sometimes cause a negative impact on adsorption [21]. Generally, for the removal of antibiotics from the polluted aquatic ecosystem, the BCs are modified and engineered accordingly with the efficiency of the desired removal. These BCs obtained as the result is divided conventionally as pristine BCs and engineered BCs.

Pristine BCs are most commonly employed for the treatment of antibiotics present in the water stretches. They are produced from the residual biomasses with slow pyrolysis at a specified condition of 700 °C and ≤ 2% oxygen supply. To get maximum adsorption efficiency through adsorbate–BC interactions, the pristine BCs are modified to obtain engineered BCs. The modification includes physical and chemical modifications.

- Physical modification involves the development of BC's pore structure for the increased hydrophobicity and aromaticity.
- Chemical modification involves improving of BC's surface functional groups for the increased polarity and porosity [7].

4 Magnetic carbon materials

Magnetic BCs were prepared by placing the biochar in the iron and cobalt solution for magnetization [4]. Magnetic carbon materials (MCM) and other composites with magnetic properties have been integrated into the biochar and used in the removal of medicaments from water and wastewater. The composite nanomaterial was prepared by a combination of clay (bentonite), chitosan powdered activated carbon, and iron magnetic nanoparticles were integrated into the biochar. These BCs were used in the removal of medicaments having both cationic and anionic properties.

Due to low cost, high adsorption, plenty of availability, and capability to introduce into the biochar, iron particles are used to produce magnetic carbon materials [6].

4.1 Induction of magnetic properties

To impart magnetic properties, several precursors use the combination of metal oxide nanoparticles. The magnetic properties can be imparted by using magnetite or maghemite in the framework of carbon-based materials [22]. The precoating process consists of combing iron salts and iron oxides with carbonaceous precursors followed by pyrolysis [23].

4.2 Magnetic precursors

To produce the MCM precursors, 32% uses renewable biomass from agricultural wastes such as pine sawdust, coconut shell, sugar bagasse, and corn stalk to produce

BCs. To achieve magnetization of the materials, the formation of non-magnetic iron oxides and iron hydroxides has to be prevented. The presence of nonmagnetic iron oxides/hydroxides will reduce the magnetic property of the BC, which will make their recovery difficult after use.

MCMs can be obtained by oxidative hydrolysis of ferrous salts, co-precipitation, and thermochemical methods: pyrolysis and calcination, mechanical milling [6].

MCMs can also be obtained through the oxidative hydrolysis of Fe_2 solution in an alkaline media. The initial concentration of pharmaceuticals may influence the adsorption. Lower adsorption rates were attained due to the higher initial concentration of SMX [22].

The magnetic properties can be obtained by adding magnetite (Fe_3O_4) and maghemite (Fe_2O_3) to the BC framework [22]. The magnetic properties of MCM were influenced by the magnetic particles embedded in it [24]. The equilibrium between the liquid- and the solid-phase MCMs was calculated using Langmuir and Freundlich isotherm models. Sometimes adsorption model was changed from Langmuir to Freundlich depending on the nature and composition of the adsorption biochar [25].

The catalytic activity of BC is lower than that of nanocarbon or modified BC [26]. Iron-based compounds are commonly used due to their enormous availability capability to introduce into BCs. Co-precipitation of metal oxides is the conventional method used to produce magnetic particles due to its simplicity, productivity, and low cost. It consists of mixing and heating. The most commonly produced nanoparticles are magnetite and maghemite. By precoating and/or postcoating reaction, incorporation of magnetic nanoparticles can be performed [6].

High-energy, ball milling technique is an highly economical and efficient way to prepare powdered $MC_{AC}M$ and $MC_{BC}M$. In the case of $MC_{AC}M$ adsorption takes place in its micropores. $MC_{BC}M$ adsorption occurs at the material's surface. By this technique iron magnetic particles are dispersed on the carbonaceous matrix of activated carbon and BC. The size of the nanoparticles was determined by the time of the milling process. Higher time is needed to produce smaller diameter nanoparticles. The mechanical energy produced in the process is used to separate the pharmaceuticals adsorbed [24]. In the practical application of MCM, rapid adsorption rates play an important role [23].

5 Agricultural biochar

The husk of grains, namely wheat husk and rice husk are high-capacity adsorbents that are agricultural by--products [27]. The high amount of tannin obtained from tree bark is a by-product from the timber industry which is a potential adsorbent. The polyhydroxyl group of tannin acts as the active part of the molecule for the adsorption process obtained from tree bark. Agricultural waste-embedded TiO_2 removes SMX through hydrophobic interaction [21].

6 Fluoroquinolone removal by sludge biochar

The sludge-derived BC is used for efficient adsorption of gatifloxacin, a fluoroquinolone antibiotic at a relatively low cost [28].

7 Ceftiofur and florfenicol removal by biochar

BCs produced from dairy fiber, pinewood, hickory wood, Brazilian pepper, bamboo, mixed wood, poplar wood, cherry pit, pine bark, peanut hull were used for the removal of ceftiofur and florfenicol. Dairy fiber and pinewood biochar were found to be the efficient sorbent materials for ceftiofur and florfenicol with > 99.98% removal efficiency [29].

Table 2: Biochar – temperature and pH dependency for adsorption.

Biochar	Temperature	Adsorption	Reference
Pine wood	425 °C	Low at high pH High at intermediate pH	[30]
Pine wood	550–650 °C	100% removal efficiency	[29]
Pine chip	300 °C	High	[30]
Hickory wood	450 °C	Lowest	[28]
Sugarcane	450–600 °C	Higher	[28]
Clay bio char	–	Higher in pH range 2.0–11.0	[32]
Magnetic biochar	–	Stable throughout pH 4.0–9.0 Decrease with increase in pH	[33]

8 Magnetic carbon biochars for sulfa antibiotics removal

Magnetic carbon-integrated biochars help in removing sulfa antibiotics SMX, SPY and sulfathiazole, ceftriaxone, amoxicillin, cephalexin, NOR, ciprofloxacin (CIP), tetracycline (TC), and erythromycin [32].

Clay-integrated BC prepared with potato stem and natural attapulgite shows higher adsorption for NOR in the pH range of 2.–11.0 [32]. The adsorption of SMX, which is an ionizable compound takes place due to hydrophobic π–π interactions rather than electrostatic interactions [22]. This could affect the adsorption process [27] due to electrostatic repulsions between BC and SMX [8].

MCMs are unstable below the pH value of 4 and at the pH value of 10, but were stable between the pH 4 and pH 9. The highest adsorption attained at pH 4.5 and the lowest adsorption takes place at the pH value of 8.5. Adsorption efficiency of naproxen is more when compared to acetaminophen due to its affinity, and polar π–π interactions dominated adsorption as well as π–π stacking, forming naphthalene aromatic ring when compared to the benzene ring in acetaminophen [22].

The adsorption rate was also dependent on the hydrophobicity character of different pharmaceuticals. Erythromycin existed predominantly in its anionic forms. So electrostatic attraction is impossible. Therefore, adsorption takes place in response to van der Waals force [33].

9 Doxycycline hydrochloride removal by peanut shell biochar

The biomass from the peanut shell can be transformed into BC for the adsorption of doxycycline hydrochloride from an aqueous environment [34].

10 Paracetamol

Grape stalk, cork bark, and Yohimbe bark bio chars were used for the removal of paracetamol. Grape stalk shows the more absorbent capacity for paracetamol removal. Π-stacking interactions between lignin syringyl and guaiacol moieties in grape stalk and aromatic ring in paracetamol together with H-bonding help the biochar to adsorp [35].

11 Ibuprofen removal by pinewood biochar

Pinewood-derived BC produced at 425 °C shows low adsorption capacity for ibuprofen at high pH condition, due to electrostatic repulsion between an adsorbent and adsorbate. The adsorption pH of ibuprofen is 3. The adsorption is at a peak when pH is 8.0 due to formation of hydrogen bonds between carboxylate anions and unionized surface phenolic groups present on BC. Pinewood BC is produced as a by-product of bio-oil. Naproxen and ibuprofen were adsorbed by van der Waals interactions, dipole–dipole attractions, π–π EDA interactions, and hydrogen bonding [30].

The nanofiltration membrane with charged surface modifying macromolecule in the biochar surface is used in the separation of ibuprofen molecule. Commercial activated carbons incorporated in the biochar can be used efficiently in the removal of ibuprofen, that is, activated carbon prepared from olive waste cake has the potential to remove ibuprofen, naproxen, diclofenac, and ketoprofen in aqueous media [36].

Acetaminophen is adsorbed by pinewood by π–π interactions [37]. Pine needles form of BCs possess more adsorption ability when used as BC [4].

12 Adsorption of erythromycin

The adsorption was also dependent on the pharmaceuticals hydrophobicity character. Erythromycin existed predominantly in its anionic forms. So electrostatic attraction is impossible. Therefore, adsorption takes place in response to van der Waals force [22].

Ball-milled magnetic BC effectively adsorbs pharmaceuticals like carbamazepine and TC [24]. KOH, H_2SO_4, and methanol-modified BC enhance TC removal [9]. Coconut shell and Fe_3O_4 are combined for efficient adsorption of carbamazepine and TC [28, 38]. Sulfonated sawdust BC is used in the removal of TC via electrostatic interactions [39].

13 Tetracycline removal by carbon nanoparticles

TC is one of the most widely used antibiotics, which was not metabolized and eliminated from the body. They can enter the surface and groundwater. Discharge from hospitals and pharmaceutical companies also enters the groundwater and aquatic streams [10].

TC was dominant in its anionic form when the pH lies between 4.0 and 8.0. Hence, adsorption is decreased due to repulsion force exerted by TC toward the BC surface [12].

TC is an amphoteric molecule with multiple ionizable functional groups [20]. Affinity of TC for MCM was higher at low pH in solutions. The adsorption decreases with an increase in pH of the solution from 3 to 10 and 4 to 9, respectively. The π–π electron donor–acceptor interactions and cation–π bonding are the interactions that take place between TC and MCM [24].

Sawdust biomass was used in the adsorption of TC antibiotics [34]. BCs impregnated with bentonite clay were good adsorbents of TC [40]. TC adsorption in a sawdust-based BC composite with the addition of zinc and iron chloride salts showed an almost threefold higher adsorption capacity in relation to TC than did pristine BC. The material produced by subjecting the feedstock to the action of a mixture of both salts exhibited a fourfold increase in Q_{max} value versus original BC [41].

14 Amoxicillin/cephalexin removal

Carbon nanoparticles obtained from vine wood, incorporated in BCs are suitable for the adsorption process of antibiotics such as amoxicillin, cephalexin, TC, and penicillin G. Highest adsorption of TC and penicillin G was 0.4 g L^{-1} at pH 2 [42].

H_3PO_4 modification of BCs derived from swine manure and rice straw increases the adsorption of TC. The highest adsorption capacity for TC is 25% more than pristine BC for modified BC obtained from manure. Electrostatic interactions, cation exchange, metal complexation, π–π EDA interactions, and hydrogen bonding were responsible for the mechanism of TC adsorption onto pristine-derived BC. But after metal cations Ca^{2+}, Mg^{2+}, Na^{+} were removed, π–π EDA interactions and hydrogen bonding were mainly responsible for the adsorption. The adsorption was observed to significantly increase with increasing pH from 5 to 9, which was explained by the larger contribution of electrostatic interactions to TC binding [2].

TC adsorption for a sawdust-based BC composite with the addition of iron chloride and zinc salts exhibited an almost threefold higher adsorption capacity in relation to TC than did pristine BC. The material made by exposing the feedstock to the action of a mixture of both salts revealed a fourfold increase in Q_{max} value versus original BC [41].

Composite BCs derived from potato stems and leaves as well as from magnetic manganese oxide (BC–FeMn) as adsorbents of three antibiotics belong to the quinolone class, that is, NOR, CIP, and enrofloxacin.

Cow manure and rice husk-derived BCs were used for oxytetracycline and TC via π–π electron donor–acceptor interactions [23]. The adsorption of TC was found to be 8.37 mg g^{-1} by rice husk which is comparatively high than other adsorbent materials in literature. Oxytetracycline antibiotic was adsorbed in maize straw too [27].

There exists a competing effect between oxytetracycline and carbaryl for BC adsorption. This competition may hinder the adsorption of carbaryl toward BC produced at a temperature of 400–600 °C [32].

Polymeric sorbents remove polar or nonpolar pharmaceuticals from water. It can be synthesized as ionic or hydrophobic (13)

The removal of antibiotics from water is done using BC composites with the addition of metals and their compounds via adsorption and degradation of an antibiotic [38] is possible. When BC is coated with titanium oxide (BC–TiO_2) as a photocatalyst for the degradation of SMX, the adsorption capacity of the composite lowers the adsorption capacity of the original BC. The cause of the decrease is related to the

Table 3: Antibiotics and mechanism of adsorption.

Biochar	Antibiotics	Mechanism of adsorption	Reference
Agricultural wastes	Sulfamethoxazole	π–π EDA interaction	[21]
Bamboo biochar	Sulfamethoxazole sulfathiazole	π–π EDA interaction Hydrogen bonding	[43]
Brazilian pepper wood	Sulfamethoxazole	Hydrophobic bonding	[28, 38]
Bamboo biochar	Sulfamethazine Sulfathiazole	Lewis acid–base interactions	[43]
Bur cucumber plan	Sulfamethazine	π–π interaction, electrostatic interaction, hydrogen bonding	[8]
Cow manure biochar	Oxytetracycline	π–π EDA interactions, specific site adsorption	[32]
Faujasite-type zeolites	Azithromycin, ofloxacin, SMX	Electrostatic interactions, hydrogen bonding	[44]
Hickory wood	Sulfamethoxazole	Surface functional groups	[28, 38]
Loblolly pine chips	Sulfamethoxazole	π–π interactions	[37]
Loblolly pine	Carbamazepine, diclofenac	Hydrogen bonding	[43]
Pine wood	Acetaminophen	Π–π interactions	[37]
Pinewood	Ibuprofen naproxen	Van der Waals interactions, Dipole–dipole attractions, π–π EDA interactions and hydrogen bonding	[30]
Polymeric substances	Polar or non-polar pharmaceuticals	Ionic or non-ionic	[13]
Sludge samples	Gatifloxacin	–	[28, 38]
Sulfonated saw dust	Tetracycline sulfamethoxazole	Electrostatic interactions	[39]
CA-Montmorillonite	Chlorpheniramine	Swelling	[40]
SD-SO_3H	SMX	Electrostatic interaction	[39]
Clay biochar (potato stem + natural attapulgite)	Norfloxacin	Electrostatic interaction, hydrophobic interaction, hydrogen bonding, π–π EDA interactions	[11]
Agricultural waste + TiO_2	SMX	Hydrophobic interaction	[21]

Table 3 (continued)

Biochar	Antibiotics	Mechanism of adsorption	Reference
Magnetic biochar (coconut shell + Fe_3O_4)	Carbamazepine, tetracycline	Surface complexation	[33]
Rice husk	Tetracycline	π–π EDA interactions	[23]

composite's adsorption parameters. The amalgamation of TiO_2 into the BC reduces the number of oxygen functional groups participating in the binding of the drug. However, comparing with the efficiency of pure titanium oxide and the BC's reduced adsorption ability, the adsorption capacity was higher for BC–TiO_2 (91%) than for pure titanium oxide (59%). A high degree of TiO_2 dispersion in the composite was found – the SMX [2, 27].

Cadmium ions have a positive effect over the adsorption of SMX onto the surface of BC. The effect on the surface of the BC get neutralized due to the interference of cadmium ions paving way to adsorption of SMX. In this process, electrostatic repulsion between anionic SMX and BC has been reduced by cadmium ions [16].

15 Ciprofloxacin removal by chemically prepared carbon biochar adsorption

Chemically prepared carbon biochar from date palm (*Phoenix dactylifera*) removes CIP. It is observed that the highest absorption of CIP was attained at initial pH 6, and wet adsorbent showed high performance than dry adsorbent. It is an endothermic reaction and is physical and spontaneous [10].

16 Synthetic sorbents for removal of pharmaceuticals

Polymeric sorbents integrated biochar remove polar or nonpolar pharmaceuticals from water. It can be synthesized as ionic or hydrophobic. Zeolites were used for water treatment processes. They are microporous materials, which had networks of SiO_4 and AlO_4. Faujasite zeolites are used for the removal of azithromycin, ofloxacin, and SMX antibiotics. They exhibit a pH tendency and the mechanism of adsorption via electrostatic interactions and hydrogen bonding [44]. SD-SO_3H BC is used in the removal of SMX by way of electrostatic interaction [39].

The BC surfaces have a net negative charge, and docusate containing a strongly acidic sulfonate group is anionic over a wide pH range. Electrostatic attraction is precluded and adsorption takes place by hydrophobic interactions. Bamboo BC, Brazilian pepperwood, sugarcane bagasse, and hickory wood were used to remove SMX. Rice straw BCs are used in the removal of SMX through π–π EDA interactions and surface complexation [16].

Hickory chips BCs were used to remove sulfonamides in aqueous solutions; bamboo BCs were produced at 450 °C and 600 °C and hickory BCs were produced at 450 °C, which showed the lowest adsorption, whereas the highest adsorption was noticed from sugarcane bagasse produced at 450°and 600 °C [28, 38].

17 Cotton gin and guayule bagasse waste biochars

BCs derived from cotton gin and guayule bagasse are used as filters to remove pharmaceutical wastes from effluent waters. The surface properties of BC vary depending on the feedstock which in turn reduces some pharmaceutical compounds. The intermediate degree of hydrophobicity enhances the removal of pharmaceuticals since most medicaments exhibit weak acids or bases that are hydrophobic in nature [45].

18 Tea waste biochar for sulfamethazine

Sulfamethazine was adsorbed by tea waste BC from the aqueous media. Tea waste BC adsorption capacity is more than pristine BC [8].

The advanced methods involved in the treatment of wastewater treatment are advanced oxidation processes. BC acts as an electron shuttle and the precursor persulfate activation takes place [46]. Loblolly pine chips are used in the adsorption of SMX through π–π interactions [37]. Loblolly pine chip BCs were also used in the adsorption of carbamazepine and diclofenac through hydrogen bonding [43].

Bur cucumber plant-converted BCs are used in the removal of sulfamethazine. Adsorption is enhanced through steam activation of bur cucumber BCs [8]. Magnetic BC was decreased with an increase in pH [47].

Due to its adsorption characteristic, BC was used to adsorb the dyes, heavy metals, pesticides, herbicides, and other pollutants. Advancement methods of BC production reported a positive increase in the removal of pollutants from wastewater due to its high porosity, surface area, and adsorption properties. BC is considered an adsorbent material enriched with nutrients to promote the sustainable agriculture [48].

BC is advantageous because it is eco-friendly, safe, cost-effective, and has high porosity, surface area, and adsorption capacity [49]. The amount of negative charges that exit the BC surface helps to adsorb heavy metal ions with positive charge using electrostatic interactions [50].

19 Conclusion

Adsorption or ozonation leads to degradation of the PPCPs. The efficiency of adsorption may turn on many factors, that is, the composition and physicochemical properties determine the adsorption. The adsorption processes are determined by the functional groups, $\pi-\pi$ EDA interactions of pollutants with the BC surface, and hydrophobic interactions. pH, initial concentration, and temperature also play an important role in the process of adsorption. The environment which got polluted with antibiotics creates many potential threats to the health and life of organisms.

Global consumption of drugs by humans and animals and untreated industrial effluents aggravates the scale of contamination. Various techniques have been in use for the removal of such pollutants. One of the best techniques in recent times is to use BC-based adsorbents. Physically and chemically modified BC and BC composites can be effective in removing antibiotics.

References

[1] Kümmerer K, Pharmaceuticals in the environment – a brief summary, In: Kümmerer K, eds, Pharmaceuticals in the Environment, Springer, Berlin, Heidelberg, 2008, 3–21.

[2] Chen Y, Shi J, Du Q, Zhang H, Cui Y, Antibiotic removal by agricultural waste biochars with different forms of iron oxide, RSC Adv., 2019, 9(25), 14143–14153.

[3] Petrie B, Barden R, Kasprzyk-Hordern B, A review on emerging contaminants in wastewaters and the environment: Current knowledge, understudied areas and recommendations for future monitoring, Water Res., 2015, 72, 3–27.

[4] Yaashikaa PR, Kumar PS, Varjani S, Saravanan A, A critical review on the biochar production techniques, characterization, stability and applications for circular bioeconomy, Biotechnol. Rep., 2020, 28, e00570.

[5] Rocha LS, Pereira D, Sousa E, Otero M, Esteves VI, Calisto V, Recent advances on the development and application of magnetic activated carbon and char for the removal of pharmaceutical compounds from waters: A review, Sci. Total Environ., 2020, 718, 137272.

[6] Siddiqui MT, Nizamuddi S, Baloch HA, *et al.*, Fabrication of advance magnetic carbon nanomaterials and their potential applications: A review, J. Environ. Chem. Eng., 2018, 7(1), 102812.

[7] Krasucka P, Pan B, Ok SY, Mohan D, Sarkar B, Oleszczuk P, Engineered biochar – a sustainable solution for the removal of antibiotics from water, Chem. Eng. Sci., 2020, 405, 126926.

[8] Rajapaksha AU, Vithanage M, Ahmad M, *et al.*, Enhanced sulfamethazine removal by steam-activated invasive plant-derived biochar, J. Hazard. Mater., 2015, 290, 43–50.

[9] Ahmed MB, Zhou JL, Ngo HH, Guo W, Adsorptive removal of antibiotics from water and wastewater: Progress and challenges, Sci. Total Environ., 2015, 532, 112–126.
[10] El-Shafey ESI, Al-Lawati H, Al-Sumri AS, Ciprofloxacin adsorption from aqueous solution onto chemically prepared carbon from date palm leaflets, J. Environ. Sci., 2012, 24(9), 1579–1586.
[11] Wang W, Zheng B, Deng Z, Feng Z, Fu L, Kinetics and equilibriums for adsorption of poly(vinyl alcohol) from aqueous solution onto natural bentonite, Chem. Eng. Sci., 2013, 214, 343–354.
[12] Rajapaksha AU, Dilrukshi Premarathna KS, Gunarathne V, Ahmed A, Vithanage M, Sorptive removal of pharmaceutical and personal care products from water and wastewater, Pharmaceut. Personal Care Prod. Waste Manage. Treat. Technol., 2019, 213–238.
[13] Akhtar J, Amin NAS, Shahzad K, A review on removal of pharmaceuticals from water by adsorption, Desalin. Water Treat., 2015, 57(27), 12842–12860.
[14] Sirés I, Brillas E, Remediation of water pollution caused by pharmaceutical residues based on electrochemical separation and degradation technologies: A review, Environ. Int., 2012, 40, 212–229.
[15] Rodarte-Morales AI, Feijoo G, Moreira MT, Lema JM, Degradation of selected pharmaceutical and personal care products (PPCPs) by white-rot fungi, World J. Microbiol. Biotechnol., 2011, 27(8), 1839–1846.
[16] Han X, Liang C, Li T, Wang K, Huang H, Yang X, Simultaneous removal of cadmium and sulfamethoxazole from aqueous solution by rice straw biochar, J. Zhejiang Univ. Sci. B, 2013, 14(7), 640–649.
[17] Islam MA, Auta M, Kabir G, Hameed BH, A thermogravimetric analysis of the combustion kinetics of karanja (*Pongamia pinnata*) fruit hulls char, Bioresour. Technol., 2016, 200, 335–341.
[18] Liu W-J, Jiang H, Yu H-Q, Development of biochar-based functional materials: Toward a sustainable platform carbon material, Chem. Rev., 2015, 115(22), 12251–12285.
[19] Styszko K, Nosek K, Motak M, Bester K, Preliminary selection of clay minerals for the removal of pharmaceuticals, bisphenol A and triclosan in acidic and neutral aqueous solutions, C R Chim., 2015, 18(10), 1134–1142.
[20] Zhu X, Liu Y, Qian F, Zhou C, Zhang S, Chen J, Preparation of magnetic porous carbon from waste hydrochar by simultaneous activation and magnetization for tetracycline removal, Bioresour. Technol., 2014, 154, 209–214.
[21] Kim JR, Kan E, Heterogeneous photocatalytic degradation of sulfamethoxazole in water using a biochar-supported TiO_2 photocatalyst, J. Environ. Manage., 2016, 180, 94–101.
[22] Reguyal F, Sarmah AK, Gao W, Synthesis of magnetic biochar from pine sawdust via oxidative hydrolysis of $FeCl_2$ for the removal sulfamethoxazole from aqueous solution, J. Hazard. Mater., 2017, 321, 868–878.
[23] Zhang H, Wang Z, Li R, *et al.*, TiO_2 supported on reed straw biochar as an adsorptive and photocatalytic composite for the efficient degradation of sulfamethoxazole in aqueous matrices, Chemosphere, 2017, 185, 351–360.
[24] Shan D, Deng S, Zhao T, *et al.*, Preparation of ultrafine magnetic biochar and activated carbon for pharmaceutical adsorption and subsequent degradation by ball milling, J. Hazard. Mater., 2016, 305, 156–163.
[25] Sforza E, Kumkum P, Barbera E, Kumar S, Bioremediation of industrial effluents: How a biochar pretreatment may increase the microalgal growth in tannery wastewater, J. Water Process Eng., 2020, 37, 101431.
[26] Huang D, Zhang Q, Zhang C, *et al.*, Mn doped magnetic biochar as persulfate activator for the degradation of tetracycline, Chem. Eng. Sci., 2019, 123532.
[27] Chen H, Gao B, Li H, Removal of sulfamethoxazole and ciprofloxacin from aqueous solutions by graphene oxide, J. Hazard. Mater., 2015, 282, 201–207.

[28] Yao H, Lu J, Wu J, Lu Z, Wilson PC, Shen Y, Adsorption of fluoroquinolone antibiotics by wastewater sludge biochar: Role of the sludge source, Water Air Soil Pollut., 2012, 224(1), 1370–1378.
[29] Mitchell SM, Subbiah M, Ullman JL, Frear C, Call DR, Evaluation of 27 different biochars for potential sequestration of antibiotic residues in food animal production environments, J. Environ. Chem. Eng., 2015, 3(1), 162–169.
[30] Essandoh M, Kunwar B, Pittman CU, Mohan D, Mlsna T, Sorptive removal of salicylic acid and ibuprofen from aqueous solutions using pine wood fast pyrolysis biochar, Chem. Eng. Sci., 2015, 265, 219–227.
[31] Jung C, Oh J, Yoon Y, Removal of acetaminophen and naproxen by combined coagulation and adsorption using biochar: Influence of combined sewer overflow components, Environ. Sci. Pollut. Res., 2015, 22, 10058–10069.
[32] Li M, Zhao Z, Wu X, Zhou W, Zhu L, Impact of mineral components in cow manure biochars on the adsorption and competitive adsorption of oxytetracycline and carbaryl, RSC Adv., 2017, 7(4), 2127–2136.
[33] Rakshit S, Punamiya P, Datta R, Sarkar D, Sorption of oxytetracycline on magnetite-water interface, Geochim. Cosmochim. Acta, 2010, 74(12), A846.
[34] Ben Y, Fu C, Hu M, Liu L, Wong MH, Zheng C, Human health risk assessment of antibiotic resistance associated with antibiotic residues in the environment: A review, Environ. Res., 2019, 169, 483–493.
[35] Villaescusa I, Fiol N, Poch J, Bianchi A, Bazzicalupi C, Mechanism of paracetamol removal by vegetable wastes: The contribution of π–π interactions, hydrogen bonding and hydrophobic effect, Desalination, 2011, 270(1–3), 135–142.
[36] Baccar R, Sarrà M, Bouzid J, Feki M, Blánquez P, Removal of pharmaceutical compounds by activated carbon prepared from agricultural by-product, Chem. Eng. Sci., 2012, 211–212, 310–317.
[37] Jung C, Park J, Lim KH, *et al.*, Adsorption of selected endocrine disrupting compounds and pharmaceuticals on activated biochars, J. Hazard. Mater., 2013, 263, 702–710.
[38] Yao Y, Gao B, Chen H, *et al.*, Adsorption of sulfamethoxazole on biochar and its impact on reclaimed water irrigation, J. Hazard. Mater., 2012, 209–210, 408–413.
[39] Ahsan MA, Islam MT, Hernandez C, *et al.*, Biomass conversion of saw dust to a functionalized carbonaceous materials for the removal of Tetracycline, Sulfamethoxazole and Bisphenol A from water, J. Environ. Chem. Eng., 2018, 6(4), 4329–4338.
[40] Li B, Zhang T, Biodegradation and adsorption of antibiotics in the activated sludge process, Environ. Sci. Technol., 2010, 44(9), 3468–3473.
[41] Zhou Y, Liu X, Xiang Y, *et al.*, Modification of biochar derived from sawdust and its application in removal of tetracycline and copper from aqueous solution: Adsorption mechanism and modelling, Bioresour. Technol., 2017, 245, 266–273.
[42] Pouretedal HR, Sadegh N, Effective removal of Amoxicillin, Cephalexin, Tetracycline and Penicillin G from aqueous solutions using activated carbon nanoparticles prepared from vine wood, J. Water Process Eng., 2014, 1, 64–73.
[43] Ahmed MB, Zhou JL, Ngo HH, Guo W, Johir MAH, Sornalingam K, Single and competitive sorption properties and mechanism of functionalized biochar for removing sulfonamide antibiotics from water, Chem. Eng. Sci., 2017, 311, 348–358.
[44] De Sousa DNR, Insa S, Mozeto AA, Petrovic M, Chaves TF, Fadini PS, Equilibrium and kinetic studies of the adsorption of antibiotics from aqueous solutions onto powdered zeolites, Chemosphere, 2018, 205, 137–146.

[45] Ndoun MC, Elliott HA, Preisendanz HE, Williams CF, Knopf A, Watson JE, Adsorption of pharmaceuticals from aqueous solutions using biochar derived from cotton gin waste and guayule bagasse, Biochar, 2021, 3(3), 89–104.

[46] Boczkaj G, Fernandes A, Wastewater treatment by means of advanced oxidation processes at basic pH conditions: A review, Chem. Eng. Sci., 2017, 320, 608–633.

[47] Daghrir R, Drogui P, Tetracycline antibiotics in the environment: A review, Environ. Chem. Lett., 2013, 11, 209–227.

[48] Varjani S, Joshi R, Srivastava VK, Ngo HH, Guo W, Treatment of waste- water from petroleum industry: Current practices and perspectives, Environ. Sci. Poll. Res, 2019a, 1–9, doi: https://doi.org/10.1007/s11356-019-04725-x.

[49] Jiang A, Cheng Z, Shen Z, Guo W, QSAR study on the removal efficiency of organic pollutions in supercritical water based on degradation temperature, Chem. Cent. J., 2018, 12, 16–23.

[50] Zhang W, Shi X, Zhang Y, Gu W, Li B, Xian Y, Synthesis of water-soluble magnetic graphene nanocomposites for recyclable removal of heavy metal ions, J. Mater. Chem. A, 2013, 1(5), 1745–1753.

Reena Kumari, Poonam Dhankhar, Vikram Dalal

Applications of biochar to remediate heavy metal–contaminated soil

Abstract: Soil contamination by heavy metals has become a serious environmental concern worldwide in the last decade. The immense number of techniques like surface caping, landfilling, vitrification, and immobilization have been used to remediate heavy metals from the contaminated soils. In recent years, biochar has been received significant attention in absorption of heavy metal pollutants, as it is a cost-effective and environment-friendly material to remediate the various types of contaminants such as pharmaceutical and organic compounds, and dyes. This chapter summarizes the preparation of biochar along with its physicochemical properties and the effect of biochar on heavy metal–contaminated soil. Furthermore, the adsorption mechanism behind the biochar interactions and heavy metal soil contaminants are described in detail. In addition to it, the possible risks associated with in situ biochar remediation strategy are outlined, which will aid future research in the development and sustainable use of biochar.

Keywords: biochar, remediation, contaminated soil, heavy metal, adsorption

1 Introduction

In recent years, soil pollution has significantly increased and it is caused due to the presence of contaminants like chemicals, compounds, or substances at undesirable concentrations in soil, which may affect the normal physicochemical activities of nontargeted organisms [1]. Several toxic compounds, such as dyes, phthalates, and biphenyl congeners are reported to inhibit the endocrine receptors [2–6]. Various enzymes like dye-degrading peroxidase, mono-2-ethylhexyl phthalate (MEHP) hydrolase, di-2-ethylhexyl phthalate (DEHP) hydrolase, and phthalate reductase have been identified to efficiently remediate the hazardous pollutants from the environment [5–8]. Inadequate use of pesticides and fertilizers to achieve high yields, accelerated industrialization and urbanization, as well as poor household management and industrial effluents, all contributed to an environmental pollution scenario. This raised a major public concern regarding the environmental and public health protection [9]. The major inorganic contaminants are heavy metals (HMs) such as cadmium, chromium, cobalt and metalloids like arsenic and selenium. Due to certain adverse effects of HMs (mobility, nondecomposed, bioaccumulation, and ecotoxicity), metal-contaminated soil and sediment have received considerable attention from the concerned authorities [10–12].

https://doi.org/10.1515/9783110734003-018

In terrestrial systems, soil is the primary source of chemical pollutants. The fundamental material needs of people for their survival and growth rely heavily on land farming and management [13]. Soil retained metals through sorption, precipitation, and complexation can be subjected to a number of processes, including plant retention, leaching, volatilization, redox, and (de)methylation [13]. Since they remained permanent in the soil and cannot be decomposed by organic activity, therefore they affect the crop yield and quality, even threaten human and animal health by direct exposure and through the food chain [10, 14]. While, sediments are the primary sink and pollutant source in aquatic environments and play a significant role in both ecosystem and ecology [13]. Once the metals are dumped into an aquatic environment, approximately 90% of them are deposited on sedimentary surfaces due to adsorption, precipitation, flocculation, and absorption into mineral grid structures [9]. However, if the environmental conditions for benthic and water are changed, bound metals may be released from sediments, making them more mobile and bioavailable [13].

It is therefore, necessary to improve the stability of contaminants as well as decrease their bioavailability and migration potential in natural matrixes. A variety of remediation methods, including physical remediation (washing, flushing, solidification, thermal desorption, and vacuum ventilation methods), chemical remediation (vitrification, leaching, immobilization, and electrokinetic methods), and biological remediation (microorganisms and plants) have been developed to remediate the HMs [15]. However, these methods exist, but with their own set of limitations such as complicated technique, poor feasibility, inefficiency, high economic cost, short duration, and high secondary risk [15, 16].

One environmental and economical approach is biochar-based remediation technology for the metal-contaminated soil [13]. Biochar is a solid carbonaceous material produced from thermochemical conversion (<700 °C) of carbonaceous biomass, such as agricultural residues, manures, forest residues, algal biomass, activated sludge, energy crops, and digestate under O_2-limiting conditions [17]. Biochar has a broader range of environmental applications due to its unique properties, such as high adsorption capability, microporosity, high specific surface area, and ion exchange capacity [17]. All these make biochar an attractive and cost-effective alternative method for soil remediation.

2 Soil pollution with heavy metals and sources

2.1 Heavy metals and their toxic effects

Although there is no precise description of HM, it has been characterized in the literature as a naturally occurring metallic element with high specific density, which is five times higher than water [18]. Their potential to cause toxicity even at low concentrations made them contaminants of urgent concern. HMs are defined as trace elements as they are found in trace concentrations over the earth's crust [18]. It has also been reported that metals such as arsenic, cadmium, chromium, cobalt, lead, nickel, mercury, selenium, and zinc are highly toxic to plants and animals, even in trace amounts [19]. Based on their role in living organisms, they are categorized as essential and nonessential elements. Metals such as manganese (Mn), nickel (Ni), iron (Fe), and zinc (Zn) are essential elements, as they are required to perform various physiological and biochemical functions of plants [20]. Other metals such as lead (Pb), arsenic (As), cadmium (Cd), chromium (Cr), and mercury (Hg) are nonessential elements with no physiological role but they entered into the living systems via the food chain [20]. In recent decades, besides their normal occurrence, the quantity of HMs has increased drastically, particularly when a number of industries release their waste material into fresh-water without proper treatment [18]. The increased amount of HMs beyond their permissible limit has become a major threat to the environment as well as to human health [19]. HMs may enter into the human body through accumulation in the food chain and become harmful when they are accumulated in the soft tissues of body [19]. For example, Cd poisoning has a number of deleterious effects on cellular molecules, primarily by inducing an oxidant-antioxidant imbalance. In addition, Cd has been involved in the pathogenesis of many cancers, myocardial infarction, peripheral artery disease, itai–itai disease, hypertension, and diabetic nephropathy [21]. The chronic exposure to arsenic results in a variety of symptoms, including hypertension, diabetes mellitus, obstetric complications, neurological effects, blood vessel, and respiratory system disorders, as well as skin lesions such as leucomelanosis, melanosis, and keratosis [22]. Furthermore, Pb is a nonessential element to the human body, and excessive Pb consumption can have a detrimental effect on the skeletal, nervous, enzymatic,

Table 1: Impact of some heavy metals on human health.

Heavy metals	Health effects
Arsenic (As)	Nausea, vomiting, diarrhea, painful neuropathy, encephalopathy, hypopigmentation/hyperkeratosis, diabetes, cancer: skin, bladder, lung, ncephalopathy
Cadmium (Cd)	Proteinuria, pneumonitis, lung cancer, osteomalacia
Chromium (Cr)	Acute renal failure, hemolysis, gastrointestinal hemorrhage, pulmonary fibrosis, lung cancer

Table 1 (continued)

Heavy metals	Health effects
Copper (Cu)	Gastrointestinal irritation/hemorrhage, blue vomitus, hemolysis, multiorgan dysfunction syndrome, wilson disease, anaemia, mental disorders, cystic, insomnia, postpartum psychosis, autism, rheumatoid arthritis, heart problem, hypertension, hyperactivity, nausea, enlargement of liver, inflammation
Mercury (Hg)	Elemental fever, diarrhea, vomiting, acute lung injury, caustic gastroenteritis, metallic taste, nausea, neurasthenia, tremor, gingivostomatitis, hypersensitivity, nephritic syndrome
Nickel (Ni)	Dermatitis, encephalopathy, myocarditis, nasopharyngeal tumors, pulmonary fibrosis, reduced sperm count
Lead (Pb)	Vomiting, abdominal pain, diarrhea, neurologic degeneration, osteoporosis, copper deficiency: anemia

immune, endocrine, and circulatory systems [23]. Table 1 represents the adverse effects of particular HMs on human health.

2.2 Sources of heavy metals

The distribution of HMs in the soil is correlated with natural processes and anthropogenic activities. Numerous research has shown that natural sources of HMs are generally of little concern to the ecosystem compared with anthropogenic activities [19]. The primary source of HMs is the parent rock materials and different natural weathering processes including volcanic eruptions, rock weathering, forest fires, sea-salt sprays, biogenic sources, and wind-borne soil particles that allow HMs to be released into the soil [11]. Heavy elements can be found as hydroxides, phosphates, oxides, sulfates, sulfides, silicates, and organic compounds. The most prevalent HMs are lead (Pb), mercury (Hg), arsenic (As), zinc (Zn), chromium (Cr), nickel (Ni), cadmium (Cd), and copper (Cu). Although the listed HMs are present in traces but pose health risks to humans and other animals [18]. Recent developments in industrial, agricultural, and urban sectors have made a significant contribution to increased contamination of soil by HMs. Anthropogenic activities like mining and smelting, disposal of municipal wastes, combustion of fossil fuel refining, sewage irrigation, application of pesticides, and fertilizer application, in the agricultural soil, have contributed to the high levels of HMs in the environment. Over the past few decades, the annual release of HMs has reached 1,35,000 t for Zn, 783,000 t for Pb, 939,000 t for Cu, and 22,000 t (metric ton) for Cd [19]. Figure 1 illustrates both natural and anthropogenic environmental sources of HMs.

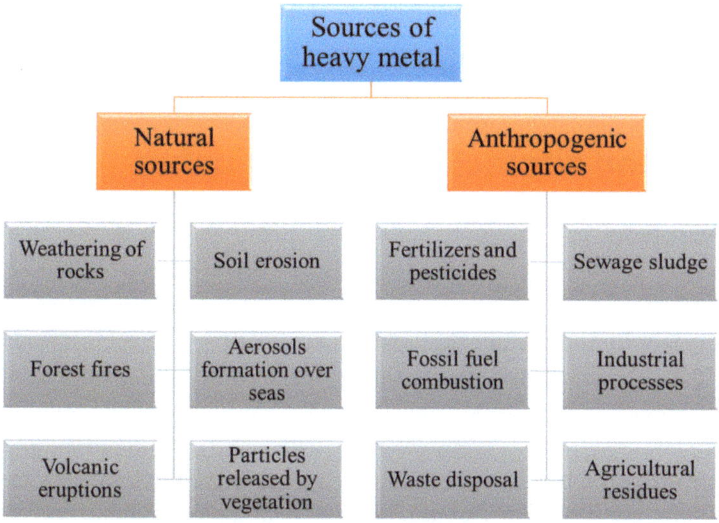

Figure 1: Heavy metal sources (natural and anthropogenic) in environment.

2.3 Remediation of heavy metal polluted soils

Various approaches have been implemented to remediate the HMs from different polluted sites. The remediation approaches were classified into two types: in situ and ex situ approaches [11]. In in situ approach, the treatment of contaminants is performed on its original site, without shifting the polluted soil itself. Whereas in ex situ approach, the contaminated soil is excavated and removed from the site for treatment. The in situ remediation provides a variety of potential technological, economic and environmental advantages over ex situ remediation. However, the choice of soil remediation method depends on the characteristics of the site, the concentration of contaminants, the types of contaminants to be removed, and the final usage of the contaminated medium [19].

The two main processes for HM polluted soils are immobilization and extraction, which can be performed in situ or ex situ (on or off-site). Metal immobilization refers to the process of stabilizing or fixing metal ions that decreases their leaching into the soil matrix and turn them into a less soluble, harmful, or bioavailable form in the soil, thus minimizing environmental risks [9]. While, during the extraction processes, metals are separated from the composition of soil. Extraction declines the volume and concentration of metals from the contaminated soil. However, immobilization approaches are more reasonable due to many advantages over-extraction in respect to cost and handling. HM remediation from soil has shown encouraging and efficient results when AC or biochar is used as an adsorbent (immobilization agent). Because of the low energy demand and potential impact on global warming, AC biochar has emerged as the preferred alternative to AC [9]. Table 2 shows the merits and demerits of different remediation techniques and their application status.

3 Manufacture and characteristics of biochar

3.1 Biochar production

Biochar is prepared mainly from crop biomass, wood chips/branches, agricultural residues (cereal straw, hazelnut, and peanut shell, wheat straw, etc.), and other woody biomass, as well as from animal manure, sewage sludge, de-inking paper sludge, and other organic wastes [16]. Major thermochemical conversion technologies include slow pyrolysis, intermediate pyrolysis, fast pyrolysis, torrefaction, gasification, and hydrothermal carbonization for biochar preparation [16, 17]. These technologies are mainly categorized according to their peak temperatures, heating speeds, reaction atmospheres, residence periods, and other parameters, of which the main products are biochar (solid fraction), bio-oil (partly condensed VM), and syngases (e.g., CO, CO_2, CH_4, and H_2) [17]. The yield of biochar is greatly dependent on the type of pyrolysis employed, which is the most common method for biochar production.

Pyrolysis is a process of thermochemical decomposition of biomass under the controlled pyrolysis temperature in the absence or presence of little oxygen in the environment [24]. Slow pyrolysis is the typical method of pyrolysis that is conducive not to the formation of fluid and gas products, but rather to the production of biochar. In this process, the biomass is pyrolyzed over a broad range of carbonization temperatures at a heating rate of approximately 0.1–1 °C/s for a residence period ranging from a few hours to several days [16]. The rise in peak temperature (400–500 °C) can result in an increase in the fixed carbon content of biochar. On the contrary, fast pyrolysis uses a higher heating rate (>200 K min^{-1}) and a shorter residence period (<2 s) which produces a small amount of biochar, high energy density bio-oil, and syngas with a low energy density [18]. Intermediate pyrolysis works at the operating conditions between fast and slow pyrolysis, resulting in a more balanced distribution of solid–liquid product yields. The biochar produced from this pyrolysis method has a brittle structure and contains a less reactive tar that is ideal for use as a solid fuel, soil modification, and fertilizer [18].

3.2 Properties of biochar

The physicochemical characteristics of biochar relevant to its soil remediation efficiency include surface area and pore distribution, CEC, pH, aromaticity, and VM content (Figure 2) [24]. The surface area is the entire surface area occupied by a

Table 2: Merits and demerits of available techniques for the remediation of heavy metal from contaminated soil.

Method	Remediation technique	Merits	Demerits
Physical remediation	Landfilling	High security, immediate clean up	High cost, requiring extra space for waste storage
	Surfacecapping	Low in cost, easy to install, high security	Restricted to small areas and specific geographic regions, loss of land
	Encapsulation	Fast to install, high security,	High cost, restricted to small, shallow contamination areas
Chemicalremediation	Solidification	High efficiency, fast to implement	High cost, treated area losing environmentalfunctions
	Stabilization	Easy to implement, affordable, immediate effects	Temporary effectiveness, metal-specific, contaminants left in soil
	Soil washing	Fast effects with high efficiency	Extreme soil disturbance
	Soil flushing	Low cost, simple to install, minimal soil disturbance	Best suited to coarse-textured soils, potentialgroundwater pollution
	Vitrification	High efficiency	Confined to a limited soil area/volume, high cost
Bioremediation	Contaminanttransformationbymicroorganisms	Low cost, minimal soil disturbance, simple to use	Low efficiency
	Contaminantremovaland/orstabilization byplants	High public acceptance, easy to use, low cost	Metal-specific, time consuming, restricted to shallow contamination, low efficiency

solid object, which is determined by the pore structure of the solid object. The surface area of biochar typically increases with an increment of pyrolysis temperature [25]. The large surface area of biochar with a highly porous structure usually leads to the higher adsorption of HM and organic contaminants. The pH value of biochar indicates its alkalinity or acidity. The majority of biochars are alkaline and with the increase in pyrolysis temperature, the content of the ash increases such that the biochar's alkaline quality increases, leading to higher pH of biochar. It has been reported that the addition of biochar to soil increases the pH thereby lowering the mobility and bioavailability of Pb, Cd, and Zn [9]. A higher pH value of biochar may substantially support its HMs adsorption by a number of mechanisms. CEC of biochar is defined as the ability to exchange the environment positively charged ions with its own cations. In general, low-temperature biochars have high CECs because they contain a lot of oxygenated functional groups, which indicates a high rate of complex formation with metal cations. Biochar with a higher CEC can help in the preservation of nutrients (e.g., N, P, and K) as well as the adsorption of pollutants when added to soil. The presence of mineral or ash content in biochar enhances its pH, which facilitates the adsorption of cationic metal ions like Pb^{2+}, Zn^{2+}, Cd^{2+}, and Ni^{2+} by increasing cation exchange. The biochar aromatization can be demonstrated by H/C and O/C ratios [25]. The aromaticity of biochar is affected by the temperature of pyrolysis. Biochar produced at a higher temperature has lower H/C and O/C ratios, which indicate a steady rise in aromaticity and a decrease in polarity at higher temperatures. Biochar with high aromaticity would be more effective for HM adsorption via improved cation-π interactions. The high VM content usually indicates that biochar contains more acidic surface functional groups, which may help in the adsorption of HMs through increased CEC and complexation capability [24].

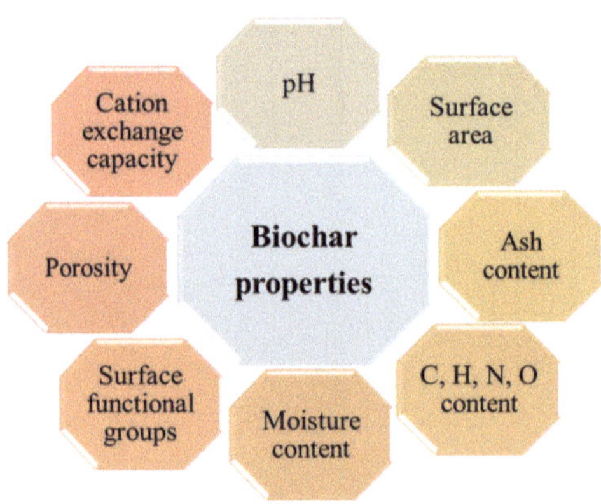

Figure 2: Physicochemical properties of biochar.

4 Biochar interaction mechanism in soil

4.1 Remediation mechanism

Biochar is a porous structure with aromatic elements along with active functional groups on its surface. These properties of biochar played a pivotal role in the remediation of various organic, inorganic, and gases in soil using different methods like electrostatic interactions, physical adsorption, and precipitation (Figure 3).

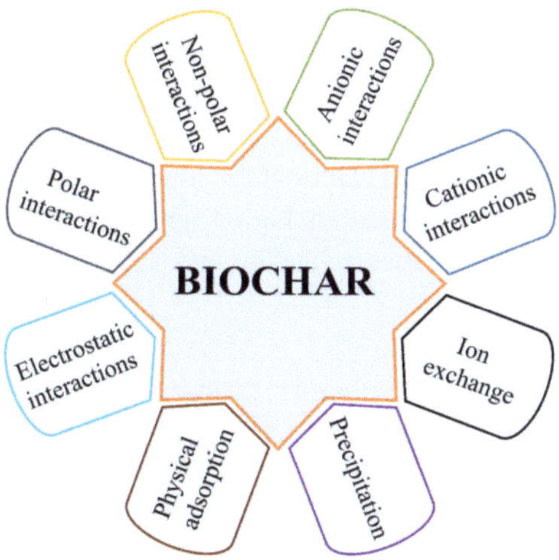

Figure 3: Various mechanisms of interactions of biochar with pollutants.

4.1.1 Physical adsorption

The interactions between adsorbent and adsorbate result in physical adsorption. It generally occurs due to intermolecular interactions and is also known as van der Waals adsorption, which is reversible in nature. The surface energy and pore volume of biochar played a significant role in the physical adsorption of HMs in Van der Waals adsorption. Zinc, cadmium, and arsenic are the ions that have been physically adsorbed on the biochar surface [26].

4.1.2 Precipitation

The combination of metal cations and biochar produces insoluble carbonates and phosphates. The pyrolysis of hemicellulose and cellulose in plant biomass at high temperatures (above 300 °C) form alkaline biochar that can precipitate metal in soil [27]. The combination of HMs and ash content like calcium, sodium, phosphorus, sulfur, potassium, and silicon of animal biochar produce insoluble minerals. It has been reported that phosphate present in cow manure biochar can immobilize lead in soil [27]. The adsorption mechanism of cow manure and rice husk biochar for zinc, copper, lead, and cadmium was compared and found that metals precipitate with carbonate or phosphates.

4.1.3 Electrostatic interactions

The electrostatic interaction between metal cations and soil particles is due to the high negative charge of biochar. The electrostatic interactions between biochar and metal depend on several factors like ionic radius and valence of HM, pH of soil, and point of zero charges (PZC) of biochar [28]. The presence of biochar in soil enhances the pH and CEC of soil, which further can enhance the electrostatic interactions between soil particles and HMs. The rice straw biochar exhibits higher remediation capacity than AC due to the electrostatic interactions between biochar and lead.

4.1.4 Complexation

The functional groups like -OH, -COOH, and -COH present on the surface of biomass helps in complex formation with HMs [29]. At lower temperatures, oxygen functional groups in biochar help in the immobilization of HMs. The oxidation of biochar surfaces can enhance the formation of oxygen functional groups. The biochar-metal complex can be produced due to cationic groups of metals and oxygen functional groups of biochar. The oxygen functional groups of biochar enhance the concentration of ligands on the soil surface, which further helps in the immobilization of more HMs due to the formation of ligand-heavy metal complexes.

4.1.5 Ion exchange

The exchange of metals, ions, or ionizable protons like Ca^{2+}, Na^{+}, Mg^{2+}, and K^{+} on the surface of biochar due to HM ions is known as ion exchange. The biochar's surface chemical properties primarily influence the ion exchange performance. The high CEC of biochar can enhance the efficiency of ion exchange between metal cations and biochar. Lee et *al.* found that pyrolysis temperatures of 250–300 °C sustained the highest CEC of biochar as high-temperature pyrolysis reduces the functional oxygen groups on its surface [30]. The efficiency of ion exchange mechanism is correlated with pH of soil. The reduction in pH of soil than PZC of biochar can cause more HMs to be attracted by biochar surface. Hydrothermal biochar exhibit a higher oxygen functional group than pyrolysis biochar.

4.2 Remediation of organic pollutants

The biochar has been used to remediate organic pollutants like agrochemicals (pesticides, fungicides, herbicides, insecticides such as atrazine, chlorpyrifos, carbofuran, pyrimethanil, and simazine), drugs/antibiotics (acetaminophen, ibuprofen, sulfamethoxazole, sulfamethazine, tetracycline, and tylosine, etc.), industries organics compounds (anthracene, catechol, m-dinitrobenzene, naphthalene, polychlorinated biphenyl, phenanthrene, p-nitrotoluene, and pyrene, *etc.*), volatile organic compounds (VOCs) (benzene, butanol, furan, hexane, and TCE, etc.), and cationic aromatic dyes (brilliant blue, methylene blue, methyl violet, and rhodanine, etc.) [31].

The removal of organic pollutants by biochar is due to chemisorption (electrophilic interactions) and physisorption like hydrogen bonds, electrostatic, and hydrophobic attractions between OH, COOH, and R-OH functional groups of biochar to pollutants [32]. The organic pollutants can be remediated using chemical transformations, biodegradation, and partitioning, [32, 33]. Several factors such as pyrolysis temperature, pH, and the ratio of biochar to pollutants affect the remediation of pollutants by biochar. High pyrolysis temperature (>500 °C) can results in loss of functional groups, which can increase the binding of nonpolar organic pollutants by biochar [32]. While at low pyrolysis temperature (<500 °C), biochar is high polar, low aromatic, and basic in nature that helps in the removal of polar organic compounds [33]. Soybean straw and rice husk-derived biochars pyrolyzed at high temperatures (>600 °C) promote remediation of the nonpolar pollutants like TCE and carbofuran from the contaminated water [33]. Polar herbicides and insecticides were removed due to the presence of functional groups in biochar at less than 350 °C. In the presence of biochar in soil, pollutants immobilize on the surface of biochar and decrease the availability of pollutants for plants. The electrostatic interactions occurred between anionic biochar and aromatic cationic dyes (methyl blue, methyl violet, and rhodanine) [31, 34]. The presence of electron acceptor and donor functional groups on the surface of polar biochar pyrolyzed at a temperature less than 400 °C plays a significant role in the adsorption of organic dyes by biochar. The functional groups present in crop residues-derived biochars exhibit high efficiency for methyl blue and methyl violet [31, 34]. The presence of net negative charge on biochar at higher pH facilitates the electrostatic interactions between methyl violet and biochar [31, 34]. The hydrogen bonds between anionic sulfamethazine (SMZ) and OH or COOH group of biochar occurred at high pH. While, at neutral and lower pH, π–π electron donor–acceptor interactions stabilize the SMZ with biochar.

4.3 Remediation of inorganic pollutants

The interactions between HMs and biochar depend on biomass feedstocks and pyrolysis conditions. HMs bind with biochar via reduction, precipitations, cation exchange, electrostatic attraction, and complexation [13]. Table 3 represents the various biochar applications for the remediation of HMs.

4.3.1 Copper (Cu)

The negative potential and functional groups in chicken manure-derived biochar showed the potential for the binding of Cu. The immobilization of Cu at biochar depends on the pH of the soil as the increment in pH can induce the complexation of Cu(II) with functional groups of biochar. It has been reported that bamboo and woodchip-derived biochar adsorb the Cu(II) and results in the formation of $Cu(OH)_2$ and $CuCO_3$. The alkaline pH of biochar is required for the immobilization of Cu [35]. While mobility of Cu is regulated by the concentration of soil dissolved organic carbon (DOC) in the soil.

4.3.2 Arsenic (As)

The phosphorus (P) content plays a vital role in the mobility of Arsenic (As) [25]. The soil treated with P-containing biochar like pine needles, soybean stover, and rice straw-derived biochar helps in the desorption of As from the soil [32]. In anaerobic conditions, rice straw-derived biochar enhances the presence of Fe-reducing bacteria such as *Caloramtor*, *Bacillus*, and *Clostridium* in soil, which can reduce As. Qiao et al. also found similar results using palm-derived biochar under anoxic conditions [36]. The alkaline and aromatic groups on biochar cause electron shuttle that can facilitate reduction of As. Choppala et *al.* found that π-electrons in functional groups of chicken manure-derived biochar cause the reduction of As [37]. Fe and Mn biochar showed the ability to immobilize the As.

Table 3: Biochar application for the remediation of different metals.

Metals	Types of feedstock	Medium	Initial concentration (mg L^{-1})	Removal efficiency (%)
Cu	Activated broiler	Water	134	80
	Activated corn stover	Water	134	98
	Bamboo	Soil	136	41

Table 3 (continued)

Metals	Types of feedstock	Medium	Initial concentration (mg L^{-1})	Removal efficiency (%)
	Chicken manure	Soil	1,805	79
	Hardwood	Solution	240	58
	Rice straw	Soil	693	46
	Sesame straw	Water	10	28
	Soybean straw	Water	134	85
As	Corn straw	Soil	73	70
	Empty fruit bunch	Solution	–	–
	Rice straw	Soil	120	–
	Sewage sludge	Soil	99	43
	Oak bark	Solution	–	70
Cd	Activated broiler	Water	236	85
	Activated corn stover	Water	236	91
	Bamboo	Soil	1.4	49
	Coconut shell	Groundwater	300	97
	Corn stalk	Soil	2	91
	Maize straw	Soil	2	51
	Soybean straw	Water	236	54
	Sesame straw	Water	10	94
	Wheat straw	Soil	5	59
Pb	Bamboo	Soil	443	73
	Coconut shell	Ground water	350	64
	Coconut shell	Ground water	350	92
	Dairy-manure	Soil	70,000	89
	Miscanthus	Soil	3,110	92
	Rice straw	Soil	527	71
	Soybean stover	Soil	1,945	95
	Vegetable waste	Soil	1,445	87
	Wheat straw	Soil	100	27

Table 3 (continued)

Metals	Types of feedstock	Medium	Initial concentration (mg L^{-1})	Removal efficiency (%)
Hg	Dairy manure	Solution	–	90
	Rice hull	Soil	129	44
	Rice husk	Soil	1,000	90
	Soybean straw	Solution	–	90
	Wheat straw	Soil	129	26
Cr	Sesame straw	Water	10	67
	Tar sand	Solution	55	96
	Walnut shell	Solution	55	80
	Wheat straw	Soil	308	60

4.3.3 Cadmium (Cd)

The behavior of Cd in the soil is highly determined by the pH of the soil [38]. The alkaline groups like OH^-, CO_3^{2-}, and PO_4^{3-} present on the surface of biochar convert Cd into $CdCO_3$, $Cd_3(PO_4)^2$, and $Cd(OH)_2$ precipitates. The strong adsorption affinity of soil calcite ($CaCO_3$) with Cd(II) is a major factor for the reduction of bioavailability of Cd, but the broad surface area and functional groups in biochar also important for the immobilization of Cd. The rice-straw-derived biochar immobilizes the Cd and decreases its concentration in the soil rhizosphere. The corn straw-derived biochar can reduce 91% of Cd in farmland soil [38]. The deprotonation of functional groups enhances the immobilization and adsorption of Cd by biochar from the soil.

4.3.4 Lead (Pb)

The immobilization of Pb in the soil depends on the π-cation electron donor–acceptor interactions between biochar and Pb(II) ion. Ahmad et al. reported that in soybean stover-derived biochar π electron-rich biochar graphene surface is responsible for interacting with π electron-deficient positively charged Pb(II) ion [39]. The sewage-derived biochar and poultry-derived biochar consist of phosphates that can form insoluble compounds like $Pb_5(PO_4)_3OH$, $Pb_5(PO_4)_3Cl$, and β-$Pb_9(PO_4)_6$ with Pb(II). The functional groups present on biochar affect the immobilization of Pb. Vegetable waste-derived biochar has the potential to immobilize the Pb due to covalent bonding between -NH_2 group of biochar and Pb.

4.3.5 Mercury (Hg)

Biochar is an important tool for the remediation of Hg, as it is highly toxic in nature in soil. In hardwood-derived biochar surface, the carboxyl groups coordinate with Hg (II) ion and form $COOHg^+$ complex, which further decreases the mobility of Hg in soil. Rice husk-derived biochar has more potential to form sulfide precipitates as it has higher sulfate than wheat straw-derived biochar. It has been reported that methylation of Hg transforms it into methylmercury (MeHg) which is highly toxic and accumulated in soil than Hg [40]. The rice root can decrease the pH of soil, which can increase the methylation of Hg. The sewage sludge-derived biochar has high organic matter that can promote the growth of microorganisms in Hg-contaminated soil to form the MeHg. Therefore, alkaline biochar is effective biochar to remediate the Hg as it can inhibit the methylation of Hg.

4.3.6 Chromium (Cr)

In the soil, chromium (Cr) is found in two valence states as Cr(III) and Cr(VI). Therefore, immobilization of Cr involves the combination of adsorption, reduction, and precipitation. The functional groups like C–O, -COOH, C = O, and -OH of biochar interact with Cr(IV) [41]. The electrons present in the functional groups of biochar transform Cr(VI) into Cr(III), which further can precipitate in $Cr(OH)_3$ or Cr_2O_3 complex.

4.4 Remediation of toxic gases

Biochar has been used to remove toxic compounds from gaseous streams. Biochar derived from hardwood bits, camphor, rice hulls, and bamboo can effectively separate H_2S from biogas. The moisture content (<80% v/w), pH (<7.0), available surface area, and chemical bonding with functional groups like OH and COOH were the key determinants of H_2S adsorption. Xu et al. reported that alkali biochar interacts with H_2S in the presence of O_2 and forms K, Na_2SO_4 [42]. TCE is a common volatile organic compound identified as a pollutant in groundwater and soil. Biochar derived at high pyrolysis temperature (700 °C) from peanut shells or soybean straw is efficient to remove TCE.

4.5 Remediation of miscellaneous

Biochar is also efficient to remove toxic compounds like phenolic compounds (4-methyl-guaiacol, 4-vinyl-guaiacol, catechol, *m*-, *p*-, *o*-cresol, guaiacol, naphthalene, syringaldehyde, and vanillin, etc.) produced during the pretreatment of lignocellulosic biomass, wastewaters from agrochemical, chemical industries, electronic, electroplating, metal processing, pharmaceutical, oil, and mines [43]. Softwood and hardwood-derived biochars have been used to treat wastewater from chemical industries, pharmaceuticals, oils, and mines [44]. Moreover, biochar derived from bamboo, cotton gin, cottonseed hull, and ground-flax shive can efficiently remove 5-hydroxymethyl furfural (HMF) and furfural. It has been reported that biochar treated with steam NaOH, HNO_3, and $KMnO_4$ exhibit 99% efficiency for removal of HMF and furfural [43].

5 Risk associated with biochar

Biochar has a broad range of applications in remediation, but it showed the risk to the environment and living organisms. Initially, it was reported that biochar remediates the HMs, dioxins, VOCs, and polyaromatic hydrocarbons (PAHs) from soil. But, it was found that usage of double-edged biochar enhance the rate of release of toxic substance and even cause a threat to secondary pollutants as well [45]. Biochar is identified to play a vital role in the utilization of greenhouse gases (GHG) like CO_2 and other non-CO_2 gases. However, the ongoing research found that under specific conditions, biochar can promote the emission of GHG [46]. Biochar declines the release of CH_4, but it can also promote the emission of N_2O and other GHG as well [47]. Biochar showed a negative effect on the survival and growth of bacteria, decreasing the growth and diversity of acidophilic fungi and earthworms in the soil [48].

6 Conclusion

Since the green revolution, rapid industrialization, and increasing usage of agrochemical-based crop production methods have significantly increased the persistence of organic and HM pollution in the food chain and surrounding environment. In recent years, soil contamination by HMs has become a worldwide environmental issue, as HMs adversely affect the quality of agricultural products and human health. The various physical, chemical, and biological methods have been employed to remediate the HM-contaminated soil. Nevertheless, these methods have significant limitations such as high economic cost, poor feasibility, inefficiency, complicated technique, short duration, and high secondary pollution problems. Consequently, effective remediation techniques are needed. Recent research has shown that biochar is a promising and efficient method for HM remediation from the contaminated soil. The variability in physical and chemical properties (e.g., surface area, pH, CEC, and microporosity) allows biochar to be more effective in specific applications. The biochar can interact with pollutants like agrochemical (herbicides, pesticides, insecticides, and fungicides, etc.), drugs, antibiotics, polyaromatic compounds, industrial organic compounds, VOCs, cationic aromatic dyes, inorganic pollutants (copper, arsenic, cadmium, lead, and chromium, *etc.*), and toxic gases due to different methods such as physical adsorption, precipitation, electrostatic interactions, complexation, and ion exchange.

In the last decade, research has been performed to understand and explore the usage of biochar to remediate HMs and other pollutants. Still, improvements in biochar are required to remediate toxic compounds and HMs, as biochar cannot completely remediate these pollutants from the soil. The distribution of microbial communities along with the properties of biochar is not explored, therefore, it needs to investigate the effect of biochar properties on the microbial communities. There is a requirement for the development of biochar which does not have any effect on the emission of GHG. It is required to use microbial organisms or enzymes in biochar so the HMs or pollutants can be easily transformed into nonpollutants.

List of Abbreviations

AC	Activated carbon
CEC	Cation exchange capacity
DEHP	Di-2-ethylhexyl phthalate
DOC	Dissolved organic carbon
GHG	Greenhouse gases
HM	Heavy metal
HMF	Hydroxymethyl furfural
MEHP	Mono-2-ethylhexyl phthalate
PAH	Polyaromatic hydrocarbon
PZC	Point of zero charges
SMZ	Sulfamethazine
TCE	Trichloroethylene
VM	Volatile matter
VOC	Volatile organic compound

References

[1] Rodríguez-Eugenio N, McLaughlin M, Pennock D, Soil Pollution: A Hidden Reality, FAO, 2018.
[2] Kumari R, Dhankhar P, Dalal V, Structure-based mimicking of hydroxylated biphenyl congeners (OHPCBs) for human transthyretin, an important enzyme of thyroid hormone system, J. Mol. Graph. Model., 2021, 105, 107870.
[3] Singh N, Dalal V, Kumar P, Molecular docking and simulation analysis for elucidation of toxic effects of dicyclohexyl phthalate (DCHP) in glucocorticoid receptor-mediated adipogenesis, Mol. Simul., 2020, 46, 9–21.

[4] Singh N, Dalal V, Kumar P, Structure based mimicking of Phthalic acid esters (PAEs) and inhibition of hACMSD, an important enzyme of the tryptophan kynurenine metabolism pathway, Int. J. Biol. Macromol., 2018, 108, 214–224.

[5] Singh N, Dalal V, Dhankhar P, Kumar P, Phthalates – A class of ubiquitous pollutant: Microbial and enzymatic degradation perspectives, In: Removal of Toxic Pollutants Through Microbiological and Tertiary Treatment, Elsevier, 2020, 487–513.

[6] Dhankhar P, Dalal V, Singh N, Gurjar BR, Sharma AK, Kumar P, Bioremediation of synthetic dyes: Dye decolorizing Peroxidases (DyPs), In: Removal of Toxic Pollutants Through Microbiological and Tertiary Treatment, Elsevier, 2020, 453–486.

[7] Dhankhar P, Dalal V, Mahto JK, Gurjar BR, Tomar S, Sharma AK, Kumar P, Characterization of dye-decolorizing peroxidase from Bacillus subtilis, Arch. Biochem. Biophys., 2020, 693, 108590.

[8] Singh N, Dalal V, Mahto JK, Kumar P, Biodegradation of phthalic acid esters (PAEs) and in silico structural characterization of mono-2-ethylhexyl phthalate (MEHP) hydrolase on the basis of close structural homolog, J. Hazard. Mater., 2017, 338, 11–22.

[9] Shakya A, Agarwal T, Potential of biochar for the remediation of heavy metal contaminated soil, In: Biochar Applications in Agriculture and Environment Management, Springer, 2020, 77–98.

[10] Bolan N, Kunhikrishnan A, Thangarajan R, Kumpiene J, Park J, Makino T, Kirkham MB, Scheckel K, Remediation of heavy metal (loid) s contaminated soils–to mobilize or to immobilize?, J. Hazard. Mater., 2014, 266, 141–166.

[11] Bastami KD, Afkhami M, Mohammadizadeh M, Ehsanpour M, Chambari S, Aghaei S, Esmaeilzadeh M, Neyestani MR, Lagzaee F, Baniamam M, Bioaccumulation and ecological risk assessment of heavy metals in the sediments and mullet Liza klunzingeri in the northern part of the Persian Gulf, Mar. Pollut. Bull., 2015, 94, 329–334.

[12] Wang Y, Wang H-S, Tang C-S, Gu K, Shi B, Remediation of heavy-metal-contaminated soils by biochar: a review, Environ. Geotech., 2019, 40, 1–14.

[13] Wang M, Zhu Y, Cheng L, Andserson B, Zhao X, Wang D, Ding A, Review on utilization of biochar for metal-contaminated soil and sediment remediation, J. Environ. Sci., 2018, 63, 156–173.

[14] Adriano DC, Wenzel W, Vangronsveld J, Bolan N, Role of assisted natural remediation in environmental cleanup, Geoderma, 2004, 122, 121–142.

[15] Guo M, Song W, Tian J, Biochar-facilitated soil remediation: mechanisms and efficacy variations, Front. Environ. Sci., 2020, 8, 183.

[16] Cheng S, Chen T, Xu W, Huang J, Jiang S, Yan B, Application research of biochar for the remediation of soil heavy metals contamination: a review, Molecules, 2020, 25, 3167.

[17] Oliveira FR, Patel AK, Jaisi DP, Adhikari S, Lu H, Khanal SK, Environmental application of biochar: current status and perspectives, Bioresour. Technol., 2017, 246, 110–122.

[18] Masindi V, Muedi KL, Environmental contamination by heavy metals, Heavy metals, 2018, 10, 115–132.

[19] Li C, Zhou K, Qin W, Tian C, Qi M, Yan X, Han W, A review on heavy metals contamination in soil: effects, sources, and remediation techniques, Soil Sediment Contam. Int. J., 2019, 28, 380–394.

[20] Tchounwou PB, Yedjou CG, Patlolla AK, Sutton DJ, Heavy metal toxicity and the environment, In: Molecular, Clinical and Environmental Toxicology, 2012, 133–164.

[21] Ghosh K, Indra N, Cadmium treatment induces echinocytosis, DNA damage, inflammation, and apoptosis in cardiac tissue of albino Wistar rats, Environ. Toxicol. Pharmacol., 2018, 59, 43–52.

[22] Rahman MM, Ng JC, Naidu R, Chronic exposure of arsenic via drinking water and its adverse health impacts on humans, Environ. Geochem. Health, 2009, 31, 189–200.
[23] Kankia H, Abdulhamid Y, Determination of accumulated heavy metals in benthic invertebrates found in Ajiwa Dam, Katsina State, Northern Nigeria, Arch. Appl. Sci. Res., 2014, 6, 80–87.
[24] Shen Z, Zhang Y, McMillan O, O'Connor D, Hou D, The use of biochar for sustainable treatment of contaminated soils, Sustainable Remediation of Contaminated Soil and Groundwater: Materials, Processes, and Assessment, 2019, 119.
[25] Wang S, Xu Y, Norbu N, Wang Z, Remediation of biochar on heavy metal polluted soils, in: IOP Conference Series: Earth and Environmental Science, IOP Publishing, 2018, 042113.
[26] Beesley L, Inneh OS, Norton GJ, Moreno-Jimenez E, Pardo T, Clemente R, Dawson JJ, Assessing the influence of compost and biochar amendments on the mobility and toxicity of metals and arsenic in a naturally contaminated mine soil, Environ. Pollut., 2014, 186, 195–202.
[27] Cao X, Harris W, Properties of dairy-manure-derived biochar pertinent to its potential use in remediation, Bioresour. Technol., 2010, 101, 5222–5228.
[28] Dong X, Ma LQ, Li Y, Characteristics and mechanisms of hexavalent chromium removal by biochar from sugar beet tailing, J. Hazard. Mater., 2011, 190, 909–915.
[29] Park JH, Choppala GK, Bolan NS, Chung JW, Chuasavathi T, Biochar reduces the bioavailability and phytotoxicity of heavy metals, Plant Soil, 2011, 348, 439–451.
[30] Lee JW, Kidder M, Evans BR, Paik S, Buchanan Iii A, Garten CT, Brown RC, Characterization of biochars produced from cornstovers for soil amendment, Environ. Sci. Technol., 2010, 44, 7970–7974.
[31] Xu R-K, Xiao S-C, Yuan J-H, Zhao A-Z, Adsorption of methyl violet from aqueous solutions by the biochars derived from crop residues, Bioresour. Technol., 2011, 102, 10293–10298.
[32] Ahmad M, Rajapaksha AU, Lim JE, Zhang M, Bolan N, Mohan D, Vithanage M, Lee SS, Ok YS, Biochar as a sorbent for contaminant management in soil and water: a review, Chemosphere, 2014, 99, 19–33.
[33] Ahmad M, Lee SS, Dou X, Mohan D, Sung J-K, Yang JE, Ok YS, Effects of pyrolysis temperature on soybean stover-and peanut shell-derived biochar properties and TCE adsorption in water, Bioresour. Technol., 2012, 118, 536–544.
[34] Li G, Zhu W, Zhang C, Zhang S, Liu L, Zhu L, Zhao W, Effect of a magnetic field on the adsorptive removal of methylene blue onto wheat straw biochar, Bioresour. Technol., 2016, 206, 16–22.
[35] Gonzaga MIS, Mackowiak C, De Almeida AQ, Wisniewski A Jr, De Souza DF, Da Silva Lima I, De Jesus AN, Assessing biochar applications and repeated Brassica juncea L. production cycles to remediate Cu contaminated soil, Chemosphere, 2018, 201, 278–285.
[36] Qiao J-T, Li X-M, Li F-B, Roles of different active metal-reducing bacteria in arsenic release from arsenic-contaminated paddy soil amended with biochar, J. Hazard. Mater., 2018, 344, 958–967.
[37] Choppala G, Bolan N, Kunhikrishnan A, Bush R, Differential effect of biochar upon reduction-induced mobility and bioavailability of arsenate and chromate, Chemosphere, 2016, 144, 374–381.
[38] Gao X, Peng Y, Zhou Y, Adeel M, Chen Q, Effects of magnesium ferrite biochar on the cadmium passivation in acidic soil and bioavailability for packoi (Brassica chinensis L.), J. Environ. Manage., 2019, 251, 109610.
[39] Ahmad M, Ok YS, Kim B-Y, Ahn J-H, Lee YH, Zhang M, Moon DH, Al-Wabel MI, Lee SS, Impact of soybean stover-and pine needle-derived biochars on Pb and As mobility, microbial community, and carbon stability in a contaminated agricultural soil, J. Environ. Manage., 2016, 166, 131–139.

[40] Zhang H, Feng X, Larssen T, Shang L, Li P, Bioaccumulation of methylmercury versus inorganic mercury in rice (Oryza sativa L.) grain, Environ. Sci. Technol., 2010, 44, 4499–4504.
[41] Lyu H, Zhao H, Tang J, Gong Y, Huang Y, Wu Q, Gao B, Immobilization of hexavalent chromium in contaminated soils using biochar supported nanoscale iron sulfide composite, Chemosphere, 2018, 194, 360–369.
[42] Xu G, Sun J, Shao H, Chang SX, Biochar had effects on phosphorus sorption and desorption in three soils with differing acidity, Ecol. Eng., 2014, 62, 54–60.
[43] Li Y, Shao J, Wang X, Deng Y, Yang H, Chen H, Characterization of modified biochars derived from bamboo pyrolysis and their utilization for target component (furfural) adsorption, Energy Fuels, 2014, 28, 5119–5127.
[44] Frankel ML, Bhuiyan TI, Veksha A, Demeter MA, Layzell DB, Helleur RJ, Hill JM, Turner RJ, Removal and biodegradation of naphthenic acids by biochar and attached environmental biofilms in the presence of co-contaminating metals, Bioresour. Technol., 2016, 216, 352–361.
[45] Lyu H, He Y, Tang J, Hecker M, Liu Q, Jones PD, Codling G, Giesy JP, Effect of pyrolysis temperature on potential toxicity of biochar if applied to the environment, Environ. Pollut., 2016, 218, 1–7.
[46] Yang X, Meng J, Lan Y, Chen W, Yang T, Yuan J, Liu S, Han J, Effects of maize stover and its biochar on soil CO2 emissions and labile organic carbon fractions in Northeast China, Agric. Ecosyst. Environ., 2017, 240, 24–31.
[47] Liu J, Shen J, Li Y, Su Y, Ge T, Jones DL, Wu J, Effects of biochar amendment on the net greenhouse gas emission and greenhouse gas intensity in a Chinese double rice cropping system, European J. Soil Biol., 2014, 65, 30–39.
[48] Chen J, Liu X, Zheng J, Zhang B, Lu H, Chi Z, Pan G, Li L, Zheng J, Zhang X, Biochar soil amendment increased bacterial but decreased fungal gene abundance with shifts in community structure in a slightly acid rice paddy from Southwest China, Appl. Soil Ecol., 2013, 71, 33–44.

Wan Ting Tee, Billie Yan Zhang Hiew,
Suchithra Thangalazhy-Gopakumar, Suyin Gan, Lai Yee Lee

Biochar as a remediation solution for pharmaceutical-contaminated wastewater

Abstract: Water pollution caused by the discharge of pharmaceutical residues is a relatively recent environmental issue affecting clean water supply. The worldwide consumption of pharmaceutical products is continually increasing at rapid speed; hence, there is a need to control their concentration in the water resources. Recently, biochar has attracted significant exploration in wastewater treatment due to its abundant availability, facile production, and cost-effectiveness. Additionally, biochar has demonstrated great potential in removing pharmaceutical contaminants from water bodies indicating that it can be an effective, economical, and environmentally friendly resource for wastewater treatment. In this chapter, the main pollution point sources of pharmaceutical compounds are first discussed. Five most prevalent compounds in the current environment classified based on their therapeutic classes are evaluated along with their associated adverse health and environmental effects. The advantages and disadvantages of conventional and advanced removal technologies for pharmaceutical contaminants from the aquatic system are assessed. The chapter further reviews the synthesis methods of biochar as well as the mechanisms involved in the adsorption of pharmaceuticals, in detail. The current and future prospects of biochar in pharmaceuticals removal are also discussed in the context of biochar feasibility and challenges encountered.

Keywords: Adsorption, biochar, biomass, pharmaceutical, pyrolysis, wastewater

1 Introduction

Industrialization has played a prominent role in the growth of global economy for the past few decades. Pharmaceutical industries in particular have contributed intensely to this development with an estimated annual production exceeding 20 million tons [1]. The production of pharmaceutical compounds is increasing exponentially to meet the growing demands for prevention and treatment of illnesses, and in sustaining the expansion of agricultural and aquacultural industries. Pharmaceutical products are commonly used in medical facilities, animal husbandry, drugs-manufacturing industries, aquafarming, and daily human life. The extensive utilization of pharmaceuticals worldwide, combined with inappropriate disposal methods as well as ineffective removal in existing wastewater treatment plants, have resulted in their existence in aquatic ecosystems [2].

https://doi.org/10.1515/9783110734003-019

The advancement of technology, especially in analytical measurement tools, has enabled the detection of pharmaceutical contaminants in aquatic environment, which are in trace concentrations, from nanograms per liter to micrograms per liter. In order to obtain an accurate detection of pharmaceutical samples in complex environment matrices, precise sample preparation and sensitive detection tools are required. Liquid or gas chromatography coupled with mass spectrometer provides the desired method to determine the pharmaceutical concentrations in aqueous systems. The evolution in measurement techniques further escalated the discovery of the ubiquitous presence of synthetic pharmaceuticals in marine life, thereby attracting attention of researchers from various fields to delve deeper into this issue.

The health and environmental impacts of pharmaceutical residues have become a major concern due to their presence in water sources, which eventually accumulate in human bodies through drinking water and food chain cycle. Pharmaceutical products are categorized into 24 therapeutic classes, whereby nonsteroidal anti-inflammatory drugs (NSAIDs), anticonvulsants, antibiotics, and lipid regulators are the four major classes reported in literature [3]. Pharmaceuticals are persistent and able to bioaccumulate in the environment due to their complex physicochemical properties. They cause detrimental effects on terrestrial and aquatic life such as extinction of species, delayed egg hatching, and reduction in reproduction rate [4, 5].

Several studies investigated the pathways of pharmaceutical pollutants to the ecosystem. Mompelat and Le Bot [3] explored the origins and routes of a variety of pharmaceutical residues. Households and non-domestic points such as hospitals, industries, and services are the major sources associated with the excretion from human beings and the improper disposal of drugs. The chemical compounds discharged from these point sources are usually directed to wastewater purification facilities and eventually released to the environment owing to the incapability of current removal technologies. Moreover, some countries discharged untreated waste streams to the water bodies directly, thus resulting in severe pollution issues [6, 7]. Other than that, animal farming and aquaculture are responsible for the direct discharge of pharmaceutical loaded waste. Landfills containing pharmaceutical waste materials from households and medical facilities can produce leachate, which may seep through soil into the groundwater. Hence, effective wastewater treatment is essential to prevent the pharmaceutical residues from impacting living organisms.

Current wastewater purification technologies are ineffective in eliminating the pharmaceutical residues. This deficiency has motivated researchers to investigate and develop advanced technologies to tackle this concerning issue. Among the investigated methods, adsorption using biochar is a promising alternative for the removal of pharmaceutical residues. Adsorption is a relatively simple and cost-effective process, while biochar is a relatively new adsorbent, thermochemically transformed from organic and inorganic wastes. The numerous microporous structures and great surface area allows the adsorption of pharmaceutical contaminants even at very low concentrations. The aim of this chapter is to provide a state-of-the-art review of this viable adsorbent in the removal of pharmaceutical contaminants from aqueous phase. The challenges and future prospects of biochar adsorption are included to address the current development and possible improvement of biochar to maximize its potential in the field of pharmaceuticals removal.

2 Background

The discovery of pharmaceuticals in the environment through published articles began around the 1940s and earlier, but it only started receiving attention from the 1990s onwards [8]. Both the literature reviews completed by Aus Der Beek, Weber [9], and Daughton [8] showed that pharmaceuticals-related environmental issues had received massive surge of attention from the years 2001 to 2010 from researchers around the world with an approximate thousandfold increment in the number of published articles as compared to the previous decade. This could be due to the advancement of analytical techniques, such as gas chromatography-mass spectrometry (GC-MS) and high-performance liquid chromatography-mass spectrometry (HPLC-MS), enabling the detection of previously undiscovered compounds in the aqueous environment [1, 8]. The presence of pharmaceutical substances, which are known as the emerging contaminants, in the environment has garnered significant interests from researchers worldwide, including major international organizations such as the United States Environmental Protection Agency (USEPA), World Health Organization (WHO), European Commission (EU), and United States Geological Survey (USGS). Nonetheless, despite the rapid global development of pharmaceuticals market, these contaminants are not yet classified under the list of hazardous waste pharmaceuticals by the USEPA (EPA, 2019).

3 Sources of pharmaceutical residues

The lack of quantitative data on the relative distribution of pharmaceutical residues in the environment affects the assessment and traceability accuracy of the emission sources. Pharmaceutical residues usually originate from several waste streams contributing to the complexity and diversity of chemicals present in the water sources. The direct and indirect discharges of pharmaceuticals into surface water and groundwater were reviewed by Mompelat and Le Bot [3]. In short, the point sources are inclusive of the release from households, industries, hospitals, aquaculture, and animal farming, as well as leachate from landfills, runoff from biosolids, and bank filtration at agricultural sites using treated wastewater for irrigation.

One of the major and primary sources for pharmaceuticals in the environment is through humans' urine and feces excreted upon consumption of medicines at either health-care facilities or households [10]. The excretions containing metabolized pharmaceuticals are discharged into the wastewater treatment facilities. Besides that, drain disposal is an allowable disposal method for the pharmaceutical waste [10]. Routine disposal of unused pharmaceuticals down the sewer, such as through sink, toilet, and floor drain, at health-care facilities has significantly contributed to the presence of the pharmaceutical compounds and their complete or partially degraded metabolites in water sources [6].

Removal of pharmaceuticals may occur to a certain degree in current wastewater treatment plants, and the final residues may be discharged in the form of sewage sludge to be incinerated, landfilled, or used as biosolids in agricultural lands [1]. Land runoff, leaching, or bank filtration, which are associated with landfill and agricultural usage, are undesirable indirect pathways for pharmaceuticals to reach at the groundwater. Besides that, sediments with various active sites were demonstrated to adsorb and retain the xenobiotic compounds in the surface water which further reduced their biodegradation rate [11].

The highly polluted effluent from drugs-manufacturing firms is another undeniable source of pharmaceuticals. Pharmaceutical containing effluents from the manufacturing sites are more concentrated in specific type and location as compared to the human excretion due to the absence of metabolism by human body which reduces the concentration [12]. Several studies have reported that the pharmaceutical industrial effluents consisted of high amounts of nonbiodegradable organic matter such as antibiotics, reproductive hormones, steroids, and many other drug-based pollutants which had adverse effects on human beings and the environment [13, 14]. One recent study on the pharmaceutical effluents in India has revealed the presence of 13 different drugs along with other pollutants in the water samples [15].

The direct release of untreated waste streams from domestic and industrial sectors to the environment is one of the biggest concerns as well. For instance, it has been reported that most of the pharmaceutical industrial effluents in Nigeria were discharged directly into the water streams [7], while in Brazil, more than half of the generated sewage was released untreated [6]. On the other hand, even though treatments were imposed prior to discharge, the inefficacy of the treatment system was evidenced by the presence of the pharmaceuticals in surface water, drinking water, and groundwater [16]. Therefore, proper disposal legislations and effective treatment methods are required to control the discharge of pharmaceuticals to the environment.

4 Health and environmental impacts

Mompelat and Le Bot [3] classified pharmaceutical products and its by-products traced in the water environment into 24 therapeutic classes as listed in Table 1, whereby NSAIDs, anticonvulsants, antibiotics, and lipid regulators are the four most prevalent classes in terms of published references. Dolar and Zokic [13] showed similar findings in their study, that is, the predominant pharmaceuticals detected in drinking water, surface water, and groundwater in the United Nation regions were diclofenac, carbamazepine, ibuprofen, sulfamethoxazole, and naproxen in descending order. Diclofenac, ibuprofen, and naproxen are classified as NSAIDs, while carbamazepine is an anticonvulsant, and sulfamethoxazole (SMX) is an antibiotic. Figure 1 shows the structures of these five predominant pharmaceuticals.

NSAIDs' diclofenac (2-[(2, 6-dichlorophenyl) amino] benzene acetic acid) is a widely used antipyretic analgesics agent to control pain and relieve inflammation resulting from illness such as arthritis, gout, and migraine [17]. Diclofenac can be taken orally or applied topically. During topical application, only 6–7% of the pharmaceutical active compounds will be absorbed while the residual on the skin and clothes will be washed off and discharged into the sewage. Conversely, for oral application, a small fraction of diclofenac will not be fully metabolized by the human body, which will then be excreted through urine and feces [18]. The presence of diclofenac in the environment poses detrimental effect to both terrestrial and aquatic animals. In the Indian subcontinent, consumption of carcasses with diclofenac residual was the major cause of the sudden extinction of several vulture species [19]. Chronic tissue damage in mussels was due to exposure to diclofenac at concentrations of nanograms per liter, which induced lipid peroxidation [20].

Carbamazepine (5 H-dibenzo[b,f]azepine-5-carboxamide) is an anticonvulsant commonly used for the treatment of epilepsy, controlling seizures, and psychiatric disorders such as depression. It has been reported as one of the pharmaceuticals with the highest global production (1,014 tons per year) and is able to persist in nature, thus resulting in its ubiquity in the environment [21]. Several ecotoxicological studies were carried out to determine the effect of carbamazepine on biological organisms mainly focusing on the aquatic environment. Ferrari and Paxeus [22] demonstrated its possible adverse effects toward bacteria, algae, microcrustaceans, and fish, while Zhang and Zhang [23] further proved its inhibition in the growth of algae.

Ibuprofen (2-[4-(2-methylpropyl)phenyl]propanoic acid) is one of the most consumed over-the-counter drugs for the treatment of pain, minor injury, inflammation, or fever with an approximate annual consumption of 200 tons per year [40]. Ibuprofen has been continuously detected in the environment due to the deficiencies of current water treatment technology. Han et al. [4] reported that ibuprofen in the aquatic system affected the cyclooxygenase pathway of fish, reducing their reproduction rate and delaying their egg hatching. The presence of ibuprofen with

Table 1: Therapeutic classes of pharmaceutical compounds with their medical purposes [3].

Therapeutic class	Compounds	Medical purpose	Reference
Antacid	Cimetidine, ranitidine	Neutralize acid in stomach to soothe heartburn and indigestion	[24]
Antianginal	Dehydronifedipine	Relieve angina symptoms and prevent future attacks; reduce risk of heart attack and stroke	[25]
Antiasthmatic	Albuterol	Aid in preventing symptoms of asthma, allergic rhinitis, mastocytosis, and bronchospasm	[26]
Antibiotic	Sulfamethoxazole, ceftriaxone	Terminate bacteria	[1]
Anticoagulant	Warfarin	Inhibit coagulation cascade in blood vessel	[27]
Anticonvulsant	Carbamazepine, primidone	Anti-seizure drugs; stabilize mood; treat mood disorder	[1]
Antidepresssant	Amitryptiline, fluoxetine	Control neurotransmitter in brain to treat major depression and relieve long-term neuropathic pain	[1]
Anti-diabetic	Metformin, glibenclamide	Treat diabetes by preventing undue rises in blood glucose	[28]
Anthelminthic	Ivermectin	Eliminate intestinal worms	[29]
Antifungal	Clotrimazole	Treat and prevent fungal infections	[30]
Antihistamine	Diphenhydramine	Minimize the condition of itchiness, sneezing, and runny nose from allergic reaction	[31]
Antihypertensive	Diltiazem, valsartan	Treat high blood pressure and prevent complications of hypertension	[32]
Antineoplastic	Cyclophosphamide, tamoxifen	Control or eliminate neoplastic cells	[1]
Beta blocker	Acebutolol, pindolol	Hinder the effect of the hormone adrenalin and the neurotransmitter noradrenalin	[1]
Birth control	Ethynylestradiol	Prevent pregnancy	[32]
Bronchodilator	Terbutaline, clenbuterol	Treat lung condition by relaxing airway muscles to overcome airflow obstruction	[33]

Table 1 (continued)

Therapeutic class	Compounds	Medical purpose	Reference
Cardiac stimulant	Digoxin	Heart stimulant	[34]
Contrast media	Lomeprol, diatrizoate	Intensify the contrast of vascular on magnetic resonance (MR)	[1]
Diuretic	Bendroflumethiazide, furosemide	Induce fluid removal from body through urination	[35]
Lipid regulator	Atorvastatin, pravastain	Control triglycerides and cholesterol in blood	[1]
Nonsteroidal anti-inflammatory drug (NSAID) and analgesic	Diclofenac, ibuprofen	Minimize pain and inflammation symptoms	[1]
physchostimulant	Caffeine, amphetamine	Stimulate central nervous system to cause excitation and increase alertness	[36]
Vasodilatator	Pentoxyfiline	Treat or prevent hypertension by dilating vascular smooth muscle to allow blood flow	[37]

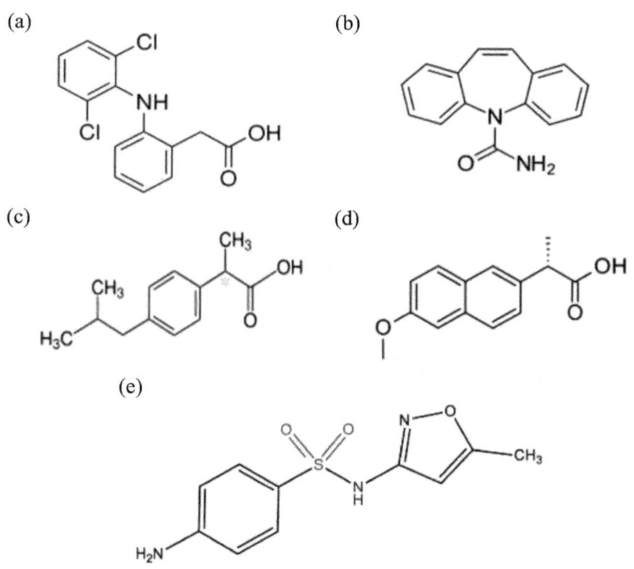

Figure 1: Chemical structures of (a) diclofenac, (b) carbamazepine, (c) ibuprofen, (d) naproxen, and (e) sulfamethoxazole [38, 39].

other drug residuals in the environment presents a potential threat to both aquatic life and human health [40].

SMX (1, 2-oxazole) is a widely used antibiotic by humans to treat bacterial infections such as bronchitis, urinary tract infections, and prostatitis. It is also extensively applied as growth promoters in farming and animal husbandry. Owing to the recalcitrant and biorefractory nature of sulfamethoxazole, it can hardly be removed in the activated sludge treatment process. In addition, its high solubility in water led to its recurring presence in surface water, drinking water, and groundwater [41]. The existence of antibiotics in the water sources even in low concentration induced the emergence of antibiotic-resistant pathogenic microorganisms in human bodies and animals as well as in the environment, which resulted in the inefficacy of the current antibiotic formulations [42].

Naproxen (2-(6-methoxy-2-napthyl)propionic acid), which belongs to the NSAIDs group, is another widely prescribed pharmaceutical utilized to lessen fever, inflammation, pain, and stiffness in our daily life [43]. It is also the preferred NSAIDs drug used to treat osteoarthritis patients with high cardiovascular risk. The derivatives of naproxen have been indicated to be more toxic than its parent compound to aquatic invertebrates and vertebrates [44]. Wojcieszynska and Guzik [5] have documented the relevant impacts of naproxen in their work. For instance, the concentration of naproxen bioaccumulated in the bile of fishes was 1,000 times higher than the surrounding water, inhibiting the egg fertilization of fishes after an exposure to 0.1 $\mu g\ L^{-1}$ over one complete life cycle. Moreover, naproxen affected the messenger ribonucleic acid (mRNA) and cyclooxygenases of zebrafish, causing gastrointestinal, renal, and teratogenic effects, pericardial oedema as well as lowering their heart rate.

The discussion above has highlighted that the five major pharmaceutical residuals found prevalently in the environment have a common attribute, which is their resistivity toward conventional removal technologies contributed mainly by their unique physicochemical properties. Major conventional and advanced removal technologies will be discussed in the following section along with their pros and cons.

5 Removal technologies

The ubiquity of pharmaceutical residuals in the environment is an indication of the ineffectiveness of current wastewater treatment plants to treat these emerging contaminants. Standard physical and chemical processes such as filtration, lime softening, coagulation-flocculation, and activated sludge process are commonly applied in wastewater treatment plants. Although these treatment methods are cost-effective, they are not efficient in removing pharmaceuticals and may produce undesirable derivatives as well [6]. A recent study on conventional wastewater treatment plant was carried out by Al-Mashaqbeh and Alsafadi [45]. By comparing the samples collected from the influent and effluent, incomplete removal of pharmaceutical components was clearly shown with less than 46% removal for carbamazepine (22.5%), sulfamethazine (37.2%), and SMX (45.8%). In addition, negative removal of diphenhydramine and thiabendazole were found that might be due to the possible recombination, accumulation, and transformation of compounds during secondary activated sludge treatment [45]. In view of the low efficiency of existing removal technologies, more advanced treatment methods have to be developed such as membrane separation, advanced oxidation process (AOP), and adsorption.

5.1 Membrane separation

Membrane separation techniques such as ultrafiltration, microfiltration, nanofiltration, and reverse osmosis have been used to remove high molecular weight compounds. These techniques have the requirements of high operating pressure and low inlet flow [6]. Since the molecular size of pharmaceuticals are considerably small, between 150 and 500 kDA, the compounds can easily pass through ultrafiltration and microfiltration. Thus, these methods are usually employed as pretreatment processes [46]. On the contrary, nanofiltration and reverse osmosis are able to achieve near complete removal of the compounds as demonstrated by Dolar and Zokic [13], resulting from the compatibility of the molecular size and weight of pharmaceuticals with the separation dimension of the membranes. Despite that, proper disposal of the membrane residue, high energy consumption, membrane fouling, and associated costs have reduced the feasibility of these processes in treating water contaminated with pharmaceuticals [46].

5.2 Advanced oxidation

AOPs include ozonation (O_3), ultraviolet (UV), photocatalytic, electrochemical, wet air oxidation, ultrasound irradiation, Fenton oxidation, and ionizing radiation technologies. Lately, these techniques have been intensively studied for the decontamination of water with pharmaceuticals. AOPs generate highly reactive species that are able to further oxidize and mineralize pharmaceuticals into carbon dioxide and inorganic ions [46]. One of the main disadvantages of AOP is the complex energy consuming devices such as UV lamps and ozonizers which result in high operating cost. More importantly, AOPs may form by-products during the partial oxidation process. These by-products may have higher toxicity than the original pollutants [47]. Enhanced AOPs such as UV/hydrogen peroxide process offers a promising technology to remove pharmaceuticals in wastewater. However, more studies are required to determine the functionalities, operating conditions, and targeted pharmaceuticals of the enhanced AOPs because of the higher complexity in terms of devices and wastewater matrices.

5.3 Adsorption

Adsorption treatment process is one of the most promising techniques for pharmaceuticals removal as it is generally effective, does not generate toxic substances, and consumes low amount of energy. The efficiency of adsorption process is strongly dependent on the physical and chemical characteristics of the adsorbents such as surface area, pore structures, point of zero charge (PZC), and surface chemical groups [48]. Moreover, the purification method is affected by process parameters such as temperature, pH of solution, contact time, dosage of adsorbent, and initial concentration of pollutant [47]. Different types of adsorbents have been studied for the adsorption of pollutants such as activated carbon, clays, biochar, chitosan, and metal organic frameworks. Among these adsorbents, activated carbon is the most well-developed adsorbent for use in wastewater treatment facilities. It has large surface area and abundant micropores, but its industrial application is limited by the cost of raw material and regeneration process, as well as deterioration of adsorbent structure with increasing operation time, resulting in declining removal efficiency [1, 6].

Recently, biochar was identified as a potential adsorbent for pharmaceuticals removal as it exhibited desirable attributes comparable with those of activated carbon, such as large amounts of micropores, high specific surface areas, facile preparation steps, and a variety of low cost synthesis feedstocks [49]. Biochar can be produced from heating the feedstock materials under oxygen-deficient environment such as pyrolysis, gasification, torrefaction, and hydrothermal carbonization. The feedstocks for the thermochemical processes are usually agricultural or forestry wastes, while the end products are biochar, syngas, and bio-oil. Syngas and bio-oil produced can be used to produce electricity as well further minimize waste disposal and energy consumption [6]. Moreover, due to its intermediate hydrophobic properties, biochar has a high potential to adsorb most pharmaceutical residual which are usually weak bases or acids with moderate hydrophobicity [50]. Thus, biochar has the potential to become a better and cheaper alternative to replace commercial activated carbons for the removal of pharmaceutical contaminants from the environment.

6 Production technologies of biochar

Biochar can be prepared from a wide range of biomasses with various chemical and physical properties. Properties such as moisture content, volatile components, fixed carbon, ash content, caloric values, and fractions of lignin, cellulose, and hemicellulos, are to be considered during the biochar preparation. Generally, a high yield of solid biochar can be obtained from biomasses with more lignin and less cellulose contents, and the porosity increases with the lignin content of the biomass. Furthermore, the biochar formed at higher temperature is expected to be a π-donor, thus forming a biochar with lower oxygen-to-carbon ratio, whereas biochar formed at lower temperature tends to be a π-acceptor.

The synthesized biochar can have different chemical and physical attributes depending on the production technology such as torrefaction, gasification, hydrothermal, flash carbonization, and slow and fast pyrolysis.

6.1 Pyrolysis

Pyrolysis can be considered as a primary biochar preparation method which is defined as a thermal decomposition process at high temperatures under the absence or minimal presence of oxygen. This process can be further classified into slow, fast, and flash pyrolysis depending on the dwelling temperature, heating rate, and residence time. Slow pyrolysis generally produces high amounts of solid biochar at low heating rates and temperatures from 300 to 700 °C, whereas fast pyrolysis utilizes fast heating rate to attain higher bio-oil yield [51]. Meanwhile, torrefaction is a mild pyrolysis process that is carried out under lower heating rates from 200 to 300 °C with negligible amount of bio-oil produced but higher amount of biochar synthesized [52]. The surface chemical composition of developed biochar does not change significantly due to incomplete carboniation when the biomass feedstock is pyrolyzed at temperatures below 300 °C. However, biochar with high carbon number and degree of graphitization can be obtained when the biomass is pyrolyzed at higher temperatures and the formation of pores and cracks on the biochar is promoted due to the release of volatile gases [53].

6.2 Hydrothermal carbonization

Hydrothermal carbonization is another popular technique for preparation of biochar and the solid product formed through this process is known as hydrochar. This process carbonizes the biomass feedstock by applying relatively mild temperatures to the biomass suspended in a liquid media under pressurized condition. The main advantage of this process is no significant amount of harmful substances are produced during the synthesis stage as water is the main solvent used. However, the highly pressurized system could create safety risk during operation. The hydrothermally produced biochar generally exhibits lower carbon content but higher oxygen content than the biomass precursor, thus such a biochar is expected to be rich in oxygenated functional groups [53].

6.3 Flash carbonization

During flash carbonization, biomass is packed into the pyrolysis chamber and a flash fire is used to heat the biomass from the bottom of the chamber. A flash fire at elevated pressure of more than 1,000 kPa is ignited and blazed toward the biomass in an upward direction while a continuous air flow is maintained concurrently from the top of the chamber in a downward direction [54]. Under the fast heating rate, the chemical reaction takes less than 30 min to transform the biomass into gaseous fuel, liquid bio-oil, and solid biochar [52]. Hirst and Taylor [55] conducted a biomass flash carbonization experiment, and the resulting carbon product was compared with conventional pyrolysis and hydrothermal carbonization processes. The biocarbon prepared by employing flash carbonization technique contained a larger amount of micropores than the two aforementioned techniques [55].

6.4 Gasification

Gasification involves the thermochemical decomposition of carbonaceous materials under an oxygen deficient environment. With the presence of gasification agents such as steam, oxygen, or air at high temperatures (750–900 °C), syngas is produced as the main product [52]. The yield of syngas is typically 85% comprising carbon dioxide, hydrogen, carbon monoxide, methane, and some other hydrocarbons. The reaction temperature is the key factor in determining the production of syngas and its composition. It was reported that as the temperature increased, the amount of hydrogen and carbon dioxide increased while the other gaseous components decreased [52]. On the other hand, biochar is one of the by-products of gasification process with a mere yield of 10% and the rest is bio-oil (5%). Biochar resulting from this process has smaller particle size and is stronger, more stable, and resistant to chemical oxidation [56]. These attributes render its wide applications in various fields including adsorbent in adsorption technique, direct carbon fuel cell of electrochemical application, additives for anaerobic digestion, and catalyst in biodiesel production [57].

7 Characterization of biochar

The effectiveness of biochar in adsorption applications strongly depends on its physicochemical, surface, and structural or molecular characteristics [52]. These characteristics can be determined through a combination of characterization methods. Currently, scanning electron microscopy (SEM), thermogravimetric analysis (TGA), x-ray diffraction (XRD), Brunauer–Emmett–Teller (BET), and Fourier transform infrared spectroscopy (FTIR) are the commonly used methods to characterize biochar. Furthermore, these methods can also be used to predict and justify the occurrence of the adsorption process and the possible mechanisms involved. Coherence results can usually be found from different analytical techniques and these can be used to reinforce and conclude the findings of researchers.

7.1 Scanning electron microscopy

SEM and transmission electron microscope (TEM) are used to identify the surface morphology of biochar. SEM images are able to provide detailed depiction of the distribution and arrangement of mesopores and micropores within the biochar. SEM and TEM can be coupled with energy dispersive x-ray spectroscopy (EDX) to further analyze the elemental composition of biochar. These methods are usually employed for the comparison of morphology and elemental properties of the biochar before and after adsorption or modification processes. For instance, Singh and Srivastava [58] incorporated iron oxide to biochar for the removal of ofloxacin. The incorporated iron oxide particles could be clearly seen as black dots attached to the carbon surface on the TEM images. The size of the iron oxide particles estimated by TEM fell within the range of 26–41 nm. On the SEM images, the granular structures on biochar were the iron and silica phases which were then confirmed by the corresponding EDX spectrograph. By comparing the EDX results, the atomic percentage of carbon increased by 4% after the adsorption of ofloxacin. This could possibly be due to the attachment of ofloxacin molecules that contained carbon atom on the biochar surface [58].

7.2 Fourier transform infrared spectroscopy

FTIR is a vibrational technique generally used to determine the functional groups present on the biochar surface. The carboxylic acid (–COOH), hydroxyl (–OH), amino, and lactonic groups are the essential functional groups on biochar that play a major role in the adsorption process [52]. A graph of transmittance (%) versus wavenumber (cm^{-1}) will be generated from the FTIR analysis for the determination of functional groups. Different functional groups are represented by the specific peaks that occur at various location in the infrared spectrometry. The shift in FTIR peaks after absorption signifies the chemical reactions involved that are responsible for the removability of the pollutants. From an FTIR analysis carried out by Paunovic and Pap [59] on its functionalized biochar, a broad peak of O–H stretching which belongs to hydroxyl group was found at wavelength 3,800–3,200 cm^{-1}. This hydroxyl group contributed to the electron donor characteristics of the functionalized biochar. Besides that, strong π-electron acceptors of symmetric and asymmetric stretching vibrations of carboxylic acid that contained CH, CH_2, and CH_3 groups were found at peaks 2,921.11 and 2,853.50 cm^{-1}, respectively. After the naproxen adsorption process, mono-substitute aromatic structures were discovered on the saturated biochar at new peaks of 566.66 and 498.63 cm^{-1}. This indicated the occurrence of π–π electron-donor-acceptor (EDA) interactions on the aromatic rings of the functionalized biochar [59].

7.3 X-ray diffractometer

X-ray diffractometer (XRD) is broadly used to analyze the structure and crystallinity of biochar. From the XRD diffractogram, the observed sharp and strong peaks signify the crystalline nature of the nanocrystals formed on the adsorbent. Ma and Li [60] successfully synthesized a magnetic sludge biochar for the removal of tetracycline (TC) and ciprofloxacin. Zinc (Zn) and iron (Fe), which are two commonly found heavy metals in the environment, were used for the modification of sludge-derived biochar (SB), whereby Zn was used as the precursor of the final magnetic SB (Fe/Zn-SB). The respective SB, Zn-SB, and Fe/Zn-SB were examined using XRD. A sharp and great diffraction peak of amorphous carbon was obtained at 26° resembling complete carbonization of the different types of biochar during the pyrolysis process. In addition, new peaks of 35.3°, 42.9°, and 56.9° were found on Fe/Zn-SB. These peaks signified the formation of magnetic media (γ-Fe_2O_3) that increased the magnetic sensitivity of the final biochar product. A diffraction peak at 20° corresponded to the generation of graphitic structure which acted as the π-electron donor for the adsorption process. This process involved the π–π conjugation between biochar surface and aromatic ring (π-electron acceptor) of TC and ciprofloxacin [60].

7.4 Thermogravimetric analysis

TGA is commonly used to examine the thermal stability of biochar by monitoring its weight loss while exerting increasing temperature onto the biochar. There are three distinct stages that can be deduced from a TGA curve as the temperature increases. The first stage involves the loss of moisture, second stage involves the fast devolatilization of organic materials, and the third stage is the continuous slight devolatilization of organic materials. Yu and Gu [61] analyzed the thermodecomposition characteristics of an activated grapefruit peel-derived biochar under 10 °C min^{-1} with a constant nitrogen gas flow of 20 mL min^{-1} and the temperature was increased from room temperature to 800 °C. At the first stage (30–180 °C), an obvious weight loss of 7% from the activated biochar was seen at 50 °C, which was associated with the evaporation of water content and organic matters with low boiling point. During the second stage, at temperatures between 180 and 600 °C, another weight loss of 7% was accounted. Decomposition of lignin and other inorganic salts might have occurred as well as the volatilization of complexed organic acids on the surface of the activated biochar. Comparatively, only 5% weight loss was found on the inactivated biochar at stage 2. This could possibly be due to the improvement of structural components such as cellulose and hemicellulose by the employed activation chemical which further enhanced the thermal degradation capability of the activated biochar. Finally, at the last stage (>600 °C), the activated biochar faced continuous weight loss corresponding to the reactions between the carbon materials formed and the volatile gas released from previous stage [61].

7.5 Brunauer–Emmett–Teller analyzer

BET is an equipment used to determine the specific surface area of biochar. The specific surface area is one of the prominent factors that governs the adsorption efficiency of a biochar adsorbent. BET is usually carried out to compare the increment of specific surface area after the raw biomass has undergone pyrolysis or other modification processes. Raw biomass and unmodified biochar typically contain a low amount of microporous structures and hence poor adsorptive capacity. The BET-specific surface area of biochar derived from spent coffee ground was increased significantly from 3.644 to 116.592 $m^2\ g^{-1}$ after chemical activation using NaOH [62]. Besides that, more than twofold increment in the BET surface area could be detected from the modification of biochar using $ZnCl_2$. The specific surface area enlarged considerably from 94.8 to 264 $m^2\ g^{-1}$ [58]. The results from the BET analysis showed the significance of chemically modifying the biochar surface to increase the availability of adsorptive sites.

8 Influencing factors on biochar physical and chemical properties

Physicochemical properties of biochar such as porosity, surface area, surface charge, pH, and functional groups are essential in determining the adsorption capability of biochar. Thus, it is important to discuss the correlated influencing factors for enhanced understanding of their impacts on biochar properties and their potential in sequestering pollutants.

8.1 Surface area and porosity

Surface area and porosity are the major physical properties of biochar that affect the removal efficiency of the adsorption process. Great surface area leads to higher availability of active sites for the uptake of pollutants while high porosity improves the mass transfer between the pollutants and biochar surface [63]. Surface area and porosity are highly intercorrelated with each other. Generally, increasing porosity will increase the surface area as well. The porous structure of biochar has different diameter sizes ranging from micropores to macropores. A micropore has less than 2 nm of diameter, a mesopore diameter is in between 2 and 50 nm while a macropore has a diameter of more than 50 nm [63]. The selection of feedstock will affect the surface area of the produced biochar. For instance, a biochar that was derived from grass had lower surface area but higher ash content as compared to one derived from woody biomass [64]. The porosity of biochar is mainly formed during the water dehydration stage of the pyrolysis process. Higher temperatures and higher heating rates during pyrolysis favor the release of volatile matters and create an internal pressure, thus inducing more formation of porous structure and subsequently increasing the biochar surface area. However, in some cases, biochars produced from high pyrolysis temperatures have lower surface area and porosity. As previously reported, this might be due to a damaged porous structure at high temperatures or the blockage of porous structure by tar that reduced the total surface area [65].

Different activation methods can also be used to enhance the surface morphology of biochar. As evident in the BET characterization section earlier, the surface area of biochar can be improved drastically through activation using chemical reagents. The surface area of spent coffee-ground biochar was increased by 30 times after treatment with NaOH [62]. The SB activated by $ZnCl_2$ and $FeCl_3$ increased about 1.8 times in both surface area and pore volume with an addition of magnetic properties to facilitate the adsorption process [58]. Furthermore, wild plum kernel biochar that was activated by KOH coupled with microwave demonstrated the successful development of porosity and rough surface as evidenced from SEM images [59]. Another notable research done by Shin and Kwak [66] has shown more than a hundredfold of increment in both surface area and total pore volume after the spent coffee waste-derived biochar was treated with NaOH. The specific surface area increased from 62 to 655.4 $m^2\ g^{-1}$ while the total pore volume increased from 0.032 to 0.456 $cm^3\ g^{-1}$ [66]. Contrarily, an acid treatment using H_2SO_4, HCl, and HNO_3 on tea-waste biochar conducted by Peiris and Nayanathara [67] has shown destructive effect on the pore wall structure based on the SEM images. Other physical activation methods such as steam, carbon dioxide, ball milling, and ultrasound irradiation have showed promising results in the enhancement of biochar surface area, pore size, and density [64].

8.2 Functional groups

Functional groups on biochar play a prominent role in the adsorption process. Oxygen-containing amino functional groups such as hydroxyl, carboxyl, phenolic, and amine are able to remove contaminants from aqueous solution through chemical bonding formation [49]. The selection of biomass and the pyrolysis temperature during preparation of biochar are pivotal in determining the functional groups on the final biochar product. Typically, when the pyrolysis temperature increases, the biochar produced will contain less functional groups. The O/C, N/C, and H/C contents in biochar that contribute to hydroxyl, carboxylic, and amino groups decrease with increasing temperature. The mass reduction in functional groups at high temperature is a result of a higher degree of carbonization [65]. In addition to that, at increasing pyrolysis temperature, the amount of acidic functional groups formed on the biochar will reduce, which decreases the cation exchange potential during pollutants removal [68]. The diminution in overall intensity of FTIR spectra at increasing temperature also indicates the occurrence of dehydration that reduces the alcohol/carboxylic –OH stretching and decreases the polar functional groups due to the disappearance of $-CH_2$ aliphatic bands [67]. On the other hand, biochar produced by hydrothermal carbonization may contain higher amounts of oxygen-containing functional groups [53].

The selection of feedstock also affects the surface chemistry of biochar due to the differences in cellulose and lignin contents in the biomass which impact the amount of carboxyl groups [63]. Moreover, the desired functional groups can be introduced by chemical activation using various acid, base, or oxidizing agents. Peiris and Nayanathara [67] have shown that acid treatment using H_2SO_4 and HNO_3 on tea-waste biochar introduced carbonyl stretching (1,690–1,720 cm^{-1}), while HCl promoted the formation of single-bonded C–O (1,200–1,250 cm^{-1}). After the activation of biochar using NaOH, FTIR peaks at 3,400 and 1,570 cm^{-1} corresponding to hydrogen bond (O–H and N–H) and aromatic bond (C = C), respectively, were determined [69]. This showed that the base activation agent promoted the formation of hydroxyl group and the generation of a stable aromatic structure. Similar FTIR results were determined by Shin and Kwak [66] after the spent coffee-waste biochar was activated by NaOH. The synergistic effect of both functional groups and surface morphologies of biochar by NaOH activation had tremendously improved the adsorption capacity of spent coffee-waste biochar in removing naproxen (14.81–192.04 $\mu mol\ g^{-1}$), diclofenac (15.73–98.74 $\mu mol\ g^{-1}$), and ibuprofen (16.20–80.02 $\mu mol\ g^{-1}$) from spiked lake water matrix [66].

8.3 Stability

Stability of biochar has to be taken into account for practical applications especially in wastewater treatment. Biochar should exhibit thermal stability, chemical stability, and insolubility under reaction conditions to prevent dissolution of organic matter from the biochar [49]. For example, possible leaching of heavy metals from biochar prepared from sewage sludge with high heavy metals may cause secondary contamination of the water matrix [70]. Moreover, high stability of biochar enables the regeneration and reuse of biochar for adsorption processes. The main influencing factors of biochar stability are the feedstock selected and the pyrolysis temperature during biochar preparation. High aromatic and lignin contents in biomass are favorable to produce biochar with higher stability. Aromatic carbon content can be enhanced through increasing pyrolysis temperature [64]. Apart from that, lignin content is positively correlated to the carbon content of biochar produced. Production technology such as hydrothermal carbonization is able to produce biochar with higher carbon content as compared to pyrolysis and gasification [49].

8.4 Surface charges and pH

Similarly, the surface charges and pH of biochar are also dependent on the feedstock type and the pyrolysis temperature. From a review carried out by Li and Dong [65], it was reported that most of the feedstocks produced alkaline biochar. Some exceptions were oak wood and corn stover which were acidic. However, when the pyrolysis temperature was increased to 600 °C, the pH of oak wood biochar increased to 9.5. Moreover, biochars from wheat straw and wastewater sludge were acidic at low temperatures (200–400 °C) but became alkaline at higher temperatures. Generally, the pH of biochar increases as pyrolysis temperature increases. In addition, biochar produced from higher pyrolysis temperature has greater ash content. Therefore, the high pH of biochar could be attributed to its high ash content [65]. Besides that, the disappearance of acidic functional groups at elevated temperature was responsible for the high pH value as well [58].

8.5 Biochar yield

Other than the physicochemical properties, the yield of biochar is very important in assessing the overall efficiency of the adsorption process. Biochar yield is highly dependent on the selection of biomass. In recent years, a wide range of biomass have been chosen as the feedstock for biochar production, especially agricultural crops and their by-products due to their abundant availability. These feedstocks mainly consist of polymers such as lignin, cellulose, and hemicellulose. Among the polymers, the volatility of hemicellulose and cellulose are higher as they can be completely pyrolyzed at 350 °C, whereas lignin with lower volatility requires temperatures up to 500 °C. The lignin content of biomass is directly proportional to the carbon content of biochar produced from pyrolysis process [64]. Generally, residues biomass such as peanut shells, paper-mill waste, and rice husk exhibited higher biochar yield than woods, manures and microalgae across the entire range of tested temperatures.[71].

Furthermore, pyrolysis conditions such as temperature, residence time, and heating rate pose significant effects on the yield of biochar [64]. As temperature is increased, the yield of biochar decreases but the surface area and carbon content of biochar will increase instead. Secondary pyrolysis reaction that involves cracking will occur at high heating rate, which results in the reduction of biochar yield. Hence, lower heating rate will encourage the production of biochar. The effect of pyrolysis residence time is interrelated with the heating temperature and heating rate. Under the same heating temperature, the yield of biochar from fast pyrolysis reduces as the residence time increases. In slow pyrolysis, biochar yield increases as the residence time prolongs. The increasing reaction time allows the repolymerization of biochar elements, thus increasing the yield of biochar [64].

8.6 Physical treatments of biochar

The desired physical and chemical characteristics of biochar can be achieved in a more economical and environmentally friendly manner by subjecting the biochar to physical treatment. Notably, the physical treatment of biochar can be classified into different modification methods. Methods such as ball milling, gas activation and microwave irradiation, and magnetism are among the most employed physical treatments on biochar. Furthermore, physical treatment utilizes clean and easily controlled reagents such as steam, oxygen, and carbon dioxide in the biochar modification, resulting in cheaper biochar which is purer than those modified via chemical modification [72].

8.6.1 Steam activation

Steam activation is the most prominent modification method among all physical activation methods. It involves partial gasification of biochar with steam, resulting in the decomposition and crystallization of carbon materials. Initially, oxygen is released from the decomposition of water vapors and forms CO with surface carbon atom while the hydrogen released forms carbon–hydrogen complexes. CO_2 and H_2 are then formed through further oxidization by steam. The devolatilization process occurs resulting in the formation of larger pores and increment in the surface area.

Steam activation on tea-waste biochar carried out by Rajapaksha and Vithanage [73] has shown significant enhancement on the surface morphologies of biochar. The surface area of steam-activated tea-waste biochar was increased from 342.2 to 576.1 $m^2\ g^{-1}$ while its pore volume increased from 0.022 to 0.109 $cm^3\ g^{-1}$. Hence, the adsorption capacity of steam-activated tea-waste biochar on sulfamethazine has increased about 81% (from 5.50 to 9.95 mg g^{-1}) as compared to raw tea-waste biochar [73].

8.6.2 Ball milling

Ball milling is a simple, solid state, mechanical–chemical process that utilizes the mechanical energy of moving balls in a machine to break down and grind biochar. The structural and adsorptive properties of biochar will be modified through the dynamic forces. Ball milling has been recognized as a green and low-cost biochar activation method. Huang and Zimmerman [74] successfully demonstrated the excellent adsorption capability of ball-milled biochar. The best removal efficiency achieved for SMX and sulfapyridine (SPY) were 83.83% and 89.6% by ball-milled biochars, while the unmilled biochars showed nearly zero adsorptivity toward the two sulfonamide antibiotics. The increment in both specific surface area and functional groups could greatly contribute to the improvement in the removability of the two antibiotics. The FTIR analysis showed the increase in surface functional groups such as –CO, aromatic C = O/C = C, and $-CH_2$ at 1,262, 1,696/1,597, and 2,920 cm^{-1}, respectively. These oxygen-containing functional groups induced the formation of hydrogen bonding due to their affinity toward the polarity of SMX and SPY. On top of that, the ball-milled biochar has been demonstrated to be effective in removing other water pollutants as well [74].

8.6.3 Microwave

Conventional heating is a time-consuming method in which uniform heating of materials with different sizes and shapes is hard to achieve. Microwave activation is a better option to activate biochar through heating. The microwave electromagnetic base irradiation energy with frequencies ranging from 300 MHz to 300 GHz is transmitted through the biochar at molecular level, and this will be converted to heat [75]. Hence, thermal energy can be transferred internally and uniformly with a lower consumption of time and energy. Thus, microwave can be incorporated in pyrolysis process as a source of heat to produce biochar while concurrently modifying its structure and properties [76]. The microwave-assisted pyrolysis was able to produce biochar yields of more than 65 wt% and the biochar product contained more functional groups and greater BET surface areas of 450–800 $m^2\ g^{-1}$ as compared to unmodified biochar [75].

8.6.4 Magnetism

Magnetic modification of biochar is another emerging physical modification method for biochar. One of the challenges faced by current adsorption technology is the incomplete separation of tiny and light adsorbent particles from the treated water source. This is to prevent the occurrence of secondary pollution by the pollutant-saturated adsorbent and to improve the cost-effectiveness of process through the reusability of adsorbents. The induced magnetic property is able to ease the separation of minute biochar particles from the treated wastewater and effluents by imposing a magnetic force. Additionally, the incorporation of iron oxides on biochar can increase its metal adsorptive capacities by increasing the surface area and functional groups as well as its cation exchange potential. A magnetic biochar synthesized by Singh and Srivastava [58] was able to be completely separated from water sample in 4 min. New functional groups were found on the FTIR spectra of the magnetic biochar such as aliphatic C–H, Fe–O, and C = O. In addition, the BET surface area and total pore volume of the magnetic biochar increased considerably. These properties facilitated the adsorption capacity of the magnetic biochar for the removal of TC and ciprofloxacin, with maximum values of 145 and 74.2 mg g^{-1}, respectively [58].

Despite all the advantages of physical activation methods that have been mentioned earlier, each of the methods has some downsides too. Ball-milled biochar is prone to surface runoff due to the increased dispersibility in water while steam activation reduces the surface functional groups of biochar. Meanwhile, microwave biochar has low reproducibility due to the difficulty in power control, whereas there can be possible reduction in vacant pores owing to blockage by introducing a magnetic medium. Nevertheless, chemically modified biochar often exhibits higher decontamination efficiency and adsorption capacity due to more functional groups embedded on its surface or the presence of specifically tailored functional groups for certain pollutant removal. Hence, it is important to modify the biochar according to the desirable properties required for the target pollutants based on the influencing factors. Biochar with abundant functional groups, high surface area, porosity, and stability is usually able to deliver high removal efficiency of pharmaceutical residues.

9 Effect of process parameters on biochar adsorption of pharmaceuticals

Pharmaceutical removal efficiency of biochar is not only controlled by the physicochemical properties of developed biochar but also affected by the process parameters during adsorption stage. Parameters such as biochar dosage, initial adsorbate concentration, temperature, pH of solution, and contact time are the most influential parameters in the adsorption process.

9.1 Biochar dosage

The effect of biochar dosage is related to the availability of binding sites for pharmaceutical adsorption. Generally, high biochar dosage results in high pharmaceutical removal owing to large amount of active sites for binding the pharmaceutical residues. This is supported by the findings reported by Singh and Srivastava [58]. From their study, ofloxacin removal was increased from 36 to 90%, whereas adsorption capacity decreased from 11 to 4 mg g^{-1} as the textile mill sludge biochar dosage was increased from 1 to 6 g L^{-1}. The removal efficiency increased sharply from 1 to 4 g L^{-1} and thereafter became nearly constant. The results indicated that ofloxacin adsorption was enhanced due to the increased binding sites as more biochar dosage was applied [58].

9.2 Solution pH

Solution pH is another critical parameter in adsorption process as the surface charge of biochar and speciation of adsorbate can be altered by the surrounding pH. The surface charges interaction between the adsorbent and adsorbate species is critical in facilitating electrostatic interaction during adsorption. According to the study conducted by Paunovic and Pap [59], the highest naproxen adsorption capacity (25 mg g^{-1}) by microwave-functionalized biochar (WpOH) was achieved at pH 5–7. Naproxen is an acidic pharmaceutical compound with acid dissociation constant (pK_a) of 4.2. This indicated that naproxen was dissociated into anionic form when $pH > pK_a$ and exhibited as a neutral molecule when $pH < pK_a$. The point of zero charge (pH_{PZC}) of WpOH was 7.2, where WpOH surface became positively charged at $pH < 7.2$ and negatively charged at $pH > 7.2$. At $pH < 4.2$, the positively charged WpOH surface did not form any electrostatic interactions with neutrally charged naproxen species. Hence, the adsorption mechanisms at this stage could be attributed to hydrogen bonding, hydrophobic, and π–π interactions. At pH 5–7, electrostatic interaction was formed between the positively charged WpOH surface and negatively charged naproxen anions. This additional adsorption mechanism led to enhanced naproxen adsorption capacity. However, the adsorption capacity decreased when pH > 7 due to electrostatic repulsion between the like charges of the WpOH surface and anionic naproxen species. Therefore, the solution pH is essential to control the surface chemistry of the adsorption process, and it is important to adjust pH of the solution for optimum removal of the pharmaceuticals.

9.3 Temperature

Temperature is also an influencing parameter in the adsorption process. In the study of SMX adsorption onto sugar bagasse biochar, the removal efficiency decreased from 92.18% to 85.98% when temperature was increased from 30 to 50 °C. This could be possibly be due to the decrease in adsorbent density causing difficulty in retaining the biochar in liquid medium, hence a reduction in surface activity was observed [39]. On the contrary, an increase in adsorption capacity was observed in the study conducted by Yu and Gu [61] on TC removal by grapefruit peel biochar. The TC adsorption capacity increased from 27.92 to 30.95 mg g^{-1} when temperature was increased from 25 to 45 °C. Higher temperatures might have resulted in higher intermolecular motion intensities, thus increasing the mobility of TC to the surface of biochar [61]. The experimental data from temperature study are useful in determining adsorption isotherms such as Langmuir, Freundlich, Temkin, and Sips models, as well as adsorption thermodynamics properties. These parameters are crucial for the design of large-scale adsorption systems.

9.4 Initial concentration of adsorbate

Initial concentration of adsorbate is another crucial parameter in the adsorption process to determine the relationship between the available sorption sites on biochar and the amount of adsorbate. From the study carried out by Paunovic and Pap [59] using microwave-functionalized biochar (WpOH), the effect of initial concentration of naproxen was studied from 3.1 to 125.3 mg L^{-1} with constant WpOH dosage of 0.4 g L^{-1}, pH 6, temperature of 22 °C, and 180 min of contact time. The results showed that this adsorption process was highly dependent on the initial naproxen concentration. Complete removal (100%) of naproxen was achieved at initial concentrations of 3.1 and 6.3 mg L^{-1}. This was due to the abundant availability of vacant sites with low amount of adsorbate. Hence, all the naproxen adsorbates were easily removed by WpOH at low initial concentration. At 15.7 mg L^{-1}, the removal efficiency dropped significantly to 60% with a continuous diminishing trend to about 23% when the initial concentration was increased to 125.3 mg L^{-1}. Owing to the fixed dosage of WpOH at 0.4 g L^{-1}, the limited sorption sites on biochar would eventually be taken up by the adsorbates as the concentration increased. Thus, the removal efficiency of naproxen decreased with increasing naproxen concentration [59]. A similar trend was observed by Singh and Srivastava [58] using a magnetic SB to remove ofloxacin from aqueous solution. The adsorption sites of biochar were occupied at low concentration, hence resulting in decreased efficiency at higher initial concentration of ofloxacin [58].

10 Equilibrium and kinetic of pharmaceuticals adsorption onto biochar

Adsorption equilibrium and kinetic studies provide useful information regarding the adsorption performance, adsorbent rate, and the mass transfer mechanisms involved, revealing the interactions between adsorbate and adsorbent. The mass transfer mechanisms involve three main steps. The first step is the external diffusion whereby adsorbate is transferred through the liquid film to the adsorbent surface. The driving force for external diffusion is the difference in concentration of the bulk solution and the adsorbent surface. The second step is the internal diffusion of adsorbate into the adsorbent pores. The third cum last step is the attachment of the adsorbate on the active sites of the adsorbent. Both equilibrium and kinetic information are important for the design and optimization of the adsorption system.

10.1 Isotherm models

Isotherm studies are important to evaluate the maximum adsorption capacity and further identify the mechanisms involved. The equilibria experimental data are usually fitted into several adsorption isotherm models such as the Langmuir, Freundlich, Temkin, and Dubinin–Radushkevich (D-R). The Langmuir model assumes the solid surface of adsorbent is uniform and contains a finite number of active sites with the same sorption activation energies. The adsorbates are assumed to occupy the sites in a monolayer manner without any interaction with adjacent sorbed adsorbate [77]. The dimensionless separation factor, R_L, can be used to determine the process favorability. The adsorption process is linear if R_L=1, irreversible if $R_L = 0$, favorable if $0 < R_L < 1$ and unfavorable if $R_L > 1$ [78]. The Freundlich model describes the logarithmic depletion of sorption energy with increasing surface coverage. It assumes multilayer adsorption occurs on heterogeneous surface and is reversible [79]. The Temkin model considers the interaction of adsorbent–adsorbate and assumes a linear reduction in free sorption energy with the surface coverage [80]. The D-R model takes into account that the adsorption process occurs on heterogeneous surface. It assumes a high degree of rectangularity in the adsorption isotherm which is related to the porous structure of the adsorbent [81]. Equations (1)–(5) are the mathematical equations of the respective isotherm models.

Langmuir:

$$q_e = \frac{q_m K_L C_e}{1 + K_L C_e} \tag{1}$$

Hall's separation factor:

$$R_L = \frac{1}{1 + K_L C_0} \tag{2}$$

Freundlich:

$$q_e = K_f C_e^{\frac{1}{n}} \tag{3}$$

Temkin:

$$q_e = \frac{RT}{b} \ln(AC_e) \tag{4}$$

Dubinin and Radushkevich:

$$q_e = q_s \exp(-K_D \varepsilon^2) \tag{5}$$

where q_m (mg g^{-1}) is the Langmuir maximum adsorption capacity, K_L (L mg^{-1}) is the Langmuir adsorption constant, C_0 (mg L^{-1}) is the highest initial concentration, K_f ((mg g^{-1})(Lmg^{-1})$^{1/n}$) is the Freundlich constant, n is the Freundlich exponent, b (J mol^{-1}) is the Temkin constant, A (L mg^{-1}) is the maximum binding constant, q_s (mg g^{-1}) is the D-R maximum sorption capacity, K_D (mol^2 kJ^{-2}) is the D-R constant related to sorption energy, ε $(= RT\ln(1 + (1/C_e)))$ is the Polanyi potential.

10.1.1 Isotherm models case studies

Various isotherm empirical models have been applied for the analysis of experimental equilibrium data related to the adsorption of pharmaceuticals from aqueous solution using biochar. These led to the vast variability of results which were highly dependent on the properties of both adsorbent and adsorbate. The most extensively used models by researchers in this field are the Langmuir and Freundlich models [82].

Ma and Li [60] synthesized a magnetic biochar from sewage sludge through hydrothermal carbonization process to remove TC and ciproflaxin from aqueous solution. The experimental equilibrium data were evaluated using the Langmuir, Freundlich, and D-R models. From the Langmuir model fitting, the calculated maximum adsorption capacity for ciproflaxin and TC were 72.3 and 145 mg g^{-1}, respectively.

The adsorption capacity of the magnetic TC was comparatively high as compared to other adsorbents for the removal of antibiotics. Among the three isotherm models, the equilibrium data were best fitted to the Freundlich model with the highest nonlinear coefficient of determination, R^2, than the Langmuir model. High desirability of the adsorption process was also shown from the values of $1/n$ that lay between 0 and 1. Hence, the Freundlich model indicated that the active sites on the magnetic biochar surface was heterogeneous. In addition, the model assumed that the multilayer physisorption of ciproflaxin and TC was the dominant adsorption mechanism which was confirmed by the free energy calculated. The reversible physisorption process also allowed the regeneration and reusability of the saturated magnetic biochar [60].

Freundlich model was also the best fitted model for TC removal using biochars derived from maple leaf, rice straw, wine manure, and sewage sludge [60, 69, 83]. This showed that TC might be more favorable to be removed by biochar with heterogeneous active sites through multilayer physisorption process. On the other hand, the equilibrium data for TC adsorption onto activated biochars from grapefruit peel and spent coffee ground were best fitted with the Langmuir model rather than the Freundlich model. However, the maximum adsorption capacities for both activated biochars were lower at 34.58 and 113.64 mg g^{-1}, respectively. This signified that the favorability of the adsorption process was not solely dependent on the adsorbate but also on the adsorbent properties.

The experimental data of the adsorption of naproxen onto a microwaved biochar derived from wild plum kernel were fitted to the Langmuir, Freundlich, and D-R models [59]. The Langmuir model was the best fitting model suggesting the occurrence of monolayer adsorption of naproxen molecules onto the active sites of the biochar. The biochar surface was assumed to be homogeneous and the actives sites were identical. The maximum adsorption capacity obtained from the model was 73.13 mg g^{-1}. From the Freundlich model, high K_f coefficient of 18.39 (mg g^{-1}) (L $mg^{-1})^{1/n}$ and $1/n$ coefficient with a value of 0.28 were calculated. These results implied that the microwaved biochar had high affinity toward the naproxen molecules. The D-R model suggested that the adsorption was dominant by physisorption process. However, due to the poor fitting of D-R model to this system (R^2 = 0.81), the adsorption mechanisms involved had to be further proven from other aspects such as kinetic models and characterization techniques.

10.2 Kinetic models

Kinetic study is crucial for the determination of time required for the adsorption process to achieve equilibrium and gaining insights into the nature of the adsorption process. Adsorption kinetic correlates the rate of adsorbate uptake by adsorbent as a function of time. Typically, an adsorption kinetic profile exhibits three stages. The first stage is fast adsorption, where abundant unoccupied active sites are available for adsorbate uptake, followed by slower adsorption rate as the sites are progressively occupied as time goes by. Finally, the kinetic profile demonstrates a plateau when the adsorbate–adsorbent system achieves equilibrium. Adsorption models such as the pseudo-first-order kinetic, pseudo-second-order kinetic, and Elovich models are often used to describe the kinetic behavior of the system. The adsorption mechanism can be determined by fitting the experimental data to the intraparticle diffusion model.

The pseudo-first-order kinetic model assumes a proportional relationship between the occupation rates of sorption sites and the amount of vacant sites available. It is usually used to predict the kinetic behaviour of adsorption process at the initial stage [84]. Meanwhile, the pseudo-second-order kinetic model considers the saturation rate of binding sites is proportional to the squared amount of unoccupied sites [85]. The Elovich kinetic model describes the occurrence of chemisorption process at the solid surface reactions on a highly heterogeneous surface [86]. Finally, the kinetic data can be further studied using the intraparticle diffusion model to determine the rate-determining steps for the adsorption of adsorbate onto the adsorbent. The trend of intraparticle diffusion plot (q_t versus $t^{0.5}$) suggests the rate-controlling step of the adsorption process. A linear plot passing through the origin implies that the rate is solely limited by intraparticle diffusion. Inversely, a linear or nonlinear plot which does not pass through the origin indicates that the rate is controlled by two or more steps [87]. Equations (6)–(9) are the mathematical expressions for the kinetic models.

Pseudo-first-order:

$$q_t = q_e\left(1 - e^{-k_1 t}\right) \tag{6}$$

Pseudo-second-order:

$$q_t = \frac{k_2 t q_e^2}{1 + k_2 t q_e} \tag{7}$$

Elovich:

$$q_t = \frac{1}{\beta}\ln(\alpha\beta) + \frac{1}{\beta}\ln(t) \tag{8}$$

Intraparticle diffusion:

$$q_t = k_p t^{0.5} + C \tag{9}$$

where k_1 (min^{-1}) is the pseudo-first-order rate constant, k_2 (g (mg min)$^{-1}$) is the pseudo-second-order rate constant, β (g mg^{-1}) is the Elovich constant, α (mg (g min)$^{-1}$) is the initial adsorption rate, k_p (mg(g min$^{0.5}$)$^{-1}$) is the intraparticle diffusion rate constant, and C (mg g^{-1}) is the y-intercept of intraparticle diffusion plot.

10.2.1 Kinetic models case studies

In current research, the pseudo-first-order, pseudo-second-order, and Elovich model seem to be the most employed models for the kinetic study of pharmaceuticals adsorption onto biochar. From the recent data collection conducted by Li and Zou [82], most of the removal of emerging contaminants using biochar complied to the pseudo-second-order kinetic model, while the Elovich model was also relatively relevant. Some studies have included the intraparticle diffusion model as well to determine the rate-limiting steps of their adsorption process.

Keerthanan and Bhatnagar [88] fitted the experimental kinetic data of steam-activated tea-waste biochar used for caffeine removal into both the pseudo-second-order and Elovich models. Initially, a rapid increase in adsorption capacity was observed and equilibrium was achieved at 1.5 h. The kinetic data were best fitted to the Elovich model with a R^2 of 0.972. This good fit suggested the occurrence of chemisorption process during the adsorption of caffeine onto the heterogeneous surface of biochar. The initial sorption rate, α, calculated from the Elovich equation was 55.66 mg (g min)$^{-1}$. This indicated that the adsorption of caffeine by the activated biochar at the beginning was a random process due to the presence of high amount of vacant active sites. As the time increased, the sorption sites on the biochar were occupied by caffeine until equilibrium was achieved. Due to the limited amount of vacant sites remaining, a desorption constant which was also known as the Elovich constant, β, was determined as 0.62 g mg^{-1}. Hence, the high α and low β values signified that the caffeine adsorption onto the steam-activated tea-waste biochar was a favorable adsorption process [88].

The adsorption kinetic data obtained by Kim and Bhatia [83] for the removal of aqueous TC using a maple leaf-derived biochar were fitted to various kinetic models. The pseudo-second-order model showed the best fit ($R^2 = 0.988$) among all models signifying chemical adsorption was the rate-limiting step. Initially, the adsorption of TC onto biochar increased consistently for 24 h, then the increment slowed down due to the reduction in active sites on the biochar. The results suggested that the chemical adsorption processes such as hydrogen bonding and metal complexation that happened on the biochar surface were faster and more dominant than the physical adsorption process. The metal complexation between TC and the insoluble chelates with cation of Ca^{2+} and Mg^{2+} was another minor influencing mechanism involved in this process. Besides that, the intraparticle diffusion linear plot did not pass through the origin, suggesting that the adsorption rate of TC onto maple leaf-derived biochar was not controlled by only one rate-limiting step [83].

The experimental kinetic data of Singh and Srivastava [58] on the removal of ofloxacin using a magnetic SB were best fitted to the pseudo-second-order model as well. The equilibrium of the adsorption process was achieved in approximately 5 h. The data were then fitted into the Weber–Morris intraparticle diffusion model (equation (9)) to determine the diffusion mechanisms involved and the rate-controlling step of the entire adsorption process. The Weber–Morris intraparticle diffusion plot suggested that the entire adsorption process could be distinguished into three stages. In the first stage, the plot with high gradient indicated the occurrence of the dominant film diffusion process where ofloxacin molecules traveled through the liquid film surrounding the biochar. The second stage implied the pore diffusion of ofloxacin molecules entering into the biochar while the final stage with a smaller gradient represented the attainment of adsorption equilibrium state [58].

11 Mechanisms of pharmaceuticals adsorption onto biochar

Adsorption process is a complicated process where multiple mechanisms could occur to sequester the pharmaceutical pollutants. The typical adsorption mechanisms governing the pharmaceutical adsorption onto biochar are illustrated in Figure 2.

One of the adsorption mechanisms of pharmaceutical removal by biochar is π–π interaction, which is a relatively weak interaction which commonly occurs among aromatic rings. The adsorbate species could adsorb onto the adsorbent surface via electron transfer between the electron acceptor and donor. The driving force of this mechanism is the π–π interaction between the protonated aniline ring with π-electron rich graphitic surface [89]. Since biochar has a graphitic surface, it is expected that the surface has higher π-electron density. Furthermore, the polycondensed and electron-rich graphitized surface in high temperature biochar could act as the π-donor, which bind the electron-withdrawing molecules within the pharmaceutical components. Pharmaceutical components typically contain long chain aromatic rings, thereby π–π interactions could possibly occur between the aromatic rings and the carbonaceous surface of biochars [82].

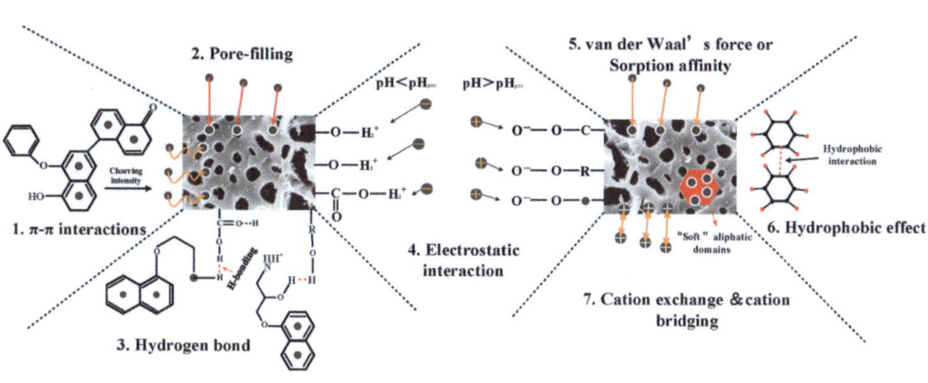

Figure 2: Typical adsorption mechanisms for pharmaceutical uptake by biochar Reprinted from Li et al. [82] with permission from Elsevier. Copyright 2019 Elsevier.

As previously mentioned, surface functional groups are important in facilitating the adsorption process. The surface functional groups of biochar determine the chemical properties of the biochar adsorbent. Functional groups such as phenolic, hydroxyl, carboxyl, and anhydride are capable of forming hydrogen bonds with the functional groups of pharmaceuticals [89]. The formation of hydrogen bonds can lead to a strong covalent bond energy and are not easy to break, thus hydrogen bonding is important in the pharmaceutical uptake by biochar. Additionally, the surface functional groups of modified biochars can facilitate π–π interactions. The hydroxyl groups situated in the benzene rings may play the role of π–electron donor, whereas the protonated carboxyl groups may function as the π–electron acceptor.

Electrostatic interaction is another dominant adsorption mechanism of pharmaceutical removal by biochar, particularly toward adsorption of ionic and ionizable pharmaceuticals. The solution pH can alter the surface charge of biochar and degree of ionization of adsorbates. An ionic bond will be formed between cations and anions generated after electron loss and gain. Ionic exchange is a reversible reaction between ions in the liquid phase and ions on the solid phase. The mechanism will be triggered when certain ions in the liquid phase are adsorbed onto the solid, thereby releasing equivalent ions back into the liquid phase in order to achieve a net zero charge [90].

Nonpolar functional groups such as C–H groups on biochar can create a mutual repulsion with water molecules and this phenomenon is known as hydrophobic interaction [91]. The hydrophobic interaction and partitioning process facilitates the adsorption of hydrophobic and neutral organic compounds onto the biochar. Biochars formed at low temperatures have amorphous structure and contain aliphatic domains that can provide an effective medium for partitioning of hydrophobic pharmaceutical compounds. Several studies showed that hydrophobic interaction was the main mechanism of their adsorption process in sequestrating organic pollutants [92].

Additionally, pore filling and intraparticle diffusion are important adsorption mechanisms for pharmaceutical adsorption on biochar. Adsorption process is highly depended on the surface properties of biochar with adsorption capability strongly associated to the micropore surface area and pore diffusion mechanism. Pore filling is a process where the pharmaceuticals occupy the microporous (<2 nm) and mesoporous (2–50 nm) structures of the biochar. The type of biochar and the polarity of the pharmaceuticals are important governing factors in this mechanism. Generally, biochar that contains little amount of volatile matter is able to achieve the highest pore filling removal efficiency under low concentrations of adsorbate [92]. The pharmaceutical adsorption of different biochars are summarized in Table 2.

Table 2: Adsorption performance of different biochars toward pharmaceutical residues.

Biochar	Biomass	Pollutant	Treatment	q_m (mg g^{-1})	Mechanisms	Reference
Iron oxide nanoparticle incorporated on mesoporous biochar	Textile mill sludge	Ofloxacin	Slow pyrolysis, chemical functionalization	19.74	π–π electron donor-acceptor, hydrogen bonding, pore filling	[58]
Engineered tea-waste biochar	Tea waste	Caffeine	Steam activation, slow pyrolysis	15.4	Electrostatic interaction, nucleophilic attraction, hydrogen bonding, π–π interaction	[88]
Spherical biochar	Glucose	Paracetamol	Hydrothermal carbonization, slow pyrolysis	286	Pore filling, hydrogen bonding, π–π, and n–π interaction	[95]
Porous biochar	Pharmaceutical sludge	Tetracycline	Chemical activation, slow pyrolysis	379.8	Hydrogen bonding, π–π conjugation, and pore filling	[69]
Alkaline modified biochar	Spent coffee ground	Tetracycline	Chemical activation, slow pyrolysis	113.6	Electrostatic interaction, hydrogen bonding, pore filling, and π–π interaction	[62]
Microwave functionalized biochar	Wild plum kernels	Naproxen	Pyrolysis, microwave, and chemical functionalization	73.14	Electrostatic interaction, hydrogen bonding, and π–π interaction	[59]
Sphere-shaped aromatized biochar	Sugarcane bagasse	Sulfamethoxazole	Hydrothermal carbonization, chemical activation	400	Hydrogen bonding and π–π interaction	[39]

(continued)

Table 2 (continued)

Biochar	Biomass	Pollutant	Treatment	q_m (mg g^{-1})	Mechanisms	Reference
Gliricidia sepium biochar	*Gliricidia sepium*	Caffeine	Slow pyrolysis	16.26	Electrostatic interaction, n–π interaction, hydrogen bonding, π–π interaction	[93]
Magnetic sludge biochar	Municipal sludge	Tetracycline	Hydrothermal carbonization, chemical activation	145	Hydrogen bonding, hydrophobic interaction, π–π interaction	[60]
		ciprofloxacin		74.2		
Grapefruit peel-based biochar	Grapefruit peel	Tetracycline	Carbonization and chemical activation	37.92	Pore filling, charge interaction, and chemical bonding	[61]
Waste sludge/leaf biochar	Waste sludge, leaf	Diclofenac	Slow pyrolysis	877	Electrostatic interaction, pore filling	[96]
Maple leaf-derived biochar	Maple leaf	Tetracycline	Slow pyrolysis	407.3	Metal complexation, hydrogen bonding, and hydrophobic interaction	[83]
Chinese herbs-derived biochar	*Flueggea suffruticosa*	Tetracycline	Slow pyrolysis, chemical activation	200	Hydrogen bonding, electrostatic interaction	[97]
Graphene oxide-coated biochar	Bamboo sawdust	Sulfamethazine	Slow pyrolysis	–	Hydrogen bonding, π–π interaction, electrostatic interaction, pore filling, cation exchange	[98]
Hickory chips biochar	Hickory chips	Sulfamethoxazole	Pyrolysis, ball-milled activation	100.3	Hydrophobic interaction, π–π interaction, hydrogen bonding, electrostatic interaction	[74]
Bamboo biochar	Bamboo	Sulfapyridine	Pyrolysis, ball-milled activation	57.9	Hydrophobic interaction, π–π interaction, hydrogen bonding, and electrostatic interaction	[74]

11.1 Notable case studies

Keerthanan and Rajapaksha [93] have investigated the influence of pyrolysis temperature on biochar formation and its application in caffeine removal. In their work, *Gliricidia sepium* biochar was synthesized at pyrolysis temperatures of 300, 500, and 700 °C. It was reported that increasing pyrolysis temperature resulted in an increased ash content (6.03–20.51%) due to the increase in the concentration of inorganic substances and combustion of organic matters [93]. Additionally, the H/C ratio of the synthesized biochar decreased at increasing pyrolysis temperature. These results indicated that the aromaticity and polarity indices of the biochar decreased at higher pyrolysis temperature due to the condensation of aromatic rings on the biochar surface. This hypothesis was supported by the FTIR analysis, with the detection of C–H bending of aromatic ring at 864 cm^{-1}, increased intensity in C–H bending peak at 1,443 cm^{-1}, and shifting of C = C stretching peak at 1,567 cm^{-1} [93]. In a caffeine adsorption study, the biochar synthesized at 700 °C had the highest caffeine adsorption capacity (16.3 mg g^{-1}) as compared to 500 and 300 °C, which was approximately 4 mg g^{-1} [93]. This might be due to the widening and formation of internal pore structure of biochar at 700 °C. This finding was validated by the BET results where the surface area of biochar synthesized at 700 °C was the highest (216.4 m^2 g^{-1}), followed by biochars synthesized at 500 °C (76.30 m^2 g^{-1}), and 300 °C (1.02 m^2 g^{-1}), whereas the pore volume of biochar increased from 0.0012 to 0.1097 cm^3 g^{-1} as the pyrolysis temperature increased.

In another study on the effect of modification and magnetization of rice straw biochar, Dai and Meng [94] reported that the modified magnetic biochar showed enhanced TC adsorption capacity up to 98.33 mg g^{-1} as compared to the biochar without modification (37.80 mg g^{-1}). One of the possible reasons is that magnetic iron oxides and oxygenated functional groups of biochar formed strong hydrogen bonding with TC and surface complexation. Furthermore, the graphitic structure and aromatic rings of biochar could facilitate strong π–π and cation–π interactions, thus enhancing the TC removal [94].

Another notable study conducted by Tran and Tomul [95] revealed an innovative spherical biochar in paracetamol removal. In the study, a spherical biochar was synthesized by hydrothermal carbonization of pure glucose followed by slow pyrolysis, whereas nonspherical biochar was synthesized via slow pyrolysis using pomelo peels as the feedstock. Figure 3 illustrates the surface morphology of the reported spherical and nonspherical biochars. It was demonstrated that the specific surface area of the spherical biochar was higher than the nonspherical biochar (1,292 > 1,033 m^2 g^{-1}). However, the pore volume of the spherical biochar (0.704 cm^3 g^{-1}) was lower than that of the nonspherical biochar (1.074 cm^3 g^{-1}). Therefore, the paracetamol adsorption onto the nonspherical biochar was dominated by pore filling. As for the spherical biochar, the amount of oxygen-containing functional groups (43%) was higher than that of the nonspherical biochar (38%), and surface interactions such as hydrogen

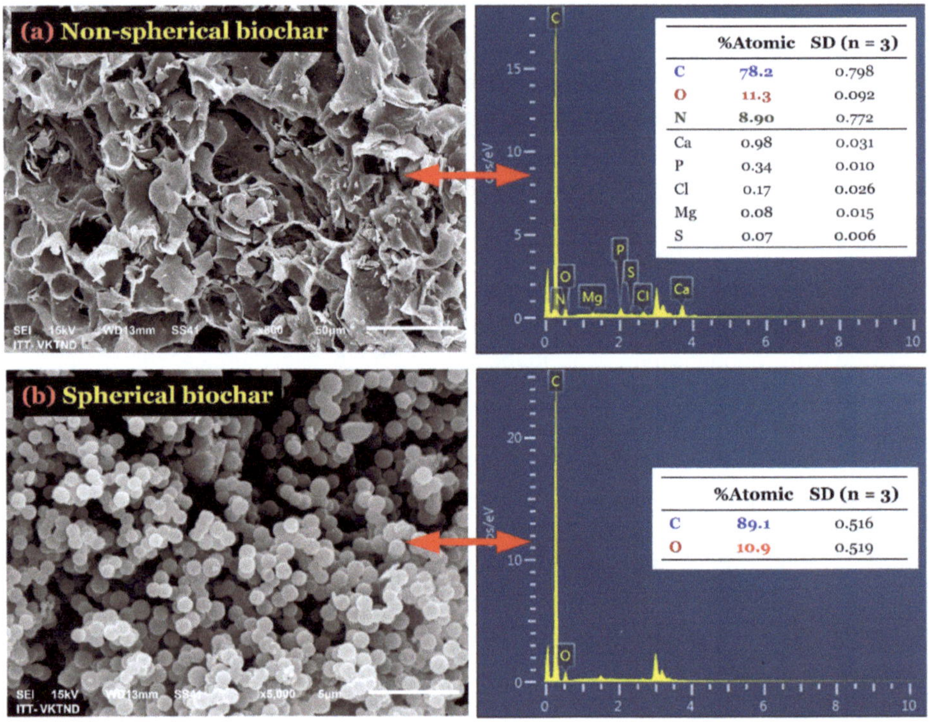

Figure 3: SEM imaging for nonspherical and spherical biochar. Reproduced from Tran et al. [95] with permission from Elsevier. Copyright 2020 Elsevier.

bonding and π–π interaction were the dominant adsorption mechanisms for paracetamol removal by the spherical biochar [95].

The remarkable work carried out by Prasannamedha and Kumar [39] demonstrated the mechanisms involved in the adsorption of SMX onto sugarcane bagasse biochar. Sugarcane bagasse, the raw material, was first carbonized hydrothermally and then activated using sodium hydroxide (NaOH) under inert atmosphere. From the FTIR results for both the inactivated and activated biochars, the major functional groups were discovered to be hydroxyl and carboxylic groups, which were responsible for the adsorption of SMX onto the carbon skeleton of the activated biochar. The FTIR peak was shifted from 1,030 to 1,113 cm^{-1} after activation, corresponding to C–O stretching of carboxylic acid and aromatic esters that were caused by the deformation of chemical structure using the reducing agent NaOH. The PZC of the activated biochar was evaluated to be pH 6.5 and the adsorption was favorable at pH 5–7.5. At PZC, the acidic functional groups of activated biochar such as carboxylic acid (–COOH) was deprotonated thus becoming an H-acceptor while the phenolic acid (–OH) was protonated and turned into an H-donor. Hence, inducing strong EDA interaction between the H-bond of SMX and O-functional groups of the activated biochar resulted in good adsorption performance. In addition, SMX has two aromatic structures bonded to sulfonamide ($-SO_2-NH$), namely, para amino phenyl and methyl oxazolyl ring. These intensified the immobilization of SMX on the activated biochar through the π–π interaction between the sulfonamide group and the graphitic region of the activated biochar [39].

Besides that, SEM images depicted that the activated biochar has very small and random pores that were formed from the evolution of volatile matter. Nitrogen physisorption carried out on the activated biochar indicated the presence of a large amount of mesopores as compared to micropores. These porous structures served as the active sites for SMX adsorption. The fitting of experimental data was best done with the Elovich model, suggesting adsorbate pore filling as the primary adsorption mechanism. This was further explained by the intraparticle diffusion plot, whereby the initial stage (<30 mins) involved the adherence of adsorbate to the active sites after boundary layer diffusion [39].

Similarly, the removal of diclofenac by chemically activated biochar generated from waste sludge/leaf has shown an astounding result of 877 mg L^{-1} [96]. The waste sludge/leaf was first pyrolyzed at 200 °C then activated by 9% of HCL solution. Compared to the unactivated biochar that exhibited only 58.4% of removal efficiency, the efficiency of chemically activated biochar increased significantly to 100%. The BET surface area of biochar increased from 3.30 to 4.17 m^2 g^{-1} while the pore diameter enhanced from 6.74 to 11.69 nm. After adsorption of diclofenac, the pore volume of activated biochar decreased from 0.00372 to 0.00017 cm^3 g^{-1}. This measurement further proved the occurrence of pore filling during adsorption onto the biochar [96].

Flueggea suffruticosa is a Chinese herbal medicine residue used as the feedstock for biochar preparation [97]. The residue was first modified by zinc chloride and then carbonized in a tubular furnace at 500 °C under a heating rate of 10 °C min^{-1} for 90 min. The resulting biochar was used to remove TC, chlortetracycline (CTC) and oxytetracycline (OTC). The highest maximum adsorption capacity of 200 mg g^{-1} was found for the removal of CTC using the modified biochar. By comparing the SEM images, the modified biochar showed a rougher and more porous surface structure than the raw biochar. Thus, the specific surface area was greatly increased to 1,556 m^2 g^{-1}. In addition, from the N_2 adsorption-desorption isotherm, the modified biochar was found to contain micropores. These great physical structural properties significantly contributed to the high removal efficiency through physical adsorption such as pore filling. Besides that, electrostatic interaction between TCs and biochar played a dominant role in the removal process of TCs. The ionic strength of the TCs and the modified biochar was determined using $CaCl_2$ salt. $CaCl_2$ had negative effects on the adsorption of TC and CTC, whereas it increased the removability of OTC. This finding suggested the presence of electrostatic repulsion between OTC and modified biochar, thus resulting in the lowest maximum adsorption capacity of 129.9 mg g^{-1} for OTC. In addition, the importance of hydrogen bonding mechanism could be seen from the FTIR spectra. The three TCs had similar functional groups and they were rich in hydroxyl, amine, and aromatic groups (O–H, C = O, and C–N bonds). These functional groups formed hydrogen bonding with the hydroxyl and carboxyl groups in the modified biochar. In a nutshell, the removal of TCs by the modified biochar was governed by both physical and chemical processes [97].

Huang and Zimmerman [74] performed a novel approach in using ball-milled biochar to remove two sulfonamide antibiotics which are SMX and SPY from water media. The removability of both antibiotics from the wastewater was notable. The ball-milled biochars were derived from bagasse, bamboo, and hickory chips at pyrolysis temperatures of 300, 450, and 600 °C, respectively. The ball-milled hickory chips biochar pyrolyzed at 450 °C provided the best removal efficiency for SMX (83.3%) while the ball-milled bamboo biochar pyrolyzed at the same temperature obtained the highest removal efficiency for SPY (89.6%). The enhancement of the surface's structural and chemical properties created by the ball milling has been discussed under the physical treatment section. The large amounts of functional groups (CH_2, C = O, C = C, and –CO) that were detected on both hickory chips and bamboo derived ball-milled biochars resulted in multiple adsorption mechanisms such as π–π interaction, hydrophobic interaction, and hydrogen bonding. Furthermore, the adsorption processes of both biochars were found to be highly pH dependent due to the electrostatic interactions between the adsorbent and adsorbate. At pH values from 3.5 to 8.5, the surface of both biochars was tested to be negatively charged. At low acidic pH, the main species of SMX and SPY were neutral. However, as the pH solution was further increased, the anionic species of SMX and SPY increased as well. This resulted in the electrostatic repulsion between the adsorbent and adsorbate. Thus, the removal efficiencies of both antibiotics at high pH were poor [74].

12 Prospects and conclusions

Biochar adsorption has been demonstrated as a promising technology for pharmaceutical removal from liquid media with benefits such as facile operation, low cost, and environmentally friendliness. In recent year, a number of patents related to different biochar applications have been filed as shown in Table 3.

Up to present (10 April 2021), there are 7,665 patents related to "biochar treatment" registered under Google Patents. This indicated that the applications of biochar have been translated into technology deemed feasible for upscaling. Despite the increasing number of patents filed and research articles published, there is still limited use of biochar in commercial wastewater purification facilities, and a majority of the reported studies have been conducted at laboratory level.

Further development of effective biochar adsorbents is critical to fully utilize the great potential of this environmentally friendly and low cost material at large-scale wastewater treatment plants. As such, adequate knowledge on the biochar

Table 3: Registered patents on biochar applications in recent years.

Year	Title of patent	Patent number
2021	Method for removing antibiotics in aquaculture wastewater	CN112340818A
2020	Lake and reservoir water body circulating purification utilization system and circulating purification method thereof	CN112499909A
2020	Process and method for preparing active carbon by hydrothermal co-carbonization of enteromorpha and municipal sludge	CN112479204A
2020	Biochar preparation device	CN112430468A
2020	Method for reducing heavy metal content of sludge-based biochar	CN112520963A
2020	Zinc-rich biochar, preparation method thereof and application of zinc-rich biochar in passivation and remediation of heavy metal contaminated soil	CN112457853A
2020	Hierarchical pore biochar and preparation method and application thereof	CN112456488A
2020	Method for preparing biochar by co-pyrolysis of excess sludge and hazelnut shells and photocatalytic modification method of biochar	CN112387269A
2020	Modified biochar for adsorbing high-concentration phenolic wastewater and preparation method thereof	CN112473622A
2020	Method for preparing magnetic biochar from kitchen garbage	AU2020103433A4
2020	Preparation method of rice straw biochar	CN112358362A
2019	Absorbent for municipal wastewater treatment	WO2021048527A1
2019	Preparation and use of zeolite and biochar composite material	WO2020257853A1
2019	Methods and systems for producing biochar	WO2020227135A1
2019	Hybrid biochar and manufacturing method therefor	WO2020226351A1
2019	Systems and methods of regenerating activated carbon	WO2020198394A1
2019	Hybrid biochar and method of preparing for the same	KR102018188B1
2019	Carbon dioxide adsorber using chicken manure biochar and fabricating method of the same	KR102203466B1
2018	Decorative biochar and manufacturing method thereof	KR20200027867A
2018	Method and apparatus for processing biomass	WO2019186216A1
2017	Torrefied biomass briquettes and related methods	WO2019118565A1
2017	Hydrochar	WO2019110860A1
2017	Apparatus and method of producing activated carbon material	WO2019104382A1

physical properties (porosity, surface area, particle size, density, and stability) and chemical properties (functional groups and surface chemistry) will form the basis in the development and selection of effective biochar in pharmaceutical removal. These physicochemical properties are highly dependent on the biomass feedstock and pyrolysis conditions. The pristine biochar can be further modified using either physical or chemical treatment to improve the properties of the biochar, thereby resulting in enhanced removal efficiency. Through these modifications, the developed biochar may acquire beneficial properties such as improved surface area, porosity, and additional functional groups for higher treatment efficiency.

Currently, most research works mainly focus on the development of biochar and its performance in pharmaceutical removal, but the sustainability of using biochar in wastewater treatment remains scarce. Hence, life cycle analysis and technoeconomic analysis should be evaluated to assess the feasibility of the engineered biochar in large-scale applications from the aspects of environment, safety, and economy. Furthermore, the fate of the exhausted biochar should be investigated as well. This information is crucial to determine the potential risks of leached chemicals from the modified biochar and its corresponding recovery methods to avoid unintended secondary contamination from the modified biochar.

References

[1] Wang JL, Wang SZ, Removal of pharmaceuticals and personal care products (PPCPs) from wastewater: a review, J. Environ. Manage., 2016, 182, 620–640, doi: 10.1016/j.jenvman.2016.07.049.

[2] Wang JL, Chu LB, Irradiation treatment of pharmaceutical and personal care products (PPCPs) in water and wastewater: an overview, Radiat. Phys. Chem., 2016, 125, 56–64, doi: 10.1016/j.radphyschem.2016.03.012.

[3] Mompelat S, Le Bot B, Thomas O, Occurrence and fate of pharmaceutical products and by-products, from resource to drinking water, Environ. Int., 2009, 35(5), 803–814, doi: 10.1016/j.envint.2008.10.008.

[4] Han S, et al., Endocrine disruption and consequences of chronic exposure to ibuprofen in Japanese medaka (Oryzias latipes) and freshwater cladocerans Daphnia magna and Moina macrocopa, Aqu. Toxicol., 2010, 98(3), 256–264, doi: 10.1016/j.aquatox.2010.02.013.

[5] Wojcieszynska D, Guzik U, Naproxen in the environment: its occurrence, toxicity to nontarget organisms and biodegradation, Appl. Microbiol. Biotechnol., 2020, 104(5), 1849–1857, doi: 10.1007/s00253-019-10343-x.

[6] De Andrade JR, et al., Adsorption of Pharmaceuticals from Water and Wastewater Using Nonconventional Low-Cost Materials: a Review, Ind. Eng. Chem. Res., 2018, 57(9), 3103–3127, doi: 10.1021/acs.iecr.7b05137.

[7] Abdulkareem AO, et al., Effect of untreated pharmaceutical plant effluent on cardiac Na+-K+-ATPase and Ca2+-Mg2+-ATPase activities in mice (Mus Musculus), Toxicol. Rep., 2019, 6, 439–443, doi: 10.1016/j.toxrep.2019.05.002.

[8] Daughton CG, Pharmaceuticals and the Environment (PiE): evolution and impact of the published literature revealed by bibliometric analysis, Sci. Total Environ., 2016, 562, 391–426, doi: 10.1016/j.scitotenv.2016.03.109.
[9] Aus Der Beek T, et al., Pharmaceuticals in the environment – Global occurrences and perspectives, Environ. Toxicol. Chem., 2016, 35(4), 823–835, doi: https://doi.org/10.1002/etc.3339.
[10] EPA, Management Standards for Hazardous Waste Pharmaceuticals and Amendment to the P075 Listing for Nicotine, Vols 84, No. 36, 2019, Federal Register.
[11] Kastner M, et al., Classification and modelling of Nonextractable Residue (NER) formation of xenobiotics in soil – a synthesis, Crit. Rev. Environ. Sci. Technol., 2014, 44(19), 2107–2171, doi: 10.1080/10643389.2013.828270.
[12] Solanki A, Boyer TH, Pharmaceutical removal in synthetic human urine using biochar, Environ. Sci. Water Res. Technol., 2017, 3(3), 553–565, doi: 10.1039/c6ew00224b.
[13] Dolar D, et al., RO/NF membrane treatment of veterinary pharmaceutical wastewater: comparison of results obtained on a laboratory and a pilot scale, Environ. Sci. Pollut. Res., 2012, 19(4), 1033–1042, doi: 10.1007/s11356-012-0782-7.
[14] Babaahmadi F, et al., Data on metal levels in the inlet and outlet wastewater treatment plant of hospitals in Bushehr province, Iran, Data Brief, 2017, 10, 1–5, doi: 10.1016/j.dib.2016.11.054.
[15] Kumari V, Tripathi AK, Characterization of pharmaceuticals industrial effluent using GC-MS and FT-IR analyses and defining its toxicity, Appl. Water Sci., 2019, 9(8), doi: https://doi.org/10.1007/s13201-019-1064-z.
[16] Campestrini I, Jardim WF, Occurrence of cocaine and benzoylecgonine in drinking and source water in the São Paulo State region, Brazil, Sci. Total Environ., 2017, 576, 374–380, doi: 10.1016/j.scitotenv.2016.10.089.
[17] Hiew BYZ, et al., Adsorptive removal of diclofenac by graphene oxide: optimization, equilibrium, kinetic and thermodynamic studies, J. Taiwan Inst. Chem. Eng., 2019, 98, 150–162, doi: 10.1016/j.jtice.2018.07.034.
[18] Bernardo M, et al., High efficacy on diclofenac removal by activated carbon produced from potato peel waste, Int. J. Environ. Sci. Technol., 2016, 13(8), 1989–2000, doi: 10.1007/s13762-016-1030-3.
[19] Swan GE, et al., Toxicity of diclofenac to Gyps vultures, Biol. Lett., 2006, 2(2), 279–282, doi: 10.1098/rsbl.2005.0425.
[20] Lonappan L, et al., Diclofenac and its transformation products: environmental occurrence and toxicity – A review, Environ. Int., 2016, 96, 127–138, doi: 10.1016/j.envint.2016.09.014.
[21] Zhang YJ, Geissen SU, Gal C, Carbamazepine and diclofenac: removal in wastewater treatment plants and occurrence in water bodies, Chemosphere, 2008, 73(8), 1151–1161, doi: 10.1016/j.chemosphere.2008.07.086.
[22] Ferrari B, et al., Ecotoxicological impact of pharmaceuticals found in treated wastewaters: study of carbamazepine, clofibric acid, and diclofenac (vol 55, pg 359, 2003), Ecotoxicol. Environ. Saf., 2003, 56(3), 450–450, doi: 10.1016/S0147-6513(03)00111-8.
[23] Zhang W, et al., Eco-toxicological effect of Carbamazepine on Scenedesmus obliquus and Chlorella pyrenoidosa, Environ. Toxicol. Pharmacol., 2012, 33(2), 344–352, doi: 10.1016/j.etap.2011.12.024.
[24] Dajani EZ, Dajani NE, Shahwan TG, Over-the-counter drugs, In: Johnson LR (Editor), Encyclopedia of Gastroenterology, Elsevier, New York, 2004, 16–23, doi: https://doi.org/10.1016/B0-12-386860-2/00529-3.

[25] Vardanyan RS, Hruby VJ, 19 – Antianginal drugs, In: Vardanyan RS, Hruby VJ (Editors), Synthesis of Essential Drugs, Elsevier, Amsterdam, 2006, 257–267, doi: https://doi.org/10.1016/B978-044452166-8/50019-4.
[26] Mosby C, Mosby's Dental Drug Reference (Eleventh Edition), Jeske AH (Editor), Mosby, St. Louis, 2014, 202–350, doi: https://doi.org/10.1016/B978-0-323-16916-5.00003-1.
[27] Fedan JS, Thrombosis-embolism☆, In: Reference Module in Biomedical Sciences, Elsevier, 2016, doi: https://doi.org/10.1016/B978-0-12-801238-3.99531-X.
[28] Furman BL, Antidiabetic agents, In: Enna SJ, Bylund DB (Editors), xPharm: The Comprehensive Pharmacology Reference, Elsevier, New York, 2007, 1, doi: https://doi.org/10.1016/B978-008055232-3.61040-2.
[29] Romero-González R, Garrido Frenich A, Martínez Vidal JL, Veterinary drugs residues: anthelmintics, In: Motarjemi Y (Editor), Encyclopedia of Food Safety, Academic Press, Waltham, 2014, 45–54, doi: https://doi.org/10.1016/B978-0-12-378612-8.00243-2.
[30] Scholar E, Antifungal agents, In: Enna SJ, Bylund DB (Editors), xPharm: The Comprehensive Pharmacology Reference, Elsevier, New York, 2007, 1–3, doi: https://doi.org/10.1016/B978-008055232-3.61020-7.
[31] Van Cauwenberge P, et al., ALLERGY | Allergic Rhinitis, In: Laurent GJ, Shapiro SD (Editors), Encyclopedia of Respiratory Medicine, Academic Press, Oxford, 2006, 80–92, doi: https://doi.org/10.1016/B0-12-370879-6/00023-5.
[32] Coleman JJ, Butt TF, Chapter 20 – Antihypertensive drugs, In: Aronson JK (Editor), Side Effects of Drugs Annual, Elsevier, 2009, 349–370, doi: https://doi.org/10.1016/S0378-6080(09)03120-1.
[33] Couetil LL, 28 – Inflammatory diseases of the lower airway of athletic horses, In: Hinchcliff KW, Kaneps AJ, Geor RJ (Editors), Equine Sports Medicine and Surgery (Second Edition), W.B. Saunders, 2014, 605–632, doi: https://doi.org/10.1016/B978-0-7020-4771-8.00028-4.
[34] Winchester HA, Chapter 2 – Emergency management of poisoning, In: Shannon MW, Borron SW, Burns MJ (Editors), Haddad and Winchester's Clinical Management of Poisoning and Drug Overdose (Fourth Edition), W.B. Saunders, Philadelphia, 2007, 13–61, doi: https://doi.org/10.1016/B978-0-7216-0693-4.50007-4.
[35] Montalvão SIGHM, Singh V, Haque S, Chapter 5 – Bioassays for bioactivity screening, In: Rocha-Santos T, Duarte AC (Editors), Comprehensive Analytical Chemistry, Elsevier, 2014, 79–114, doi: https://doi.org/10.1016/B978-0-444-63359-0.00005-7.
[36] Favrod-Coune T, Broers B, The health effect of psychostimulants: a literature review, Pharmaceuticals (Basel, Switzerland), 2010, 3(7), 2333–2361, doi: 10.3390/ph3072333.
[37] Shukla AC, Steven JM, McGowan FX, CHAPTER 16 – Cardiac physiology and pharmacology, In: Coté CJ, Lerman J, Todres ID (Editors), A Practice of Anesthesia for Infants and Children (Fourth Edition), W.B. Saunders, Philadelphia, 2009, 361–395, doi: https://doi.org/10.1016/B978-141603134-5.50020-2.
[38] Karimi-Maleh H, et al., Recent advances in using of chitosan-based adsorbents for removal of pharmaceutical contaminants: a review, J. Clean. Prod., 2021, 291, doi: https://doi.org/10.1016/j.jclepro.2021.125880.
[39] Prasannamedha G, et al., Enhanced adsorptive removal of sulfamethoxazole from water using biochar derived from hydrothermal carbonization of sugarcane bagasse, J. Hazard. Mater., 2021, 407, 124825, doi: https://doi.org/10.1016/j.jhazmat.2020.124825.
[40] Illes E, et al., Hydroxyl radical induced degradation of ibuprofen, Sci. Total Environ., 2013, 447, 286–292, doi: 10.1016/j.scitotenv.2013.01.007.
[41] Binh VN, et al., Antibiotics in the aquatic environment of Vietnam: sources, concentrations, risk and control strategy, Chemosphere, 2018, 197, 438–450, doi: 10.1016/j.chemosphere.2018.01.061.

[42] Oh WD, et al., Surface-active bismuth ferrite as superior peroxymonosulfate activator for aqueous sulfamethoxazole removal: performance, mechanism and quantification of sulfate radical, J. Hazard. Mater., 2017, 325, 71–81, doi: 10.1016/j.jhazmat.2016.11.056.
[43] Changanaqui K, et al., Blue LED light-driven photoelectrocatalytic removal of naproxen from water: kinetics and primary by-products, J. Electroanal. Chem., 2020, 867, doi: https://doi.org/10.1016/j.jelechem.2020.114192.
[44] Jallouli N, et al., UV and solar photo-degradation of naproxen: TiO2 catalyst effect, reaction kinetics, products identification and toxicity assessment, J. Hazard. Mater., 2016, 304, 329–336, doi: 10.1016/j.jhazmat.2015.10.045.
[45] Al-Mashaqbeh O, et al., Removal of Selected Pharmaceuticals and Personal Care Products in Wastewater Treatment Plant in Jordan, Water, 2019, 11(10), doi: https://doi.org/10.3390/w11102004.
[46] Xu Y, et al., Advances in technologies for pharmaceuticals and personal care products removal, J. Mater. Chem A, 2017, 5(24), 12001–12014, doi: 10.1039/c7ta03698a.
[47] Kyzas GZ, et al., New approaches on the removal of pharmaceuticals from wastewaters with adsorbent materials, J. Mol. Liq., 2015, 209, 87–93, doi: 10.1016/j.molliq.2015.05.025.
[48] Akhtar J, Amin NAS, Shahzad K, A review on removal of pharmaceuticals from water by adsorption, Desal. Water Treat., 2016, 57(27), 12842–12860, doi: 10.1080/19443994.2015.1051121.
[49] Wang XQ, et al., Recent advances in biochar application for water and wastewater treatment: a review, Peerj, 2020, 8, doi: 10.7717/peerj.9164.
[50] Ndoun MC, et al., Adsorption of pharmaceuticals from aqueous solutions using biochar derived from cotton gin waste and guayule bagasse, Biochar, 2021, 3(1), 89–104, doi: 10.1007/s42773-020-00070-2.
[51] Li D-C, Jiang H, The thermochemical conversion of non-lignocellulosic biomass to form biochar: a review on characterizations and mechanism elucidation, Bioresour. Technol., 2017, 246, 57–68, doi: https://doi.org/10.1016/j.biortech.2017.07.029.
[52] Yaashikaa PR, et al., A critical review on the biochar production techniques, characterization, stability and applications for circular bioeconomy, Biotechnol. Rep., 2020, 28, e00570, doi: https://doi.org/10.1016/j.btre.2020.e00570.
[53] Chen Y-D, et al., Production, properties, and catalytic applications of sludge derived biochar for environmental remediation, Water Res., 2020, 187, 116390, doi: https://doi.org/10.1016/j.watres.2020.116390.
[54] Jeyasubramanian K, et al., A complete review on biochar: production, property, multifaceted applications, interaction mechanism and computational approach, Fuel, 2021, 292, 120243, doi: https://doi.org/10.1016/j.fuel.2021.120243.
[55] Hirst EA, Taylor A, Mokaya R, A simple flash carbonization route for conversion of biomass to porous carbons with high CO2 storage capacity, J. Mater. Chem A, 2018, 6(26), 12393–12403, doi: 10.1039/c8ta04409k.
[56] Prabakar D, et al., Advanced biohydrogen production using pretreated industrial waste: outlook and prospects, Renew. Sustain. Energy Rev., 2018, 96, 306–324, doi: https://doi.org/10.1016/j.rser.2018.08.006.
[57] You SM, et al., A critical review on sustainable biochar system through gasification: energy and environmental applications, Bioresour. Technol., 2017, 246, 242–253, doi: 10.1016/j.biortech.2017.06.177.
[58] Singh V, Srivastava VC, Self-engineered iron oxide nanoparticle incorporated on mesoporous biochar derived from textile mill sludge for the removal of an emerging pharmaceutical pollutant, Environ. Pollut., 2020, 259, 113822, doi: https://doi.org/10.1016/j.envpol.2019.113822.

[59] Paunovic O, et al., Ionisable emerging pharmaceutical adsorption onto microwave functionalised biochar derived from novel lignocellulosic waste biomass, J. Colloid Interface Sci., 2019, 547, 350–360, doi: https://doi.org/10.1016/j.jcis.2019.04.011.

[60] Ma Y, et al., Hydrothermal synthesis of magnetic sludge biochar for tetracycline and ciprofloxacin adsorptive removal, Bioresour. Technol., 2021, 319, 124199, doi: https://doi.org/10.1016/j.biortech.2020.124199.

[61] Yu H, et al., Activation of grapefruit derived biochar by its peel extracts and its performance for tetracycline removal, Bioresour. Technol., 2020, 316, 123971, doi: https://doi.org/10.1016/j.biortech.2020.123971.

[62] Nguyen V-T, et al., Alkaline modified biochar derived from spent coffee ground for removal of tetracycline from aqueous solutions, J. Water Process Eng., 2021, 40, 101908, doi: https://doi.org/10.1016/j.jwpe.2020.101908.

[63] Madadi R, Bester K, Fungi and biochar applications in bioremediation of organic micropollutants from aquatic media, Mar. Pollut. Bull., 2021, 166, 112247, doi: https://doi.org/10.1016/j.marpolbul.2021.112247.

[64] Pandey D, Daverey A, Arunachalam K, Biochar: production, properties and emerging role as a support for enzyme immobilization, J. Clean. Prod., 2020, 255, doi: https://doi.org/10.1016/j.jclepro.2020.120267.

[65] Li HB, et al., Mechanisms of metal sorption by biochars: biochar characteristics and modifications, Chemosphere, 2017, 178, 466–478, doi: 10.1016/j.chemosphere.2017.03.072.

[66] Shin J, et al., Competitive adsorption of pharmaceuticals in lake water and wastewater effluent by pristine and NaOH-activated biochars from spent coffee wastes: contribution of hydrophobic and π-π interactions, Environ. Pollut., 2021, 270, 116244, doi: https://doi.org/10.1016/j.envpol.2020.116244.

[67] Peiris C, et al., The influence of three acid modifications on the physicochemical characteristics of tea-waste biochar pyrolyzed at different temperatures: a comparative study, RSC Adv., 2019, 9(31), 17612–17622, doi: 10.1039/C9RA02729G.

[68] Singh SV, et al., Pyrolysis temperature influences the characteristics of rice straw and husk biochar and sorption/desorption behaviour of their biourea composite, Bioresour. Technol., 2020, 314, doi: https://doi.org/10.1016/j.biortech.2020.123674.

[69] Liu H, Xu G, Li G, Preparation of porous biochar based on pharmaceutical sludge activated by NaOH and its application in the adsorption of tetracycline, J. Colloid Interface Sci., 2021, 587, 271–278, doi: https://doi.org/10.1016/j.jcis.2020.12.014.

[70] Wang JL, Wang SZ, Preparation, modification and environmental application of biochar: a review, J. Clean. Prod., 2019, 227, 1002–1022, doi: 10.1016/j.jclepro.2019.04.282.

[71] McBeath AV, Wurster CM, Bird MI, Influence of feedstock properties and pyrolysis conditions on biochar carbon stability as determined by hydrogen pyrolysis, Biomass Bioener., 2015, 73, 155–173, doi: https://doi.org/10.1016/j.biombioe.2014.12.022.

[72] Kazemi Shariat Panahi H, et al., A comprehensive review of engineered biochar: production, characteristics, and environmental applications, J. Clean. Prod., 2020, 270, 122462, doi: https://doi.org/10.1016/j.jclepro.2020.122462.

[73] Rajapaksha AU, et al., Steam activation of biochars facilitates kinetics and pH-resilience of sulfamethazine sorption, J. Soils Sediment, 2016, 16(3), 889–895, doi: 10.1007/s11368-015-1325-x.

[74] Huang J, et al., Ball milled biochar effectively removes sulfamethoxazole and sulfapyridine antibiotics from water and wastewater, Environ. Pollut., 2020, 258, 113809, doi: https://doi.org/10.1016/j.envpol.2019.113809.

[75] Wang B, Gao B, Fang J, Recent advances in engineered biochar productions and applications, Crit. Rev. Environ. Sci. Technol., 2017, 47(22), 2158–2207, doi: 10.1080/10643389.2017.1418580.
[76] Krasucka P, et al., Engineered biochar – A sustainable solution for the removal of antibiotics from water, Chem. Eng. J., 2021, 405, doi: https://doi.org/10.1016/j.cej.2020.126926.
[77] Langmuir I, The adsorption of gases on plane surfaces of glass, mica and platinum, J. Am. Chem. Soc., 1918, 40(9), 1361–1403, doi: 10.1021/ja02242a004.
[78] Hall KR, et al., Pore- and Solid-Diffusion Kinetics in Fixed-Bed Adsorption under Constant-Pattern Conditions, Indust. Eng. Chem. Fund., 1966, 5(2), 212–223, doi: 10.1021/i160018a011.
[79] Freundlich HMF, Over the Adsorption in Solution, J. Phys. Chem., 1906, 57, 385–471.
[80] Temkin M, Pyzhev V, Kinetics of Ammonia Synthesis on Promoted Iron Catalysts, Acta Physico., 1940, 12, 217–222.
[81] Dubinin MM, Radushkevich LV, The equation of the characteristic curve of activated charcoal, Proc. Acad. Sci. Phys. Chem. Sec., 1947, 55, 331–333.
[82] Li L, et al., Biochar as a sorbent for emerging contaminants enables improvements in waste management and sustainable resource use, J. Clean. Prod., 2019, 210, 1324–1342, doi: https://doi.org/10.1016/j.jclepro.2018.11.087.
[83] Kim JE, et al., Adsorptive removal of tetracycline from aqueous solution by maple leaf-derived biochar, Bioresour. Technol., 2020, 306, 123092, doi: https://doi.org/10.1016/j.biortech.2020.123092.
[84] Lagergren S, About the theory of so-called adsorption of soluble substances, Kungliga Svenska Vetenskapsakademiens, Handlingar, Band, 1898, 24, 1–39.
[85] Ho YS, McKay G, Sorption of dye from aqueous solution by peat, Chem. Eng. J., 1998, 70(2), 115–124, doi: http://dx.doi.org/10.1016/S0923-0467(98)00076-1.
[86] Kumar PS, et al., Adsorption behavior of nickel(II) onto cashew nut shell: equilibrium, thermodynamics, kinetics, mechanism and process design, Chem. Eng. J., 2011, 167(1), 122–131, doi: http://dx.doi.org/10.1016/j.cej.2010.12.010.
[87] Weber WJ, Morris JC, Kinetics of adsorption carbon from solutions, Sanit. Eng. Div., Proc. Am. Soc. Civ. Eng., 1963, 89, 31–60.
[88] Keerthanan S, et al., Engineered tea-waste biochar for the removal of caffeine, a model compound in pharmaceuticals and personal care products (PPCPs), from aqueous media, Environ. Technol. Innov., 2020, 19, 100847, doi: https://doi.org/10.1016/j.eti.2020.100847.
[89] Zhang P, et al., Microwave-assisted chemical modification method for surface regulation of biochar and its application for estrogen removal, Process Saf. Environ. Protect., 2019, 128, 329–341, doi: https://doi.org/10.1016/j.psep.2019.06.009.
[90] Wu L, et al., MgO-modified biochar increases phosphate retention and rice yields in saline-alkaline soil, J. Clean. Prod., 2019, 235, 901–909, doi: https://doi.org/10.1016/j.jclepro.2019.07.043.
[91] Xie L, et al., Role of molecular architecture in the modulation of hydrophobic interactions, Curr. Opin. Colloid Interface Sci., 2020, 47, 58–69, doi: https://doi.org/10.1016/j.cocis.2019.12.001.
[92] Ambaye TG, et al., Mechanisms and adsorption capacities of biochar for the removal of organic and inorganic pollutants from industrial wastewater, Int. J. Environ. Sci. Technol., 2020, doi: 10.1007/s13762-020-03060-w.
[93] Keerthanan S, et al., Caffeine removal by Gliricidia sepium biochar: influence of pyrolysis temperature and physicochemical properties, Environ. Res., 2020, 189, 109865, doi: https://doi.org/10.1016/j.envres.2020.109865.

[94] Dai J, et al., Effects of modification and magnetization of rice straw derived biochar on adsorption of tetracycline from water, Bioresour. Technol., 2020, 311, 123455, doi: https://doi.org/10.1016/j.biortech.2020.123455.
[95] Tran HN, et al., Innovative spherical biochar for pharmaceutical removal from water: insight into adsorption mechanism, J. Hazard. Mater., 2020, 394, 122255, doi: https://doi.org/10.1016/j.jhazmat.2020.122255.
[96] Zhang H, et al., Production of biochar from waste sludge/leaf for fast and efficient removal of diclofenac, J. Mol. Liq., 2020, 299, 112193, doi: https://doi.org/10.1016/j.molliq.2019.112193.
[97] Wang H, et al., Adsorption of antibiotics from water by using Chinese herbal medicine residues derived biochar: preparation and properties studies, J. Mol. Liq., 2021, 325, 114967, doi: https://doi.org/10.1016/j.molliq.2020.114967.
[98] Huang D, et al., Sorptive removal of ionizable antibiotic sulfamethazine from aqueous solution by graphene oxide-coated biochar nanocomposites: influencing factors and mechanism, Chemosphere, 2017, 186, 414–421, doi: https://doi.org/10.1016/j.chemosphere.2017.07.154.

A. Saravanan, S. Jeevanantham, S. Karishma

Biochar as sustainable strategy for remediation and regeneration of heavy metal–contaminated soil

Abstract: Rapid development in industrialization and urbanizations leads to several undesirable changes in the environment. Heavy metal pollution was considered as one of the major issues which cause threatens to plants, animals and humans. Metal processing, mining, paint industries and energy utilizing sectors are the key sources for the heavy metal pollution. In humans, continuous exposure of heavy metal leads to improper function of crucial organs such as heart, brain, lungs, liver, and kidney. Different researches have been carried out toward the reduction of these toxic heavy metals in the environment by different conventional methodologies, but each method has certain limitations. Currently, the use of biochar has gained more attention for the remediation of heavy metals from the contaminated soil. Biochar was prepared from the different organic sources such as agricultural waste, and plant and animal wastes. Application of biochar to the contaminated soil reduces the bioavailability and mobility of heavy metals in soil and their accumulation in the plant. Along with the heavy metal remediation, biochar can also regenerate the soil to retain their functional properties and reduce the waste generation from agricultural practices. There are different factors such as properties of biochar, soil pH, type of heavy metals and their concentration in soil, redox potential of soil, and plant and rhizosphere microorganisms that affect the biochar function (i.e., heavy metal remediation in soil). In this chapter, preparation of biochar from different organic sources, their potential and mechanism for remediation of heavy metals from the contaminated soil, and the difficulties in remediation using biochar will be discussed in detail. The current advancement in biochar preparation and the future scope for biochar application for various remediation practices are also explained in this chapter.

Keywords: biochar, heavy metal, soil contamination, remediation, bioavailability of heavy metals

1 Introduction

Heavy metals are the elements having higher specific gravity (>5.0) and the atomic weight ranging from 63.5 to 200.6. Heavy metals are considered as one of the major environmental pollutants due to their nonbiodegradability, flexibility, and accumulation capabilities in plants and animals. Various industrial practices and human anthropogenic activities increase the heavy metals concentration in the environment. The

https://doi.org/10.1515/9783110734003-020

accumulation of heavy metals in plants leads to reduced plant growth and productivity. In soil, the heavy metal pollution affects the soil properties such as soil pH, mineral contents, and soil microorganisms. From the soil and water ecosystems, the toxic heavy metals can enter the humans through the food chain. Once it enters into humans, it causes various harmful diseases. However, some of the metal ions are required as micronutrients for various metabolic activities in humans; they can also cause harmful effects due to prolonged accumulation in the human body. Arsenic (As), antimony (Sb), cadmium (Cd), cobalt (Co), copper (Cu), lead (Pb), mercury (Hg), and so on are considered as major toxic heavy metals that cause harmful diseases including anemia, skin inflammation, liver damages, pulmonary disorders, gastrointestinal disorders, kidney damage, brain damage, renal edema, mental retardation, nervous system damage, and cancer to humans due to accumulation of heavy metals in human body [1].

Different remediation strategies have been utilized for the reduction of heavy metals in soil and water environments. Among them, using biochar is one of the most effective heavy metal remediation approaches which can remove the heavy metals from the contaminated soil by diverse mechanisms including adsorption, ion exchange, complexation, and precipitation processes. Biochar is a carbon-rich product which was derived from the organic materials (biomasses) such as agricultural residues, compost manure, biomass energy crops, industrial bioresidues, and sewage sludge. The effectiveness of heavy metal remediation in soil strongly depends on the properties of the biochar. The properties of the biochar are influenced by various factors such as mineral and elemental contents in organic biomass and pyrolytic temperature. After application of biochar to the soil, heavy metal remediation potential can be interrupted by several factors including soil pH, types and concentration of heavy metals present in soil, redox potential of soil, plant root system, and rhizosphere microorganisms [2]. Heavy metal remediation through adsorption mechanism was carried out by using biochar as adsorbent material. The surface properties of biochar such as surface area, pore structure, size and volume, and availability of functional groups on biochar surfaces play a major role in the heavy metal adsorption. The adsorption mechanism is carried out by biochar either through forming electrostatic interactions (physical adsorption) between the biochar surface and heavy metals or by forming hydrogen and ionic bonds (chemical adsorption) between the functional groups on biochar surfaces to the heavy metals. Generally, the biochar produced by the conventional approaches has poor surface properties and functional groups presence on the biochar surfaces which results in reduced heavy metal remediation from soil. In recent times, the surface properties and availability of functional groups on biochar surfaces could be increased by different physical, chemical and biological pretreatment approaches. These modification processes incorporate necessary functional groups to the biochar as well as increase the surface properties such as pore size, structure, volume, and surface area of the biochar [3].

Heavy metal remediation through complexation process was done by immobilizing the heavy metals in soil using the biochar. The oxygen-containing functional group, such as –OH, –COOH, and –C = O, has the capability to form complexes straightforwardly with the toxic heavy metals in soil. Thereby, these functional groups could increase the adsorption capability of biochar. Usually, biochar is composed of higher percentages of cations than anions, which influence the cation exchange capability of biochar. After biochar application to the soil, the cations present on the biochar surfaces could be exchanged by the cationic heavy metals in the soil. Similar to the complexation process, the mineral contents in the biochar could form precipitates with the heavy metals in the soil [4].

This chapter deals with the heavy metal pollution in environment, especially in soil ecosystem and the different remediation mechanisms using biochar to reduce the heavy metal concentration in soil. The sources from which biochar is prepared and their mechanism of heavy metal remediation, such as adsorption, complexation, ion exchange, and precipitation, are explained in detail. The factors that influence the heavy metal remediation from soil are discussed in detail. The recent advancement in the biochar production with excellent heavy metal remediation capabilities is explained, and the current status and necessity for future research in heavy metal remediation using biochar are also discussed in detail.

2 Sources of biochar

Biochar can be synthesized from various sources such as agricultural residues, compost, manure, biomass energy crops, industrial bi-residues, and sewage sludge.

2.1 Wood-based biochar

Wood is the traditional biomass form employed for biochar production, which is primarily derived from forest plantations, woodlands, and its associated industries. Enormous quantities of wood-based biomass is derived from diverse industrial activities that generate biomass in the form of chips, shavings, saw dust and bark. In spite of valuable products being produced by wood pyrolysis, it synthesizes biochar as a by-product that possesses unique physical and chemical properties for environmental remediation. Specific surface area of the biochar is one of the pivotal factors in the removal of heavy metals through adsorption. During pyrolysis process, rich organic matter in wood-based biochar volatizes imparting complex porous structure or network. Besides surface area, surface functional groups also play a unique role in the metal remediation. Wood-based biochar contains surplus organic functional groups that can interact with the metal ions via electrostatic interaction, ion exchange, chelation, and other adsorptive mechanisms. A significant amount of oxygen is also present in the woody biochar that aids in the removal of specific metal ions.

2.2 Agriculture-based biochar

Agricultural processes produce massive amounts of residues, which are buried as waste or burned, posing a major environmental threat. Usage of agro-residues as feedstocks for biochar processing is seen as a viable option for waste management and creation of potential pollutant-removing properties through char synthesis. Crop residues such as shells, leaves, stem, bark, stalks, and bagasses are left unutilized after harvestation. Some agro-based residues – pomelo peel, wheat bran, rice husk have been utilized for the adsorption of heavy metal ions. Higher inorganic mineral content was observed in agro-based biochar relative to woody biochar, thus contributing to metal ions from aqueous solutions through ion exchange mechanisms.

3 Mechanism of heavy metal remediation by biochar

Biochar remediates heavy metals from the environment through two different mechanisms, namely,
- Physical adsorption
- Chemical adsorption

The heavy metal remediation mechanisms using biochar was shown in Figure 1.

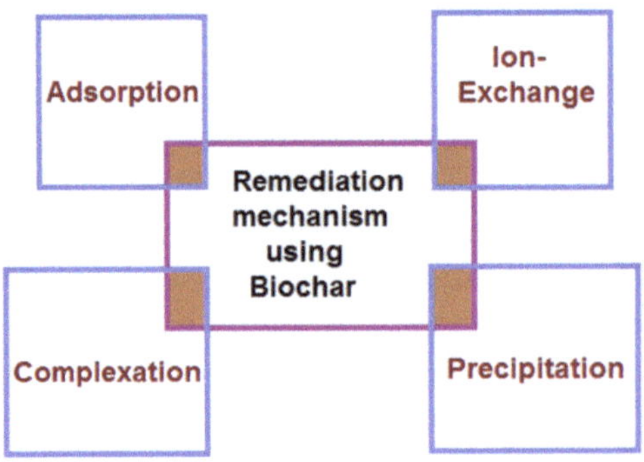

Figure 1: Heavy metal remediation mechanisms using biochar.

3.1 Physical adsorption

Physical mechanism involves the adsorption of heavy metals onto the porous surfaces of biochar. Surface characteristics of biochar such as pore structure, size, and surface area influences the physical adsorption of heavy metals. Macroporous, mesoporous, and microporous structures are present in the biochar which retains the heavy metal ions. Physical sorption is caused by weak interactive forces such as Vander Waal's forces and hydrogen bonds between biochar and heavy metals. Physisorption usually occurs along with other chemical interaction processes. In most of the effluent treatments, physical adsorption has an advantage of biochar regeneration and reusage following the sorption process [5]. Since physically bound heavy metals are soluble in nature, the adsorbed metals are only persistent for a limited time and are reversible, and may not be useful in reducing complete bioavailability in environment.

3.2 Chemical adsorption

In addition to physical adsorption, chemical adsorption is a significant mechanism which helps remove all heavy metals by biochar. Chemical interaction mechanisms include:

i) Electrostatic interaction
ii) Complexation
iii) Ion exchange
iv) Precipitation

3.2.1 Electrostatic interaction

Biochar's bonding to heavy metals is caused by its electrostatic potential or charges, which can be determined by characterization studies. Tomczyk et al. observed electrostatic interaction to be a possible mechanism of copper removal by wood waste biochar [6]. Heavy metal concentration and pH indirectly influences the charges on the biochar surface. Higher concentration of metal ions enhances the electrostatic mechanism of adsorption. Biochar is supplied with negative charge at high pH, which can make it easier to remove positively charged heavy metal ions. Negatively charged metal ions, on the other hand, experience the same phenomenon conversely [7].

3.2.2 Complexation

Ligands and functional groups on the surface of the biochar could form complexes with heavy metal ions thereby immobilizing them. Functional groups such as hydroxyl (–OH), carboxyl (–COOH), and carbonyl (–C = O) which possess oxygen could easily favor complex formation with heavy metals. They include the necessary binding sites, which improve biochar's basic metal adsorption capacity. Dairy manure-based biochar containing hydroxyl and carboxyl groups form complexes with lead (Pb) and cadmium (Cd) ions playing a crucial role in their adsorptive removal [8]. Biochar developed at low temperatures contains a lot of oxygen relative to that at high temperatures, which can tightly bind with heavy metal ions.

3.2.3 Ion exchange

Biochar has high percentage of cations relative to anions, which contributes to its cation exchange ability. Cations such as Mg^{2+}, Mn^{2+}, and Ca^{2+} present in the biochar could exchange these ions with heavy metal ions in the solution. Most of the animal waste biochar such as crayfish shell biochar are rich in calcium (Ca^{2+}) which could be exchanged with positive heavy metals such as cadmium (Cd^{2+}) or copper (Cu^{2+}) [9]. Heavy metal ions that have been ion exchanged on biochar may be reversibly exchanged and biologically taken up by plants for their metabolism. Significantly pH changes affect desorption, and this mechanism is predominant in treatment of heavy metal effluents.

3.2.4 Precipitation

Biochar consists of high minerals that can precipitate with heavy metal ions in aqueous solutions. Minerals can directly involve in precipitation of heavy metals or can create an alkaline environment that aids in the process. Plant-based biochar, such as rice straw, grass, and eucalyptus leaves, contains carbonate, silica oxide, chlorine, and phosphorous mineral groups which forms precipitation with Cd^{2+} ions. Hydroxyl (OH^-) and Cl^- groups also coprecipitate and further precipitate with Cd^{2+} ions forming $Cd_2(OH)_3Cl$ [10]. Mobility of metal ions in the precipitated complex depends on the precipitate formed and the pH of the environment. High complex precipitates are not easily accessible thus resulting in complete removal of heavy metals from the environment.

4 Factors affecting the heavy metal remediation

Biochar is a powdered charcoal that is prepared from the organic biomass through different processes such as pyrolysis, gasification, torrefaction, and flash and hydrothermal carbonization. There are different factors that decide the properties of biochar including temperature, condition for pyrolysis, and composition of organic materials. Correspondingly, there are different factors that strongly influence the heavy metal remediation using biochar; they are presented in Figure 2.

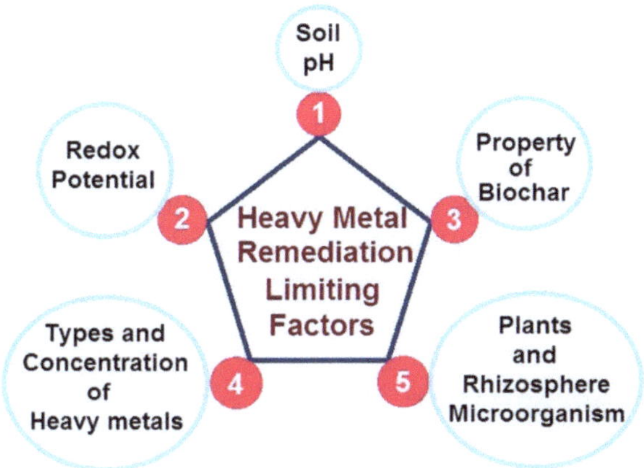

Figure 2: Factors affecting the heavy metal remediation using biochar.

4.1 Properties of biochar

Different varieties of organic biomasses were utilized for the preparation of biochar. Therefore, the composition of each biomass varies significantly with others and that results in changes in the biochar properties. They key mechanism of biochar for environmental remediation is stabilization of heavy metals in soil. The changes in biochar properties affect the heavy metal stabilization in soil. Temperature is one of the crucial factors that influence the properties of biochar by changing the medium pH, surface area, cation exchange capacity, functional groups presence, and mineral content of the organic materials [11]. The increase in pyrolysis temperature decreases the oxygen (O) and hydrogen (H) concentration in the organic materials which results in changes in the availability of functional groups on biochar surfaces. Mineral content of the organic materials are involved in the remediation of heavy metals from the environment by forming metal precipitates with heavy metals. Thereby, the mineral content reduces the bioavailability and mobility of heavy metals in the environment. Therefore, the organic material utilized for biochar preparation should have sufficient mineral content and essential elements for effective heavy metal remediation. The changes in these contents affect the properties of biochar, which results in reduced heavy metal removal. Therefore, the biochar should be prepared with the sufficient characteristics.

4.2 Soil pH

Generally, biochar prepared from the organic biomasses are alkaline in nature. The application of biochar to the soil increases the soil pH and results in the hydrolysis of heavy metals in soil. The increase in pH influences the immobilization of heavy metals in soil by forming metal precipitates. Soil pH can also influence adsorption and transform of heavy metals into oxidizable and residual states. The changes in the soil pH influences the heavy metal complexation process in soil, it also reduces the possibility of desorption of heavy metals in soil. The soil pH is one of the important factors responsible for the soil microbial activity for the regeneration of soil fertility. The soil pH can affect the soil enzyme activity by changing the nutrient availability in soil, concentration of substrate in soil, and microbial diversity in the soil [12].

4.3 Heavy metal type and concentration

The effectiveness of heavy metal remediation using biochar also depends on the types and concentration of heavy metals in the soil. If the biochar has sufficient functional groups, they could remediate the heavy metal contaminated soil through adsorption process. The adsorption might be due to the electrostatic interaction (physical adsorption) or by forming covalent or ionic bonds between the heavy metals and the functional groups on biochar surfaces (chemical adsorption). Different heavy metals are present in soil including zinc (Zn), lead (Pb), cadmium (Cd), and chromium (Cr). Among them, Cu and Cd are immobilized in the soil by forming complexes with the organic matters available in soil. Other heavy metals like Zn, Pb, and Cr are abundantly available in soil due to various anthropogenic activities; they affect the soil properties, plants, and microorganisms in soil. The higher concentration of these heavy metals in soil reduces the organic contents of the plant, and reduces the microbial populations such as bacteria, fungi, and earthworms. They can also disrupt various enzymes such as dehydrogenase, phosphatase, arylsulfatase, and β-galactosidases. Compared with animals and microorganisms in soil, plant has the capability to tolerate higher concentration of heavy metals and also the potential to reduce the toxicity of heavy metals as well as to translocate the heavy metals from soil and accumulate it themselves. The biochar used for remediation of heavy metals will either remove the heavy metals through adsorption or by immobilizing those heavy metals in the soil and influencing the bioremediation (microbial degradation) and phytoremediation (remediation by plants) process [13].

4.4 Redox potential

The redox potential of soil is affected by various factors such as flooding and drought conditions. Redox potential affects the heavy metal remediation using biochar through influencing soil pH, redox chemistry of different metals like Mn and S, microbial community, and their activities in soil. The redox potential can increase the soluble heavy metal contents in soil under aerobic condition with the application of biochar. Apart from increase in soluble heavy metal contents, the changes in soil redox potential enhance the release of various toxic heavy metals such as Mo, Co, and As into the soil. Therefore, the soil redox potential affects the heavy metal remediation using biochar [14].

4.5 Plants and rhizosphere microbes

The application of biochar to soil for remediation of toxic heavy metals directly interacts with the plant roots and rhizosphere microorganisms. The plant root system has specific mechanisms to immobilize and translocate or degrade the toxic heavy metals in soil; adding biochar to soil interferes with the remediation mechanism of plant root system. Therefore, the plant root system will alter the physiochemical properties of biochar, which results reduced heavy metal bioremediation. Applications of biochar in the soil have direct effect in heavy metal remediation mechanism by both the plant root system as well as biochar. The low molecular weight compound called root exudates are secreted by the plant root to immobilize the heavy metals for phytoremediation process. Many studies reported that the root exudates affects heavy metal remediation using biochar by reducing their immobilization efficiency and adsorption capability of biochar. The root exudates in soil influence the changes in the biochar surfaces by protonation, which will reduce the adsorption capacity of biochar for cationic heavy metals, but it increases the anionic heavy metal adsorption capacity [15].

The rhizosphere of plant has specific physical and chemical properties that will influence the heavy metal remediation. The microorganisms present in the rhizosphere can dissolve the heavy metals in soil through their different metabolic activities. For example, by secreting the low molecular weight acids, the microorganisms can dissolve the heavy metals in soil. These microorganisms utilize the carbon in the soil as energy source for their growth and metabolic activities. Biochar is one of the major carbonaceous substances that can be utilized by the microorganisms as energy source after application to the soil. Therefore, microorganisms alter the characteristics of biochar in the soil, which results in reduced heavy metal remediation by biochar. The biochar can also be completely degraded by the microorganisms if the biochar is available in soil for longer periods [16].

5 Difficulties in heavy metal remediation by biochar

Heavy metal remediation using biochar is an effective remediation approach due to its diverse removal mechanism of heavy metals from soil. Though biochar effectively reduces the heavy metal concentration in soil, it can also have some difficulties to execute its remediation mechanism.

- There is no complete removal of all heavy metals, only specific heavy metals can be removed from soil using biochar.
- It is still in the laboratory research phase and is not applicable for industry-scale application.
- The efficiency of heavy metal remediation using biochar strongly depends on the properties of biochar, soil pH, types and concentration of heavy metals, redox potential of soil, plant root system, and rhizosphere microorganisms.
- Under aerobic conditions, biochar increases the availability of heavy metals in soil.
- After heavy metal adsorption using biochar, either the toxic heavy metals should be separated from biochar or their toxicity should be reduced after adsorption. It is difficult to carry out these processes. It also needs higher cost investment.

6 Advancement in biochar preparation

Biochar produced by the conventional methodologies has lower adsorption efficiencies for toxic heavy metal removal. Recently, there are different modification approaches by which the biochar surface properties were improved for efficient removal of toxic heavy metals even at higher concentration in soil. The pretreatment methods used for the synthesis of biochar with excellent surface properties are shown in Figure 3. Different physical, chemical, and biological modifications methods have been utilized. Through physical modification process, the biochar properties including pore structure, surface area, and pore volume are improved. Physical modification of biochar is carried out by using stream and different gaseous agents such as oxygen, nitrogen, and carbon dioxide. Chemical modification of biochar increases the availability of functional groups on biochar surfaces, porosity, and cation exchange capability of biochar. Chemical modification biochar is done by using acids such as sulfuric acid, hydrochloric acid, and phosphoric acid, and alkali such as potassium carbonate, zinc chloride, sodium hydroxides, and potassium hydroxide. Biological modification of biochar enhances the hydrophobicity, surface area, and cation and anion exchange capabilities of biochar. Biological biochar modification is carried out by either anaerobic digestion or by using microorganisms like bacteria. Therefore, the biochar produced through these physical, chemical, and biological modification processes have excellent adsorption capabilities for toxic heavy metal remediation from the contaminated soil.

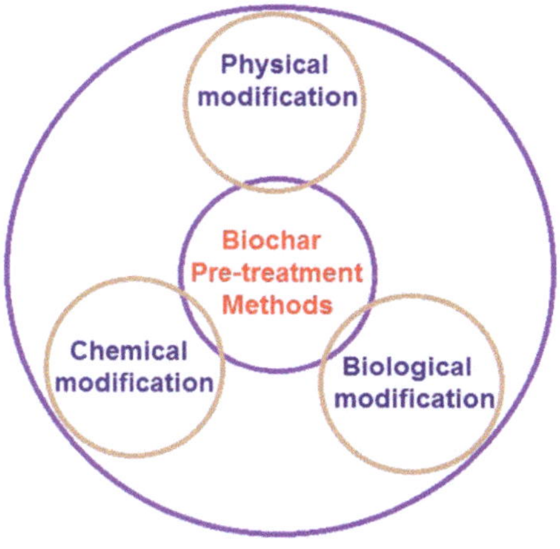

Figure 3: Biochar pretreatment methodologies.

7 Conclusion and future scope

Biochar is most commonly used and one of the efficient alternative heavy metal remediation approaches for contaminated soil and water. Biochar is most probably synthesized from the organic wastes including agricultural residues, compost manure, biomass energy crops, industrial bio-residues, sewage sludge, and wood-based biomasses. Biochar removes heavy metals from the soil through different mechanisms including physical adsorption and chemical adsorption, such as electrostatic interactions, complexation, ion exchange, and precipitation. Different factors influence the heavy metal remediation using biochar, such as properties of biochar, soil pH, types and concentration of heavy metals, redox potential, plant root systems, and rhizosphere microorganisms. However, these factors affects the heavy metal remediation efficiency, biochar removes heavy metals effectively from the contaminated soil. In future, research needs to be carried to produce biochar with efficient surface properties and higher stability for remediation of heavy metals and other toxic pollutants from the environment. A cost-effective, regenerative pretreatment for heavy metal reutilization for various applications needs to be developed. Studies need to be carried out to explain the mechanism of heavy metal remediation in the industrial scale similar to the in-situ laboratory phase of research using biochar.

References

[1] Carolin CF, Kumar PS, Saravanan A, Joshiba GJ, Naushad M, Efficient techniques for the removal of toxic heavy metals from aquatic environment: a review, J. Environ. Chem. Eng., 2017, 5, 2782–2799, doi: https://doi.org/10.1016/j.jece.2017.05.029.

[2] Rechberger MV, Kloss S, Rennhofer H, Tintner J, Watzinger A, Soja G, Lichtenegger H, Zehetner F, Changes in biochar physical and chemical properties: accelerated biochar aging in an acidic soil, Carbon, 2017, 115, 209–219.

[3] Michal K, Isabel H, Thomas DB, Barbara C, Jadwiga SZ, Patryk O, Activated biochars reduce the exposure of polycyclic aromatic hydrocarbons in industrially contaminated soils, Chem. Eng. J., 2017, 310, 33–40.

[4] Jiang A, Cheng Z, Shen Z, Guo W, QSAR study on the removal efficiency of organic pollutants in supercritical water based on degradation temperature, Chem. Central. J., 2018, 12, 16–23.

[5] Zhang A, Li X, Xing J, Xu G, Adsorption of potentially toxic elements in water by modified biochar: a review, J. Environ. Chem. Eng., 2020, 8(4), 104196, doi: https://doi.org/10.1016/j.jece.2020.104196.

[6] Tomczyk A, Boguta P, Sokolowska Z, Biochar efficiency in copper removal from Haplic soils, Int. J. Environ. Sci. Technol., 2019, 16, 4899–4912, doi: https://doi.org/10.1007/s13762-019-02227-4.

[7] Dai Z, Meng J, Shi Q, Xu B, Lian Z, Brookes PC, Xu J-M, Effects of manure and lignocelluloses derived biochars on adsorption and desorption of zinc by acidic types of soil with different properties, Eur. J. Soil. Sci., 2016, 67, 40–50, doi: https://doi.org/10.1111/ejss.12290.

[8] Chen Z-L, Zhang J-Q, Huang L, Yuan Z-H, Li Z-J, Liu M-C, Removal of Cd and Pb with biochar made from dairy manure at low temperature, J. Integr. Agric., 2019, 18(1), 201–210, doi: https://doi.org/10.1016/S2095-3119(18)61987-2.

[9] Zhang D, Zhang K, Hu X, He Q, Yan J, Xue Y, Cadmium removal by MgCl2 modified biochar derived from crayfish shell waste: batch adsorption, response surface analysis and fixed bed filtration, J. Hazard. Mater., 2021, 408, 124860, doi: https://doi.org/10.1016/j.jhazmat.2020.124860.

[10] Dai W, Xu M, Zhao Z, Zheng J, Huang F, Wang H, Liu C, Xiao R, Characteristics and quantification of mechanisms of Cd^{2+} adsorption by biochars derived from three different plant-based biomass, Arab. J. Chem., 2021, 14, 103119, doi: https://doi.org/10.1016/j.arabjc.2021.103119.

[11] Zhang C, Shan B, Zhu Y, Tang W, Remediation effectiveness of Phyllostachys pubescens biochar in reducing the bioavailability and bioaccumulation of metals in sediments, Environ. Pollut., 2018, 242, 1768–1776.

[12] Hong YK, Kim JW, Lee SP, Yang JE, Kim SC, Heavy metal remediation in soil with chemical amendments and its impact on activity of antioxidant enzymes in Lettuce (Lactuca sativa) and soil enzymes, Appl. Biol. Chem., 2020, 63, 42, doi: https://doi.org/10.1186/s13765-020-00526-w.

[13] Komkiene J, Baltrenaite E, Biochar as adsorbent for removal of heavy metal ions [Cadmium(II), Copper(II), Lead(II), Zinc(II)] from aqueous phase, Int. J. Environ. Sci. Technol., 2016, 13, 471–482, doi: https://doi.org/10.1007/s13762-015-0873-3.

[14] El-Naggar A, Shaheen SM, Hseu Z, Wang S, Ok YS, Rinklebe J, Release dynamics of As, Co, and Mo in a biochar treated soil under pre-definite redox conditions, Sci. Total Environ., 2019, 657, 686–695.

[15] Alozie N, Heaney N, Lin C, Biochar immobilizes soil-borne arsenic but not cationic metals in the presence of low-molecular-weight organic acids., Sci. Total Environ., 2018, 1188.
[16] Wang J, Shi L, Zhai L, Zhang H, Wang S, Zou J, Shen Z, Lian C, Chen Y, Analysis of the long-term effectiveness of biochar immobilization remediation on heavy metal contaminated soil and the potential environmental factors weakening the remediation effect: a review, Ecotoxicol. Environ. Safe., 2021, 207, 111261, doi: https://doi.org/10.1016/j.ecoenv.2020.111261.

Vijyendra Kumar, Muskaan Kedia, A B Soni, Dharm Pal,
Bidyut Mazumdar, Gamini Sahu

Role of biochar in the removal of organic and inorganic contaminants from wastewater

Abstract: We all know how the water resources are continuously being deteriorated by the number of organic and inorganic materials like dyes, metals, pesticides and by-products from factories. Discharge of these pollutants has a negative impact on our ecosystem. To remove these pollutants there are several technologies, but we need a technology that is both economic and efficient. Here comes biochar (carbonized biomass or charcoal formed by the pyrolysis process of biomass) to rescue us from our problem. Biochar is eco-friendly and readily available which adds to its attention at present and in the future.

Keywords: biochar, organic and inorganic contaminants, methods, technology

1 Introduction

In today's world, the rapid growth in industrial technology is a boon as well as a bane for us. If we see the water system, around 0.14% of the total water source is available worldwide for us and yet its demand is increasing by 1% per year [1]. Around 80% of all industrial water is sent to nature without any treatment. This results in major environmental concern as the water contains toxic metal ions, organic and inorganic materials, and also cancer-producing agents [2]. To curb this situation, many advanced technologies are put forth. Efficient materials like activated carbon, which adsorbs and degrades catalytically, and filters have been used to remove contaminants. Another example is multiwalled carbon nanotubes (MWCNTs/Pd) composite; it acts as a catalyst and removes organic matter from the wastewater. This material has a good catalytic power, but at the same time, fabrication of MWCNT composite is expensive and complex [3].Other materials used for the same purpose are hazardous too, therefore these are not sufficient.

Biochar-based technology, due to their low cost, high efficiency, and simple procedure, is most popular these days for treatment of wastewater. Biochar (BC), a carbon-rich pyrogenic material, has many properties which can be exploited for our usage such as [4]:

i. Large specific surface area
ii. High porosity
iii. Exceptional ion exchange capacity
iv. Abundant oxygen functional group

https://doi.org/10.1515/9783110734003-021

v. Can easily be generated from biomasses like crop straw, industrial by-product, and municipal waste [5].
vi. Feedstock is abundant and of low cost.

Thus, BC is a great alternative in water treatment technologies. They remove organic dyestuffs, phenols, and pharmaceutically active compounds through adsorption or catalysis

2 Different methods for the preparation of biochar

It was found that the raw BC failed to remove the high contents of the contaminants and also could not remove the selective pollutants from the wastewater. Thus, BC-based hybrid composites were formed by the combination of foreign materials to overcome this problem. An example of this is using chitosan and pyromellitic dianhydride in combination with BC; this brought more functional groups to the BC's surface and resulted in high adsorption of Cu (II) ions [6].

There are four types of modification techniques generally employed on raw BC which depends on the operation procedure:

a) Heat/gas (like steam, ozone, and CO_2) [7]
b) Acidic/alkaline alteration (such as HCl, H_2SO_4, and KOH or K_2CO_3) [8].
c) Decoration of organic and inorganic material (such as chitosan, Fe_3O_4, MgO, and MnO_x) [9]
d) Nanomaterial (like carbon nanotubes, graphene and layered double hydroxide, and ZnS nanocrystalline) coating [10].

2.1 Biochar from various feedstocks

BC is the strong material acquired by pyrolysis of biomass in an oxygen lacking climate at temperature < 800 °C. Since we have different kinds of biomass as shown in Figure 1[11–13] (i.e., different chemical composition and structure), different types of pyrolysis products are yielded.

Some of the BC produced by different types of feedstock and their properties are given below in Table 3 [11–13].

The biomass feedstock can be divided into the following:

i. Residue from agriculture and forest
ii. By-products of industries and municipality waste
iii. Nonconventional matters

Agricultural and forest residue are the mostly used types of biomass for the formation of BC. They are widely used because of the following benefits [14]:

i. Available at low price
ii. Rich in natural resources
iii. Available in abundance such as rice husk, straws, and wood chips
iv. Large content of cellulose and lignin
v. Results in higher carbon and oxygen content in BC

The combination of a high cellulose content and a porous structure causes pyrolysis and evaporation of the biomass (cellulose), allowing for quick impurity removal [15]. Thus, structure of biomass plays a significant part in the formation of porous BC. Large amounts of carbon and oxygen results in high yield of BCs along with its associated functional groups, which acts as vital sites for elimination of impurities [16, 17].

Table 1: Comparative analysis of two different types of agricultural biomass.

Material	Woody biomass	Straw biomass
Structure	These have rigid and physically condensed structure	Loose packing
Result of pyrolysis	Improves stability of porous structure	Easily broken after pyrolysis, resulting in compact structure with vague pores distribution
Example of feedstock	Pine wood	Straw crops of wheat, oat, barley etc.

Industrial by-products and municipal waste includes biomass from sources like the paper industry, sewage treatment industry, and solid waste from human life and food industry [18, 19]. These biomasses are a mixture of organic and inorganic compounds. BC is produced from this type of biomass, such as coking wastewater treatment sludge and press cake from anaerobic digester. This type of BC has a high amount of ash and nonorganic components. High ash means high phosphate adsorption capacity. BC containing metal ions like Mg^{2+}, Ca^{2+}, and K^{+} have excellent ion exchange capacity for removing heavy metal ions [18, 20].

Nonconventional materials involve wastes like tire rubber, bones, plastics, animal manure, and cardboards. As the structure and composition of these biomasses differ from one another, the structure and composition of BC resulting from them also differs. The following are the properties of BC produced by different nonconventional materials [16, 21–24].

Table 2: Comparison of three types of biomass.

Biochar raw material	Pig manure	Waste cardboard	Tire rubber
Ash content	Ash content 65.48%	Lower than pig manure	High carbon and volatile content
Other properties	Uneven and rough surface with nonuniform pores	Fibrous structure without proper pores	Strong affinity toward azo dye, heavy metal ions

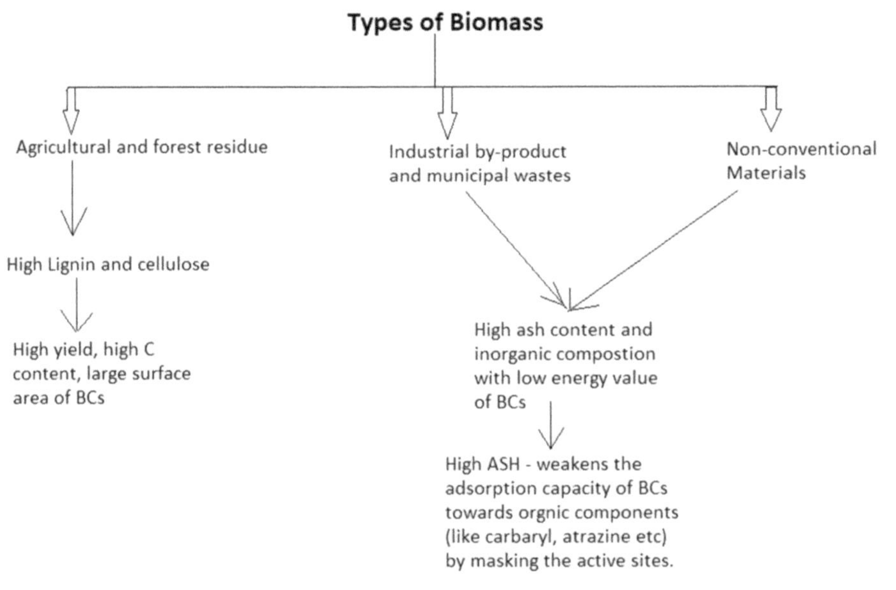

Figure 1: Types of biomass.

2.2 Hybridized biochar

The commonly obtained BC (pristine BC) has some defects like limited function and low anti-interference ability, which leads to poor water treatment. To overcome these defects, many new strategies are developed by which their functions can be enhanced through modifications. Strategies like pretreatment of biomass and modification of BC's surface has resulted in novel BC-based materials, for example magnetic biochar composite (MBC) and nonmetal oxides.

Table 3: Biochar-based materials and its applications in organic dye removal.

Feedstock	Pyrolysis condition (°C, residence time)	Scheme applied	Properties and benefits	Impurity	Effect on completion of removal
Rice straw	600 (1 h)	Moderate pyrolysis completed in N_2	a) Cheap b) High ash content c) High rough surface as compare to activated carbon	Violet crystal	2 times > activated charcoal
Cabbage	600 (1 h)	Moderate pyrolysis completed in N_2	a) Cheap b) High ash content c) High rough surface as compare to activated carbon	Violet crystal	4.8 times > activated charcoal
Wood chips	600 (1 h)	Moderate pyrolysis completed in N_2	a) High carbon content b) Low cost than activated carbon	Violet crystal	adsorption performance comparable with activated charcoal
Residue of *Spirulina platensis* biomass	450 (2 h)	Moderate pyrolysis completed in closed space	a) High carbon content b) high ash content than activated carbon	Methylene blue	comparable to activated charcoal
Paper sludge and wheat husk	500 (20 min)	Chemical coprecipitation of iron monoxide nanoparticles on BCs	a) Magnetic biochars b) High surface area c) High porous volume when compared to pure BCs	Malachite green	Better removal of MG than activated charcoal
Coconut shell	450 (2 h)	Calcination along with wet impregnation	a) TiO_2 crystals spread homogeneously over the surface of biochar	Reactive brilliant blue KN-R	More effective than 100% BCs
Chicken feather	450 (1 h)	Pyrolysis of tetrabutyl titanate-treated biomass	a) Covered with TiO_2 particles b) High N_2 content c) Large surface area d) Cheaper	Rhodamine B	Rate of deterioration higher than pure BC

2.2.1 Magnetic biochar composite

Separation of pristine BC from wastewater was a difficult task because of their small sizes [25]. This separation can be enhanced by converting BCs to MBCs. Preparation of these MBCs includes pretreatment of biomass and post-loading of magnetic matter.

Pretreatment of biomass involves treatment of biomass with Fe^{2+}/Fe^{3+} or chemical coprecipitation of Fe^{2+}/Fe^{3+} onto biomass before pyrolysis [26].

A magnetic sludge-derived BC (MSDBC) is produced by treating $FeCl_2$–$FeCl_3$ with pretreated sludge [27]. This has the following properties:

i. Good magnetic property (saturation magnetization (M_s) = 8.49 emu g^{-1})
ii. Stable in low iron leaching

Another MBC can be formed by hydrothermal carbonization of a mixture of biological sludge and ferric sludge. Resulted composite shows the following properties:

i. Excellent magnetic property (saturation magnetization (M_s) = 18.8 emu g^{-1})
ii. Magnetic stability

Thus, now using the ferromagnetic property MBC can be easily separated using a permanent magnet in the magnetic field.

Obtained MBC has following advantages when compared to pure BC using the same method [28]:

i. Small diameter
ii. Small pore size
iii. Large specific surface area
iv. High carbonization degree

Post loading of magnetic matter involves putting magnetic matter directly on BC. Co precipitation process helps in the formation of MBCs. First and foremost, immaculate BC is set up by paper slime and wheat husk through pyrolysis at 500 °C. From that point forward, the dried BC is blended in with a solution of $FeCl_2·4H_2O$, $FeCl_3·6H_2O$, ethanol, and HCl and mixed, followed by ammonia arrangement treatment [13].

Another MBC is manufactured by using preprepared nano-Fe_3O_4 particles through a self-assembly procedure. The result reveals:

i. Fe_3O_4 nanoparticles are well dispersed on the BC with small agglomeration.
ii. Fe_3O_4/BC has saturation magnetization of 42 emu g^{-1}.

MBCs can likewise be created by ball processing BC with iron particles. Their characteristics were:

i. Fe_3O_4 particles uniformly dispersed on the BC
ii. Fe_3O_4/BC has saturation magnetization of 19.0 emu g^{-1}
iii. Around 200 nm in the main diameter of Fe_3O_4/BC.

2.2.2 Nonmetallic oxides/hydroxides biochar composites

Combining pristine BC with various nonmetallic oxides/hydroxide**s** results in BC's extra functions and strengths. Thus, it is seen as an advisable idea. This nonmetallic oxide/hydroxide composite BC can be produced either by pretreatment of biomass using a precursor or by immobilization of nonmetallic oxide/hydroxide after pyrolysis.

Pretreatment of biomass using a precursor: This method involves modification of biomass using a selected precursor and then pyrolysis under optimum conditions [29].

A nitrogen-doped copper BC was formed by this method using $CuCl_2{\cdot}2H_2O$ as the precursor. It shows the following properties:
1. Irregular-shaped grooved surface morphology
2. Decrease in surface area
3. Increase in catalytic capacity

Another enhanced BC is periclase (MgO) nanocomposites (PE-MgO/BC) [30]. In this, a process called combined electrochemical modification (CEM) was employed. $MgCl_2$, as a precursor, was used as an electrolyte and the biomass as the host material for production of BC composite. This CEM method [31]:
i. Provides MgO nanocomposites binding sites
ii. Increases the porosity of BC
iii. Decreases the time consumption for preparation
iv. Produced composite biochar has the highest surface area (56.42 m^2/g) and the maximum adsorption capacity

Other metal salts like $AlCl_3$, $MnCl_2$, $ZnCl_2$, and $CaCl_2$ can also be used in place of $MgCl_2$, and $CuCl_2$.

Immobilization of nonmetallic oxide/hydroxide after pyrolysis – BC composites can also be produced by this method. In this, one of the common techniques is wet impregnation, which immobilizes the nonmetallic oxide/hydroxide onto the BC [32].

Using an ultrasound-assisted wet impregnation method, a series of new TiO_2/BC hybrid materials such as TiO_2/MSP550, TiO_2/MSP700, TiO_2/SWP550, and TiO_2/SWP700 can be fabricated. These BCs are produced by first pyrolysis of softwood pellets and miscanthus straw pellets at 550 and 700 °C, respectively. Then the resulting BCs are treated with titanium isopropoxide (TTIP) suspension through ultrasound-assisted technology. Finally, composite BC are obtained by calcination at 400 °C in an oxygen-limited atmosphere. The properties of the TiO_2/MSP and TiO_2/MSW are [44]:

i. Modified physicochemical properties
ii. irregular plate-like structure
iii. High specific surface area
iv. High photocatalytic activity

Another common method for immobilization is sol–gel method. A novel TiO_2, supported on coconut shell BC, calcined at 450 °C, is fabricated by this technique [33]. This composite has high decolorization efficiency at strong acidic and basic conditions. A sonochemical/sol–gel method is used to synthesize ZrO_2–BC. Other than these two methods, we can also immobilize nanometallic oxides/hydroxides by the deposition of preprepared metal oxide nanoparticles onto the BC [34]. Example of this is combining CeO_2 with pristine BC to yield CeO_2–H@BC, which has 10 wt% of CeO_2 loadings.

3 Removal of organic and inorganic pollutants

3.1 Elimination of organic dyes

Organic dyes, like azodyes and anthraquinone dyes present in water as a contaminant, are harmful and carcinogenic to the human body. They are generally produced during manufacture of textiles, paper, dyes, rubber, and so on [35]. They are:
i. Chemically stable
ii. Soluble in water
iii. Colorful
iv. Have complex aromatic structure

Removal of these undesirable contaminants can be achieved by techniques such as photocatalysis, adsorption, membrane separation, and advanced oxidations. But among them, adsorption and catalysis are the most effective, low cost and easy operation techniques hence are widely used. BC is used as adsorption/catalysis matrix as it is cost-effective and has a number of vital binding sites. Let us see the performance of BC and BC-based hybrid materials in eliminating the organic dye impurity.

By BC: BCs are used as adsorbent for the organic dye removal. BC derived from sewage sludge, which acts like an adsorbent, is found to remove Methylene Blue (MB), a component of dyeing water, up to 100% in 10 h [36]. Its adsorption capacity is around 29.85 mg g^{-1} at pH 7. Electrical contact, n–π interaction, hydrogen bond interaction, and ion exchange are all involved in MB adsorption on the adsorbent.

Another BC derived from *Spirulina platensis* algae biomass is used as adsorbent for the removal of Congo red (CR) and MB adsorption. Adsorption performance for CR was 82.6% and for MB 92.6%, almost the same as activated carbon.

Table 4: Biochar (Spirulina platensis) performance for the removal of CR and MB.

Impurity	MB	CR
Adsorption capacity (mg g^{-1})	57.80	51.28
pH	>7 (alkaline)	2
Adsorbent dosage (mg L^{-1})	2,000	2,000
Initial concentration of dye (mg L^{-1})	90	90

The electrical interaction between the adsorption site and the cationic dye MB/ anionic dye CR causes this discrepancy.

BC produced by other biomasses are also used for organic material removal. For a greater ash composition and surface area, there are more active areas, resulting in a higher affinity for organic dye removal.

By BC-based hybrid materials: As the adsorption capacity of original BC is sometimes low to meet the standards, we use the hybrid BC composites.

For example [37],

Table 5: Hybrid biochar performance with organic pollutant (MB).

At 25 °C	Pristine biochar	Hybrid biochar composite (Al_2O_3 nanoparticle-coated BCs)
MB adsorption capacity (mg g^{-1})	26.8	35.0

Similarly, Fe-modified BC has high surface area and porous volume in comparison with original BC, thus it is used for the removal of malachite green organic dye.

Also, magnetic hybrid composite BCs can be efficiently removed from the solution, thus removing any source of secondary pollution due to residual adsorbent. The enhanced properties due to metallic nanomaterials like surface area, porosity, functional group, and extra functionalities add to the benefits in removing organic dyes.

Catalytic degradation can also be employed in place of adsorption for the treatment of dyeing wastewater. Catalysts like TiO_2, ZnO, and CdS can be used for photocatalytic degradation of dye impurities [38]. Decolorization of reactive Brilliant Blue KN-R is done photocatalytically by the TiO_2-supported coconut shell BC [33].

Table 6: Decolorization rate in removing KN-R shown.

TiO_2/BC composite (with xenon lamp)	Pristine BC (without illumination)
81.09%	65.48

TiO_2 particles generate holes, h^+, and electrons, e^-, when they adsorb energy greater than or equal to band gap. These holes and electrons on reaction with H_2O/OH and O_2 produce OH radical and O_2^- superoxide radical, respectively. These radicals have high oxidizability which efficiently removes KN-R. Experiments have proven that TiO_2/BC composites can be reused again.

Another example is of CeO_2–H@BC nanocomposite derived by hydrothermal method for Reactive Red 84 (RR84) degradation [34]. Presence of ceria particles increases:

1. Number of cavitation bubbles on catalyst surface
2. Mass transfer rate of RR84 molecules

This leads to more production of OH radicals. Also due to the presence of vacant d-orbitals in the metal, more reactive oxidizing agents are easily generated, thus sonocatalytic degradation is enhanced. Highest degradation efficiency is 98.5% by 1 g L^{-1} of CeO_2–H@BC, 450 W ultrasonic power, pH 6.5, and 10 mg L^{-1} RR84 concentration.

Other catalysts are also used to fabricate hybrid BC composite, which can be used for organic dye removal. Hybridization has two main effects on the BC:

1. Increase in physicochemical properties
2. Addition of new functionalities.

3.2 Expulsion of phenol and compound intermediates

Phenolic toxins (for instance, pentachlorophenol (PCP), bisphenol A (BPA)) are delivered by natural blend, paper and plastic assembling, farming, and so forth [19, 23]. Chemical intermediates (e.g., diethyl phthalate (DEP), 2-mercaptobenzothiazole (MBT)) are used in preparation of dyestuffs and pesticides [39]. These compounds found in wastewater are highly poisonous for the human body and need to be degraded. But due to high stability of aromatic and heteroaromatic structures, they are difficult to biodegrade [40]. So, we need to develop efficient materials to remove them.

3.2.1 By biochar ozonation technique

This is employed for elimination of phenolic contaminants from solution. In this, a number of BCs are synthesized through pyrolysis of sludge at different temperatures, which acts as a catalyst in the ozonation [19]. Some of the synthesized BC are BC500, BC700, and BC900. About 95% of phenol is removed in 30 min by BC500 and BC900. Degradation of phenol is due to the generation of superoxide radical O_2^- (strong oxidizing agent) by carbonyl groups present on the BC surface.

Another BC framed by sewage slime is utilized as an activator to actuate persulfate for the reactant debasement of BPA [41]. Metal in the sludge precursor and the singlet oxygen formed catalytically by the ketone structure of BC are the main reasons behind the BPA degradation. Their removal efficiency is greater than 90%.

Chemical intermediates can be removed by the BC containing oxygen, nitrogen, and sulfur atoms. BC such as P300, P500, W300, and W500 are used as catalysts for the DEP degradation [42].

Table 7: Biochar and DEP.

Biochar	Percentage DEP degraded
P300	58.9%
P500	72.3%
W300	52.3%
W500	60.9%

Due to the high persistent free radical (PFR) production at high temperature, P500 has a higher degradation efficiency than P300. The distinction in DEP debasement between P500 and W500 is because of the distinction in their biochar carbon lattice and broken up natural matter.

3.2.2 By biochar-based hybrid material

Pristine BC has lower phenol removal capacity than hybrid BC composites [44]. TiO_2/BC hybrid material is used for phenol degradation by photolysis. TiO_2/MSP700 removes 64% of phenol, whereas pristine BC removes 58% of phenol. Only 10% loss in phenol is observed after five cycles of photooxidation reaction by TiO_2/MSP700.

BC photocatalyst free from any metals like g-C_3N_4–C is used to degrade MBT catalytically. g-C_3N_4–C has a higher degradation efficiency of 90% than pure g-C_3N_4 of 49% [43]. Photo-generated electrons (e^-) get transferred to the BC surface after visible light radiation, which reacts with oxygen to produce active superoxide radical O_2^-, thus helping in increased performance of g-C_3N_4–C.

3.3 Removal of pharmaceutically active compounds

With expanding well-being mindfulness and higher well-being norms, the utilization of different drugs, like anti-infection agents, calming medications, and analgesics, has expanded a lot in ongoing years, which likewise determined more genuine ecological issues because of the expanded arrival of wastewater which contains enormous pharmaceutical active compounds [45].

The accumulation of antibiotics in water results in drug-resistant bacteria, which will then be killed by more powerful antibiotics and get trapped in a vicious circle [46]. These pharmaceutical compounds can also affect humans by entering the food chain. Therefore, it is necessary to remove them.

3.3.1 By biochar

BCs have shown their usefulness in removing pharmaceuticals compounds also. A BC formed by pyrolysis of malt rootlets is used to activate sodium persulfate (SPS), which removes the sulfamethoxazole (SMX), an antibiotic, by its oxidation [47]. Degradation efficiency of 250 µg L^{-1} SMX is around 94% in 30 min when degraded by 250 mg L^{-1} SPS and 90 mg L^{-1} BC.

Another useful BC is that generated by wetland plants. The BC generated acts as a catalyst for peroxydisulfate (PDS) activation and the degradation of SMX in water [48]. Pretreating the biomass with ammonium nitrate improved the N content in BC in this study. BCs with more nitrogen, enormous explicit surface area, and profoundly graphitic nanosheet structures were framed at high strengthening temperatures (> 700 °C). The addition of N provided a large number of adsorption sites for organic pollutants to bind to through electrostatic force. 100% SMX removal is achieved by N-BC900 in 20 min.

3.3.2 By Biochar-based hybrid material

To eliminate antibiotic medication TET tetracycline, a drug dynamic compound, straw BC doped with iron and zinc is utilized [49].

The resultant composite has:

i. Rough surface but uniform crystal particles
ii. High surface area
iii. Huge pore size to bind TET atoms through hydrogen bond or π–π electron-donor- acceptor (eda)

Most extreme adsorption limit of Fe/Zn BC is 102.0 mg g^{-1}.

For the reactant debasement of ofloxacin (OFX), a cobalt oxide immobilized BC (BC–Co_3O_4) is employed. In 10 min, 90% of OFX is removed by the BC–Co_3O_4/oxone system [50].

4 Economic analysis

For additional turn of events and progression of BC-based materials, financial aspects assume a significant part. Low cost in the outline and fabrication is always desired. Therefore, it is important to gather sufficient cost data on BC-based products. Cost of BC-based materials depends on:

i. Raw material collection and transportation
   a. Price of raw material
   b. Transportation cost
ii. Local transport price
iii. Haul distance
   a. Biomass availability
iv. Fabrication strategies
v. Fabrication condition and scale
vi. Energy consumption
vii. Equipment cost and depreciation
viii. Recycling
ix. Country of production

Cost also depends on the type of biomass. Generally, cost of biomass from municipalities and industries is less than that from agriculture and forests [54, 55].

Table 8: Different stocks and their cost.

Biomass	Cost in US (in $) per ton
Wood waste	78
Corn stover feedstock	75
Food waste	55
Sewage sludge	45
Garden and green waste	34

5 Potential risks

Although BC has many advantages for wastewater treatment, its potential risk cannot be ignored. There are two sources of environment risks:

a) From production process of BC-based hybrid material
b) Liberation of harmful materials from BC-based materials

Various organic chemicals like dioxins, PCDD/DF-furans, and polycyclic aromatic hydrocarbons (PAH) are yielded by pyrolysis at low temperatures. Yield of BC is high at this low pyrolysis temperature but along with it the degree of toxicity is also high. If BC is produced in large amounts, then we have to bear with the accompanying pollution too. Also some of the fabrication methods of BC results in the effluent of strong acid/alkali/oxidant [51].

Secondary pollution also occurs due to the presence of heavy metals like Cu, Cd, and Pb in the BC. BC containing these toxic elements, when treated with water, releases these harmful elements to the water that can cause severe problems.

Furthermore, nanomaterials (such as titanium dioxide, magnesium oxide, and oxide of graphene) in some BC can be leached out and recycled after many uses, causing unintentional environmental damage [52].

Thus, future BC studies should focus on improving the manufacture cycle of BC-based items and improving their steadiness when used, to diminish or eliminate the conceivable liberation of destructive contaminations into the setting.

6 Conclusion and future perspectives

So far we studied BC, their uses, different methods to produce them, and their application for the wastewater treatment. Various types of BC and BC-based hybrid composites are fabricated accompanying the low cost, easy availability, and wide variety of biomasses. Organic contaminants present in wastewater get bound on the active sites of the BC by hydrogen bonds, electrostatic adsorption, or n–π conjugate action. Inorganic pollutants are removed via ion exchange, surface-complexion, precipitation, or ionic contact. The performance of these BCs can be compared with the activated charcoal and CNTs. The presence of porous structure and functional group allows BC to combine with other functional materials and thus we have different nanometallic oxides/hydroxides. For example, we have seen BC combine with magnetic nanoparticles. They have a high removal capacity of contaminants via Fenton-like oxidation reaction. These hybrid BCs, owing to their advanced properties, find use in many applications.

We saw the formation of BC from various kinds of feedstock which includes agricultural residue, municipal waste, industrial by-products, and nonconventional materials, also, the formation of hybridized BC-based materials, MBCs and nonmetallic oxides/hydroxides by various methods. Usage of BC and BC-based hybrid composites in the removal of organic dyes, phenols, chemical intermediates, and pharmaceutically active compounds were discussed in detail. Comparison of the BC in the removal efficiency helped in finding the best BC for removing a particular impurity.

We also discussed the cost estimation of various BCs and their hybrids. The cost of BC depends on various factors, mainly on type of biomass and the place. BC, despite their vast applications, still has more potential which can unlock other problems.

Many studies in the field of biochar have been completed to this point, but there are still some peculiar challenges in biochar planning and uses.

- Amount of energy required for the BC preparation is quite high, thus a cut in preparation cost is the need of the hour.
- Properties of BC vary according to their raw material source and synthesis conditions. As a result, precise regulation of BC with desired compositions and structures remains elusive.
- The majority of the literature focuses on metallic oxides and hydroxide or other nanomaterial alteration technologies. However, precisely controlling the configurations of hybrid matter at the minute level to achieve improved performance optimization is challenging.
- Despite the fact that several experiments have shown the superior efficiency of BC-based products in the elimination of organic compounds from aqueous solutions, there is still a shortage of accurate knowledge regarding realistic wastewater treatment.
- Study of synchronized elimination of organic and inorganic matter from impure water should also be looked upon.
- Along with all these, the effect of BC and their hybrids on aquatic life should not be ignored.

References

[1] Chen B, Ma Q, Tan C, Lim TT, Huang L, Zhang H, Carbon based sorbents with three-dimensional architectures for water remediation, Small, 2015, 11(27), 3319–3336.
[2] Li J, Wang X, Zhao G, Chen C, Chai Z, Alsaedi A, Hayat T, Wang X, Metal-organic framework-based materials: superior adsorbents for the capture of toxic and radioactive metal ions, Chem. Soc. Rev., 2018a, 47, 2322–2356.
[3] Cano OA, Gonzalez CAR, Paz JFH, Madrid PA, Casillas PEG, Hernandez ALM, Perez CAM, Catalytic activity of palladium nanocubes/multiwalled carbon nanotubes structures for methyl orange dye removal, Catal. Today, 2017, 282, 168–173.
[4] Cha JS, Park SH, Jung SC, Ryu C, Jeon JK, Shin MC, Park YK, Production and utilization of biochar: a review, J. Ind. Eng. Chem., 2016, 40, 1–15.
[5] Zhu Q, Wu J, Wang L, Yang G, Zhang X, Adsorption characteristics of Pb2+ onto wine lees-derived biochar, Bull Environ Contam Toxicol, 2016, 97, 294–299.
[6] Deng J, Liu Y, Liu S, Zeng G, Tan X, Huang B, Tang X, Wang S, Hua Q, Yan Z, Competitive adsorption of Pb(II), Cd(II) and Cu(II) onto chitosan-pyromellitic dianhydride modified biochar, J. Colloid Interface Sci., 2017, 506, 355–364.
[7] Rajapaksha AU, Vithanage M, Ahmad M, Seo DC, Cho JS, Lee SE, Lee SS, Ok YS, Enhanced sulfamethazine removal by steam-activated invasive plant-derived biochar, J. Hazard. Mater., 2015, 290, 43–50.
[8] Li YC, Shao JG, Wang XH, Deng Y, Yang HP, Chen HP, Characterization of modified biochars derived from bamboo pyrolysis and their utilization for target component (furfural) adsorption, Energy Fuels, 2014b, 28, 5119–5127.

[9] Han Z, Sani B, Mrozik W, Obst M, Beckingham B, Karapanagioti HK, Werner D, Magnetite impregnation effects on the sorbent properties of activated carbons and biochars, Water Res., 2015, 70, 394–403.
[10] Li C, Zhang L, Gao Y, Li A, Facile synthesis of nano ZnO/ZnS modified biochar by directly pyrolyzing of zinc contaminated corn stover for Pb(II), Cu(II) and Cr(VI) removals, Waste Manag., 2018c, 79, 625–637.
[11] Sewu DD, Boakye P, Woo SH, Highly efficient adsorption of cationic dye by biochar produced with Korean cabbage waste, Bioresour. Technol., 2017, 224, 206–213.
[12] Nautiyal P, Subramanian KA, Dastidar MG, Experimental investigation on adsorption properties of biochar derived from algae biomass residue of biodiesel production, Environ. Process, 2017, 4, 179–193.
[13] Kulaksiz E, Gozmen B, Kayan B, Kalderis D, Adsorption of Malachite Green on Fe-modified biochar: influencing factors and process optimization, Desalin. Water Treat., 2017, 74, 383–394.
[14] Singh R, Prakash A, Balagurumurthy B, Singh R, Saran S, Bhaskar T, Hydrothermal liquefaction of agricultural and forest biomass residue: comparative study, J. Mater. Cycles Waste Manag., 2015, 17, 442–452.
[15] Kumar S, Loganathan VA, Gupta RB, Barnett MO, An assessment of U(VI) removal from groundwater using biochar produced from hydrothermal carbonization, J. Environ. Manag., 2011, 92, 2504–2512.
[16] Lonappan L, Rouissi T, Das RK, Brar SK, Ramirez AA, Verma M, Surampalli RY, Valero JR, Adsorption of methylene blue on biochar microparticles derived from different waste materials, Waste Manag., 2016, 49, 537–544.
[17] Sewu DD, Jung H, Kim SS, Lee DS, Woo SH, Decolorization of cationic and anionic dye-laden wastewater by steam activated biochar produced at an industrial-scale from spent mushroom substrate, Bioresour. Technol., 2019, 277, 77–86.
[18] Takaya CA, Fletcher LA, Singh S, Anyikude KU, Ross AB, Phosphate and ammonium sorption capacity of biochar and hydrochar from different wastes, Chemosphere, 2016, 145, 518–527.
[19] Zhang F, Wu K, Zhou H, Hu Y, Sergei P, Wu H, Wei C, Ozonation of aqueous phenol catalyzed by biochar produced from sludge obtained in the treatment of coking wastewater, J. Environ. Manag., 2018a, 224, 376–386.
[20] Ho SH, Chen YD, Yang ZK, Nagarajan D, Chang JS, Ren NQ, High-efficiency removal of lead from wastewater by biochar derived from anaerobic digestion sludge, Bioresour. Technol., 2017, 246, 142–149.
[21] Sanchez-Olmos LA, Medina-Valtierra J, Sathish-Kumar K, Cardenas MS, Sulfonated char from waste tire rubber used as a strong acid catalyst for biodiesel production, Environ. Prog. Sustain. Energy, 2017, 36, 619–626.
[22] Meng J, Liang S, Tao M, Liu X, Brookes PC, Xu J, Chemical speciation and risk assessment of Cu and Zn in biochars derived from co-pyrolysis of pig manure with rice straw, Chemosphere, 2018, 200, 344–350.
[23] Peng P, Lang YH, Wang XM, Adsorption behavior and mechanism of pentachlorophenol on reed biochars: pH effect, pyrolysis temperature, hydrochloric acid treatment and isotherms, Ecol. Eng., 2016, 90, 225–233.
[24] Inyang M, Dickenson E, The potential role of biochar in the removal of organic and microbial contaminants from potable and reuse water: a review, Chemosphere, 2015, 134, 232–240.
[25] Reguyal F, Sarmah AK, Gao W, Synthesis of magnetic biochar from pine sawdust via oxidative hydrolysis of FeCl2 for the removal of sulfamethoxazole from aqueous solution, J. Hazard. Mater., 2017, 321, 868–878.

[26] Chaukura N, Murimba EC, Gwenzi W, Synthesis, characterisation and methyl orange adsorption capacity of ferric oxide biochar nano-composites derived from pulp and paper sludge, Appl. Water Sci., 2016, 7, 2175–2186.
[27] Wang J, Liao ZW, Ifthikar J, Shi LR, Chen ZQ, Chen ZL, One-step preparation and application of magnetic sludge derived biochar on acid orange 7 removal via both adsorption and persulfate based oxidation, RSC Adv., 2017a, 7, 18696–18706.
[28] Zhang H, Xue G, Chen H, Li X, Magnetic biochar catalyst derived from biological sludge and ferric sludge using hydrothermal carbonization: preparation, characterization and its circulation in the Fenton process for dyeing wastewater treatment, Chemosphere, 2018b, 191, 64–71.
[29] Cai XX, Li J, Liu YG, Yan ZL, Tan XF, Liu SB, Zeng GM, Gu YL, Hu XJ, Jiang LH, Titanium dioxide-coated biochar composites as adsorptive and photocatalytic degradation materials for the removal of aqueous organic pollutants, J. Chem. Technol. Biotechnol., 2018, 93, 783–791.
[30] Jung KW, Ahn KH, Fabrication of porosity-enhanced MgO/ biochar for removal of phosphate from aqueous solution: application of a novel combined electrochemical modification method, Bioresour. Technol., 2016, 200, 1029–1032.
[31] Fang C, Zhang T, Li P, Jiang RF, Wang YC, Application of magnesium modified corn biochar for phosphorus removal and recovery from swine wastewater, Int. J. Environ. Res. Public Health, 2014, 11, 9217–9237.
[32] Wang Z, Shen D, Shen F, Li T, Phosphate adsorption on lanthanum loaded biochar, Chemosphere, 2016b, 150, 1–7.
[33] Zhang S, Lu X, Treatment of wastewater containing Reactive Brilliant Blue KN-R using TiO2/ BC composite as heterogeneous photocatalyst and adsorbent, Chemosphere, 2018, 206, 777–783.
[34] Khataee A, Gholami P, Kalderis D, Pachatouridou E, Konsolakis M, Preparation of novel CeO2-biochar nanocomposite for sonocatalytic degradation of a textile dye, Ultrason. Sonochem., 2018, 41, 503–513.
[35] Mahmood S, Khalid A, Arshad M, Mahmood T, Crowley DE, Detoxification of azo dyes by bacterial oxidoreductase enzymes, Crit. Rev. Biotechnol., 2016, 36, 639–651.
[36] Fan S, Wang Y, Wang Z, Tang J, Tang J, Li X, Removal of methylene blue from aqueous solution by sewage sludge-derived biochar: adsorption kinetics, equilibrium, thermodynamics and mechanism, J. Environ. Chem. Eng., 2017, 5, 601–611.
[37] Wang X, Bayan MR, Yu M, Ludlow DK, Liang X, Atomic layer deposition surface functionalized biochar for adsorption of organic pollutants: improved hydrophila and adsorption capacity, Int. J. Environ. Sci. Technol., 2017c, 14, 1825–1834.
[38] Khan MM, Lee J, Cho MH, Au@TiO2 nanocomposites for the catalytic degradation of methyl orange and methylene blue: an electron relay effect, J. Ind. Eng. Chem., 2014, 20, 1584–1590.
[39] Abdul G, Zhu XY, Chen BL, Structural characteristics of biochar-graphene nanosheet composites and their adsorption performance for phthalic acid esters, Chem. Eng. J., 2017, 319, 9–20.
[40] Michałowicz J, Duda W, Phenols-sources and toxicity, Pol. J. Environ. Stud., 2007, 3, 347–362.
[41] Huang BC, Jiang J, Huang GX, Yu HQ, Sludge biochar-based catalysts for improved pollutant degradation by activating peroxymonosulfate, J. Mater. Chem. A., 2018, 6, 8978–8985.
[42] Fang GD, Liu C, Wang YJ, Dionysiou DD, Zhou DM, Photogeneration of reactive oxygen species from biochar suspension for diethyl phthalate degradation, Appl. Catal. B., 2017, 214, 34–45.
[43] Zhu Z, Fan WQ, Liu Z, Yu Y, Dong HJ, Huo PW, Yan YS, Fabrication of the metal-free biochar-based graphitic carbon nitride for improved 2-Mercaptobenzothiazole degradation activity, J. Photochem. Photobiol. A, 2018c, 358, 284–293.

[44] Lisowski P, Colmenares JC, Masek O, Lisowski W, Lisovytskiy D, Kaminska A, Lomot D, Dual functionality of TiO2/biochar hybrid materials: photocatalytic phenol degradation in the liquid phase and selective oxidation of methanol in the gas phase, ACS Sustain. Chem. Eng., 2017, 5, 6274–6287.
[45] Jain K, Sharma P, Jayalakshmi M, A comparative study of consumption behavior of pharmaceutical drugs. In: International conference on recent developments in science, engineering and technology, vol 799, 2018, 27–33.
[46] Bougnom BP, Piddock LJV, Wastewater for urban agriculture: a significant factor in dissemination of antibiotic resistance, Environ. Sci. Technol., 2017, 51, 5863–5864.
[47] Kemmou L, Frontistis Z, Vakros J, Manariotis ID, Mantzavinos D, Degradation of antibiotic sulfamethoxazole by biochar activated persulfate: factors affecting the activation and degradation processes, Catal. Today, 2018, 313, 128–133.
[48] Zhu S, Huang X, Ma F, Wang L, Duan X, Wang S, Catalytic removal of aqueous contaminants on N-doped graphitic biochars: inherent roles of adsorption and nonradical mechanisms, Environ. Sci. Technol., 2018b, 52, 8649–8658.
[49] Zhou Y, Liu X, Xiang Y, Wang P, Zhang J, Zhang F, Wei J, Luo L, Lei M, Tang L, Modification of biochar derived from sawdust and its application in removal of tetracycline and copper from aqueous solution: adsorption mechanism and modelling, Bioresour. Technol., 2017, 245, 266–273.
[50] Chen LW, Yang SJ, Zuo X, Huang Y, Cai TM, Ding DH, Biochar modification significantly promotes the activity of Co3O4 towards heterogeneous activation of peroxymonosulfate, Chem. Eng. J., 2018b, 354, 856–865.
[51] Dong CD, Chen CW, Hung CM, Synthesis of magnetic biochar from bamboo biomass to activate persulfate for the removal of polycyclic aromatic hydrocarbons in marine sediments, Bioresour. Technol., 2017, 245, 188–195.
[52] Arabyarmohammadi H, Darban AK, Van Der Zee S, Abdollahy M, Ayati B, Fractionation and leaching of heavy metals in soils amended with a new biochar nanocomposite, Environ. Sci. Pollut. Res., 2018, 25, 6826–6837.
[53] Brown TR, Wright MM, Brown RC, Estimating profitability of two biochar production scenarios: slow pyrolysis vs fast pyrolysis, Biofuels Bioprod. Biorefin., 2011, 5, 54–68.
[54] Shackley S, Hammond J, Gaunt J, Ibarrola R, The feasibility and costs of biochar deployment in the UK, Carbon Manag., 2014, 2, 335–356.

Index

https://doi.org/10.1515/9783110734003-022

Bei Fragen zur Produktsicherheit wenden Sie sich bitte an:
If you have any questions regarding product safety,
please contact:

Walter de Gruyter GmbH
Genthiner Straße 13
10785 Berlin
productsafety@degruyterbrill.com